Communications
in Computer and Information Science 424

Stanisław Kozielski   Dariusz Mrozek
Paweł Kasprowski   Bożena Małysiak-Mrozek
Daniel Kostrzewa (Eds.)

Beyond Databases, Architectures and Structures

10th International Conference, BDAS 2014
Ustroń, Poland, May 27-30, 2014
Proceedings

 Springer

Volume Editors

Stanisław Kozielski
Dariusz Mrozek
Paweł Kasprowski
Bożena Małysiak-Mrozek
Daniel Kostrzewa

Silesian University of Technology, Gliwice, Poland

E-mail: {stanislaw.kozielski, dariusz.mrozek, kasprowski,
bozena.malysiak-mrozek, daniel.kostrzewa}@polsl.pl

ISSN 1865-0929   e-ISSN 1865-0937
ISBN 978-3-319-06931-9   e-ISBN 978-3-319-06932-6
DOI 10.1007/978-3-319-06932-6
Springer Cham Heidelberg New York Dordrecht London

Library of Congress Control Number: 2014937803

Typesetting: Camera-ready by author, data conversion by Scientific Publishing Services, Chennai, India

Printed on acid-free paper

Springer is part of Springer Science+Business Media (www.springer.com)

Preface

Databases are present in almost every IT system and collect data describing many areas of human life and activity. Collecting and processing an increasing number of data makes it necessary to use computer architectures and data structures providing high performance and high availability of database systems. Moreover, a vast majority of data require appropriate processing and searching algorithms in order to make the data readable for users and to discover the knowledge that is hidden inside the data.

Beyond Databases, Architectures and Structures (BDAS) is a series of conferences that intends to give the state of the art of the research that satisfies the needs of modern, widely understood database systems, architectures, models, structures, and algorithms focused on processing various types of data. The aim of the conference is to reflect the most recent developments of databases and allied techniques used for solving problems in a variety of areas related to database systems, or even go one step forward—beyond the horizon of existing databases, architectures, and data structures.

The 10th BDAS Scientific Conference was a continuation of the highly successful BDAS conference series started in 2005, traditionally held in Ustroń, Poland. The idea of organizing the conference on databases came from the Institute of Informatics at the Silesian University of Technology in Gliwice, Poland, in 2004. The idea probably matured much earlier in the minds of many people working at the institute. However, in 2004 three people from the Institute—Dr. Bożena Małysiak (today Bożena Małysiak-Mrozek), Dr. Paweł Kasprowski, and Dr. Dariusz Mrozek—who are still members of the BDAS Organizing Committee, met together and made a decision to organize BDAS. Prof. Stanisław Kozielski, head of the Institute of Informatics, agreed to be the Program Committee Chair of the conference and holds this position to this day.

For many years BDAS attracted hundreds or even thousands of researchers and professionals working in the field of databases. Among attendees of our conference were scientists and representatives of IT companies. Several editions of BDAS were supported by our commercial, world renowned partners developing solutions for the database domain, such as IBM, Microsoft, Sybase, Oracle, and others. BDAS annual meetings have become an arena for exchanging information on the widely understood database systems and data processing algorithms.

BDAS 2014 was the 10th jubilee edition. We decided that this edition should be special in several ways. Therefore, we made significant changes. First of all, BDAS changed its name while retaining the acronym. The old name of the conference, Databases: Applications and Systems, evolved to Beyond Databases, Architectures and Structures, to emphasize the scientific nature of the conference. For the first time, BDAS 2014 was organized under the technical co-sponsorship of the IEEE Poland Section. We also initiated a cooperation with Springer, which resulted in the publication of this book.

Today's BDAS is focused on all aspects of databases. It is intended to have a broad scope, including different methods of data acquisition, processing, and storing. This book consists of 56 carefully selected papers that are assigned to 11 thematic groups:

- Query languages, transactions and query optimization
- Data warehousing and big data
- Ontologies and Semantic Web
- Computational intelligence and data mining
- Collective intelligence, scheduling, and parallel processing
- Bioinformatics and biological data analysis
- Image analysis and multimedia mining
- Security of database systems
- Spatial data analysis
- Applications of database systems
- Web and XML in database systems

The first group is related to query languages, database transactions, and query optimization. Papers assembled in this group discuss hot topics of query selectivity estimation, join ordering in distributed databases, distributed transaction processing, NoSQL, and object-relational mapping. The next group of papers concerns issues related to data warehousing and big data. The group consists of four papers presenting research devoted to the ETL process in stream data warehouses, modeling XML data warehouses, and trajectory data warehouses, and finally, data quality aspects and barriers in big data. The third group consists of six papers devoted to ontologies and the Semantic Web. The research devoted to computational intelligence and data mining is presented in 11 papers gathered in the fourth group. These papers show a wide spectrum of applications of various exploration methods, such as decision rules, clustering, support vector machines, fuzzy numbers, artificial immune algorithms, to solve many real problems. The next group of papers is related to collective intelligence, scheduling, and parallel processing. The group consists of four papers on global decision making, multi-agent systems, scheduling resource utilization in a computing system, and parallel data processing with multi-core CPU and GPU architectures. This group of papers is followed by the group devoted to bioinformatics and biological data analysis. Two papers assembled in the group describe methods for fast and accurate classification of biological sequences and gene ontology term similarity analysis in graph database environment. The third paper introduces possibilities of eye movement analysis. The book also contains three papers devoted to image analysis and multimedia mining, including descriptions of methods for hand shape classification, content-based image indexing, and storing facial expressions and emotions. Some aspects of the security of databases systems, including backup systems, DDoS attacks, and user authentication, are discussed in three successive papers. The next five papers show how databases may be used for spatial data analysis and processing. The following eight papers present different usages of databases starting from maritime fleet management, through determination of nuclei spin, lean architecture governance, medical applications, and ending with monitoring systems. Some aspects of data visualization and integration are also covered in this section. Finally, the last three papers consider applying Web 2.0 concepts in the development of the energy planning portal and the use of XML to relational data transformations.

We would like to thank all Program Committee members and additional reviewers for their effort in reviewing the papers. Special thanks to Piotr Kuźniacki - builder and for nine years administrator of our website www.bdas.pl. The conference organization would not have been possible without the technical staff: Maria Woźniak, Dorota Huget, and Jacek Pietraszuk.

We hope that the broad scope of topics related to databases covered in this proceedings volume will help the reader to understand that databases have become an important element of nearly every branch of computer science.

April 2014

Stanisław Kozielski
Dariusz Mrozek
Paweł Kasprowski
Bożena Małysiak-Mrozek
Daniel Kostrzewa

Organization

BDAS 2014 was organized by Institute of Informatics, Silesian University of Technology, Poland.

Technical Program Committee

Stanisław Kozielski (Chair)	Silesian University of Technology, Poland
Werner Backes	Sirrix AG Security Technologies, Bochum, Germany
Andrzej Chydziński	Silesian University of Technology, Poland
Tadeusz Czachórski	IITiS, Polish Academy of Sciences, Poland
Yixiang Chen	East China Normal University, Shanghai, P.R. China
Po-Yuan Chen	China Medical University, Taichung, Taiwan, University of British Columbia, Canada
Sebastian Deorowicz	Silesian University of Technology, Poland
Krzysztof Goczyła	Gdansk University of Technology, Poland
Marcin Gorawski	Silesian University of Technology, Poland
Jarek Gryz	York University, Ontario, Canada
Andrzej Grzywak	Silesian University of Technology, Poland
Brahim Hnich	Izmir University of Economics, Turkey
Xiaohua Tony Hu	Drexel University, Philadelphia, USA
Zbigniew Huzar	Wroclaw University of Technology, Poland
Tomasz Imielinski	Rutgers University, New Brunswick, USA
Pawel Kasprowski	Silesian University of Technology, Poland
Przemysław Kazienko	Wroclaw University of Technology, Poland
Jerzy Klamka	IITiS, Polish Academy of Sciences, Poland
Andrzej Kwiecień	Silesian University of Technology, Poland
Antoni Ligęza	AGH University of Science and Technology, Poland
Bożena Małysiak-Mrozek	Silesian University of Technology, Poland
Marco Masseroli	Politecnico di Milano, Italy
Zygmunt Mazur	Wroclaw University of Technology, Poland
Tadeusz Morzy	Poznan University of Technology, Poland
Mikhail Moshkov	King Abdullah University of Science and Technology, Saudi Arabia
Dariusz Mrozek	Silesian University of Technology, Poland
Mieczysław Muraszkiewicz	Warsaw University of Technology, Poland
Tadeusz Pankowski	Poznan University of Technology, Poland

Organizing Committee

Additional Reviewers

Nurzyńska Karolina
Pluciennik-Psota Ewa
Respondek Jerzy
Sikora Marek
Sikorski Andrzej
Sitek Paweł
Świderski Michał
Świtoński Adam

Traczyk Tomasz
Tutajewicz Robert
Waloszek Wojciech
Werner Aleksandra
Wyciślik Łukasz
Zawiślak Rafał
Zghidi Hafed
Zielosko Beata

Sponsoring Institutions

Technical co-sponsorship of the IEEE Poland Section.

Table of Contents

Ontologies and Semantic Web

Computational Intelligence and Data Mining

Collective Intelligence, Scheduling, and Parallel Processing

Bioinformatics and Biological Data Analysis

Image Analysis and Multimedia Mining

Security of Database Systems

Spatial Data Analysis

Applications of Database Systems

Web and XML in Database Systems

Using the Model of Continuous Dynamical System with Viscous Resistance Forces for Improving Distribution Prediction Based on Evolution of Quantiles

Dariusz Rafal Augustyn

Silesian University of Technology, Institute of Informatics,
16 Akademicka St., 44-100 Gliwice, Poland
draugustyn@polsl.pl

Abstract. The paper considers the problem of prediction of a probability distribution. We take into account an extrapolation model based on evolution of quantiles. We may use any concrete model which allows to track and extrapolate boundaries of buckets of an equi-height histogram. This histogram with $p+1$ boundaries is equivalent to p-quantiles. Using such baseline extrapolation model we may obtain lines of locations of bucket boundaries that may intersect in future. To avoid intersections and to extend (in time) correctness of the results, we propose to use a model of continuous dynamical system with viscous resistance forces for obtaining improved lines of locations. The proposed model allows to obtain lines with unchanged shapes or very similar ones (comparing to the results from the baseline extrapolation model) but without any intersections. This approach will be helpful when a previously used baseline extrapolation model is too much time limited. The work was inspired by the problem of prediction of an attribute value distribution used for query selectivity estimation. However, the proposed method may be applied not only in query optimization problem.

Keywords: prediction of distribution, evolution of quantiles, equi-height histogram, continuous dynamical system, viscous resistance, optimization method.

1 Introduction

Obtaining the optimal query execution plan requires to early estimate the size of data that satisfying a query condition. This is done by the query optimizer which estimates a query selectivity parameter which is the number of table rows satisfying a query condition divided by the total number of rows. For a range query condition, the selectivity is a definite integral of probability density function (PDF) which describes a distribution of table attribute values. So the PDF estimator is needed like a histogram [7]. There are many histogram types but equi-width and equi-height one are commonly used in Database Management

S. Kozielski et al. (Eds.): BDAS 2014, CCIS 424, pp. 1–9, 2014.

Systems. An equi-height histogram with p buckets is equivalent to p-quantiles representation of a distribution.

The representation of PDF (commonly a histogram) are built (or re-built) during so-called update statistics activity. This activity needs to gather values (all or a sample) from a database table. Because of a high workload of this process (including data access) which highly utilizes IO, it is performed rarely. We may easily obtain a new approximate representation by using some model of prediction which rather utilizes CPU (not IO). Of course, this requires some historical data about representations gathered during several statistics updates performed previously.

These are many approaches to the problem of finding a prediction models which may allow to extrapolate distribution. For example, there is an approach which is based on tracking the evolution of K statistics moments [1]. Here, we collect the values of k-th non-central moments ($k = 1 \ldots K$) obtained in previous update statistics activities. Using any extrapolation model (e.g. [1,6]) we may predict new values of all k-th moments in future. Using maximum entropy principle [3] we may reconstruct PDF estimator subject to obtained values of predicted moments.

In this paper we use the approach based on the idea of evolution of quantiles [5]. Here we collect $p+1$ values of bucket boundaries of an equi-height histogram during some previous update statistics activities, i.e. we track the evolution of p-quantiles. Also in this method, any extrapolation model may be used for prediction the future values of bucket boundaries.

In this paper we propose the new method of improving the results obtained by applying any baseline extrapolation model applied for evolution of quantiles.

We propose to apply the model of nonlinear continuous dynamical system with viscous resistant forces for simulating movements of bucket boundaries. This allows to extend rightness of predicted values for further future (comparing to the results obtained from the baseline extrapolation model).

2   Motivating Example

Let us assume that we have a baseline extrapolation model which allows to predict values of p-quantiles. This means that we can extrapolate in time the boundaries of an equi-height histogram buckets. Such histogram has p buckets. Let us denote x_i for $i = 1 \ldots p+1$ as boundaries of buckets that the j-th bucket has the following boundaries (x_j, x_{j+1}). It is obvious that:

$$\underset{i=1 \ldots p}{\forall} \; x_{i+1} > x_i. \tag{1}$$

The model is valid when satisfies the condition (1).

Let us assume the results of applying the extrapolation model shown in Fig. 1 ($p = 4$, so boundaries of histogram buckets determine quartiles). The extrapolation model returns locations of boundaries $x_i(t_j)$ where $i = 1 \ldots 5$ and $t_j = 0, 0.4, 0.8, \ldots, 18.8$. For simplicity there are continuous lines presented in

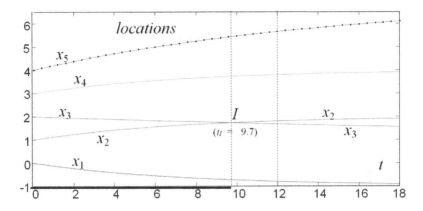

Fig. 1. The results of applying the exemplary model for extrapolation of locations of boundaries of equi-height histogram, used for predicting the 4-quantiles. Evolution of locations along $[0\ 18]$.

Fig. 1 (for $t \in [0, 18]$) but the model gives discrete values (shown only for x_5). This exemplary model is valid only for time belonging into $[0\ 9.7)$. At the $t_I = 9.7$ the lines of locations of x_2 and x_3 intersect (see the point I in Fig. 1).

The problem which appears is how to extend the model validity for higher values of time, i.e. how to move (towards the future) or eliminate the point of intersection without significant changes of shapes of lines of locations x_i. For example, let us assume that we want to extend validity of the model for the time interval $[0\ 12]$.

3   Proposed Method – Presentation

We may notice that a movement of boundaries should resemble a laminar flow (a streamline flow). It occurs when a fluid flows in parallel layers, with no disruption between the layers. Inspired by this observation, to solve the problem (no crossing lines of locations) we propose to use a mechanical model of a body movement inside a liquid.

Let us assume that we already know (from the extrapolation model) locations $x_i(t_j)$ of the i-th boundary where $i = 1 \ldots N, t_j = t_1, \ldots, t_k$ ($N = 5, t_1 = 0$). We can reconstruct approximate velocities $v_i(t)$ of the i-th boundary as follows:

$$v_i(t_j) = \frac{x_i(t_{j+1}) - x_i(t_j)}{t_{j+1} - t_j} \tag{2}$$

for $t_j = t_1, \ldots, t_{k-1}$ (see Fig. 2). We can also reconstruct approximate acceleration $a_i(t_j)$ of the i-th boundary as follows:

$$a_i(t_j) = \frac{v_i(t_{j+1}) - v_i(t_j)}{t_{j+1} - t_j} \tag{3}$$

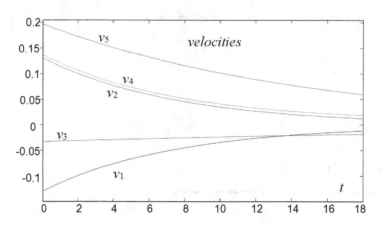

Fig. 2. Evolution of velocities

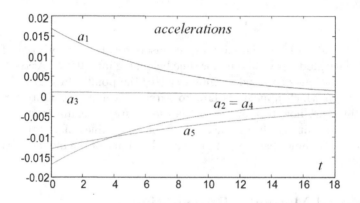

Fig. 3. Evolution of accelerations

for $t_j = t_1, \ldots, t_{k-2}$ (see Fig. 3). In the exemplary model $t_{k-2} = 18$.

Let us assume that a_i are external forces obtained from the extrapolation model. Because the accelerations are known only in discrete moments of time, to obtain $a_i(t)$ for continuous time interval $t \in [0\ t_{k-2}]$ we use the cubic spline interpolation method [8].

We can describe the continuous dynamical model of movement of the i-th border according to the second Newton's law:

$$m_i \frac{d^2 x_i'}{dt^2} = a_i(t). \tag{4}$$

Assuming unitary mass m_i, a movement of each i-th boundary is described by the two state equations as follows:

$$\begin{cases} \frac{dx_i'}{dt} = v_i' \\ \frac{dv_i'}{dt} = a_i(t) \end{cases} \tag{5}$$

where x_i' and v_i' (given by the dynamical model) are approximations of x_i and v_i (given by the extrapolation model). In our example (Fig. 1) we have the dynamical model having $2 \cdot (p + 1) = 10$ state equations.

To solve the dynamical model, i.e. to obtain $x_i'(t)$ and $v_i'(t)$ for $t \in [0, t_{k-2}]$, we use Range-Kutty method of integration [4] for numerical solving the system of state equations [10]. The dynamical model with the initial conditions $\boldsymbol{x}(0) = [x_1(0), \ldots, x_5(0)] = [-1, 0, 1, 2, 3]$ (Fig. 1) and $\boldsymbol{v}(0) = [v_1(0), \ldots, v_5(0)] = [-0.1298, 0.1298, -0.0331, 0.1358, 0.1974]$ (Fig. 2) is equivalent to the baseline extrapolation model. For the assumed absolute tolerance and relative one both equal 10^{-7}, $\boldsymbol{x}'(t)$ and $\boldsymbol{v}'(t)$ are almost the same like $\boldsymbol{x}(t)$ and $\boldsymbol{v}(t)$ (Fig. 1 and 2), so they are not presented.

To avoid an intersection the lines of locations let us introduce an additional fluid resistance, i.e. a drag force dependent on velocity. The resultant force (for $i = 1 \ldots p+1$) is a sum of the external force ($a_i(t)$ given from the extrapolation model) and the drag force $f_i = -b_i v_i'$. Thus the movement is now described by the following 2nd order differential equation:

$$m_i \frac{d^2 x_i'}{dt^2} = a_i(t) - b_i v_i'. \tag{6}$$

Let us define the following function (with very high values near the locations of neighbour's boundaries):

$$b(i, x) = \begin{cases} \frac{b_0}{|x - x_2'|} & \wedge \ i = 1 \\ \frac{b_0}{|x - x_{i+1}'|} + \frac{b_0}{|x - x_{i-1}'|} & \wedge \ i = 2 \ldots p - 1 \\ \frac{b_0}{|x - x_{p-1}'|} & \wedge \ i = p \end{cases} \tag{7}$$

where b_0 is constant (in time) and it is a parameter allowing to set the value of impact of the fluid resistance force on the resultant one. Finally, the coefficient b_i (in (6)) is defined as follows:

$$b_i = b(i, x_i'). \tag{8}$$

Such definition of the coefficient b_i (given by (7) and (8)) causes that the resultant force for the i-th boundary strongly increases when the i-th boundary significantly approaches to the one of its neighbours (see Fig. 4).

All above allows to formulate equations of a dynamical model with non-linear fluid resistances. For example, in this model, the movement of the 3rd boundary is defined as follows:

$$\frac{d^2 x_3'}{dt^2} = a_3(t) - b_0 \left(\frac{1}{|x_3' - x_2'|} + \frac{1}{|x_3' - x_4'|} \right) \frac{dx_3'}{dt}. \tag{9}$$

Assuming $b_0 = 0$ (zero resistant force) we again obtain the model which is equivalent to the extrapolation model (so locations are not presented). Assuming $b_0 = 1$ we obtain the dynamical model where there is no intersection of location

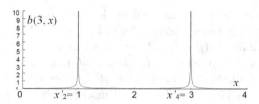

Fig. 4. The function designated for obtaining flow resistance coefficient for $i = 3$ and $t = 0$ (when location values of neighbours equal $x_2' = 1$ and $x_4' = 3$)

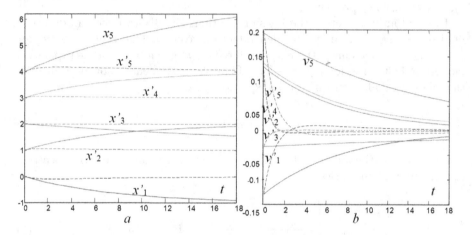

Fig. 5. Boundaries movement (dashed lines) in a highly viscous medium (for $b_0 = 1$): a) locations, b) velocities

lines (Fig. 5a) but the shapes of lines are significantly different from those obtained from the baseline extrapolation model. A viscous medium almost stops the movements of boundaries (Fig. 5) – velocities v_i' quickly come closer to zero.

Observing both Fig. 1 and 5a we may expect that we may find such value of b_0 (between 0 and 1) that the distance between x_2' and x_3' is arbitrarily small but the x_3' is still greater than x_2'.

Let us assume that those adjacent boundaries that their locations intersect in the extrapolation model should be spaced at a distance ϵ by applying the dynamical model with the proposer value of b_0. ϵ is the smallest acceptable bucket length of an equi-height histogram. This distance should be obtained at the end of extended time interval ([0 12] in our example). For our example this means that $|x_3'(t_{end} = 12) - x_2'(t_{end} = 12)| \approx \epsilon$. We may solve this as some optimization problem. The parameter of optimization procedure is b_0. Let us assume the following formula:

$$Q = \min_{i=1\ldots p} \left((|x_{i+1}'(t_{end}) - x_i'(t_{end})| - \epsilon)^2 \right) \qquad (10)$$

as a score function (we also used some penalty function to satisfy the conditions $x'_{i+1} > x'_i$ for $i = 1 \ldots p$).

For assumed $\epsilon = 0.01$ and tolerance 10^{-4}, starting from $b_0 = 0$, using Nelder-Mead simplex direct search method of optimization [9], by minimizing Q, we obtained the optimal value of $b_0 = 0.00737$.

Lines of locations are almost indistinguishable for boundaries numbered by $i = 1, 4, 5$ (Fig. 6), i.e. $x'_i \cong x_i$ for $i = 1, 4, 5$. Lines for boundaries no 2 and 3 (given by the proposed dynamical model) approximately hold their shapes (especially when the lines for $x'_2(t)$ and $x'_3(t)$ are distant i.e. for small value of time). For higher value of time boundaries no 2 and 3 repel one another (Fig. 6) so the final distant between them at $t_{end} = 12$ equals about 0.01 (Fig. 7). Of course, their lines of locations do not intersect before t_{end}. Thus the time domain of the

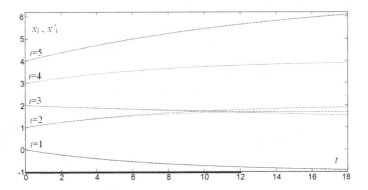

Fig. 6. Boundaries movements: x_i (given by the baseline extrapolation model) – solid lines, x'_i (given by the proposed dynamical model with optimal value of b_0) – dashed lines

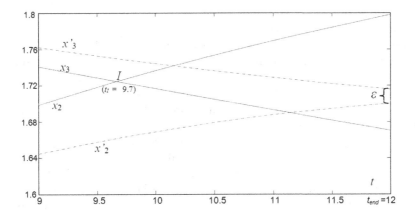

Fig. 7. x'_2, x'_3, x_2, x_3 in zoomed part of Fig. 6 for $t \in [9 \; 12]$

dynamical model validity is wider (comparing to the domain of the extrapolation model), i.e. it is determined at least on time interval $[0 \ 12]$.

4   Proposed Method – Algorithm

Let us assume that we have an extrapolation model designated to predict values of locations of $p+1$ boundaries of an equi-height histogram. The extrapolation model is valid only till t_I, i.e. till the time of the first intersection of locations lines. We are going to extend the validity of the model till t_{end} where $t_{end} > t_I$. We want to have such location lines that during the time belonging to $[0 \ t_{end}]$ each distance between any two neighbouring lines equals at least ϵ. We may assume that ϵ is significantly less than the minimal initial distance, e.g. $\epsilon = 0.01 \cdot \min_{i=1...p} (x_{i+1}(t = 0) - x_i(t = 0))$.

The below described steps of the proposed method allow to obtain the new extended model:

1. Reconstruct velocities vectors $v_i(t_j)$ and accelerations ones $a_i(t_j)$ (eq. (2) and (3)) using locations vectors $x_i(t_j)$
2. Build the continuous dynamical model (eq. (6)) using interpolated accelerations as an external forces and introducing fluid resistance forces (eq. (7) and (8))
3. Obtain the optimal parameter b_0 (eq. (7) and (9)) by minimizing the score function Q (eq. (10)) which is based on the assuming that the minimal distance between two neighbouring location should approach ϵ at t_{end}
4. Obtain the improved locations vectors $x'_i(t_j)$ using the dynamical model with fluid resistance forces and the optimal value of coefficient b_0 (obtained in the previous step).

5   Conclusions

In the paper we propose the technique which supports any method of distribution extrapolation based on evolution of quantiles. The proposed method uses results from any extrapolation model which predicts values of boundaries buckets of an equi–height histogram. The method allows to extend a time interval of prediction correctness. It will be helpful when a previously used baseline extrapolation model is too much time limited. The method is based on nonlinear continuous dynamical system with viscous resistance forces that describe movements of locations of bucket boundaries.

Although the proposed method is inspired by some aspect of query optimization it may find any different application (not only in query selectivity estimation based on extrapolated representation of attribute values distribution).

In the presented prototype module we used Matlab, but we consider to implement this functionality in Java (we consider to prepare implementations of a numerical method of integration and a method of obtaining the minimum of an univariate function). This may allow to prepare the implementation either some extrapolation model (for example the one from [1,6]) or the viscous-resistance-based dynamical model (proposed in this paper). Both might be enabled e.g. in Oracle DBMS by using $ODCIStat$ extension [11,2].

References

1. Augustyn, D.R.: Zastosowanie predykcji rozkładu wartości atrybutu w celu poprawy dokładności estymacji selektywności zapytań. Studia Informatica 34, 23–42 (2013)
2. Augustyn, D.R.: Applying advanced methods of query selectivity estimation in oracle DBMS. In: Cyran, K.A., Kozielski, S., Peters, J.F., Stańczyk, U., Wakulicz-Deja, A. (eds.) Man-Machine Interactions. AISC, vol. 59, pp. 585–593. Springer, Heidelberg (2009)
3. Buck, B.: Maximum Entropy in Action: A Collection of Expository Essays. Oxford science publications, Clarendon Press (1992)
4. Dormand, J., Prince, P.: A family of embedded runge-kutta formulae. Journal of Computational and Applied Mathematics 6(1), 19–26 (1980)
5. Eickhoff, M., McNickle, D., Pawlikowski, K.: Analysis of the time evolution of quantiles in simulation. Int. J. on Simulation: Systems, Science & Technology 7, 44–55 (2006)
6. Haber, R., Keviczky, L.: Nonlinear system identification. In: Mathematical Modelling. Kluwer Academic Publishers (1999)
7. Ioannidis, Y.: The history of histograms (abridged). In: Proceedings of the 29th International Conference on Very Large Data Bases, VLDB 2003, vol. 29, pp. 19–30. VLDB Endowment (2003)
8. MathWorks®: Cubic spline interpolation – MATLAB csapi (2013), http://www.mathworks.com/help/curvefit/csapi.html
9. MathWorks®: Find minimum of unconstrained multivariable function using derivative-free method – MATLAB (2013), http://www.mathworks.com/help/matlab/ref/fminsearch.html
10. MathWorks®: Solve nonstiff differential equations; medium order method – MATLAB ode45 (2013), http://www.mathworks.com/help/matlab/ref/ode45.html
11. Oracle®Corporation: Using Extensible Optimizer, (July 10, 2005), http://download.oracle.com/docs/cd/B28359_01/appdev.111/b28425/ext_optimizer.htm

The exIWO Metaheuristic – A Recapitulation of the Research on the Join Ordering Problem

Daniel Kostrzewa[1,2] and Henryk Josiński[3]

[1] Silesian University of Technology, Department of Industrial Informatics,
Krasińskiego 8, 40-019 Katowice, Poland
Daniel.Kostrzewa@polsl.pl
[2] Future Processing, Gliwice, Poland
dkostrzewa@future-processing.com
[3] Silesian University of Technology, Institute of Informatics,
Akademicka 16, 44-100 Gliwice, Poland
Henryk.Josinski@polsl.pl

Abstract. The authors summarize the several years research on the join ordering problem presenting a method based on the exIWO metaheuristic which is characterized by both the hybrid strategy of the search space exploration and three variants of selection of individuals as candidates for next population. The nub of the problem was recalled along with details of adaptation of the exIWO including representation of a single solution and transformation of an individual. Results of the experiments show that the exIWO algorithm can successfully compete with the SQL Server 2008 DBMS in optimization of join order in database queries.

Keywords: database query optimization, join ordering problem, query graph, exIWO algorithm.

1 Introduction

Through several editions of the BDAS conference (2008-2011) the authors presented successive development phases of the exIWO metaheuristic (expanded Invasive Weed Optimization algorithm, previously called the modified version of the IWO method) on the basis of its application for solving the join ordering problem (JOP) which is a significant part of a database query optimization process in case of both centralized and distributed data. The JOP is the NP-hard problem [1] and algorithms that compute its good approximate solutions fall into two classes: (i) augmentation heuristics that build an evaluation plan step by step according to certain criteria, (ii) randomized algorithms that perform some kind of "random walk" through the space of all possible solutions seeking a solution with minimal evaluation cost [7]. The exIWO metaheuristic belongs to the latter class. The overview of bibliography describing the methods for solving the JOP would be unusually spacious, so it has been omitted with regard to the space limit.

The authors of the original IWO algorithm [5] from University of Tehran were inspired by observation of dynamic spreading of weeds and their quick

S. Kozielski et al. (Eds.): BDAS 2014, CCIS 424, pp. 10–19, 2014.
© Springer International Publishing Switzerland 2014

adaptation to environmental conditions. The exIWO retains the basic IWO idea, but additionally is characterized by both the hybrid strategy of the search space exploration and three variants of selection of individuals as candidates for next population. Main goal of the present paper is a recapitulation of the results obtained by means of the final exIWO version for the JOP.

The organization of this paper is as follows – section 2 briefly describes the join ordering problem. Adaptation of the exIWO algorithm to the JOP is discussed in section 3. Section 4 deals with procedure of the experimental research along with its results. The conclusions are formulated in section 5.

2   Analysis of the Optimization Problem

The input of the optimization problem is given as the *query graph* (or *join graph*), consisting of all relations (generally speaking, sets of records, because they can be results of previously performed selections and projections or outcomes of subqueries executed in local databases in distributed environment) that are to be joined as its nodes and all joins specified as its edges [7], according to the join expressions formulated in the query (e.g., "R.a = S.a AND S.b = T.b"). Distinctive shapes of a join graph are the following: (i) a *star* graph, in which one of the sets of records appears in each join expression (is a central vertex of the graph presented in Fig. 1a), (ii) a *chain* graph, where 2 sets of records (constituting extreme vertices of the graph presented in Fig. 1b) appear only once in join expressions, whereas the rest of sets – twice. Remaining shapes of query graphs belong to the *irregular* group (Fig. 1c).

From the practical point of view, the objective of the JOP is to find the order of joins which guarantees that the user receives the query result as quickly as possible. The search space comprises join orders (execution plans) depicted, as a rule, in form of a *join processing tree*, which consists of the following elements (Fig. 1d): *leaves* representing sets of records, *inner vertices* corresponding to the join results, and edges connecting both join arguments with join result and, in this way, illustrating data flow from the leaves to the vertex called root which the final result is assigned to. The tree in Fig. 1d shows one of the feasible join orders for each of the query graphs from Fig. 1a, 1b, 1c. The tree belongs to the *left-deep processing trees* [7] – the subset of the complete search space characterized by a sequential system of inner vertices. In other words, every vertex of the tree is connected with at least one of leaves.

Analysis of query execution plans generated by the SQL Server 2008 system leads to the observation that the query optimizer of the aforementioned DBMS constructs exclusively join processing trees with a sequential system of vertices. Therefore, only this type of join processing trees will be taken into account in further considerations. Nevertheless, it is worthwhile to mention that the complete search space includes also *bushy trees* [7] where at least one vertex is connected with two inner vertices. This feature can be described as a parallel system of vertices. Because the shape of possible processing trees can be arbitrary, the cardinality of this set is much higher than the cardinality of the left-deep subspace – for n relations, there are $\binom{2(n-1)}{n-1}(n-1)!$ different solutions, whereas there are

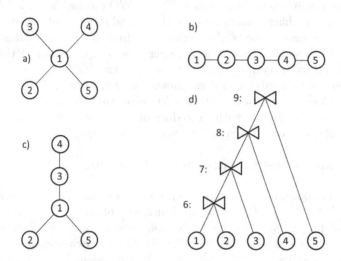

Fig. 1. Shapes of a join graph: a) star, b) chain, c) irregular, and d) the corresponding join processing tree

$n!$ ways to allocate n relations to the leaves of a left-deep tree [7]. However, it must be taken into account that some of the processing trees represent unfeasible solutions because they do not respect the conditions resulting from the join expressions.

According to the model used by the authors a value of the goal function related to the given join processing tree is interpreted as a total estimated cost of joins execution. A single join implementation is assumed to be based on the *nested loops* algorithm which (i) is implemented in commercial database management systems (DBMS) and (ii) does not impose any assumption on the data characteristics. The fundamental components of the cost model presented in [2] are formulas for estimation of cardinality of the result set of a join and length of a single record from the result set as well as formulas for estimation of a join execution cost and optionally (that is to say, in case of distributed data) of a data transfer cost. The complete solution of the JOP in the latter situation requires additionally determination of workstations where particular joins will be executed.

3   The exIWO Algorithm

The simplified pseudocode describes the exIWO using terminological convention consistent with the "natural" inspiration of the authors of the original IWO version. Consequently, the words *"individual"*, *"plant"*, and *"weed"* are treated as synonyms.

```
Create the first population.
For each individual from the population:
  Compute the value of the fitness function.
While the stop criterion is not satisfied
  For each individual from the population:
    Compute the number of seeds.
    For each seed:
      Draw the dissemination method.
      Create a new individual.
      Compute the value of its fitness function.
  Select individuals for a new population.
Return the best individual.
```

The optimization process starts with a random initialization of individuals of the first population. Cardinality of a population will remain constant in all algorithm iterations. An individual represents a single join processing tree. Configuration of the individuals' genes presented in Fig. 2 reflects the order of joins described by the tree in Fig. 1d. Leaves are numbered from 1 to 5, whereas inner vertices obtained numbers from 6 to 8: $6 = join(1, 2); 7 = join(3, 6); 8 = join(4, 7)$. The final set of records is produced by $join(5, 8)$.

1	2	3	6	4	7	5	8
W		W		W		W	

Fig. 2. An individual as a representation of a join processing tree

It is worthwhile to remark that the proposed form of an individual is also adapted to solutions constructed in case of distributed data because a gene representing a single join consists of a triple including identifiers of records sets which constitute join arguments as well as identifier W of workstation where the join will be executed. Join processing trees with the parallel system of vertices can be also mapped into the presented form of an individual.

The number of seeds S_w produced by a single weed depends on the value of its fitness function. However, as it was aforementioned, the goal function is rather interpreted as cost K_w (and therefore minimized) which allows to determine the number of seeds according to the following formula:

$$S_w = S_{min} + \left\lfloor (K_{max} - K_w) \left(\frac{S_{max} - S_{min}}{K_{max} - K_{min}} \right) \right\rfloor \tag{1}$$

where S_{max}, S_{min} denote maximum and minimum admissible number of seeds generated, respectively, by the best population member (cost K_{min}) and by the worst one (cost K_{max}).

According to the terminological convention the hybrid strategy of the search space exploration proposed by the authors of the present paper can be called "dissemination of seeds". It consists of three methods randomly chosen for each seed: spreading, dispersing and rolling down. Probability values assigned to the particular methods (p_{spr}, p_{disp}, p_{roll}) should fulfill the condition $p_{spr} + p_{disp} + p_{roll} = 1$. The draw procedure is based on the pseudorandom number generator of the uniform distribution on the interval $[0, 1)$.

The *spreading* consists in random disseminating seeds over the whole of the search space – independently of the point of the search space which represents a parent plant.

The *dispersing* is a method based on the idea proposed in the original IWO version. The degree of difference between the individual and his offspring can be interpreted as the distance between the parent plant and the place where the seed falls on the ground. The distance is described by normal distribution with a mean equal to 0 and a standard deviation truncated to nonnegative values. The standard deviation is decreased in each algorithm iteration as follows: $\sigma_{iter} = \left(\frac{iter_{max} - iter}{iter_{max}} \right)^m (\sigma_{init} - \sigma_{fin}) + \sigma_{fin}$, where $iter$ denotes the current iteration ($iter \in [1, iter_{max}]$). Consequently, the distance is gradually reduced. The number of iterations $iter_{max}$ is estimated based on the stop criterion defined as the execution time limit. The symbols σ_{init}, σ_{fin} represent, respectively, initial and final values of the standard deviation, whereas m is a nonlinear modulation factor which determines the shape of the curve of the standard deviation.

An alternative approach, proposed by the authors of the present paper, describes the distance by the Student's t-distribution with a fixed number of degrees of freedom, truncated to nonnegative values. In addition, the value produced by the Student's t random variate generator is multiplicated by the scale coefficient γ_{iter}. The scale coefficient is decreased from γ_{init} to γ_{fin} similarly to the standard deviation of the normal distribution according to the following formula: $\gamma_{iter} = \left(\frac{iter_{max} - iter}{iter_{max}} \right)^m (\gamma_{init} - \gamma_{fin}) + \gamma_{fin}$.

From the practical point of view, the distance between plants, rounded to the nearest integer value, is interpreted as the number of transformations of the parent individual. Under assumptions concerning both – the form of the individual and restriction related to the case of centralized data – the only applied transformation is based on the exchange of two records sets identifiers randomly chosen from different genes (Fig. 3). In this way any identifier cannot be used as an argument of more than one join. The resultant join order $join(2, join(4, join(3, join(1, 5))))$ is accepted for the query graphs from Fig. 1a and 1c. In case of the chain graph from Fig. 1b the solution will be rejected because it is incompatible with the conditions resulting from the join expressions. A change of workstation where a single join will be executed can be proposed as additional transformation for distributed data.

The *rolling down* can be interpreted as a movement of a seed towards a "better" location with respect to the fitness function. The term "neighbours" stands for individuals located at the distance equal to one (transformation) from the

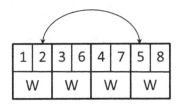

Fig. 3. Transformation of an individual

current plant. The best adapted individual is chosen from among the determined number of neighbours, whereupon its neighbourhood is analyzed in search of the next best adapted individual. This procedure is executed k times giving the opportunity to select the best adapted individual found in the last iteration as a new one. The parameter k represents also the number of neighbours taken into consideration in a single iteration of the rolling down. Thus, the method enables exploration of the vicinity of the parent individual's location in the search space.

Candidates for next population are selected in a deterministic manner according to one of the following methods: global, offspring-based and family-based. Set of candidates for the *global* selection consists of all parent plants and all their newly created descendants. By contrast, the *offspring-based* selection is limited solely to the descendants and thus should decrease the risk of stagnation at non-optimal points in the search space [6]. If the best individual so far was grown in the current population, then despite the fact that it cannot be retained in the next population it will be stored with an eye to the final optimization result. According to the rules of the *family-based* selection [8], each plant from the first population is a protoplast of a separate family. A family consists of a parent weed and its direct descendants. Only the best individual of each family survives and becomes member of the next population.

The complexity of the exIWO algorithm can be estimated as follows:

$$T_{exIWO} = O(N_w \cdot iter_{max} \cdot S_{max}) \tag{2}$$

where N_w denotes the cardinality of each population; the remaining symbols were defined in aforementioned formulas.

4   Experimental Research

The principal objective of the experiments was to compare values of the execution time for two groups of plans: the first group was composed of plans built by means of the exIWO algorithm, whereas the second one contained plans created by the query optimizer of the SQL Server 2008 system. The exIWO plans were executed "precisely" in the SQL Server 2008 DBMS owing to a set of *hints* that enable to override any execution plan the query optimizer might select for a query, and in this way to enforce the join order found by the exIWO and to apply the appropriate join algorithm.

With a view to easy automatic generation of test queries based on particular query graphs 90 test databases were created and each of them consisted of 50 tables including arbitrary number of rows (Fig. 4). Appropriate indexes were built on all columns specified as primary or foreign keys.

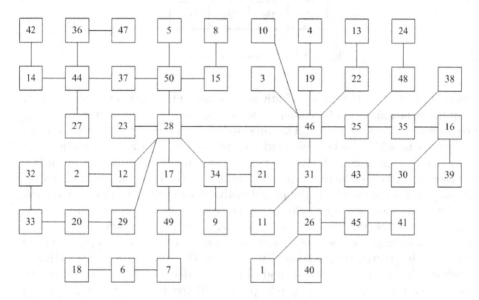

Fig. 4. Example of the irregular test query shape

The findings of the initial experiments determined the most appropriate values of the exIWO parameters for the JOP separately for each type of the query graphs. In this stage of the research the aforementioned test databases were not used. The values of the exIWO parameters were collected in Tab. 1.

The test databases found their application in the next stage of the research. Each of trials included following phases:

1. Generation of a test query.
2. Construction of the test query execution plan by means of the exIWO. It was assumed that minimum computation time amounts to 5 [s], while the real instant of program termination is equivalent to the instant of creation of the first population after the minimum computation time. And so the approximately constant time of optimization was obtained. The described process was executed 10 times. In this way 10 different variants of the plan were constructed, which results from the nondeterministic nature of the exIWO.
3. Enforcement of the test query execution by the SQL Server 2008 DBMS according to the plan constructed by the exIWO. Application of the hint OPTION(FORCE ORDER) enabled to enforce the join order determined by the exIWO. Furthermore, owing to the hint OPTION(LOOP JOIN) the nested loops algorithm was used for each join complying with assumption

Table 1. Basic parameters of the exIWO used for the join ordering problem

Description	Value (chain)	Value (star)	Value (irregular)
Population cardinality N_w	100	100	100
Execution time (stop criterion)	5 [s]	5 [s]	5 [s]
Initialization of the first population	random	random	random
Maximum number of seeds sowed by a weed S_{max}	2	5	2
Minimum number of seeds sowed by a weed S_{min}	1	1	1
Probability of the dispersing p_{disp}	0.9	1	0.2
Probability of the spreading p_{spr}	0.1	0	0.75
Probability of the rolling down p_{roll}	0	0	0.05
Type of distribution of the distance between a parent plant and an offspring (dispersing)	normal	normal	Student's t
Initial value of standard deviation σ_{init} (dispersing)	5	5	–
Final value of standard deviation σ_{fin} (dispersing)	1	1	–
Initial value of scale coefficient γ_{init} (dispersing)	–	–	5
Final value of scale coefficient γ_{fin} (dispersing)	–	–	1
Number of degrees of freedom (dispersing)	–	–	2
Nonlinear modulation factor m (dispersing)	3	3	3
Number k of examined neighbours and neighbourhoods (rolling down)	–	–	3
Selection method	global	global	global

valid for the exIWO. Each of 10 variants of the plan constructed for the given test query, as it was described in point 2, was executed 10 times, thus producing together 100 values of the execution time t_{exIWO}.

4. Construction of the test query execution plan by means of the query optimizer embedded in the SQL Server 2008 DBMS which had the possibility to use histograms providing estimates of the distribution of data in columns. Joins were performed using the nested loops algorithm.

5. Tenfold query execution enabled to evaluate the execution time t_{DBMS}.

The cache memory buffers were flushed before each execution of the test query performing the following sequence of commands:

CHECKPOINT

DBCC DROPCLEANBUFFERS.

The workstation used for experiments is described by the following parameters: Intel Core i5 M 460 2.53 GHz processor, RAM 4 GB 1066 MHz.

The relation between execution times t_{exIWO} and t_{DBMS} for different query graph types was illustrated in Fig. 5.

Fig. 5 should be interpreted as follows: Y-coordinate of a single marked point informs how big is the share of queries which were executed according to the exIWO plans in a shorter time than on the basis of the DBMS plans, while X-coordinate expresses the interval containing the percentage difference between the execution times t_{DBMS} and t_{exIWO} computed as $(t_{DBMS} - t_{exIWO})/t_{DBMS}$.

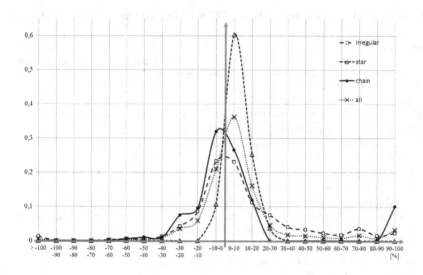

Fig. 5. Relation between execution times for exIWO plans and SQL Server 2008 plans

100%. Negative values of X-coordinate relate to the queries for which $t_{exIWO} > t_{DBMS}$. The "all" line was created on the basis of summation of numbers of queries, which were used for preparation of "star", "chain" and "irregular" lines.

Analysis of the Fig. 5 leads to the following remarks: (i) the exIWO plans are better for star query graphs, (ii) in case of irregular query graphs the line is to a greatest degree symmetrical about Y-axis what suggests that both types of plans are generally of similar quality, (iii) the exIWO plans are slightly worse for chain query graphs but, nevertheless, the last marked point on the "chain" line shows that execution time for some of exIWO plans turned out to be over tenfold shorter than for their SQL Server 2008 counterparts.

An interesting observation resulting from analysis of data not presented in the paper concerns the functioning of the SQL Server 2008 query optimizer. In most cases the estimated cost of a plan generated by the SQL Server optimizer for the given query was over 25% smaller than the value of the goal function of the exIWO plan calculated for the same query. However, this relations was not reflected in real values of the execution time. Preliminary analysis gives rise to a presumption that the SQL Server optimizer estimates the join selectivity inaccurately which finally results in underestimated cost of the execution plan.

5   Conclusion

Both the NP-hard characteristic of the join ordering problem and the phenomenon of combinatorial explosion impose the application of heuristic algorithms producing approximate but acceptable solutions in a short time. Following this idea, the authors adapted the exIWO metaheuristic for solving the JOP and the usefulness of the method was confirmed in numerous experiments – for ca. 64% of queries (58 from 90) the exIWO plans turned out to be faster in realization than the plans constructed by the SQL Server 2008 DBMS.

The exIWO metaheuristic can successfully compete with a commercial DBMS in optimization of join order in database queries. However, it is worthwhile to mention that the exIWO was applied efectually by the authors in other significant continuous as well as discrete optimization problems, to which belong: minimization of the selected numerical functions (Griewank, Rastrigin, Rosenbrock) [3], several variants of the Traveling Salesman Problem [4] and feature selection. The last-mentioned issue was an important part of the research directed at the following topics: gait-based person re-identification and automatic recognition of handwritten digits (the relevant papers are currently in press).

References

1. Ibaraki, T., Kameda, T.: Optimal nesting for computing N-relational joins. ACM Transactions on Database Systems 9(3), 482–502 (1984)
2. Kostrzewa, D., Josiński, H.: The Comparison of an Adapted Evolutionary Algorithm with the Invasive Weed Optimization Algorithm Based on the Problem of Predetermining the Progress of Distributed Data Merging Process. In: Cyran, K.A., Kozielski, S., Peters, J.F., Stańczyk, U., Wakulicz-Deja, A. (eds.) Man-Machine Interactions. AISC, vol. 59, pp. 505–514. Springer, Heidelberg (2009)
3. Kostrzewa, D., Josiński, H.: The Modified IWO Algorithm for Optimization of Numerical Functions. In: Rutkowski, L., Korytkowski, M., Scherer, R., Tadeusiewicz, R., Zadeh, L.A., Zurada, J.M. (eds.) SIDE 2012 and EC 2012. LNCS, vol. 7269, pp. 267–274. Springer, Heidelberg (2012)
4. Kostrzewa, D., Josiński, H.: Using the Expanded IWO Algorithm to Solve the Traveling Salesman Problem. In: 5th International Conference on Agents and Artificial Intelligence (ICAART 2013), pp. 451–456 (2013)
5. Mehrabian, R., Lucas, C.: A novel numerical optimization algorithm inspired from weed colonization. Ecological Informatics 1(4), 355–366 (2006)
6. Michalewicz, Z., Fogel, D.B.: How to Solve It: Modern Heuristics. Springer (2004)
7. Steinbrunn, M., Moerkotte, G., Kemper, A.: Heuristic and randomized optimization for the join ordering problem. The VLDB Journal 6, 191–208 (1997)
8. Tao, G., Michalewicz, Z.: Inver-over Operator for the TSP. In: Eiben, A.E., Bäck, T., Schoenauer, M., Schwefel, H.-P. (eds.) PPSN 1998. LNCS, vol. 1498, pp. 803–812. Springer, Heidelberg (1998)

Colored Petri Net Model of X/Open Distributed Transaction Processing Environment with Single Application Program

Marek Iwaniak and Włodzimierz Khadzhynov

Department of Electronics and Computer Science
Technical University of Koszalin, Poland
marek.iwaniak@tu.koszalin.pl, hadginov@weii.tu.koszalin.pl
http://weii.tu.koszalin.pl

Abstract. This work presents the Colored Petri Net (CPN) model of distributed transaction environment based on X/Open standards TX and XA. Due to the complexity of the model the slicing technique has been applied. Obtained reduction of the model allowed to perform the reachability analysis of each slice. Slices of the model have been extended with necessary input and output interfaces and presented as UML components. Possible usage of received components in future research work has been described.

Keywords: Colored Petri nets, distributed transactions, 2PC.

1 Introduction

When designing a distributed system the designers need to take a lot of difficult design decisions about the data distribution. To meet the system requirements system designers must face problems such as resource allocation, database schema decomposition, data replication and distributed transaction processing. A framework which assists the system designers in solving those problems is needed. Petri Net (PN) [4] is a powerful mathematical modeling language for the description of distributed systems. We want to develop a modeling framework which will allow the system designer to create a conceptual model of problem being solved and then transform it into Petri Net model for a more detailed analysis. It is potentially desired to generate code snippets from verified Petri Net model.

In order to achieve this goal we started to looking for the ways to integrate Petri Nets into distributed databases domain. Initially we have found the work [7] in which Two Phase Commit protocol (2PC) was studied with the usage of Ordinary Petri Nets for the purpose applying it in authors further goal of creating the virtual data ware house (VDW). To verify approach presented in this work we created detailed Petri Net models of Three Phase Commit protocol algorithm [1] to find out number of Ordinary Petri Nets limitations and many unwanted states in reachability analysis. To create more precise model and to eliminate those

S. Kozielski et al. (Eds.): BDAS 2014, CCIS 424, pp. 20–29, 2014.

unwanted states we decided apply Colored Petri Nets. In [2] we introduced CPN model of the Two Phase Commit protocol algorithm with two participants. By adaptation of initial marking we were able to trace different execution of protocol during reachability analysis. All previous research allowed us to analyse commit algorithms implementations at a detailed level while maintaining an appropriate level of abstraction.

Other approach in the field of Petri Nets application in distributed databases domain is [6] where CPN based framework that allows the study of different algorithms integrated inside a model of a DTP system was presented. This framework allows to define DTP environment of specified size and distribution used with various commit and deadlock algorithms and simulate system performance for specified algorithms combinations.

In this work we have looked beyond algorithms at distributed transaction environment as whole. This paper presents Colored Petri Net model of exemplary distributed transaction environment based on X/Open Distributed Transaction Processing (DTP) standard [8]. In section 2 fundamental information about Colored Petri Net is presented. Section 3 briefly describes DTP model, its components and interfaces. Section 4 presents proposed CPN model with reachability analysis of CPN slices. In summary we transform CPN slices into abstract software components which will be used in our further work on proposing new modeling framework for the distributed systems.

2   Colored Petri Nets

The classical Petri Nets were proposed by Carl Adam Petri in 1962 [5]. PN is a bipartite graph which contains two types of nodes connected by arcs. Nodes of the net are: places - represented by circles and transitions - represented by rectangles. Arcs directed from places to transitions are called transitions input arcs. Arcs directed from transitions to places are called transitions output arcs. The places can store tokens which are represented by black dots. Amount of tokens stored in all of the places is called Petri Net Marking and it represents each state of modelled process. The transitions represents events in modelled process and causes PN marking change. Each arc is described with weight of tokens. For a specific PN marking dynamic events can occur. If the input places of the transition have an amount of tokens equal to the input arc weight then the transition comes into enabled state which means it can be fired. After firing of the transition the tokens are taken away from all input places and new tokens are inserted into the output places. The number of tokens taken away from each of the input places and inserted into output places are equal to weights of input and output arcs. This process is presented in Fig. 1.

2.1   Coloured Petri Net

In Colored Petri Net (CPN) we can define token types which are distinguished by colors. CPN places can hold tokens of all defined colors at the same time, arc

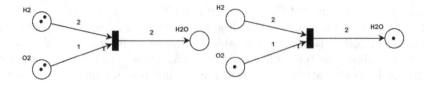

Fig. 1. The chemical reaction as PN: a) transition ready to be fired; b) transition after firing

weight can be described by any color from defined set of colors. In PN we use integer values to define arcs weight and tokens amount. In CPN we use algebraic notation so we can hold each color tokens quantity separately for example 2r (red), 3b (black). Colored Petri Net can be described by four-tuple CPN=(C,P , T , D) where: C is a collection of defined token colors, P is a collection of places $|P|$=m, T is a collection of transitions $|T|$=n, D=[D+ - D-] is incidence matrix of dimension m x n. Incidence matrix describes CPN structure and it is being created by subtraction of pre-incidence matrix from post-incidence matrix. D- is pre-incidence matrix of dimension m x n, it contains elements $d\text{-}ij = w(i,j)$ which describe the weight of transition j input arc, that is the arc directly connecting place i with transition j. D+ is post-incidence matrix of dimension m x n, it contains elements $d+ij = w(i,j)$ which describe the weight of transition j output arc, that is the arc directly connecting transition j with place i.

2.2   Reachablity Analysis

In the reachablity analysis we study subsequent marking changes that happen after transitions firing for assumed initial marking which is initial tokens distribution and represents initial state of PN. Typically reachability analysis is presented by the reachability tree which is a directed graph built with such elements as: the root - representing initial marking, nodes - representing each reachable marking, arcs - representing fired transition, leaves - representing final or repeated markings. We begin reachability analysis by checking enable/ready transition for assumed initial marking. After transition has been fired, a new marking is created. Determination of another marking can be simplified with usage of incidence matrix D which describes our Petri Net structure.

3   DTP Architecture

One of the main goals of most database management systems is to ensure data consistency in realm of concurrency. Each data change takes a whole database into entirely new database state. Many data changes can happen in one unit of time and some of them are erroneous. This matters are taken care by databases engines main feature which is transaction processing and its ACID properties (Atomicity, Consistency, Isolation and Durability). To ensure ACID properties across many databases in distributed database systems the global transaction are performed with usage of commit protocols. Mostly used is Two Phase Commit protocol which was implemented in industrial standard X/Open Distributed Transaction Processing [8]. X/Open DTP architecture model features following components:

- Application Program (AP) - Application program defines transaction and its boundaries. Within transaction boundaries AP performs a sequence of operation that involves many resource managers such as databases.
- Transaction Manager (TM) - manage global transactions , coordinates the decision to commit or rollback them. Many large commercial middleware can handle the role of transaction manager. Bitronix and LibreXA (LIXA) are worth mention open-source transaction managers.
- Resource Managers (RM) - typically it is a database engine capable of handling XA interface. Most commercial and big open-source DBMS supports this feature.

X/Open DTP architecture components communicate with each other with following three interfaces:

- AP-RM - The AP-RM interface give the AP access to resources managed by RM. This could be standard SQL or RM specific native interface.
- AP-TM - The AP-TM interface allows the AP to coordinate global transaction management with the TM. The AP do not take part in global transaction commitment. Details of the AP-TM interface has been described in the TX (Transaction Demarcation) specification [9]. Functions provided by TM are called tx_()
- TM-RM - The TM-RM interface XA (eXtended Architecture) [10] - describes interfaces between TM and RM. Functions provided by each RM for the TM are called xa_(). Function provided by TM for RM are called.

The components of X/Open DTP model and mentioned interfaces was shown on Fig. 2.

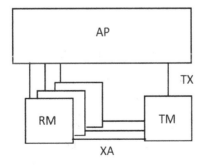

Fig. 2. DTP architecture overview

4   CPN Model of DTP Environment with Single AP

To create our CPN model we have used PIPE modeling tool. Strong sides of PIPE are its constant developed, intuitive model creation and animation mode in which we can fire enabled transitions and observe marking changes. Most of analysis and simulation modules of PIPE works only for Ordinary Petri Nets with one color define and are useless for CPN analysis. Fig. 3 presents our base CPN model and its places are described by Tab. 1. DTP environment was built with AP,TM and two RM components. Set of Colors we have used is following C = {Abort, Commit, Init, Ready}. Weights of arcs and quantity of colors in places presented in further figures and tables are ordered as shown in the C set. As we can see the CPN is complex and due to many arcs crossing may be difficult to read. It's also hard to present incidence matrix as it size is overcoming 20 x 20.

Table 1. Description of places presented on Fig.3

AP		TM		RM	
AP_INIT	AP initial state	TM_INIT	TM initial state	RM_INIT	RM initial state
AP_WAIT	AP waiting for TM	TM_WAIT	TM waiting for RMs readiness	RM_WAIT	RM is waiting for AP and TM
AP_DONE	AP done work with RMs	TM_READY	TM ready to begin commitment	RM_READY	RM is ready form commit
AP_COND	AP waiting for transaction result	TM_WAITV	TM wait for votes	RM_COMMIT	RM close with success
AP_CLOSE	AP close with success	TM_CLOSE	TM close with success	RM_FAIL	RM failed at any point
AP_FAIL	AP failed at any point	TM_FAIL	TM failed at any point		

4.1   Petri Net Slicing and Reachability Analysis

In case of complex Petri Nets the slicing technique may be applied. It assumes that we can study each slices properties and after which we can give a verdict that if all slices have some property than whole PN has it too [8]. Sliced model is presented on Fig. 4. We sliced our model of the way to get slices representing original X/Open components. Each slice has been wrapped with abstract component boundaries. During the slicing some of important arcs outer places were lost. As it was described in [3] we have restored necessary places (acting as input interfaces) and transitions (acting as output interfaces) outside each components

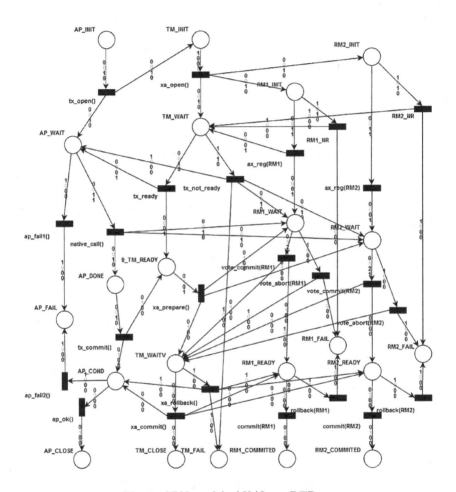

Fig. 3. CPN model of X/Open DTP

boundaries. This places have become component interfaces ports. Fig. 5 presents these ports with usage UML component notation. Due to paper limitation we will take a look only at AP slice in detail. Tab. 2 presents incidence matrix of AP component, the n is number o RMs. Let us assume we want to check components behavior in case of transaction with 2 RMs. We define initial marking M0 as collection of color sets, where we put one Init token into AP_INIT place for process initiation, one Ready token into tx1_TMin place to reflect all RMs readiness for global transaction and one Commit token into tx2_TMin place to reflect successful commitment of global transaction. After defining initial marking M0 we can begin reachability analysis by looking for enabled transition. First transition fired will be T0, by checking incidence matrix at T0 row we know that we must take away Init token from AP_INIT place and add one to AP_WAIT place and one to tx1_TMout. Refer to Tab. 3 for full reachability of analysis of this scenario and its initial marking.

Table 2. Incidence matrix of AP slice

D											RM
				AP				TM			
D1	INIT	WAIT	DONE	COND	CLOSE	FAIL	Out1	In1	Out2	In2	Rm
T0	0,0,-1,0	0,0,1,0					0,0,1,0				
T1		-1,0,0,0				1,0,0,0					
T2		0,0,-1,-1							0,0,1,0		0,0,n,0
T3			0,0,-1,0	0,1,0,0						0,1,0,0	
T4		-1,0,0,0				1,0,0,					
T5				0,-2,0,0	0,1,0,0						
T6		1,0,0,0						-1,0,0,0			
T7		0,0,0,1					0,0,0,1				
T8				1,0,0,0						-1,0,0,0	
T9				0,1,0,0						0-,1,0,0	

Table 3. Reachability of AP slice

M.	T.F.											XA
					AP				TM			
		INIT	WAIT	DONE	COND	CLOSE	FAIL	Out1	In1	Out2	In2	Rm
M0		0,0,1,0						0,0,0,1		0,1,0,0		
M1	T0		0,0,1,0					0,0,0,1		0,1,0,0		
M2	T7		0,0,1,1							0,1,0,0		
M3	T2			0,0,01						0,1,0,0	0,0,2,0	
M4	T3				0,1,0,0				0,1,0,0	0,1,0,0	0,0,2,0	
M5	T9				0,2,0,0				0,1,0,0		0,0,2,0	
M6	T5					0,1,0,0			0,1,0,0		0,0,2,0	

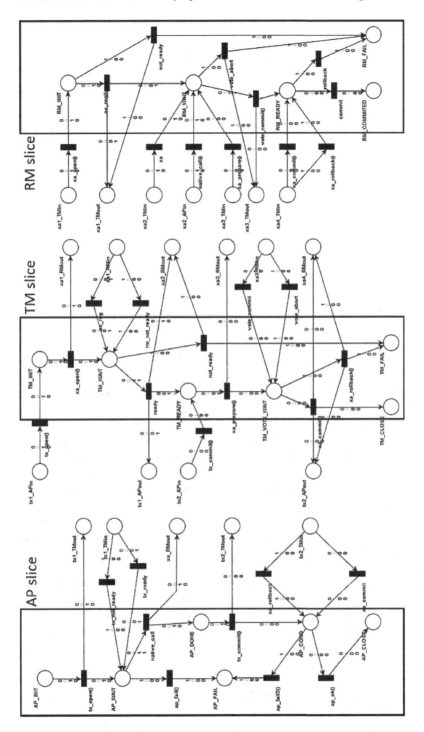

Fig. 4. The sliced CPN model of DTP

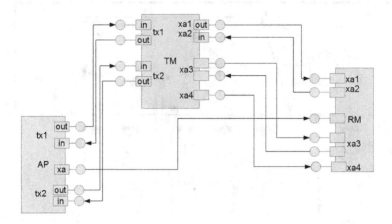

Fig. 5. Sliced model as UML component diagram

5   Summary

When designing a entity relation diagram the designer can use a tools that allow transformation from Conceptual model into database engine specific Physical model. This transitions allow the designer to see the entities he designed at physical level without their implementation. We are convinced that development of Petri Net based tools for distributed databases design which allow such a transitions of the model is needed.

In this work Colored Petri Net model of DTP environment with single AP was presented. Complex and detailed CPN model was sliced into smaller networks which represents each of DTP core components. With the addition of necessary input and output interfaces these slices became components. Each of received components can be described by its incidence matrix. These matrixes describes precisely state changes of the process encapsulated by component boundaries in reaction for specific input. Usage of incidence matrixes in reachability analysis of selected scenario described in [2] was presented in this paper with the AP component as the example.

In [6] authors were focused on creating the tool that verifies different commit and deadlock protocols cooperation in exemplary DTP system. They created a very complex and precise CPN model of DTP environment. They also divided their CPN model into core DTP components. Managing of such complex Petri Net divided into components was possible because they use Renew tool and its functionality of outer places which allows to connect separate Petri Nets. In this paper we have shown how complexity of the model can be reduced when using simpler modeling tool like PIPE which has no functionality of connecting different Petri Nets by outer places.

In future works we plan to take a closer look at applying CPN for DTP sessions trees and graphs of transaction execution modelling. This will require to take a look in detail on the inside of transition called "native_call()" in AP component and its cooperation with many RM components. Received incidence matrixes of DTP components may be very useful in future works, as they represent precisely components behavior. We are also considering a usage of Renew tool instead of PIPE tool, as it seems much more powerful in analysis and design of complex Petri Nets.

References

1. Iwaniak, M., Khadzhynov, W.: Usage of petri nets for distributed transactions modeling. SI: Studia Informatica 33(2A(105)), 255–270 (2012)
2. Iwaniak, M., Khadzhynov, W.: Reproduction of two-phase commit protocol with multiple participants with the usage of colored petri net. SI: Studia Informatica 34(2A(115)), 57–69 (2013)
3. Masri, A., Bourdeaud'huy, T., Toguyéni, A.: Performance evaluation of distributed systems: a component-based modeling approach based on object oriented petri nets (2010)
4. Murata, T.: Petri nets: Properties, analysis and applications. Proceedings of the IEEE 77(4), 541–580 (1989)
5. Petri, C.: Kommunikation mit Automaten. Ph.D. thesis, Institut fur instrumentelle Mathematik, Bonn, Germany (1962)
6. Polo Martín, M.J., Quintales, L.A.M., Moreno García, M.N.: A framework for the modelling and simulation of distributed transaction processing systems using coloured petri nets. In: Cortadella, J., Reisig, W. (eds.) ICATPN 2004. LNCS, vol. 3099, pp. 351–370. Springer, Heidelberg (2004)
7. Sarkar, B.B., Chaki, N.: Transaction management for distributed database using petri nets. International Journal of Computer Information Systems and Industrial Management Applications (IJCISIM) 2, 69–76 (2010)
8. X/Open Company Limited: Distributed Transaction Processing: Reference Model, version 3 (1996)
9. X/Open Company Limited: Distributed Transaction Processing: The TX (Transaction Demarcation) Specification (1995)
10. X/Open Company Limited: Distributed Transaction Processing: The XA Specification (1991)

Granular Indices for HQL Analytic Queries

Michał Gawarkiewicz, Piotr Wiśniewski, and Krzysztof Stencel

Faculty of Mathematics and Computer Science
Nicolaus Copernicus University
Toruń, Poland
{garfi,pikonrad,stencel}@mat.umk.pl

Abstract. Database management systems use numerous optimization techniques to accelerate complex analytical queries. Such queries have to scan enormous amounts of records. The usual technique to reduce their run-time is the materialization of partial aggregates of base data. In previous papers we have proposed the concept of metagranules, i.e. partially ordered aggregations of the fact table. When a query is posed, the actual aggregation level will be determined and the smallest fit metagranule (materialized aggregation) will be used instead of the fact table. In this paper we extend that idea with metagranular indices, i.e. indices on metagranules. Assume a user issuing an aggregate query to a fact table with a selective `HAVING` or small `LIMIT-ORDER BY` clause. The database engine can not only identify the best metagranule but it can also use the index on that metagranule in order not to scan its full content. In this paper we present the proposed optimization method based on metagranular indices. We also describe its proof-of-concept prototype implementation. Finally, we report the results of performance experiments on database instances up to 350GiB.

Keywords: analityc queries, ORM, databases.

1 Introduction

Processing analytical queries is time-consuming, since they require scanning countless records. A database admin can avoid such enormous scans by materializing properly pre-aggregated data. On the other hand, he/she must also take into account the space occupied by such redundant aggregates and the overhead on updates they bring about. Therefore, an administrator has to find an equilibrium between gains (faster queries) and costs (space and processing overhead). This activity may be supported or even performed by automated tools.

The problem how to choose a proper set of indices for an application is an interesting research topic [3]. Analyses of database workloads lead to a number of index sets that can speed up query processing. The maintenance cost of each of these sets is then computed and compared with the profits. The set with biggest difference between profits and costs is suggested to a database admin [6,10,13,5,4]. A number of index advisors has been developed in commercial databases. Since the database workload changes over time, advisors have to be

S. Kozielski et al. (Eds.): BDAS 2014, CCIS 424, pp. 30–39, 2014.
© Springer International Publishing Switzerland 2014

rerun over and over again. Therefore, online index advisors [2] and benchmarks to compare them have emerged [14]. Further research on the choice of indices led to adaptive indices, e.g. *database cracking* [11] and *adaptive merging* [9]. Since both methods prove useful, a hybrid approach has also been proposed [12]. There is also a benchmark to compare adaptive indices [8].

In our research we focus on the object-relational mapping middleware (ORM) and its inherent optimisation potential. We showed that ORM could optimize analytic queries by materialization of partial aggregates [7]. We also described a programming interface to define aggregations worth materializing (called meta-granules) and a query rewriting that facilitates using them. In this paper we extend that approach by considering usage of indices on metagranules. Assume a query rewritten so that it uses materialized aggregates instead of base data. If this query contains a HAVING or ORDER-BY clause (preferably combined with LIMIT), an index on the used metagranule can further accelerate the query. Queries are analysed using a method similar to the one presented in [1]. The proposed method collects queries from an application, and then examines them to find indices that will potentially accelerate the application. In our proof-of-concept prototype, we use Hibernate ORM. Thus, the queries considered are HQL (Hibernate QL) queries. In this paper, we report results of experimental evaluation of this prototype on database instances up to 350 GiB. The results attest that the proposed method can significantly increase the efficiency of applications that use various ORMs. This paper makes the following contributions:

- we propose a novel method to advise and automatically generate indices on metagranules;
- we describe a query rewriting method that aids using these indices;
- we show our proof-of-concept prototype implementation of these ideas;
- we show performance evaluation of this prototype using database instances of the size up to 350GiB.

The paper is organized as follows. In section 2 we motivate the idea of granular indices. Section 3 reminds our approach to materializing partial aggregates in ORMs. In section 4 we describe the concept of metagranules and discuss the cost aspect of storing them. Section 5 presents the algorithm that analyses queries collected from an application. Section 6 reports the result of experimental evaluation of the proposed method. Section 7 concludes.

2 Motivation

Assume a business analysis application with a database of the schema as presented on Figure 1. Its user poses the query shown on Listing 1.1. In the absence of auxiliary data structures such query has to scan all rows of the fact table (invline). However, if the aggregation of the expression (inl.price * inl.qty) by customer's id is materialized, the query can be answered significantly faster. Further acceleration can be achieved if we create an index on the summed column of such materialized aggregate. In such a case, the query engine

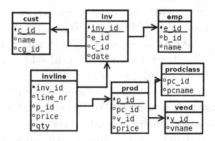

Fig. 1. The schema of a sample business database

Listing 1.1. The query to find twenty best customers

```
SELECT cust.cid, SUM(inl.price * inl.qty)
  FROM cust JOIN inv USING (c_id)
       JOIN inl USING (inv_id)
  GROUP BY cust.c_id
  ORDER BY SUM(inl.price * inl.qty) DESC
  LIMIT 20;
```

simply collects twenty tail entries from this index (provided it is stored in the ascending order).

The algorithms presented in this paper can *automatically* detect such an optimisation opportunity and suggest creating the corresponding materialized aggregate and its index. However, one has to be aware that maintaining them is costly. Therefore, we also report the results of an experimental analysis of costs and profits induced by these redundant data structures (see section 6).

3   Partial Aggregation

In this paper we focus on applications that use an object-relational mapping system to store their persistent data. Since contemporary ORMs have relatively limited functionality, an application programmer that wants to code analytic processing has two equally terrible options. He/she can accept notably low performance or bypass ORM mechanisms by directly addressing the database with SQL queries [7]. We proposed an extension to Hibernate that allows materializing aggregates without breaking the architecture of an application. Listing 1.2 shows an example class augmented with annotations @DWDim and @DWAggr that indicate aggregates to be materialized. Our extension interprets such annotations and (1) creates a table that stores SUM(quantity*price) grouped by inv_id, date, c_id, (2) augments the class Invoice with methods to query this table and (3) adds triggers that will synchronize this table with the base data.

Listing 1.2. The class Invoice with annotations on metagranules

```
@Entity
```

Listing 1.3. The query for twenty best customers using a finer aggregation

```
SELECT cust.c_id, SUM(sum_qty_x_price)
   FROM mg_inv
   GROUP BY cust.c_id
   ORDER BY SUM(sum_qty_x_price) DESC
   LIMIT 20;
```

```
@Granule(Dim = "id,_date,_customer"
            Agr = "Sum(invLines.quantity*invLines.price)")
@Granule(Dim = "date,....")
public class Invoice {
     @DWDim
     private Long id;
     @DWDim
     private Date date;
     @DWDim
     private Customer customer;
     @DWAggr(function="SUM(quantity*price)")
     private List<InvoiceLine> invLines;
}
```

Usually more than one subset of dimensions is used to aggregate data. In order to pass the set of all interesting aggregations we can use the annotation @Granule [15] as shown on Listing 1.2. This annotation indicates that the table mg_inv with columns inv_id, date, c_id, sum_qty_x_price is to be created. The first three columns are dimensions, while the fourth column is SUM(quantity*price) over these three dimensions.

4   Metagranules

Annotations presented in section 3 may impose a significant overhead on the database. Storing and maintaining numerous materialized aggregates requires space and time to synchronize data. However, we can observe that some analytic queries can be rewritten so that they used more finely aggregated data. The query from Listing 1.1 can be executed as the query from Listing 1.3. Although the original query aggregates over the customer id only, it can benefit from using finer aggregation over invoice id, date and customer id.

This query will not run in a fraction of seconds. For a 100 GiB database the query should finish in less than one minute. This is probably acceptable. The idea of materializing only a subset of desired aggregates has been presented [15]. Grouping levels called *metagranules* constitute a partial order. Figure 2 shows an example of such order. It contains a possible set of metagranules for the database schema from Figure 1.

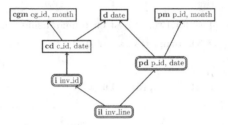

Fig. 2. The partial order of metagranules

Metagranules represent the aggregates used by the application. Some of them are chosen to be actually materialized. We call them *proper metagranules.* In Figure 2 their symbols have double border.

In the article [15] we presented methods to rewrite queries so that they use the most fit proper metagranule, i.e. the maximal metagranule smaller or equal to the desired metagranule. A smaller metagranule contains more records. Thus the query based on a smaller metagranule will finish later. For some metagranules there could be more than one metagranule that satisfies the abovementioned conditions. The metagranule d has two such proper metagranules: i and pd. Eventually, the algorithm chooses the one with smaller number of records. In [15] we performed experiments on a database instance of size 100 GiB. They confirmed the validity of this approach. This idea can be converted into an algorithm as presented in section 5.

The choice of the best set of proper metagranules constitutes another interesting problem. The more metagranules are proper the faster are the queries. On the other hand each proper metagranule induces a noteworthy cost of overhead, since it occupies space and slows down updates. In section 6 we show results of tests against database instances of the sizes 100GB and 350GB. These results reveal the impact of the introduction of new proper metagranules.

If a metagranule is proper, it can also have indices. Therefore, we have adapted the algorithms presented in [1] so that they can be used to suggest metagranular indices. Modified algorithms are discussed in section 5.

5   Processing HQL Queries

In this section we present the algorithms used in the proposed method. We start from the analysis of HQL queries to collect all metagranules possibly worth materializing. Assume that we have already collected the potential database workload, i.e. a set of queries from an application code. Initially the set of metagranules is empty. Then for each query the algorithm performs the following steps:

1. Build the abstract syntax tree (AST) for the query.
2. Locate the WHERE node and if it exists, find all database columns in its subtree, but not in aggregate functions.
3. Add all but aggregated columns to the set M of metagranule dimensions.

Table 1. Row counts in the tested database instances

Table	Small DB	Medium DB	Big DB
customer	360 000	600 000	1 200 000
invoice	59 995 970	199 948 701	601 628 100
invoiceline	629 949 439	2 099 416 306	6 316 455 713
product	12 000	24 000	36 000
product day aggr.	29 404 062	59 387 618	88 920 000

4. Find all non-aggregated columns in the subtrees of GROUP, ORDER and HAVING add them to the set M.
5. Find the aggregating function F in the AST.
6. Add M as the dimensions and F as the aggregate of the new metagranule.

Now we focus on the identification of possible metagranular indices for a query. The corresponding algorithm performs the following steps:

1. Find the metagranule for the query.
2. Build the abstract syntax tree (AST) of the query.
3. For each of the WHERE, ORDER and HAVING nodes in the AST, find all references to database columns that are also dimensions of the metagranule.
4. Add them as the columns of a potential metagranular index.

Finally, let us consider a query that can be potentially rewritten to use metagranules and metagranular indices. The corresponding algorithm performs the following steps:

1. Identify the metagranule for this query. If there is none, stop.
2. Build the abstract syntax tree (AST) of the query.
3. Remove from GROUP all columns that are dimensions of the metagranule.
4. If the GROUP node has no children, remove it.
5. Replace columns and aggregates that are members of the metagranule with columns from the table that stores the aggregated data of the metagranule.
6. Remove tables from the FROM node that are no longer referenced.
7. Add the table that stores the data of the metagranule to the FROM node.

6   Performance

We used a computer with Intel i5 3570 (ivy bridge) and 8GiB RAM. The storage was Raid 0 over 4xBlack Caviar 1TB controlled by Adaptec 2405. The operating system was Ubuntu 13.10. We used PostgreSQL 9.1. We tested against three database instances: small (base data of size 34 GiB), medium (130 GiB) and big (345 GiB). Each of them was tested in three variants. The variant *plain* contained just the base data. The variant *aggr* stored aggregated data on sales in metagranule tables mg_inv (aggregated by invoice) and mg_pd (aggregated by day). The instances in the variant *plain* have corresponding sizes 41 GiB, 134 GiB and 391 GiB. In the *index* variant, the medium and big databases have an

Table 2. The execution times of the queries for five best invoices on a given day

Variant	Medium DB	Big DB
plain	820 s	6 410 s
aggr, based on mg_inv	26 s	84 s
index (date) on mg_inv	0.4 s	0.8 s
index (date,sum_qty_x_price) on mg_inv	0.3 s	0.3 s

index on `date` and `sum_price_x_qty`. These indices increased the size of these databases to 145GiB and 418 GiB. Below we report and discuss the execution times of three analytic queries. For the sake of readability we present them in SQL, since there can be interested readers who are not familiar with HQL.

Listing 1.4. A query for five the best invoices on a given day

```
SELECT name, SUM(qty * price) AS sum_qty_x_price
  FROM invline JOIN inv USING (inv_id)
    JOIN cust USING (c_id)
  WHERE date = '2013.11.10'
  GROUP BY name, inv_id
  ORDER BY sum_qty_x_price DESC LIMIT 5;
```

Listing 1.5. The optimized query for five the best invoices on a given day

```
SELECT date, name, sum_qty_x_price
  FROM mg_inv JOIN cust USING (c_id)
  WHERE date = '2013.11.10'
  ORDER BY sum_qty_x_price DESC LIMIT 5;
```

Listing 1.6. A query for total sales for a given month sliced into days

```
SELECT date, SUM(qty * price)
  FROM invline JOIN inv USING(inv_id)
  WHERE EXTRACT(year FROM date) = 2013 AND
    EXTRACT(month FROM date) = 8
  GROUP BY date;
```

The first query finds five biggest invoices on a given day. Listing 1.4 shows its initial (user submitted) form. According to the methods discussed in this paper, this query gets rewritten to the form shown on Listing 1.5. The index suggesting algorithm presented in section 5 picked the indices (date) and (date,sum_qty_x_price) on the metagranule table mg_inv. Tab. 2 shows execution time of this query for various database instances and their variants. We can see that using a materialized aggregation significantly accelerates the query. If the system is severely loaded with such queries, it will be worth building the metagranular index.

The second query finds total sales for each day of a given month as shown on Listing 1.6. The result of this query can be computed from stored aggregates by

Table 3. Execution times of the query for total sales for a given month sliced into days

Variant	Small DB	Medium DB	Big DB
plain	365 s	2 494 s	> 2h
aggr, based on mg_inv	14 s	49 s	147 s
aggr, based on mg_pd	25 s	18 s	27 s

invoices or by products on particular days. Tab. 3 presents the execution times of this query for various database instances and their variants without indices.

Listing 1.7. A query for eight best vendors of a given month

```
SELECT vname, SUM(qty * il.price) AS sum_qty_x_price
  FROM vend JOIN prod USING(v_id)
    JOIN invline il USING (p_id)
    JOIN inv USING(inv_id)
  WHERE EXTRACT(year FROM date) = 2013 AND
    EXTRACT(month FROM date) = 8
  GROUP BY v_id, vname
  ORDER BY sum_qty_x_price DESC LIMIT 8;
```

Listing 1.8. The rewritten query for eight best vendors of a given month

```
SELECT vname, SUM(sum_qty_x_price) as sum_qty_x_price
  FROM vend JOIN prod p USING(v_id)
    JOIN mg_pd USING(p_id)
  WHERE EXTRACT(year FROM date) = 2013 AND
    EXTRACT(month FROM date) = 8
  GROUP BY p.v_id, vname
  ORDER BY suma DESC LIMIT 8;
```

Table 4. The execution times of the query for eight best vendors of a given month

Variant	Small DB	Medium DB	Big DB
plain	365s	1 521s	> 1h
aggr, based on mg_pd	10s	49s	31s

The third example query lists eight best vendors of a given month, i.e. vendors whose products had biggest sales. Listing 1.7 shows this query in the original form. Listing 1.8 presents the rewritten version that uses the materialized aggregation mg_pd. Tab. 4 reports executions times of these two version in various database instances.

It is obvious that one cannot store all desired aggregations. However, our experiments show that even a reasonably small subset of materialized aggregations can make the performance of big analytic queries acceptable.

Table 5. The impact of the presented methods in case of the big database

Version	Size		Insertion of 100 000 invoices		The query from Listing 6	
	GiB	Ratio	Time	Ratio	Time	Ratio
plain	345	100%	4.63s	100%	6 410s	100%
aggr	391	113%	556.97s	12 029%	84s	1.31%
index	418	121%	858.59s	(154%) 18 544%	0.3s	(0.36%) 0.0046%

Table 6. Time necessary to insert records on 100 000 invoices

Variant	Medium DB	Big DB
plain	5, 39 s	4, 63 s
aggr	326, 82 s	556, 97 s
index	598, 31 s	858, 59 s

We have also assessed the overhead imposed by proper metagranules. We have analysed the time needed to insert records on 100 000 invoices into three analysed database variants of three different sizes. The results are shown in Tab. 6. For a database variant with stored aggregations, appropriate triggers synchronize derived data. This causes also corresponding updates in granular indices, if they exist.

Tab. 5 shows how the presented methods impact the size of the database and the execution times of 100 000 insertions of invoices and the query from Listing 1.4. This summary contains data on the big database instance that stores 6 billions records in the table `invline`. In parentheses there are ratios of corresponding values from the last two rows.

7 Conclusions

In this paper we extend our previous research on metagranules, i.e. a partial order of possible materialized aggregates that accelerate analytic queries. We show algorithms that find possibilities to build indices on stored aggregations to further improve the efficiency of querying. Performed experimental evaluation proves that using metagranules and their indices can significantly improve the efficiency of an application. However, the overhead imposed by them is noteworthy. Therefore, one cannot materialize all desired aggregations. A quantitave model to balance the cost and profits of metagranules is needed to choose the optimal set of metagranules and indices. The development, tuning and verification of such model is a topic for our further research.

References

1. Boniewicz, A., Gawarkiewicz, M., Wiśniewski, P.: Automatic selection of functional indexes for object relational mappings system. International Journal of Software Engineering and Its Applications 7, 189–195 (2013)

2. Bruno, N., Chaudhuri, S.: An online approach to physical design tuning. In: ICDE, pp. 826–835 (2007)
3. Chaudhuri, S., Narasayya, V.R.: An efficient cost-driven index selection tool for Microsoft SQL Server. In: Proceedings of the 23rd International Conference on Very Large Data Bases, VLDB 1997, pp. 146–155. Morgan Kaufmann Publishers Inc., San Francisco (1997), http://dl.acm.org/citation.cfm?id=645923.673646
4. Choenni, S., Blanken, H., Chang, T.: Index selection in relational databases. In: Proc. International Conference on Computing and Information, pp. 491–496 (1993)
5. Choenni, S., Blanken, H.M., Chang, T.: On the automation of physical database design. In: Proceedings of the 1993 ACM/SIGAPP Symposium on Applied Computing: States of the Art and Practice, SAC 1993, pp. 358–367. ACM, New York (1993), http://doi.acm.org/10.1145/162754.162932
6. Finkelstein, S., Schkolnick, M., Tiberio, P.: Physical database design for relational databases. ACM Trans. Database Syst. 13(1), 91–128 (1988), http://doi.acm.org/10.1145/42201.42205
7. Gawarkiewicz, M., Wiśniewski, P.: Partial aggregation using Hibernate. In: Kim, T.-H., Adeli, H., Slezak, D., Sandnes, F.E., Song, X., Chung, K.-I., Arnett, K.P. (eds.) FGIT 2011. LNCS, vol. 7105, pp. 90–99. Springer, Heidelberg (2011)
8. Graefe, G., Idreos, S., Kuno, H., Manegold, S.: Benchmarking adaptive indexing. In: Nambiar, R., Poess, M. (eds.) TPCTC 2010. LNCS, vol. 6417, pp. 169–184. Springer, Heidelberg (2011), http://dl.acm.org/citation.cfm?id=1946050.1946063
9. Graefe, G., Kuno, H.: Self-selecting, self-tuning, incrementally optimized indexes. In: Proceedings of the 13th International Conference on Extending Database Technology, EDBT 2010, pp. 371–381. ACM, New York (2010), http://doi.acm.org/10.1145/1739041.1739087
10. Hammer, M., Chan, A.: Index selection in a self-adaptive data base management system. In: Proceedings of the 1976 ACM SIGMOD International Conference on Management of Data, SIGMOD 1976, pp. 1–8. ACM, New York (1976), http://dl.acm.org/citation.cfm?id=509383.509385
11. Idreos, S., Kersten, M.L., Manegold, S.: Database cracking. In: CIDR 2007, Third Biennial Conference on Innovative Data Systems Research, Asilomar, CA, USA, January 7-10, pp. 68–78 (2007) (Online Proceedings)
12. Idreos, S., Manegold, S., Kuno, H., Graefe, G.: Merging what's cracked, cracking what's merged: adaptive indexing in main-memory column-stores. Proc. VLDB Endow. 4(9), 586–597 (2011), http://dl.acm.org/citation.cfm?id=2002938.2002944
13. Rozen, S., Shasha, D.: A framework for automating physical database design. In: Proceedings of the 17th International Conference on Very Large Data Bases, VLDB 1991, pp. 401–411. Morgan Kaufmann Publishers Inc., San Francisco (1991), http://dl.acm.org/citation.cfm?id=645917.758359
14. Schnaitter, K., Polyzotis, N.: A benchmark for online index selection. In: Proceedings of the 2009 IEEE International Conference on Data Engineering, ICDE 2009, pp. 1701–1708. IEEE Computer Society, Washington, DC (2009), http://dx.doi.org/10.1109/ICDE.2009.166
15. Winiewski, P., Stencel, K.: Query rewriting based on meta-granular aggregation, pp. 457–468, http://csp2013.mimuw.edu.pl/proceedings/PDF/paper-40.pdf

Performance Analysis of .NET
Based Object–Relational Mapping Frameworks

Aleksandra Gruca and Przemysław Podsiadło

Institute of Informatics, Silesian University of Technology,
Akademicka 16, 44-100 Gliwice, Poland
aleksandra.gruca@polsl.pl

Abstract. Object-relational mapping is a technology that connects relationships with object-oriented entities, which aims to eliminate duplicate layers together with costs of maintenance and any errors arising from their existence. A lot of tools and technologies were designed in order to support and implement idea of object-relational mapping. In this paper we present the performance comparison of two most common object-relational mapping interfaces for .NET framework: Entity Framework and NHibernate. In the .NET developers community, there is a lot of discussion today about the similarities and differences of the both technologies. To address this issue, we compared the features and performance of both tools. We analysed the performance of Entity Framework and NHibernate for two different databases (MS SQL Server and PostgreSQL), different query languages (lambda expressions and LINQ for Entity Framework and HQL and Critera API for NHibernate) and compared the results with the standard SqlClient queries. The results show that there is no significant difference between these both tools and we proved that common opinion that NHibernate performs better than Entity Framework is incorrect.

Keywords: Object-relational mapping, ORM, .NET framework, Entity Framework, NHibernate, MS SQL Server, PostgreSQL, performance analysis.

1 Introduction

Most of the software applications which are nowadays developed are based on object-oriented programming languages and relational databases. Such approach allows creating applications which are built on the logical elements that interacts with each other, while the relational databases technology is responsible for preserving data structures in tabular forms with established relations. These two technologies, although they seems to be divergent, interact with each other in both theoretical and practical aspects. The object-oriented programming and relational databases are complement on data structures and interface middleware between data and the user. Theoretical background between these two technologies is obvious correspondence between the diagrams of database tables and the structure of the classes defined in the application.

S. Kozielski et al. (Eds.): BDAS 2014, CCIS 424, pp. 40–49, 2014.

Object-relational mapping (ORM) is a technology that connects relationships with object-oriented entities, which aims to eliminate duplicate layers together with costs of maintenance and any errors arising from their existence. A lot of tools and technologies were designed in order to support and implement idea of object-relational mapping but most of them is not suitable for commercial use and there are also some voices in the discussion that undermine the usability of object-relational mapping interfaces in larger projects [7].

The aim of the research presented in this paper is to analyse the performance and to compare two most popular and .NET based ORM interfaces. At the beginning stage of every software application project, when designing data access layer, a selection of the object-relation mapping interface is performed. Such decision must be considered very carefully, as subsequent costs of changing the interface even in partially implemented application, often outweigh the potential gains. Typically such a decision in made on the basis of the skills of a project team members, which often limits the number of choices and may have a bad influence on the process of designing application architecture. Therefore, information on the performance of data interfaces would be helpful in deciding on the appropriate technology.

1.1 Related Work

In the Internet one can find many articles, blogs and discussions comparing different ORM interfaces and various aspects of their practical application. In general, the discussion on ORM interfaces goes in two directions. First problem is how using any ORM interface influences the performance of database query and whole development process. Another discussion is which ORM framework is the best solution in terms of performance and other criteria that may be important for application development and maintaining. Currently, two most popular object oriented technologies are .NET and Java. There are a dozens different ORM tools for both technologies and their comparative list can be found at http://c2.com/cgi/wiki?ObjectRelationalToolComparison. Unfortunately, most of the websites that take up this subject is outdated or incomplete and there is a very little number of scientific publications focused on comparative analysis of ORM tools.

To address the above problems we analysed performance of two most popular open-source ORM interfaces based on .NET technology: Entity Framework [6] and NHibernate [1]. In the research performed in 2010 [4] the both tools were analysed, however over last years, both interfaces were extensively developed and improved, and especially Entity Framework got a lot of positive reviews since the previous release. Therefore it is reasonable to analyse if and how their performance has changed with new releases and also in comparison to the performance of the standard SqlClient queries.

The list of basic evaluation criteria for ORM tool suitability and general comparison of several ORM mapping tools based on proposed criteria can be found in [10], however the presented analyses are not focused on the tools performance. The up-to-date comparison for Java based ORM tools can be found in the article [3]. According to the authors best knowledge there is no such comparison for the current versions of selected .NET based ORM tools. In addition to the work presented in [4] we compared the performance of different query languages for both interfaces, that is lambda expressions and LINQ for Entity Framework and HQL and Critera API for NHibernate. To make our conclusion more general, we used two different relational databases.

The results of the research performed in this paper not only show the differences among analysed technologies: NHibernate and Entity Framework but also analyse whether introduced delay is within acceptable limits, if compared to the standard SqlClient mechanism. Comparative analysis were performed for Microsoft SQL Server [9] and PostgreSQL [11] databases, however the main aim of this work is comparison between interfaces, not the database systems, therefore each database has different structure, schema and data.

2 Relational Databases and Object-Oriented Programming

Persistence is ability of an object to outlive the application process in which it resides. This term is often used in conjunction with the problem of storing objects in databases as to obtain persistence, the object (its current state, properties and relations), needs to be somehow preserved in non-volatile storage such as a hard drive.

The paradigm of object-oriented programming supports the building of applications out of elements that have both data (in a current state) and behaviour. The main issue, when translating the logical representation of the objects into a form that can be stored in database, is the fact that objects do not only contain the data but also have ability to share the data and communicate with other objects. The relational paradigm, however, is based on mathematical principles and the most important issue is the data itself, the relationships and data structure.

The term object-relational impedance mismatch refers to technical and conceptual issues that arise when we try to combine object and relational artefacts. The mismatch occurs, because object-oriented programming paradigm and relational databases have different concepts of data. The impedance mismatch is manifested in several different forms [2]:

- Inheritance and Polymorphism - inheritance is rather unimportant element of the classical relational model, which focuses on the structural relationships between data. There are methods allowing mapping the inheritance hierarchies, but there is no simple method to represent polymorphism, which is crucial for object-oriented programming.
- Associations - classes that exist in object-oriented programming are associated in behavioural way, which is unimportant from the relational point of view. One can easily imagine two classes with sets of mutual references which in relational model would be represented as "many-to-many" relation which would require intermediate table.

- Data types - different database systems and object-oriented programming languages have specific data types and sometimes it is unclear how to map them. The examples of such problematic data types are *DateTime/datetime* type which can be represented in many different ways depending on implementation.
- Granularity - in object-oriented systems classes may exists on different levels of granularity, starting from classes that model business objects and ending on basic data types such as *DateTime* or *Integer*. Therefore, the idea that every class should map to its own table cannot be realized in practice, so the application designer must decide which class should be represented by its own table.
- Identity - identity is one of the most crucial issues in the relational model. In theory, each table should have its own primary key, which allows the record to be unique, even if other attributes have identical values. In case of object-oriented systems, there are two methods of identity verification of two objects. Reference comparison verify whether object references refer to the same area in memory, while value comparison compares objects attribute-by-attribute. Each of these identity types is different.
- Data navigation - naive implementation of an object graph navigation may result in an exponential explosion of application queries to the database. One can easily imagine the problem with "one-to-many" mapping when the function retrieving attribute values creates connection to the database for each item. Therefore, the main problem is how to create the mapping which will minimize number of queries and thus loads several entities via JOINs and selects the targeted entities before starting to walk network of objects.

3   Object-Relational Mapping Tools

Object-relational mapping is a technique for converting data between incompatible object and relational systems. The basic idea standing behind ORM tools is to delegate the management of persistence to the external mechanism. With the use of ORM tools it is possible to work at code-level with objects representing a domain model being separated from the structure of relational database. Therefore, the ORM technology establishes a bidirectional link with the data in a relational database and objects in code, based on a configuration and by executing SQL queries (dynamic most of the time) on the database [8].

The basic features that should be covered by any ORM mapping tool are ability to use inheritance, create hierarchies between entities and use polymorphism. Another important thing is to handle any type of relations (1-1, 1-n, n-m), support for transactions, aggregates (equivalent to SQL's SUM, AVG, MIN, MAX, COUNT) and support for grouping (SQL's GROUP BY). Additional features that should be taken into account when selecting ORM tool are a list of supported databases, ability of use query language, flexibility (customization of queries, SQL joins support, support for the data types specific to the database management system, etc.), optimization and many more numerous criteria. Detailed list and description of features that may be considered when selecting

ORM tool can be found in [8,10]. While the most of these criteria can be verified just by analysing specification and documentation of specific ORM tool, there is still a lot of uncertainty when considering performance of ORM tools. Below, we analyse some features of two selected open-source ORM interfaces for .NET Framework: Entity Framework and NHbiernate.

3.1 Comparison of Entity Framework and NHibernate

Entity Framework. Entity Framework is a set of technologies in ADO.NET supporting object-relational mapping in .NET Framework. The first version of Entity Framework was introduced in 2008 year and received a lot of criticism due to many errors and technical misgivings with design and implementation. Since the first release, Entity Framework has advanced significantly, mainly due to involvement the community feedback during design decisions. Official builds of Entity Framework are available as a fully supported Microsoft products both standalone as well as part of Visual Studio.

NHibernate. Hibernate is one of the most popular object-relational mapping tools. It is non-commercial tool, developed from 2001 and originally designed for Java language. Since 2005 there is also available version for .NET environment, called NHibernate. To achieve persistence, there is no requirement for a .NET objects to implement any NHibernate interface. To communicate with database, NHibernate uses metadata from XML mapping files instead of additional interfaces and attributes.

In Tab. 1 we present comparison of the most important features of both analysed ORM mapping tools [4,5]. The big difference between Entity Framework and NHibernate from a developer perspective is that the former offers an integrated set of services whereas the latter requires the combination of several open-source libraries. This problem can be easily solved by using external tools such as NuGet (http://www.nuget.org/), however sometimes these external libraries are not always up-to-date when new NHibernate version is released [5].

4   Experimental Analysis

4.1   Configuration of Testing Environment

Computational analyses were performed on the PC computer with processor Intel Core i5 M 560 2.67GHz, 4GB RAM and operating system Windows 8 64-bits. As testing environment we used .NET Framework, version 4.5.5, Entity Framework version 5.0 [6] and NHibernate, version 3.3.1 [1]. Analysed databases were Microsoft SQL Server 2012, version 11.0.3 with modified Nortwind database [9] and PostgreSQL, version 9.2 with modified Dellstore database [11].

To compare performance between two ORM interfaces we prepared eight different types of queries. For each ORM query the running time was measured and compared with the same operation performed with standard SqlClient method.

Table 1. Comparison of selected features of Entity Framework and NHibernate

	Entity Framework	NHibernate
Programming Language	C# and VB	C# and VB
Code-based mapping	Automated	Automated (from v3.2) or external Fluent NHibernate
Bidirectional relationships support	Yes	Yes
Transaction support	Yes	Yes
Blocking mechanisms	Optimistic and pessimistic	Optimistic and pessimistic
Optimization	Buffering and lazy loading	Buffering and lazy loading
Custom types and collections	No	Yes
Private fields mapping	No	Yes
Automatic migration	build-in Code-First Migrations	external Fluent Migrations
Queries	LINQ to Entities, Entity SQL, SQL	HQL, Criteria API, QueryOver, SQL
Database support	MS SQL, SQL Azure PostgreSQL, MySQL Oracle, SQLite, DB2, Firebird	MS SQL, SQL Azure, PostgreSQL, MySQL Oracle, SQLite, DB2, Firebird

Each query was executed 100 times and the average time was computed. Presented results do not include the data context initialization which is very expensive to create and can be performed in several different ways for each technology.

For testing purposes we selected queries that are the most representative and frequently used in software applications. We analysed basic CRUD (Create, Read, Update, Delete) functionality, as well as more advanced queries using popular operators such as joining, grouping, ordering and queries with subqueries. As we wanted to analyse if and how the performance of ORM tools evolved during the last several years, we compared our work with the results presented in article [4]. Therefore several of the queries and data types used were similar to these presented in [4] in terms of operators used and data types compared.

Each query was prepared in several versions: SQL with the use of SqlClient, HQL with the use of criteria API and with LINQ to Entites in two versions: as standard LINQ query and with lambda expressions. The queries used in our research for MySQL and PostgreSQL databases, are presented in Tab. 2 and Tab. 3 respectively.

4.2   Results for MSSQL Database

Analysis of the results for MSSQL database shows how small is the difference between two analysed ORM interfaces. This is obviously in opposition to the popular theory that performance of NHibernate is better than performance of Entity Framework. We can even notice that in some cases, Entity Framework performs a little better than NHibernate. It is also worth to notice that for a selected interfaces the time of the query execution is almost identical in most of

Table 2. Query list for MSSQL database

Q1	insert into Products values(SqlAddNew2,14,2,1kg per pack,10,null,null,null,0)
Q2	update Products set ProductName=SqlUpd where ProductName=SqlAddNew2
Q3	delete from Products where ProductName=SqlUpd
Q4	select * from Products p join Categories c on p.CategoryID = c.CategoryID join Suppliers s on p.SupplierID = s.SupplierID where ProductName =Producut 21099
Q5	select * from Products p join Categories c on p.CategoryID = c.CategoryID join Suppliers s on p.SupplierID = s.SupplierID where ProductID =21067
Q6	select * from Products p join Categories c on p.CategoryID = c.CategoryID join Suppliers s on p.SupplierID = s.SupplierID where UnitPrice =2000
Q7	select * from Products p join Categories c on p.CategoryID = c.CategoryID join Suppliers s on p.SupplierID = s.SupplierID where ProductName like %Producut% and ProductID = 21067 and UnitPrice > 20000 and UnitsOnOrder<1
Q8	select * from Products p join Categories c on p.CategoryID = c.CategoryID join Suppliers s on p.SupplierID = s.SupplierID where ReorderLevel!=13
Q9	select * from Products p join Categories c on p.CategoryID = c.CategoryID join Suppliers s on p.SupplierID = s.SupplierID where p.ProductID<50
Q10	select * from Products p join Categories c on p.CategoryID = c.CategoryID join Suppliers s on p.SupplierID = s.SupplierID where p.ProductID<50 and s.SupplierID <10
Q11	select * from Products p join Categories c on p.CategoryID = c.CategoryID join Suppliers s on p.SupplierID = s.SupplierID where ProductID > 21070 order by ProductID desc
Q12	select count(ProductID) from Products group by CategoryID
Q13	select * from Products p join Categories c on p.CategoryID = c.CategoryID join Suppliers s on p.SupplierID = s.SupplierID where s.SupplierID in (select SupplierID from Suppliers where Country = Germany)

Table 3. Query list for PostgreSQL database

Q1	insert into products(category, title, actor, price, special, common_prod_id) VALUES (2,'Sqladd2', 'Przemek Podsiadlo', 100, null, 1)
Q2	update products set title ='Sqlupd' WHERE title='Sqladd2'
Q3	delete from products WHERE title='Sqladd2'
Q4	select * from products p join categories c on p.category =c.category where title ='AIRPLANE BERETS'
Q5	select * from products p join categories c on p.category =c.category where prod_id =5600
Q6	select * from products p join categories c on p.category =c.category where price >20
Q7	select * from products p join categories c on p.category =c.category where title like '%AIRPLANE%' and prod_id = 6066 and price>10
Q8	select * from products p join categories c on p.category =c.category where special <>0
Q9	select * from products p join categories c on p.category =c.category where p.prod_id<50
Q10	select * from products p join categories c on p.category =c.category where p.prod_id<50 and c.category <10
Q11	select * from products p join categories c on p.category =c.category where prod_id > 9990 order by prod_id desc
Q12	select count(prod_id) from products group by category
Q13	select * from products p join categories c on p.category =c.category where c.category in (select category from categories where categoryname = 'Action')

the cases, therefore there is no additional overhead associated with creating the query with the use of lambda expressions for Entity Framework or Criteria API for NHbiernate.

When comparing query execution time to the standard SqlClient, the results are in general in accordance with expectations. The advantage of the standard SqlClient can be seen especially in CRUD queries when operation does not return any data and we take advantage from *ExecuteNonQuery* method. In case of more advanced queries, there are some examples, when the performance of ORM tool is even better then simple SqlClient, which is explained by better optimization for ORM based queries when referencing to database and mapping the results.

Known disadvantage of SqlClient is shallow mapping, when the mapped object from one-to-many relation have the instance of related objects, but this instance does not have any collection of the objects related to it, until another query to the database is executed. In case of using ORM tools, this problem is solved in automated manner. In addition, the mechanism of lazy loading will not load any object until it is referenced.

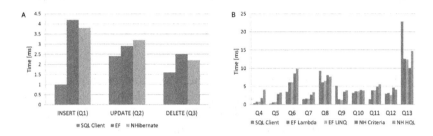

Fig. 1. MS SQL Server results: (A) – CRUD queries, (B) – advanced queries

Detailed comparison of our results with the results presented in [4] was impossible due to different testing environment and database version. However, even in a such condition we could make some general conclusions. In the results presented in [4], there is quite noticeable difference for GROUP BY between Entity Framework and NHibernate due to non-optimal manner in which group-by statements are translated into native SQL form by Entity Framework. This problem is obviously solved in the current version of Entity Framework as we do not observe such differences between both tools.

4.3 Results for PostgreSQL Database

In case of PostgreSQL, when comparing Entity Framework and NHibernate performance, we can see that NHibernate queries perform a little bit faster. This is the result of using external data provider NpgSQL http://npgsql.projects.pgfoundry.org which is developed by rather small community and with the limited resources. The influence of using NpgSQL can be also seen when analysing the differences in performance of the both interfaces and standard SQL client which are not big as in a case of other databases. Instead of NpgSQL one can use Devart's dotConnect for PostgreSQL http://www.devart.com/dotconnect/, however it is a commercial product.

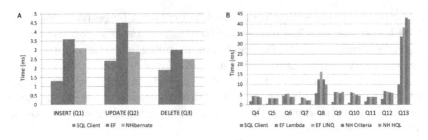

Fig. 2. PostgreSQL results: (A) – CRUD queries, (B) – advanced queries

5 Conclusions

Object-relational mapping interfaces are being more and more popular in the object-oriented applications, allowing to persist model objects to a relational database and to retrieve them. ORM technology provides a clean separation between physical structure of data and its object representation. There is a number of benefits to use ORM tools such as reduction of code, better application design and maintainability, and ability to take a full advantage of object oriented programming paradigm. Therefore it is necessary to measure the performance of ORM tools in order to recommend better and faster solution for developers.

In this paper we analysed the performance of the two most popular object-relational mapping interfaces for .NET platform. Both tools are open-source software, however they are mature and well-established products. We compared performance of the selected tools with standard SqlClient for several different, most frequently used types of queries.

The performance of standard SqlClient was in general better than performance of ORM tools, which is in accordance with intuition, as most ORM tools are designed to handle variety of data-use scenarios, far more than it is need for any single application. However, the differences between simple SqlClient approach and well-designed ORM tools is not a significant one, and the delay in execution time probably would not be noticed by the users. The analysis was performed for two different databases. As the database structures and queries were different, the results cannot be compared directly, however the observed tendency is similar in both cases, regardless of the structure and queries used.

It is also worth to notice, that there is a very small difference for NHibernate between HQL and Criteria API queries. The same observation can be made for Entity Framework when comparing performance of lambda expressions and LINQ. This allows developers to select the more suitable method. In case of using LINQ or HQL we have more readable and understandable code, especially for static and complex queries. However, when there is need to create dynamic queries, it is much easier to do it using Criteria API or lambda expressions.

The results presented in this research clearly shows that the widely held opinion that NHibernate performs better than Entity Framework is incorrect. Such opinion resulted from the drawbacks and errors that were present in the very first versions of this software, however it seems that nowadays Entity Framework gains more and more popular trust in .NET developers community. The integration of Entity Framework with the .NET Framework and extensive support for it in Visual Studio makes it more and more popular tool. In addition, .NET Framework can automatically generate edmx file, context and models which are related to database structure, while in case of NHibernate there is still need to use external and frequently commercial tools.

In the conclusion, performed analysis does not give the clear answer which ORM interface or query language should be recommended if considering only the tool performance. In the authors opinion this is desirable situation, as it gives more flexibility to the developers team. There is a lot of other criteria that need to be taken into account when recommending object-mapper solution, some of them were mentioned in section 3.

Acknowledgements. The work was supported by Polish Ministry of Science and Higher Education (0161/IP2/2011/71) and partially by Ministry of Science and Higher Education, as a Statutory Research Project (8686/E-367/S/2013).

References

1. NHibernate 3.3.1 (2013), `http://nhforge.org/`
2. Barnes, J.: Object-Relational Mapping as a Persistence Mechanism for Object-Oriented Applications. Master's thesis, Macalester College, Saint Paul, Minnesota, USA (2007)
3. Bhatti, S., Abro, Z., Rufabro, F.: Performance evaluation of java based object relational mapping tool. Mehran University Research Journal of Engineering and Technology 32(2), 159–166 (2013)
4. Cvetković, S., Janković, D.: A comparative study of the features and performance of ORM tools in a .NET environment. In: Dearle, A., Zicari, R.V. (eds.) ICOODB 2010. LNCS, vol. 6348, pp. 147–158. Springer, Heidelberg (2010)
5. Doomen, D.: Entity framework 5/6 vs nhibernate 3 – the state of affairs (2013), `http://www.dennisdoomen.net/2013/03/entity-framework-56-vs-nhibernate-3.html` (accessed December 2013)
6. Entity Framework team: Entity framework 5.0 (2013), `http://entityframework.codeplex.com/`
7. Fowler, M.: Ormhate (2013), `http://martinfowler.com/bliki/OrmHate.html` (accessed December 2013)
8. Marguerie, F.: Choosing an object-relational mapping tool, `http://madgeek.com/Articles/ORMapping/EN/mapping.htm` (accessed December 2013)
9. Microsoft Corporation: Microsoft SQL server 2012 (2013), `http://www.microsoft.com/en-us/sqlserver/`
10. Płuciennik-Psota, E.: Object relational interfaces survey. Studia Informatica 33(2A), 299–310 (2012)
11. The PostgreSQL Global Development Group: PostgreSQL 9.3.2 (2013), `http://www.postgresql.org`

Standardization of NoSQL Database Languages

Małgorzata Bach and Aleksandra Werner

Silesian University of Technology,
Gliwice, Poland
{malgorzata.bach,aleksandra.werner}@polsl.pl
http://www.polsl.pl

Abstract. NoSQL database systems have been becoming more and more popular and accepted by a database users thus their rapid development is nowadays observed. Because of this fact, modern database engines and their categories in the form of the Venn diagram are mentioned in the paper. Besides, the possibilities of using declarative languages that are modeled on SQL - the language for relational databases – in NoSQL, are presented. For this purpose selected NoSQL technologies are given in more details and their query languages are described. Moreover, the NoSQL language commands' equivalents of SQL standard are provided in this document.

Keywords: NoSQL, key-value databases, column family databases, document-oriented databases, graph databases, declarative language.

1   Introduction

The rapid growth of the NoSQL databases can be observed nowadays, although their conception has appeared quite recently. The term "NoSQL" was first used in 1998 by Carlo Strozzi to name his lightweight relational open-source database without SQL interface. The term appeared again in 2009 during the conference referring to open, distributed, no relational databases, took place in San Francisco. Cassandra, Voldemort, Dynamite, HBase and other dynamically developing databases were presented at that conference and the acronym "NoSQL" was voted to describe the class of the databases that differ from the classical relational model.

Some scholars, among them Carlo Strozzi [15] suggest that, as the current NoSQL movement departs from the relational model, it should be called more appropriately "NoREL" (i.e. Not RELational systems) or "NoJOIN" (i.e. systems without JOINing) [17].

It should be emphasized that the departure from relational model is not equivalent to belonging to the NoSQL one. For example the Hierarchical and Network database models, prevailing in the 60's, used neither relational model nor SQL language, nevertheless they are not to be included to the NoSQL trend now. This is because the development of the new movement has been related to the

S. Kozielski et al. (Eds.): BDAS 2014, CCIS 424, pp. 50–60, 2014.

expansion of the Internet, Web 2.0 network and social portals which have generated problems with computer systems' performance and scalability. That's why it is used to satisfy the above mentioned requirements [18].

NoSQL translated as a "Not Only SQL" expression may imply the departure from SQL language, but the authors of the new conception meant the departure from relational model [1]rather than the language.

However, the fact is that at the beginning of the appearance of this technology, the majority of NoSQL haven't offered the possibility of a usage of the declarative languages, such as SQL. Queries to the NoSQL databases have been usually created on the very low level of abstraction so data operating has required specialized knowledge of programming. Various and often unintuitive languages that should be learnt for the NoSQL systems usage, can be a serious problem for a really great number of their potential users.

For the tens of years with the relational databases domination, users used to operate the declarative language, but not the imperative one. As it is easier to specify **what**, in contrast to **how**, data should be retrieved. Therefore, more and more people talk about the need of language standardization and about the necessity of creating more user-friendly languages that access the non-relational data.

This paper presents and evaluates the work that has been achieved in the NoSQL area so far. For selected examples of the NoSQL solutions, the possibilities of using declarative structures in data processing are presented.

The article can be regarded as a guide in the stated range for all who have to deal with the need of choice **whether**, and if so, **what** NoSQL solution to choose.

2   The Database Market

Nowadays, it is difficult to find an area where databases do not have the application. However, the variety of applications makes it difficult to give a single solution (i.e. one/single data model) that would be appropriate for all situations.

This diversity is reflected on the diagram of Venn, presented in the Fig. 1. It can be seen that the database market is varied. There is a place for relational and non-relational databases. Database systems can also be divided into transactional (operational) and analytical. Solutions called NewSQL are also added in the Fig. 1. It is a class of modern relational database management systems that try to provide the scalability performance comparable to NoSQL systems. It is applied for online transaction processing (read-write) workloads while still maintaining the ACID [2] guarantees of a traditional relational database systems.

In the last years a variety of NoSQL databases has been developed mainly by practitioners and web companies to fit their specific requirements regarding: scalability, performance and feature-set. Because of the diversity of these

---

[1] SQL is commonly identified with Relational Database Management System – RDBMS.

[2] **ACID** (*Atomicity, Consistency, Isolation, Durability*) – a set of properties that guarantee the database transactions are processed reliably.

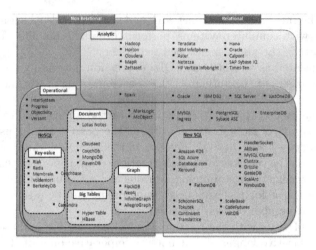

Fig. 1. The Database Market (based on: [14])

approaches, the classification of NoSQL systems is very difficult. Nevertheless, the very wide range – often niche – applications, NoSQL databases implementations can generally be classified into one of four main categories (based on data model):

- Key-Value Stores,
- Column Family Databases (Wide Column Store, BigTable Clones),
- Document-Oriented Databases,
- Graph Databases.

In the next chapters, the possibility of using declarative languages in data processing is discussed with respect to several selected NoSQL systems.

3 Cassandra

Apache Cassandra system was created especially for the Facebook network. The project started in 2007 in order to improve the process of searching users' messages (so called: InboxSearch problem). This solution is a very interesting hybrid of the model taken from the Google BigTable and the model of replication and partitioning from the Amazon Dynamo. Mentioned duality is marked in the Fig. 1, by placing Cassandra across the key-value and column-oriented databases [1].

3.1 Data Model

The data model of Cassandra looks like the model in the Google BigTable. It is said to be the prototype of all databases implementing column families. However there are some differences between both models – also in terms used.

There is a special nomenclature in the Apache Cassandra system:

- *Column* – is the smallest portion of information. The name, value and date of last modification are specified for each column. Only the newest version of data is stored in the Cassandra unlike in the Google BigTable, where a lot of data versions are stored.
- *Column Family* – is a set of rows and each row can have any number of columns. Thus the following rows can have different collection of columns. This structure can be compared with a table in a relational database, but it must be remembered that a significant difference is the lack of a rigid structure.
- *Keyspace* – is used to group column families together so a kind of namespace is created. It is said to be similar to a schema in a relational database.
- *Super Column* – is a structure that allows to group many columns.
- *Super Column Family* – the set of rows containing super columns.

3.2   CQL   Cassandra Query Language

At the beginning, the Cassandra using was considered to be rather difficult, so it wasn't very popular and doesn't gain developers' approval. Especially those of small applications. One of the attempts to make the system easier to use was the introduction (in 2011) of the conception of CQL (Cassandra Query Language) similar to the SQL. Currently, CQL-3 version is available. For example, a keyspace creation is made by a command:

```
CREATE KEYSPACE test WITH strategy_class = 'SimpleStrategy'
AND strategy_options:replication_factor = 1;
```

where parameters STRATEGY_OPTIONS:REPLICATION_FACTOR and STRATEGY_CLASS determine the number of replicates and the strategy of their placement, respectively. The sample statement that creates column family can be written as follows:

```
CREATE COLUMNFAMILY users (KEY text PRIMARY KEY, full_name text,
email text, state text, gender text, profile text,
birth_year int);
```

Instead of CREATE COLUMNFAMILY, the CREATE TABLE statement can be used. Data can be inserted by a suitable INSERT command – for example:

```
INSERT INTO users (KEY, full_name, email, state, gender,
birth_year)
VALUES (robnow', 'Robert Nowakowski ', 'bobjones@gmail.com',
'TX','M', '1980');
```

Data retrieval from a Cassandra table is realized analogously to the SQL language by a SELECT statement. For example:

```
SELECT * FROM users
WHERE birth_year='1975';
```

The syntax of SELECT statement is simplified in comparison with the original SQL language in RDBMS. There is no possibility of joining column families. The lack of JOIN servicing is one of the directives of a NoSQL idea. Filter conditions (WHERE phrase) can refer only to a key or the indexed columns. Besides, the only implemented aggregate function is COUNT. Additionally, it is not possible to group rows and sorting (ORDER BY) can take place only with a reference to the column that is a part of the composite key (*Composite Primary Keys*).

4 Hypertable

Hypertable is the open-source project, inspired by the Google BigTable system. It was started in 2007 by the engineers sponsored by Baidu, Rediff.com and Zvents Inc.

Hypertable runs on the basis of a distributed file system, such as for example the Apache Hadoop DFS, GlusterFS or Kosmos File System (KFS). It is written in C++ language.

4.1 Data Model

In a database that implements Hypertable family columns, as it was previously described in Cassandra system, the data is represented in the form of tables with the various structure of the rows. The key-value pairs are associated with the individual table cells. A key contains the ID of the row and the column. It means that there is the exact address information for each cell. Depending on the configuration used it is possible to store a great number of versions of each cell that differ in the timestamps.

4.2 HQL Hypertable Query Language

HQL is a declarative language similar to SQL that simplifies the work with Hypertable system.

Logical table grouping is achieved by namespaces that can be compared with the hierarchy of folders in the file system. For example the following commands:

```
CREATE NAMESPACE "/test";
USE "/test";
CREATE NAMESPACE "subtest";
```

cause the space 'test' and subspace 'subtest' are created. If the name begins with the '/' sign, it is treated as an absolute path, otherwise - it is treated as a subspace with respect to the current one.

Table is created by executing the command `CREATE TABLE`:

```
CREATE TABLE User (full_name, email, state, gender, profile,
birth_year
ACCESS GROUP default (full_name, email, state),
ACCESS GROUP profile (profile));
```

Hypertable does not support data types but values are treated as an opaque of bytes sequences, thus declaration of the particular columns cannot be found in the example. The possibility of access groups declaration is one of the most interesting features of the Hypertable system, because it has an influence on data storage. All data from the columns that belong to one group, are located together on a disk, which can reduce the number of Input/Output operations. In order to insert new data, the command `INSERT` is used:

```
INSERT INTO User VALUES
("row1", "full_name", "Robert Nowakowski"),
("2009-08-02 08:30:00", "row1", "email", bobjones@gmail.com);
```

Data is in a form of tuples' list that are separated by a comma. Each tuple represents the cell and can have one of two forms:

```
(row, column, value) or (timestamp, row, column, value)
```

In the first case, timestamp of a cell is automatically added, in the second one – it is explicitly given by the user. Queries are implemented by a `SELECT` statements – e.g.:

```
SELECT full_name FROM User WHERE name = "Robert Nowakowski";
SELECT * FROM User WHERE
'2008-07-28 00:00:02' < TIMESTAMP < '2008-07-28 00:00:07';
```

It is possible to define the conditions for rows, cells or timestamps, but – similarly to Cassandra system – there are a lot of limitations in comparison with the `SELECT` statement in RDBMS. For example there is no way of grouping, sorting or using the aggregate functions [4].

5   Neo4J

Neo4j is an open-source graph database that stores data in a graph, the most generic of data structures, capable of clear representing any kind of data in a highly accessible way. It is implemented in Java and is one of the older NoSQL systems. This system has been used in the production environments for 10 years. The community edition of the database is licensed under the free GNU General Public License (GPL) v3 [10].

5.1   Data Model

Neo4j is a graph database, that is, it stores data as nodes and relationships. Both nodes and relationships can hold properties in a key/value form. Property values can be either a primitive or an array of one primitive type. Nodes are often used to represent entities, but depending on the domain the relationships may be used for that purpose as well. Both – the nodes and relationships, have internal unique identifiers that can be used for the data search. The semantics can be expressed by adding directed relationships between nodes.

5.2   Cypher Query Language

Cypher is a declarative graph query language that allows to query and update of the graphs. Being a declarative language, Cypher focuses on the clarity of expressing **what** to retrieve from a graph, not **how** to do it, in contrast to imperative languages like Java, and scripting languages like Gremlin and the JRuby, which anyway can also be used in the Neo4j. Compared to the previously described languages, the syntax of Cypher commands is the least similar to the classic SQL [5,3].

For example, the equivalent SQL query:

```
SELECT *
FROM User
WHERE full_name = 'Mike'
```

is the following command in Cypher:

```
START User=node:User(full_name = 'Mike')
RETURN User
```

START clause specifies the starting point on the graph, from which the query is executed. Thus, the role of this phrase is something between FROM and WHERE clause of the SQL SELECT statement.
Cypher commands can embrace several parts, namely:

- START: Starting points in the graph, obtained via index lookups or by element IDs.
- MATCH: The graph pattern to match, bound to the starting points in START (it's equivalent to the SQL JOIN clause).
- WHERE: Filtering criteria.
- RETURN: What data set should be return. It is equivalent to the SQL SELECT clause.
- ORDER BY: Sorts the output.
- CREATE: Creates nodes and relationships.
- DELETE: Removes nodes, relationships and properties.
- SET: Set values of the properties.
- FOREACH: Performs updating actions once per each element in a list.
- WITH: Divides a query into multiple, distinct parts (the WITH clause is used to pipe the result from one query to the next one and to separate reading from updating of the graph).

The Cypher command mentioned above not only allows for data searching, but also their insertion, modification or deletion. Therefore, it is not only an equivalent of the SQL SELECT statement, but the UPDATE, INSERT and DELETE statements as well.

6   SPARQL

Neo4j supports Semantic Web technology which means that we can use RDF (Resource Description Framework) – directed, labeled graph data format. Each RDF data repository has implemented its own query language, making it difficult to move data between different documents. It caused a serious need to develop a common query language for semantic web. So, W3C organization has involved in the issue and in 2008 recommended their SPARQL (SPARQL Protocol And RDF Query Language) product as a language and protocol standard for RDF files.

The conception and syntax of SPARQL language is similar to the SQL and allows to query data set restricted by a criteria specified by the RDF predicates. RDF is a triple (entity1, property, entity2) that captures both entity attributes and relationship between entities as statement: **entity1** has **property** related to **entity2** (entity2 can be defined as a value of the property).

The SPARQL query comprises:

1. Prefix declaration where URI addresses of data, ontologies or other documents are defined.
2. Part that describes the form of a query (SELECT, CONSTRUCT, ASK, DESCRIBE).
3. Part that consists of a query pattern in the form of RDF triples.
4. Query modifiers (FILTER, ORDER BY, OPTIONAL etc.) rearranging query results.

There are four various types of SPARQL query that differ mainly in the form of returning result:

– SELECT form is used to create a list of URIs in the form of a table, which satisfy the pattern-matching requirements specified in the query [7].
– CONSTRUCT query returns an RDF graph, which is created by taking the results of the equivalent SELECT query and filling in the values of variables that occur in the CONSTRUCT template [9].
– ASK form is used to test whether or not a query pattern has a solution (is there any result for a given query pattern); it returns a boolean True/False value depending on whether or not the query pattern has any matches in the dataset [8].
– DESCRIBE query form returns all triples which contain all URIs which satisfy the pattern-matching requirements specified in the query – so it returns one RDF graph that describes a resources found. The implementation of this return form is up to each query engine.

7   Other Solutions

It is not possible to describe exactly all declarative languages working in the NoSQL products in one paper. Therefore in the consecutive paragraphs of this point, only selected – interesting in authors' opinion – projects that have been developed so far in this area are mentioned and/or labeled.

The OrientDB – graph-document database, written in Java language, supports SQL language, but in comparison with other NoSQL implementations, offers extended syntax with graph operators. There are, e.g. `ORDER BY` and `GROUP BY` phrases (although the current release supports only one field to group by) and the results can be extracted by the aggregate functions' usage. Besides, OrientDB allows a subqueries [6].

ArangoDB, earlier known as AvocadoDB, is a multi-purpose open-source database with the flexible data model for documents, graphs and key-values. This database is equipped with the declarative SQL-like query language called AQL (ArangoDB Query Language) [2].

It is worth mentioning here about the project known under the draft name UnQL (Unstructured Query Language) that was founded by Damien Katz and Richard Hipp. According to D. Katz "UnQL stems from our belief that a common query language is necessary to drive NoSQL adoption in the same way SQL drove adoption in the relational database market." [11]

So, the idea was to create a language that would allow the handling of documentary databases and semi-structural ones as well as all types of data stored in the JSON (JavaScript Object Notation) format. The assumption about the syntax of the language was the similarity to the SQL. It was due to personal experiences with SQL of both project initiators, who believed that the relational query language was enough to extend with new concepts typical for a not relational databases. Unfortunately, after very intensive work connected with UnQL, there was the significant slowdown in 2011 and the latest information about the project was in the first half of 2012.

A lot of scientific centers run researches that concern the intermediate systems between relational and non-relational databases. The prototype of such a system that translates queries formulated in SQL form used in MongoDB and Cassandra systems, is described for example in [13].

Also, the Quest Software company has developed the Toad for Cloud Databases software to migrate data between SQL and NoSQL and to perform queries to multiple databases. A question formulated in SQL is in the next step converted to the corresponding APIs that is necessary to return data from a database with the specific NoSQL solution. Toad for Cloud Databases cooperates with the Apache HBase, Amazon SimpleDB, Azure Table, MongoDB, Apache Cassandra, Hadoop and all databases that use Open Database Connectivity (ODBC) [16,12].

8   Summary

The described researches focused on the area of NoSQL databases – particularly on the analysis of declarative languages availability in the NoSQL solutions.

Authors attempted to assess the capability of developing standards in this area. Such solutions seem to be necessary because of several years of NoSQL databases presence in the modern systems. More and more companies use this model of a database rather than typical relational one or join it with the existing RDBMS (as both types cooperate together). Hybrid solutions and single non-relational, are increasingly popular, but the necessity of using different API for accessing various databases makes programming difficult. Besides, the problems with solutions portability appeared. Accordingly, the interface standardization of NoSQL database access is inevitable. Without the standardization, the study of a new NoSQL system will always be associated with learning a new programming language that can effectively discourage potential users.

As mentioned in previous chapters, NoSQL solutions market is quite diverse. There are used a lot of data models, and even within the same model differences are significant, as it can be seen for example with respect to Cassandra and HiperTable systems.

Thus, for sure, it will be very difficult to develop a single standard for all categories. In the authors' opinion, the first step of this process should be taken for each group of databases separately – i.e. key-value databases, databases implementing column family, graph and document ones.

References

1. Apache Cassandra^{TM} 1.1 Documentation (November 10, 2013), http://www.datastax.com/doc-source/pdf/cassandra11.pdf
2. ArangoDB Documentation (November 04, 2013), http://www.arangodb.org/documentation
3. Cypher queries (October 10, 2013), http://docs.neo4j.org/chunked/milestone/rest-api-cypher.html
4. HQL Reference (October 10, 2013), http://hypertable.com/documentation/reference_manual/hql/
5. Intro to NOSQL, and Cypher vs SQL: a declarative graph query language (September 14, 2013), http://www.meetup.com/Friends-of-Neo4j-Stockholm/events/87662782
6. OrientDB (October 15, 2013), https://github.com/orientechnologies/orientdb/wiki/SQL-Query
7. SELECT query form (October 10, 2013), https://code.google.com/p/tdwg-rdf/wiki/Beginners6SPARQL#6.4.3._SELECT_query_form
8. SPARQL 1.1 Query Language (October 10, 2013), http://www.w3.org/TR/2013/REC-sparql11-query-20130321
9. SPARQL by Example (October 10, 2013), http://www.cambridgesemantics.com/pl/semantic-university/sparql-by-example
10. The World's Leading Graph Database (September 10, 2013), http://www.neo4j.org/
11. Welcome to the UnQL Specification home (October 10, 2013), http://www.unqlspec.org/display/UnQL/
12. Welcome to Toad for Cloud Databases Community (October 10, 2013), http://toadforcloud.com/index.jspa

13. Cur'el, O., Hecht, R., Le Duc, C., Lamolle, M.: Data Integration over NoSQL Stores Using Access Path Based Mappings (November 10, 2013), http://hal.inria.fr/docs/00/73/83/56/PDF/finalDEXA.pdf

14. Kaskade, J.: Making Sense of Big Data (October 10, 2013), http://www.slideshare.net/infochimps/making-sense-of-big-data

15. Lith, A., Mattson, J.: Investigating storage solutions for large data: A comparison of well performing and scalable data storage solutions for real time extraction and batch insertion of data (August 10, 2013), http://publications.lib.chalmers.se/records/fulltext/123839.pdf

16. Mewald, M.: Quest zaprezentowało narzędzie do zarządzanie bazami NoSQL (October 10, 2013), http://webhosting.pl/Quest.zaprezentowalo.narzedzie.do.zarzadzanie.bazami.NoSQL

17. Strauch, C.: NoSQL Databases (November 12, 2013), http://oak.cs.ucla.edu/cs144/handouts/nosqldbs.pdf

18. Tiwari, S.: Professional NoSQL. John Wiley & Sons (2011)

Research on the Stream ETL Process

Marcin Gorawski[1,2] and Anna Gorawska[1]

[1] Silesian University of Technology,
Institute of Computer Science,
Akademicka 16, 44-100 Gliwice, Poland
{Marcin.Gorawski,Anna.Gorawska}@polsl.pl
[2] Wrocław University of Technology,
Institute of Computer Science,
Wybrzeże Wyspiańskiego 27, 50-370 Wrocław, Poland
Marcin.Gorawski@pwr.wroc.pl

Abstract. Continuously growing importance of information assisted by rapid development of systems that collect and process huge volumes of data has become a great problem in terms of processing and analyzing data. The response to current and future needs of market is a data warehouse assisted by process of data extraction. Mentioned stream ETL process enables loading real-time data without interrupting processing or conducting analysis that supports decision-making processes. This paper presents first implementation of the stream ETL process which origins from model and concept of a Stream Data Warehouse. In the first part of this paper the concept of the Stream Data Warehouse and its major components, including stream ETL, will be presented. The second part contains description of a developed stream ETL engine, as well as results of performed accuracy and efficiency analysis. Finally, paper concludes with description of future research issues that will be addressed in further research on the presented solution.

Keywords: data warehouse, data stream processing, ETL process, load balancing, Stream Data Warehouse.

1 Introduction

The increasing phenomenon of an information overload is a direct result of continuous trend to decrease cost of data distribution with the development of data processing platforms being not sufficiently fast. Therefore, sending data package is not a major problem, while processing increasing number of data volumes is a huge challenge. Moreover, processing the most current data, based on which companies make decisions to adapt operational processes, corporate strategies and business models, has become one of the top priorities. As the access to the latest information translates directly into profits, it is desirable to conduct predictive analysis to detect needs, risks and trends in the market.

In order to minimize side effects of the information overload data volumes are often stored in a data warehouse (DW) [2, 5, 8, 9, 24, 35]. The aim of

S. Kozielski et al. (Eds.): BDAS 2014, CCIS 424, pp. 61–71, 2014.

this treatment is to increase efficiency of advanced analytical and statistical calculations. This is possible because data warehouses are usually designed for analytical processing of large data sets that are available to users in a read-only mode. The data warehouse stores information that are structurally optimized for a certain area of real problems, which gives possibility to perform advanced analysis of large volumes of data. In addition, DW enables to integrate data of non-uniform format, providing the same level of access to all resources.

Thus, DWs has become preferred tools with an ETL process (Extract, Transform, Load) [1–4, 6, 22, 23, 29, 30, 32–34] - one of the elementary components of DW. Currently, the biggest weakness of a classic approach to the ETL process is that in order to perform a necessary condition to obtain exclusive access to the DW's resources has to be fulfilled. Thus, data updates in the classic DW take place only when unconditional access to DW is not required. From this perspective, the classic approach is inadequate and does not meet current requirements of international markets.

2   Stream Data Warehouse

Since establishment of the data warehouse concept a number of ideas that have responded to the problems of the last decade, were created. New solution is the Stream Data Warehouse (StrDW), which was presented in [7, 10, 13, 21] along with many other models and solutions. The Stream Data Warehouse is a solution that combines idea of the data warehouse with stream data processing model benefits. Therefore, it is more complex and sophisticated tool.

Data Stream Processing Model. Data stream processing model [5, 11, 12, 14, 16–18, 21, 25–27, 31] is one of solutions to the problem of handling huge volumes of data. The foundation of systems based on this model is assumption of data streams existence, i.e. infinite collections of data that supply system in an unpredictable manner. Moreover, the order and content of data volumes is independent of the system. Streams are used only for transmission of data in a form of tuples. According to [7] data stream can also be defined as interrelated thematic elements (records) belonging to unlimited set of tuples and timestamps. A tuple is the smallest possible portion of data containing information about a particular phenomenon or state of tested situation at a specific point in time. However, a feature that mostly distinguishes this particular model from others is the fact that tuples are processed on the fly and then deleted.

Stream Data Warehouse's Architecture. According to the definition given in [7], the Stream Data Warehouse creates a platform for processing a variety of stream data sets (consisting flat and spatial data) stored in a Stream Materialized Aggregate List (StrDW[MAL])[7, 20, 21] in a form of an analytic information. The StrDW concept assumes existence of many collaborating engines: Stream ETL (StrETL), Stream ETL Recovery, StrDW[MAL], StrSOLAP, StrSOLAM and Data Privacy Preservation system.

With respect to the data stream processing model the StrDW is connected to stream data sources (StrDS). It is assumed that StrDS systems do not have ability to store or archive generated tuples. In consequence, in case of interruption or connection failure undelivered data will be lost without possibility of its recovery or restoration.

Stream ETL Process. According to the concept of StrDW [7] an StrETL engine is actually a network of interconnected components that allow to make operations like in the classic ETL process but on data derived from data streams. Elements forming structure of the StrETL engine are:

- Remote Buffer Framework (RBF),
- Remote Integrator Framework (RIF),
- ELT-RT,
- Fault Tolerant Integrator (FTI).

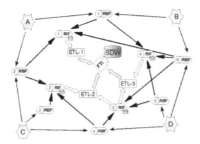

Fig. 1. An example of StrETL architecture

Fig. 1 presents an example of a network creating the StrETL engine, where objects marked in large letters from A to D are data streams. Each source is plugged into at least one RBF component, which task is to immediately receive data from the tethered StrDS.

In the next phase, tuples are processed by RIF modules, which obtain data from linked RBF's, store it and then allow to be downloaded by ETL-RT modules. A great advantage of RIF components is that they store tuples obtained from many RBF along with identifiers of the input streams. Thus, in case of engines failure RIF modules can be switched between source RBFs.

Finally, the last module within the StrETL engine is a Fault Tolerant Integrator (FTI) that detects errors, analyzes redundant data streams and merges them into one so processed data can be loaded into the StrDW[MAL]. Thus, in an ETL-RT module insertion will be held by RIF module.

3   Architecture of the Stream ETL Engine

The StrETL-RT engine, codenamed StrETL-RT, is powered by data streams which source is a sensors system. It is assumed that sensors are spread over a

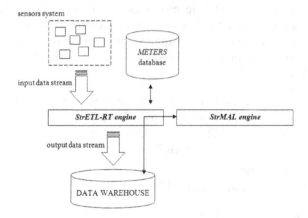

Fig. 2. Architecture of the StrETL-RT engine

closed geographical area. Therefore, apart from data streams the engine is also supplied by an external relational database called METERS in which information about the sensors system is stored, e.g. the number of elements of the system, its technical information as well as detailed data on geographical location.

One of the main objectives of this research was creation of an interface between StrETL-RT and StrMAL engines. As it is shown on the Fig. 2, the StrETL-RT engine is also supplied by requests from the StrMAL engine. With the bidirectional connection between engines existence of two streams is assumed: current and historical. The first one is sent by the StrETL-RT engine and contains real-time data derived from source systems. It is worth mentioning that in current stream only newest and properly transformed data is being sent. Simultaneously, the same tuples are transported to the last layer. Therefore, the StrMAL engine can access the most recent data, without connecting to a data warehouse database (DW[DB]). The historical stream contains data that is stored in the DW[DB] and cannot no longer be found in the current stream.

In order to emphasize the data stream processing model and avoid problems described in the first part of this paper, the StrETL-RT engine has got an exclusive access to the DW[DB]. Therefore, the StrMAL engine is obligated to direct its requests through the StrETL-RT engine.

Created StrETL-RT engine can be divided into three logical layers: identification and extraction, analysis and transformation and finally load balancing and loading layer, which will be described in forthcoming sections. The StrETL-RT engine's architecture and internal structure has been outlined in the Fig. 3, where each item marked as Si symbolizes single data stream. Source system generates data in form of a stream with respect to presented schema: sensors identifier;timestamp;measured value, while DW database requires additional information. All of them are stored in the METERS database and ought to be extracted in order to maintain tuples integrity with a DW[DB] format.

In general, StrETL-RT engine's layers run in separate processes in which multiple threads perform different operations.

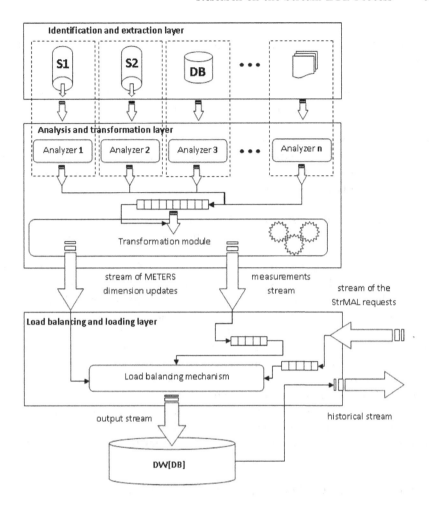

Fig. 3. The StrETL-RT engine's architecture

Identification and Extraction Layer. Extraction of data from autonomous sources is the most critical stage of the ETL process, thus the biggest challenge. Correct retrieval of data from multiple sources determines the efficiency of the whole process.

The identification and data extraction layer is responsible for downloading new data volumes, i.e. tuples which have not occurred in the DW[DB] yet. In case of a database, extraction is incremental, i.e. selection includes only those data volumes that has been entered or updated after the end of the last extraction. When it comes to extraction from stream sources it is not necessary to check whether the data has already been loaded into the DW. The data stream consists only the latest data. According to the data stream definition, tuples derived from sensors are directly transmitted to the StrETL-RT engine's input without prior archiving.

When the data is readout from a newly added sensor with unknown identifier, the METERS database is queried by the StrETL-RT engine, whether the object with a predetermined ID is a part of the sensors system. If the answer is positive then information about it is derived from the database. Consequently, only data characterizing the new sensor is loaded, so that in the next step measurements from this particular sensor may also supply the DW[DB].

Analysis and Transformation Layer. At the final stage of the ETL process volumes of data already adapted to the format of the DW are processed. Therefore, it is assumed that the data appearing at the input of this layer is properly formatted. In this phase, data is written to the relevant facts or dimensions tables of the DW[DB].

In the analysis and transformation layer tuples derived from the sensors StrDS by the previous layer are subjected to grammatical analysis in order to determine their structure and compliance with a specified grammar. Thus, raw data volumes are subjected to adaptation to the current DW database format. Described stream tuples input schema forced to create a LL(1) type grammar, which enables correct lexical and syntactic input analysis. First, the analysis is performed, i.e. each tuple is decomposed into elements and then an intermediate representation is formed, which is subjected to further transformation. Tuples undergo transformations after which such prepared data is sent to the next layer, which is responsible for communication with the data warehouse. In addition, all tuples, except those updating the list of sensors, are sent simultaneously to the input of the StrMAL engine as a current stream.

In this layer operations performed on the data meet the requirements for data quality. First goal is to ensure their accuracy and completeness. In addition, some operations cause designation of the relation necessary to maintain data integrity within the DW.

Load Balancing and Loading Layer. The last layer receives transformed tuples from sensors and updates of the METERS table (information about newly added sensors) from the previous layer, as well as inquiries from the StrMAL engine. In this layer all mentioned requests (from which appropriate queries will be created) are scheduled, which means that execution order is being determined. Such alignment is referred as load balancing [6] and is a key mechanism in the described layer. In case of developed engines, queries supplying this layer must be queued so as to meet the requirements for data quality [28] in the context of DW resources and quality of service. The issue of QoS is understood as minimization of a response time to queries addressed to the StrETL-RT engine by the StrMAL. Thus, the load balancing algorithm should allow loading the most recent data to DW and to answer requests from the StrMAL engine as soon as possible.

The main task assigned to the last layer is updating the DW[DB] with tuples coming from the analysis and transformation layer. At the entrance of the last layer there are three streams connected: stream updating METER dimension, measurements and one coming from the StrMAL engine. The last of these streams provides two-way communication between the engines. Assuming independent

StrETL-RT engine's action, i.e. without connecting to the StrMAL engine, a set input streams decreases to the first two. It is assumed that query inserting a new measurement to the DW is type of insert, and requests from the StrMAL type of select. Thus, a insert query queue is called inserts and a queue of select queries was named selects. The order of the requests depends on the chosen algorithm. For the StrETL-RT engine two algorithms were created: with and without connecting to the StrMAL engine.

The first load balancing algorithm enables scheduling requests, regardless of the engine's StrMAL connection availability, while the second one supports both queues: inserts and selects. Operation of both algorithms proceeds with emphasis to data quality preservation problem. In both created algorithms inquiries updating METER dimension are always supported in the first place. Only then scheduling of other requests can be started.

4   Test Results

Engines StrETL-RT and StrMAL are first implementation of models derived from the concept of the StrDW. As the engines StrETL-RT and StrMAL are unique solutions it is not possible to compare them with any other software on the market. Consequently, this section will present results of studies conducted only for the StrETL-RT engine without active connection with the StrMAL engine.

Measured time metrics are representing duration of the phenomenon - tuple's processing. Each of the designated metrics refers to a single tuple and shows another dimension of its processing in the StrETL-RT engine. Among measured time metrics are:

- tuple processing time - time measured from the moment tuple enters the engine until it is loaded into the DW[DB],
- tuple processing time in the analysis and transformation layer - time interval counted from the moment of entering the second layer to its departure to the next one.

All of performed tests were conducted with emphasis to a changing number of sensors in the source system. The main reason for these studies was need to verify engine's performance under changing load conditions. The consequence of increasing the number of sensors was enlargement of a input data set. Therefore, load concept will be used in the context of the number of registered sensors.

Each of conducted tests was associated with a collection of several thousand measurements obtained from multiple sensors. Consequently, presentation of all results would not be possible. Graphs presented in this section show only sample results, for no more than 100 consecutive measurements, showing a general trend. In addition, results contain only correct tuples - no measurement for fault tuples, which had to be rejected, had been shown.

In most cases, for 4 or 10 sensors, the processing time varied between 0 and 20 milliseconds. This means that in theory tuples are processed without any unnecessary delays.

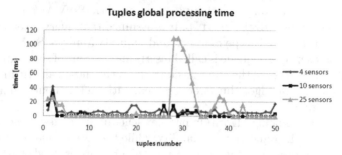

Fig. 4. Tuples global processing time

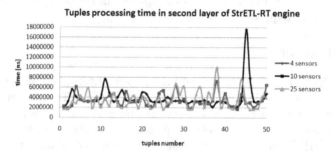

Fig. 5. Tuples processing time in the analysis and transformation layer

Similarly as in previous measurements, the number of registered sensors is proportional to the time efficiency of the layer. However, at the end of the presented set of samples a rapid spike in values was observed. The probable cause of this phenomenon is the explosion of data on the inputs of the engine. However, all measurements are collocated with tuples free of errors. As a result, processing times characterizing fault tuples are not included in this sample set. Therefore, the observed peek may also have a direct connection with the processing of a long sequence of incorrect tuples.

5   Summary

This paper presents first attempt of collaborating the data stream processing model with concepts of the ETL process, as well as an introduction to study of the StrDW. Presented experimental results have shown that the StrETL-RT engine is working properly and is capable of efficient data processing even in critical situations. The main goal of future research is to implement all components of the StrDW together with the StrSOLAP engine as the first implementation of the prototype presented in [7].

The future research on the StrETL engine alone will focus on implementing all components that creates the StrETL environment. Furthermore, a problem of process resumption will be addressed as well as user identity identification

[15]. Parallel processing, especially on CUDA architecture [19], is next planned area of interest in terms of performing transformation and analysis within the StrETL process more efficiently.

References

1. Albrecht, A., Naumann, F.: Managing ETL processes. In: Proceedings of the International Workshop on New Trends in Information Integration, NTII 2008, Auckland, New Zealand, August 23, pp. 12–15 (2008)
2. Athanassoulis, M., Chen, S., Ailamaki, A., Gibbons, P.B., Stoica, R.: MaSM: efficient online updates in data warehouses. In: Proceedings of the ACM SIGMOD International Conference on Management of Data, SIGMOD 2011, Athens, Greece, June 12-16, pp. 865–876. ACM (2011)
3. Bergamaschi, S., Guerra, F., Orsini, M., Sartori, C., Vincini, M.: A semantic approach to ETL technologies. Data and Knowledge Engineering 70(8), 717–731 (2011)
4. Berkani, N., Bellatreche, L., Khouri, S.: Towards a conceptualization of ETL and physical storage of semantic data warehouses as a service. Cluster Computing 16(4), 915–931 (2013)
5. Gorawski, M., Morzy, T., Wrembel, R., Zgrzywa, A.: Advanced data proceedings and analysis techniques. Control and Cybernetics 40, 581–583 (2012)
6. Gorawski, M.: Architecture of parallel spatial data warehouse: Balancing algorithm and resumption of data extraction. In: Software Engineering: Evolution and Emerging Technologies. Frontiers in Artificial Intelligence and Applications, vol. 130, pp. 49–59. IOS Press (2005)
7. Gorawski, M.: Advanced data warehouses. Habilitation. Studia Informatica 30(3B), 386 (2009)
8. Gorawski, M.: Multiversion spatio-temporal telemetric data warehouse. In: Grundspenkis, J., Kirikova, M., Manolopoulos, Y., Novickis, L. (eds.) ADBIS 2009. LNCS, vol. 5968, pp. 63–70. Springer, Heidelberg (2010)
9. Gorawski, M., Bańkowski, S., Gorawski, M.: Selection of structures with grid optimization, in multiagent data warehouse. In: Fyfe, C., Tino, P., Charles, D., Garcia-Osorio, C., Yin, H. (eds.) IDEAL 2010. LNCS, vol. 6283, pp. 292–299. Springer, Heidelberg (2010)
10. Gorawski, M., Chrószcz, A.: The design of stream database engine in concurrent environment. In: Meersman, R., Dillon, T., Herrero, P. (eds.) OTM 2009, Part II. LNCS, vol. 5871, pp. 1033–1049. Springer, Heidelberg (2009)
11. Gorawski, M., Chrószcz, A.: Query processing using negative and temporal tuples in stream query engines. In: Szmuc, T., Szpyrka, M., Zendulka, J. (eds.) CEE-SET 2009. LNCS, vol. 7054, pp. 70–83. Springer, Heidelberg (2012)
12. Gorawski, M., Chrószcz, A.: StreamAPAS: Query language and data model. In: 2009 International Conference on Complex, Intelligent and Software Intensive Systems, CISIS 2009, Fukuoka, Japan, March 16-19, pp. 75–82. IEEE Computer Society (2009)
13. Gorawski, M., Chrószcz, A.: Optimization of operator partitions in stream data warehouse. In: DOLAP 2011, Proceedings of the ACM 14th International Workshop on Data Warehousing and OLAP, Glasgow, United Kingdom, October 28, pp. 61–66. ACM (2011)

14. Gorawski, M., Chrószcz, A.: Synchronization modeling in stream processing. In: Morzy, T., Härder, T., Wrembel, R. (eds.) Advances in Databases and Information Systems. AISC, vol. 186, pp. 91–102. Springer, Heidelberg (2013)

15. Gorawski, M., Chrószcz, A., Gorawska, A.: Customer unification in E-commerce. In: Yin, H., Tang, K., Gao, Y., Klawonn, F., Lee, M., Weise, T., Li, B., Yao, X. (eds.) IDEAL 2013. LNCS, vol. 8206, pp. 142–152. Springer, Heidelberg (2013)

16. Gorawski, M., Gorawska, A.: AGKPStream a operatory strumieniowe. Studia Informatica 33(2A), 181–195 (2012)

17. Gorawski, M., Gorawska, A.: Stream join operators. In: 10th Students Science Conference Man-Civilization-Future. Oficyna Wydawnicza Politechniki Wroclawskiej (2012)

18. Gorawski, M., Gorawska, A., Pasterak, K.: Evaluation and development perspectives of stream data processing systems. In: Kwiecień, A., Gaj, P., Stera, P. (eds.) CN 2013. CCIS, vol. 370, pp. 300–311. Springer, Heidelberg (2013)

19. Gorawski, M., Lorek, M., Gorawska, A.: CUDA powered user-defined types and aggregates. In: 27th International Conference on Advanced Information Networking and Applications Workshops, WAINA 2013, Barcelona, Spain, March 25-28, pp. 1423–1428. IEEE Computer Society (2013)

20. Gorawski, M., Malczok, R.: On efficient storing and processing of long aggregate lists. In: Tjoa, A.M., Trujillo, J. (eds.) DaWaK 2005. LNCS, vol. 3589, pp. 190–199. Springer, Heidelberg (2005)

21. Gorawski, M., Malczok, R.: Indexing spatial objects in stream data warehouse. In: Nguyen, N.T., Katarzyniak, R., Chen, S.-M. (eds.) Advances in Intelligent Information and Database Systems. SCI, vol. 283, pp. 53–65. Springer, Heidelberg (2010)

22. Gorawski, M., Marks, P.: Influence of balancing used in a distributed data warehouse on the extraction process. In: Draheim, D., Weber, G. (eds.) TEAA 2005. LNCS, vol. 3888, pp. 84–98. Springer, Heidelberg (2006)

23. Gorawski, M., Marks, P.: Resumption of data extraction process in parallel data warehouses. In: Wyrzykowski, R., Dongarra, J., Meyer, N., Waśniewski, J. (eds.) PPAM 2005. LNCS, vol. 3911, pp. 478–485. Springer, Heidelberg (2006)

24. Gorawski, M., Marks, P.: Checkpoint-based resumption in data warehouses. In: Sacha, K. (ed.) Software Engineering Techniques; Design for Quality. IFIP, vol. 227, pp. 313–323. Springer, Boston (2006)

25. Gorawski, M., Marks, P.: Fault-tolerant distributed stream processing system. In: 17th International Workshop on Database and Expert Systems Applications (DEXA 2006), Krakow, Poland, September 4-8, pp. 395–399. IEEE Computer Society (2006)

26. Gorawski, M., Marks, P., Gorawski, M.: Collecting data streams from a distributed radio-based measurement system. In: Haritsa, J.R., Kotagiri, R., Pudi, V. (eds.) DASFAA 2008. LNCS, vol. 4947, pp. 702–705. Springer, Heidelberg (2008)

27. Gorawski, M., Pasterak, K.: Schedulery strumieniowe w AGKPStream. Studia Informatica 33(2A), 197–210 (2012)

28. Henschen, D.: 2013 analytics and info management trends. Information Week Report ID: R6061112 (2013)

29. Jörg, T., Deßloch, S.: Towards generating ETL processes for incremental loading. In: Proceedings of the 2008 International Symposium on Database Engineering & Applications. ACM International Conference Proceeding Series, vol. 299, pp. 101–110. ACM (2008)

30. Kakish, K., Kraft, T.A.: ETL evolution for real-time data warehousing. In: Proceedings of the Conference on Information Systems Applied Research (2012) ISSN 2167-1508
31. Mathioudakis, M., Koudas, N.: Twittermonitor: trend detection over the twitter stream. In: Proceedings of the 2010 International Conference on Management of Data, pp. 1155–1158. ACM (2010)
32. Vassiliadis, P.: A survey of extract-transform-load technology. International Journal of Data Warehousing and Mining (IJDWM) 5(3), 1–27 (2009)
33. Vassiliadis, P., Simitsis, A.: Near real time ETL. In: New Trends in Data Warehousing and Data Analysis, Annals of Information Systems, vol. 3, pp. 1–31. Springer US (2009)
34. Waas, F., Wrembel, R., Freudenreich, T., Thiele, M., Koncilia, C., Furtado, P.: On-demand ELT architecture for right-time BI: Extending the vision. International Journal of Data Warehousing and Mining (IJDWM) 9(2), 21–38 (2013)
35. Wrembel, R.: On handling the evolution of external data sources in a data warehouse architecture. In: Integrations of Data Warehousing, Data Mining and Database Technologies, pp. 106–147 (2011)

XML Warehouse Modelling and Querying

Fatma Abdelhedi, Landry Ntsama, and Gilles Zurfluh

IRIT SIG/ED - 118 route de Narbonne, 31062 Toulouse, France
{fatma.abdelhedi,landry.ntsama,gilles.zurfluh}@irit.fr
http://www.irit.fr/-Equipe-SIG

Abstract. Integrating XML documents in data warehouse is a major issue for decisional data processing and business intelligence. Indeed this type of data is increasingly being used in organisations' information system. But the current warehousing systems do not manage documents as they do for extracted data from relational databases. We have therefore developed a multidimensional model based on the Unified Modeling Language (UML), to describe an XML Document Warehouse (XDW). The warehouse diagram obtained is a Star schema (StarCD) which fact represents the documents class to be analyzed, and the dimensions correspond to analysis criteria extracted from the structure of the documents. The standard XQuery language can express queries on XML documents, but it is not suitable for analyzing a warehouse as its syntax is too complex for a non IT specialist. This paper presents a new language aimed at decision-makers and allows applying OLAP queries on a XDW described by a StarCD.

Keywords: XML warehouse, OLAP, XQuery, Multidimensional Modelling.

1 Introduction

Due to the increasing volume of information contained in databases and web, decision-makers are facing problems concerning the extraction, the aggregation and the analysis of data containing heterogeneous formats. In such an available mass of information, only transactional data extracted from relational databases are mainly exploited. Complex data contained in XML documents [9] are on the other hand barely exploited, even not at all. Yet, all these data represent a significant source of information for the decision-making process.

Thus, in order to integrate XML documents into a warehouse, it is first of all advisable to have a multidimensional model. We have extended the classical star diagram model [2], proposing a model where facts and dimensions are extracted from XML-Schemas [11]) that describe the documents in the source. The hierarchical structure of the documents is then partially preserved, making the dimensional schema easy to understand for the decision-makers.

We consider our model completely defined. Therefore, this paper is mainly dedicated to an OLAP query language that allows decision-makers to analyse an XDW described by a StarCD. This language is indented for non IT specialists,

S. Kozielski et al. (Eds.): BDAS 2014, CCIS 424, pp. 72–81, 2014.

and its relatively simple syntax is based on the principles developed for the complex object query language [1].

2   Related Works

Our work is developed within the context of warehousing of "documents-centric" XML documents [6]. We aim to define mechanisms that allow the decision-makers to simply manipulate complex objects organized according to a hierarchical structure (XML-Schema).

In this context, the authors of [3] propose to describe an XDW using the class diagram model of UML. In their warehouse schema, the documents are represented by a fact (**xFact**)linked to virtual dimensions. The authors emphasize the modelling aspect, without defining a concrete approach for the OLAP analysis. But they briefly present an algorithm based on XQuery to analyze a warehouse. The proposed data model in their paper enables to describe an XDW using objects classes linked to each other, and organized and defined following the user requirements. So the documents structure does not appear in the schema, as it is represented in the document source.

The authors in [5] propose a data model identical to the one above to describe an XDW. They extend the query language MDX which is basically an OLAP query language for the classical data warehouse. This extended language (XML-MDX) integrates a set of operators defined to manipulate an XML cube, allowing particularly the analysis of textual contents. The XQuery is used during the creation of the XML cube, for the calculation of the measures and definition of the dimensions. As in [3], the document structure is not represented in the warehouse schema. Furthermore, the query language designed for the OLAP analysis turns out to be complex because it requires that the decision-maker has computing knowledge, yet he is not an IT specialist.

In [7], the authors propose a new multidimensional model for the OLAP analysis of XML documents. This model called "the galaxy model" allows to define a warehouse schema as a dimensions graph. The dimensions of a galaxy are linked to each other by one or several nodes that express the compatibility between them. The fact is chosen among dimensions during the warehouse querying process. So a dimension in a galaxy schema represents as well an axis and a subject of analysis. The authors extend the classic multidimensional algebra which they adapt to their model, joining new operators into it, particularly for the textual data analysis. In their paper, the proposed data model for the XDW does not keep the initial structure of the documents. The elements in the XML documents are unstructured in the warehouse schema, so we could reasonably think that it establishes an obstacle for the query expression by the decision-makers.

To our knowledge, there is currently no modelling standard defined for XML documents. Furthermore, the works presented above only answer partially the problem we address. In our approach, we assume that the decision-makers know the structure of the document they need to analyze. Thus they can retrieve this structure in XDW schema and easily express their analysis OLAP queries.

3   Contribution

3.1   Warehouse Model Formalization

We have developed a multidimensional model that describes a XDW, under the shape of a star schema named StarCD. This model is based on the UML formalism [4] to represent Facts and Dimensions. It is built from the analysis of an XML document source described by a unique XML-Schema (XSchema), and it supports the creation of the analytics queries by decision-makers.

A StarCD is defined as follows:
$StarCD = (F, D)$ where:

- F corresponds to the fact of the schema, and it represents also the root element of the XSchema that define the analysed document class.
- $D = \{D_1, \ldots, D_n\}$ is a set of dimensions associated to the fact.

The fact is characterized by a set of measures, being textual or numeric type. The measures are described either by attributes or classes linked to the fact by aggregation links. The hierarchical structure is then identical to the one represented in the XSchema of the source. The fact is defined as follows:
$F = (M, Agg)$ where:

- $M = \{M_1, \ldots, M_p\}$ is a set of measures.
- $Agg = \{ag_1, \ldots, ag_p\}$ is the set of aggregation functions associated to each measure, with ag included in $\{SUM, COUNT, AVG, MAX, MIN\}$.

The dimensions are characterized by parameters, organized into hierarchies. Those hierarchies indicate the granularity level going from the top level parameter (All) to the bottom level one, corresponding to the parameter linked to the fact into the star schema. Dimensions are defined as follows:
$D = (P, H)$ where:

- $P = \{p_1, \ldots, p_s\}$ is the set of parameters of the dimension D.
- $H = \{h_1, \ldots, h_s\}$ is the set of hierarchies in which parameters are organized, having h defined as follows:
  - $h = \{p_1, p_2, \ldots, All\}$ where p1 is the parameter linked to the fact (corresponding to the bottom level parameter).

In the StarCD, dimensions are described as a set of classes, each class representing a parameter that enables to aggregate some fact instances. Each dimension is a tree structure whose root is linked to the fact by an association link "By". The weak attributes correspond to the attributes of classes in the schema.

The presented model introduces the **reverse hierarchy** concept. In that type of hierarchy, the top level parameter is actually linked to the fact in order to keep the XSchema structure in the star schema presented to the decision-maker.

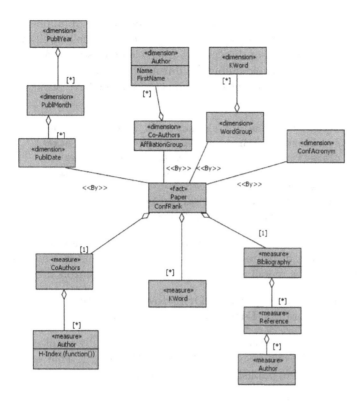

Fig. 1. XML Warehouse schema(StarCD)

Example 1: The StarCD presented in Fig. 1 represents a warehouse schema created from a collection of scientific papers; fact and measures are in the bottom of the schema, and dimensions are in the top.

This StarCD respects the hierarchical tree structure defined in the XSchema, from which it has been created. Thus, we can see an example of reverse hierarchy in the dimension Co-Authors. The class Co-Authors is directly linked to the fact because it is a first level parameter in the XSchema that describes the document source. Thus in the schema, the parameters in this dimension are reversed.

3.2 Multidimensional Expression Language

The query language presented in this section is inspired by the object query language OQL [1]. It allows decision-makers to express complex OLAP queries, and also allows the manipulation of the objects presented in the multidimensional schema StarCD. Its structure is built to make the navigation in the tree structure of an XML file easy by means of simple queries, and by using the following syntax:

```
Analyse <measure>
From <variables that define the warehouse classes>
By <dimensions>
```

The clause **Analyse** contains the measures and their applied aggregation functions. The clause **From** indicates the elements in the StarCD which will be used either as measures, or dimensions. Each element is associated to a variable ("alias") that helps to avoid ambiguity during the hierarchies' course. The clause **By** specifies the variable of the dimensions to use. The BNF structure of a query is as follows:

```
<Query_spec> ::= <Analyse_clause><From_clause><By_clause>
<Analyse_clause> ::= Analyse <Operator> (<Meseure_name>)
<From_clause> ::= From <alias> in <XML_Element_Name>
(. <XML_Element_Name>)*(,<alias> in <XML_Element_Name>
(. <XML_Element_Name>)*)*
<By_clause> ::= By <alias>(, <alias>)*
<Operator> ::= COUNT | SUM | AVG | MAX | MIN
```

In [8], the authors defined a "closed core" of OLAP operators. We picked some of those operators dedicated to complex analysis, to express multidimensional operations from the model above presented. Those operators are:

- Granularity-related operators: **Drill-down, Roll-up**
- Structural operators: **DRotate, Switch, Nest**

In order to illustrate the impact of those operators, we will consider for example the query R1 expressed using the StarCD in Fig. 1. That query calculates the number of papers published by publication's date (parameter PubliDate):

```
R1:    Analyse count(p)
       From p in Paper, d in p.PubliDate
       By d
```

Granularity-related Operators. Granularity-related operators allow analyzing data by aggregating or desegregating them according to the level of detail required. They change the granularity level of the parameter used for data calculation. This change of level can be done upwards (**Roll-up**) or downwards (**Drill-down**).

Example 2
Roll-up: This operator aggregates measures by changing the parameter of a dimension. Hence, the query R2 will count the number of papers ordered by publication date, increasing the date level to month:

```
R2:    Analyse count(p)
       From p in Paper, d in p.PubliDate, m in d.PubliMonth
       By m
```

Drill-down: This operator increases the level of detail of measures, that is, it represents the element used in the calculation at a lower level of granularity in the clause **From**. So the query R3 will count the number of papers by publication date, but decreasing the aggregation level on date to be measured in days. The queries R3 and R1 are equivalent then:

```
R3:    Analyse count(p)
       From p in Paper, d in p.PubliDate
       By d
```

On reverse hierarchies, these two operators above are also reversed.

Structural Operators. The rotation operator **DRotate** changes the dimension used for the calculation of a query. It means, in our case, a change in the parameter we need to use in the clause **From**.

Example 3: The query R4 calculates the number of papers by group of authors (Co-Authors). A rotation has been performed on the query R3 to change the dimension PubliDate by Co-Authors:

```
R4:    Analyse count(p)
       From p in Paper, co in p.Co-Authors
       By co
```

The operator **Switch** exchanges the positions of two or more variables of a parameter in a dimension. It only implies a visual effect since the measures values do not change. The operator Nest imbricates a parameter into another.

Example 4: Let's consider the following query that calculates the number of papers by authors, publication dates and key-word (parameters Author, PubliDate and KWord). The query is expressed as follows:

```
Analyse count(p)
From p in Paper,
     k in p.WordGroup.KWord,
     a, in p.CoAuthors.Author,
     d in p.PubliDate
By a.FirstName, a.LastName, k, d;
```

Switch will exchange the positions of variables a.FirstName and a.LastName, the clause By then becomes:

```
By a.LastName, a.FirstName, k, d
```

Nest will imbricate the variable d (for PubliDate) into the variable k (for KWord). The clause By is again transformed and becomes:

```
By a.LastName, a.FirstName, k.d
```

Likewise the Switch operator, Nest also has a merely visual effect; therefore, the previously calculated values do not change.

We consider that an XML structure can be defined as a tree structure graph; thus, the presented language allows navigating in that tree, and helps decision-makers to identify the depth of the elements during the query processing.

Query Expression. We propose below two queries examples defined from the StarCD.

Query 1: Calculate the number of papers by conference
Measure: **count**(Paper)
Parameter: ConfAcronym

```
Analyse count(p)
From p in Paper, c in p.ConfAcronym
By c
```

This query calculates for each distinct value of ConfAcronym, the associated number of papers.

Query 2: Calculates the number of papers by key-word and by year of publication
Measure: **count**(Paper)
Parameter: KWord (dimension WordGroup) and PubliYear (dimension PubliDate)

```
Analyse count(p)
From p in Paper,
     k in p.WordGroup.KWord,
     d in p.PubliDate.PubliYear
By k, d
```

The query groups the papers by key-words, and after by year of publication. For each couple, it then counts the number of papers in each corresponding group.

3.3   Experimentation

The materialized warehouse is supplied from a scientific document collection, sharing the same XML structure (XSchema). It contains as much facts as the source has documents. The dimensions are stored in distinct documents, and are linked to the fact by references (see Fig. 3). Indeed, we do believe that saving facts and dimensions in separated xml files insures the singularity of the values of the dimensions, and also helps to suppress the redundancy in the warehouse.

The physical XDW have been created into the database eXist-DB, which is a native XML database that enables to easily manipulate XML collections with the query language XQuery [10]. In that context, the queries expressed by a decision-maker are then translated into XQuery in order to be applied to the warehouse. Hence, a translator has been created to realise that operation. It performs an automatic translation of any analytic query into an XQuery query, with transparency for the user. The architecture of the translator is shown in Fig. 2. This translator takes the OLAP query written by the decision-maker, and uses the warehouse schema to build the translated query.

Once the query is seized by the user, the translator proceeds to perform both a lexical and a syntactic analysis. After that, the query is sent to the code generator, which performs a semantic analysis (type verification and optimisation based on the source language) before finally translating the query into XQuery. We have defined the following translation method:

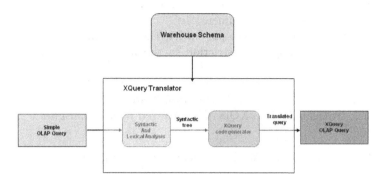

Fig. 2. Translator Architecture

```
<Paper   ConfRank="1">

    <Measures>
        <Coauthors>
                <Author H-Index="01" />
                <Author H-Index="02" />
                <Author H-Index="03" />
        </Coauthors>

        <KWord>Based on XML documents</KWord>
        <KWord>Between data warehouses</KWord>
        <KWord>Exchange of data cubes</KWord>

        <Bibliography>
            <Reference>
                <Author H-Index="19" />
            </Reference>
        </Bibliography>
    </Measures>

    <Dimensions>
        <ConfAcronym ref = "A1" />

        <PubliDate ref = "P1" />
```

Fig. 3. XML Fact Structure

- The clause **Analyse:** the measures in this clause correspond to the results to be returned. Those values will appear in the clause *return* of the XQuery language.
- The clause **From:** a clause *for* will be defined in XQuery for each variable in the clause From. Only the variables defined as dimensions (variables in the clause By of the analytic Query) will have an equivalent clause *for*.

– The clause **By:** The variables in this clause have several roles:
  • They allow to define the dimensions
  • They allow to define the display order (the clause *order by* in XQuery)
  • They allow to apply a restriction to the facts that are linked to the chosen dimensions (The clause *let* in XQuery)

Example 5: We present in Tab. 1 and Tab. 2 two analysis queries examples, showing their translations into XQuery, as realized by the translator.

Table 1. Query 1 Calculate for each conference the average number of authors)

Analysis Query	Translated Query

Analyse avg(count(a))
From p in Paper,
    a in p.CoAuthors.Author,
    c in p.ConfAcronym
By c
	for $a in //DConfAcronym/ConfAcronym
	let $doc := //Paper[Dimmensions
	    /ConfAcronym/@ref = $a/@id]
	    /Measures/MCoauthors/MAuthor
	return
	<group>
	    <Acronym>{$a/text()}</Acronym>
	    <avgAuthor>{avg(count($doc))}</avgAuthor>
	</group>

Table 2. Query 2 Calculates the number of papers by key-word and by year of publication

Analysis Query	Translated Query

Analyse count(p)
From p in Paper,
    k in p.WordGroup.KWord,
    d in p.PubliDate.PubliYear
By k, d
	for $a in //DWordGroup/WordGroup,
	    $b in distinct-values($a/KWord),
	    $c in //DPubliDate/PubliDate,
	    $d in $c/PubliYear/@year
	let $doc := //Paper[Dimensions
	    /WordGroup/@ref = $a/@id
	    and Dimensions/PubliDate
	    /@ref = $c/@id]
	return
	if (exists($doc))
	then <group>
	    <Kword>{$b}</Kword>
	    <Year>{data($d)}</Year>
	    <nbPaper>{count($doc)}</nbPaper>
	    </group>
	else()

The queries presented above are illustrating the simplicity to express an OLAP query using our language, compared to the same expression written in XQuery. Then the decision-maker has the benefits of the power of XQuery, being preserved from its complexity due to the transparency given by the translator.

4   Conclusions

In this paper, we proposed a new OLAP query language for the XML document warehouses. This language is applied on a multidimensional model defined from the object modelling formalism UML, and helps the decision-maker to be preserved from the complexity of the existing query languages or algebra. This language is translated into XQuery in order to be applied on a XDW, so we developed a translator that translate automatically any simple OLAP query into XQuery and apply it on a materialized XDW. We used the JavaCC technology (Java Compiler Compiler) to build the translator, because it allows combining the advantages of Java and the Language Theory.

We intend to integrate some new operators (such as Push, Pull or HRotate) to manage more complex queries. Furthermore, to validate our approach we are developing a prototype that will enable a semi-automatic modelling process for a document warehouse, as well as its querying through its StarCD.

References

1. Barry, D.K., Cattell, R.G.G.: The Obkect Database Standard. Morgan Kaufmann publisher (1997)
2. Golfarelli, M., Maio, D., Rizzi, S.: The dimensional fact model: a conceptual model for data warehouses. International Journal of Cooperative Information Systems 07 (1998)
3. Nassis, V., Rajagopalapillai, R., Dillon, T.S., Rahayu, W.: Conceptual and systematic design approach for xml document warehouses (2005)
4. Object Management Group (OMG): Unified Modelling Language (UML), http://www.uml.org/
5. Park, B.-K., Han, H., Song, I.-Y.: XML-OLAP: A multidimensional analysis framework for XML warehouses. In: Tjoa, A.M., Trujillo, J. (eds.) DaWaK 2005. LNCS, vol. 3589, pp. 32–42. Springer, Heidelberg (2005)
6. Pérez, J.M., Llavori, R.B., Aramburu, M.J., Pedersen, T.B.: Integrating data warehouses with web data: A survey. IEEE Trans. Knowl. Data Eng. 20(7), 940–955 (2008)
7. Ravat, F., Teste, O., Tournier, R., Zurfluh, G.: A conceptual model for multidimensional analysis of documents (2007)
8. Ravat, F., Teste, O., Zufluh, G.: Algèbre OLAP et langage graphique (2006)
9. W3C-Consortium: Extensible Markup Language (XML), http://www.w3.org/XML/
10. W3C-Consortium: W3C XML Query (XQuery), http://www.w3.org/XML/Query/
11. W3C-Consortium: XML Schema Part 0: Structures, 2nd edn., http://www.w3.org/TR/2004/REC-xmlschema-1-20041028/

Unifying Mobility Data Warehouse Models Using UML Profile

Marwa Manaa and Jalel Akaichi

Computer Science Department, ISG-University of Tunis, Le Bardo, Tunisia
{manaamarwa,j.akaichi}@gmail.com

Abstract. Many applications were interested in studying objects mobility which allowed the onset of a variety of trajectory data warehouses. As a new paradigm, launched by the evolution of classical ones to take into account mobility data provided by pervasive systems. Mobility data warehouse models was adopted by users in various fields such as those related to marketing, agriculture, health care, etc. However, the proposed conceptual models suffer from dispersed points of view that have to be unified in order to offer generic conceptual support for experts and clerical users. The purpose of this work is to propose a unified conceptual model able to unify different points of view through a generic UML description profile of facts and dimensions well adapted to new concepts imposed by mobility. Thanks to the proposed unified model, users will be able to build themselves their trajectory data warehouses regardless the case of study involving mobile objects.

Keywords: Mobility data warehouse, uml profile, unified model.

1 Introduction

In recent years, the modeling and the analysis of data out coming from moving objects activities has attracted a particular interest which is motivated by the development of pervasive systems and positioning technologies. Moving objects activities generate a huge volume of data which permits the emergence of a new concept called trajectory data. Indeed, according to authors in [13] **trajectory data is the record of the evolution of the position of a moving object traveling in space during a time interval in order to achieve a given goal**. For analysis purposes, trajectory data have to be extracted, transformed and loaded into a new comer, in the decision making field, called Trajectory Data Warehouse (TDW). TDW takes birth by integrating spatio-temporal data related to objects motion and activities into a unique repository useful for analysis and/or mining. In fact, classical Data Warehouses (DWs) [2] have shown their little support for managing data generated by moving object. Following these purposes, classical DWs have to be extended to be able to support the data warehousing changing from the representation till querying of data related to moving objects and their trajectories in order to extract knowledge from raw data captured by mobile devices. Whatever the type of the DW, classical, spatial, temporal, spatio-temporal,

S. Kozielski et al. (Eds.): BDAS 2014, CCIS 424, pp. 82–91, 2014.
© Springer International Publishing Switzerland 2014

or trajectory, the design of these DWs remains a difficult task that have to be performed by experts. However, it would be motivating to allow experts and ordinary users to define and to build themselves their model through a simple and flexible process. Traditionally, the design of a DW is based on an adequate representation of facts on one hand and dimensions of analysis on the other hand. In [14], the author has shown, through a proposition leaning on a graphical representation offering a visual help to the user, that a unified representation can be envisioned. We think that the model proposed in [14] is very interesting. However, it does not take into account the new requirement of mobile object imposed by the emergence of trajectory data and warehouse.In fact, classical DWs do not allow the analysis of the moving object and its trajectory. Various TDW models oriented applications were proposed such as those related to marketing, agriculture, health care, etc. Nonetheless, those models do not yet have a consensus for their design since it is a new paradigm and does not represent trajectories in a unified way. The goal of this paper is to propose a unified model which can be used to apprehend the modeling of facts and dimension members in a unified way. It represents the basic structure for a TDW. For validation purposes, we realize UML profile matching to the proposed unified model that will guide users to model their own specific TDW model. The remainder of this paper is organised as follow. The next section presents an overview of trajectory data and warehouses proposed in the literature. Section 3, defines the new unified model. Section 4, discusses validation. Finally, we conclude and we propose future works.

2   State of the Art

2.1   Trajectory Data Models

Thanks to the development of localisation system and mobile networks such as wireless or 3G networks, a lot of data based on mobile sources can be collected. Mobile sources are often called moving object which is an entity like person, animal, or natural phenomenon. Moving object moves in space and whose position evolves along the time. Whatever the case study is, the movement of a moving object is the evolution of their position in a spatial and temporal ways referring to the concept of trajectory. The trajectory notion has evolved over time and various trajectory models were proposed in the literature: in the beginning, the author in [6] introduced a model to represent trajectories in the form of lines associated with activities to represent entities movement. Later, the authors in [7] and [16] extended this model by representing trajectories across geospatial lifelines which defines periods of time during which entity occupies in space, and the evolution of this entity during this period is modeled with a geospatial lifeline. Later, author in [2] modeled trajectory with the position of the object in different timestamps using linear interpolation. This model is unreal because it can not be applied in road networks. Then, authors in [13] enhanced the modeling of trajectory with stops and moves. The stop is a point, it has some duration and it is temporal. Whereas move is not a point. It is a part of a trajectory between two consecutive stops, between the starting point and the first stop, or between the last stop and the end point [13]. Also, authors in [5] proposed the approach of modeling trajectory with abstract data types (TAD). Besides, this approach presents trajectory with moving points and moving regions while it is difficult to represent infinite types in a database. After presenting different research works in the trajectory data area, we propose to summarize different models resulting from studies cited above using the following table:

Table 1. Trajectory data models

Model	Modeling of Trajectory
Hagerstrand [6]	lines associated with activities.
Parent et al. [12]	spatio-temporal points.
Theriault et al. [16]	geospatial lifelines.
Tryfona et al. [17]	points, lines and regions.
Güting et al. [5]	moving points, moving lines and moving regions.
Braz [2]	positions in different timestamps (linear interpolation).
Spaccapietra et al. [13]	stops and moves.

In general, we can claim trajectory data models can not express the exact and real semantic of trajectories and none of the models take into account constraints and operations that are specific to trajectories. Consequently, these approaches can not allow trajectories analysis.

2.2 Trajectory Data Warehouse Models

Many works were proposed for the conceptual modeling of trajectory [6,7,16,2,15,13,5,17,12,5]. But, some works were proposed for the conceptual modeling of TDWs [10,2,11,4,1,8]. A DW is a collection of subject-oriented, integrated, non volatile and time varying data that allows the transformation of an information system into strategic information for decision-making purpose with a rapid execution time. Yet, modern location-aware devices deliver huge quantities of spatio-temporal data covering moving objects which must be quickly processed. Proposed works are classified into classes: The trajectory oriented approach(TOA), the cell oriented approach (COA), moves and stops, and episodes. Moves and stops [13] were explained above. Episodes is a sequence of location-based points that has the same moving dynamic. TOA is to find a minimum rectangular delimitation for a portion of trajectory. COA is to divide the area of the trajectory in cells and consequently cells will divide the space in partitions and the time in intervals. In [10], authors proposed a TDW based on classical DW using standard DW technologies. Authors in [2] proposed a generic TDW ,discussed the possibility of the usage of data mining tasks in order to discover knowledge resident in TDW and used a solution which has as proposal, the use of linear interpolation. Nevertheless, this solution is not exact, it can give a solution to the distinct count problem in the same cell, but can duplicate the count of the trajectories where a same trajectory can be counted multiple times during roll-up operations. In addition, the linear interpolation is not suitable in the reality. Because it does not take into consideration barriers and it can not be applied in a road network. Many recent decision problems related to various professions such commercial representatives, transporters, and ecologists concerned with mobility. To better satisfy professionals requirements, authors in [11] proposed a generic conceptual TDW for mobile professional's. The conceptual TDW schema contains a high number of tables which will increase the response time to queries asked. Also, the authors in [9] extended a conceptual spatial multidimensional model by incorporating a trajectory as a first class concept. Recently, authors in [4] proposed the framework called St-Toolkit for the design and the implementation of DWs that support spatio-temporal concepts. Another model were proposed in [1] related to the modeling of the displacement of the herd. The authors proposed two different models. The first one is based on entity relationships model. The second model based on the model introduced in [15] and takes into account the evolution of space and time. Very recently, a framework for TDWs was presented in [8]. Authors in this paper proposed two alternative approaches COA and TOA. The choice of a method between TOA and COA is a trade off among the selected granularity level and the number of trajectory approaches. After presenting different works in TDW area, we propose to summarize different models in the following table:

Table 2. Trajectory data warehouse models

Models	Application domain	Schema of TDW	Class
Model of Orlondo et al	A generic TDW	Star	COA
Model of Oueslati and Akaichi	TDW for mobile professional	Snowflake	Move and stop
Model of Campora et al	traffic data	Star	Episode
Model of Arfaoui and Akaichi	TDW for the displacement of the herd	star	Move and stop
Model of Marketos et al	A generic TDW	Star	COA and TOA
Model of Braz	A generic TDW(use of data mining tasks)	Star	COA

Various TDWs were proposod. We can conclude that there is not a consensus methodology for their modeling. Also, each model is specific for the concerned field.

3   The Unified Model

TDW design process consists in different steps as shown in the following algorithm 1. First, we create the fact table. Then, we define different dimensions. The design of the TDW can be finished at this step and we get a *star schema* for this TDW. Moreover, we can also make hierarchies for the dimensions by defining the members. This step results another type of structure which is the *snowflake schema*. Moreover, we can add one or many facts, and we get a *constellation schema*. Our model is a conceptual model, it provides a highest level of abstraction and it is completely independent from software and hardware.

Algorithm 1. Trajectory data warehouse construction Algorithm

---

Input: fact−type, number−of−fact, number−of−temporal−dimension, number−of
  −spatial−dimension, number−of−object−dimension
Output: Realization of the TDW model: fact, dimensions, and members
if *fact−type=star* **then**
  Createfact(fact−key, accurate−fact−attributes, holistic−fact−attributes)
  CreateDimensions(number−of−temporal−dimension,   number−of
  −spatial−dimension, number−of−object−dimension)
  Updatefact(Spatial−Dimensions, Temporal−Dimensions, Object−Dimension,
  fact−key)
end
else if *fact−type=snowflake* **then**
  Createfact(fact−key, accurate−fact−attributes, holistic−fact−attributes)
  CreateDimensions(number−of−temporal−dimension,   number−of
  −spatial−dimension, number−of−object−dimension)
  Createmembers(temporal−dimension, spatial−dimension, object−dimension)
  Updatefact(Spatial−Dimensions, Temporal−Dimensions, Object−Dimensions,
  fact−key)
end
else if *fact−type=constellation* **then**
  for $i = 1 \rightarrow NF$ **do**
    Createfact(fact−key, accurate−fact−attributes, holistic−fact−attributes)
    CreateDimensions(number−of−temporal−dimension,
    number−of−spatial−dimension, number−of−object−dimension)
    Updatefact(Spatial−Dimensions, Temporal−Dimensions, Object−Dimensions,
    fact−key)
  end
end

TDW can be applied in many contexts. We consider an application scenario related to the field of delivery system. Logistic management is an important class of applications that may take advantage of analyzing trajectory data. In fact, applications that manage truck delivery of goods are of special interest due its importance for the global economy. For this kind of application, the moving object is the truck driver who cares of bringing goods to different points of interest in a road network. While moving, it describe a trajectory which is composed of stops and moves. According to authors in [3], truck travels may be used to make sure that delivery plans are being correctly executed or they may analyze to evaluate the performance of the delivery system.

3.1   Modeling Facts

A fact table is the primary table in a dimensional model which contains measures corresponding to an event or a situation. In our example the trajectory is the fact, showing the trajectory toward several criteria like date and point of interest. In fact, measures can be obtained by the intersection of different criteria called dimensions. A TDW should include a spatial, a temporal, and a moving object dimension. A fact has the following structure: fact−name[(T),(fact−key), (temporal−reference−attributes), (spatial−reference−attributes), (moving−object−reference−attributes), (accurate−fact−attributes), (holistic−fact−attributes)] where:

- fact−name: is the name of the type
- T: is a mark for the trajectory fact type
- fact−key: is the concatenation of the values of temporal−reference− attributes, spatial−reference−attributes and moving−object−reference− attributes.
- temporal−reference−attributes: is a list of attribute names, each attribute is a reference to a member instance in a time dimension.
- spatial−reference−attributes: is a list of attribute names, each attribute is a reference to a member instance in a space dimension.
- moving−object−attributes: is a list of attribute names different from spatial or temporal attribute names, each attribute is a reference to a member instance in object dimension.
- accurate−fact−attributes: is a list of attribute names including distributive and algebraic measures.
- holistic−fact−attributes: is a list of attribute names including holistic measures.

Rule 1. (Well−formed trajectory fact type). This rule proposed by author in [14]. It requires that a trajectory fact type is well−formed if each referenced dimension is either degenerated or points to a legal entry in their dimension. We extended this rule by adding that a trajectory fact type is well−formed if each referenced dimension is either degenerated or points to a legal entry in their correspondent dimension whatsoever spatial, temporal or the object dimensions. The above rule does not take into account the spatio-temporal aspect of mobility and does not distinguish between dimensions.

Rule 2. (accurate−fact−attribute) A measure is accurate if the super aggregates can be computed from sub-aggregates such (sum, max) or can be computed from a finite set of auxiliary measures.

Rule 3. (holistic−fact−attribute) A measure is holistic if the super aggregates cannot be computed from the sub−aggregates, not even using any finite number of auxiliary measures. Then, is necessary to compute the measure in an approximated way.

3.2   Modeling Dimensions

Information connected to facts can be analyzed in terms of entities. These entities called dimensions. Dimensions and dimension members have the same structure: member−name[(M), (member−key),(list−of−reference− attributes), (list−of−property−attributes)] where:

- member−name: is the name of the memeber
- M: is a mark for the member
- member−key: is a list of attribute names, the concatenation of these attributes identifies each instance of the type
- reference−attributes: is a list of attribute names where each attribute is a reference to the dimension member.
- property−attributes: is a list of attribute names where each attribute is a property for the member. Only the member−key is mandatory.

Rule 4. In each TDW there must be at least a temporal, spatial and moving object dimensions.

4   Unified Model UML Profile Realization

The secret of a great TDW process and valuable analysis is a good conceptual modeling schema that offer a highest level of abstraction and completely software and hardware independent. To implement our approach we choose the StarUML open source platform that uses the xml language to create the profiles UML. StarUML proposed a set of generic UML profiles, allows users to add concepts an to change the representation of these concepts. In fact, An UML profile is a string value surrounded by two quotation marks << >> as we see in the following table:

<?xml version= "1.0" encoding= "UTF-8"? > <PROFILE version= "1.0" >
<HEADER> <NAME>UnifiedTDWProfile¡/NAME¿
<DISPLAYNAME>UnifiedTDWProfile< /DISPLAYNAME>
<DESCRIPTION>Unified TDW Modeling Profile< /DESCRIPTION>
<AUTOINCLUDE>True< /AUTOINCLUDE> < /HEADER>

4.1   Classes' Stereotypes

In this section we propose stereotypes for classes used in the TDW UML profile in this table:

Table 3. Classes stereotypes description and presentation

Stereotype name	Class type	Description	Icon
<<Fact>>	Class	this stereotype indicates that the class represents the variable to be analyzed	
<<TemporalDimension>>	Class	this stereotype indicates that the class represents a temporal dimension	
<<SpatialDimension>>	Class	this stereotype indicates that the class represents a spatial dimension	
<<ObjectDimension>>	Class	this stereotype indicates that the class represents moving object dimension	
<<MemberLevel>>	Class	this stereotype indicates that the class represents a member of dimension	

We give an example for <<Fact>> which is an extension for UML class as showing in this table:

<STEREOTYPE> <NAME>Fact< /NAME> <DESCRIPTION>fact
stereotype< /DESCRIPTION> <BASECLASSES>
<BASECLASS>UMLClass< /BASECLASS> < /BASECLASSES>
<RELATEDTAGDEFINITIONSET>Fact< /RELATEDTAGDEFINITIONSET>
<NOTATION>< /NOTATION> <ICON minWidth="30"
minHeight="20">Image4.bmp< /ICON>
<SMALLICON>Image4.bmp< /SMALLICON>< /STEREOTYPE>

4.2   Attributes' Stereotypes

Also, we propose stereotypes for attributes used in the modeling as we see in this table:

Table 4. Attributes stereotypes description

Stereotype name	Class type	Description
<<fact−key>>	Attribute	indicates that this attribute is the fact key
<<temporal−reference−attributes>>	Attribute	indicates that this attribute is a reference to a temporal dimension
<<spatial−reference−attributes>>	Attribute	indicates that this attribute is a reference to a spatial dimension
<<moving−object−attributes>>	Attribute	indicates that this attribute is a reference to a temporal dimension
<<member−key>>	Attribute	indicates that this attribute is the member key
<<reference−attributes>>	Attribute	indicates that this attribute is a reference to the dimension member
<<property−attributes>>	Attribute	indicates that this attribute is a property for dimension

We give an example for <<fact−key>> which is an extension for UML attribute in this table:

<STEREOTYPE> <NAME>fact−key< /NAME> <DESCRIPTION>extension
of UML attribute< /DESCRIPTION> <BASECLASSES>
<BASECLASS>UMLAttribute< /BASECLASS> < /BASECLASSES>
<RELATEDTAGDEFINITIONSET>fact−key<
/RELATEDTAGDEFINITIONSET> < /STEREOTYPE>

4.3   Measures' Stereotypes

Then, we propose stereotypes for measures as we see in the following table:

Table 5. Measures stereotypes description

Stereotype name	Class type	Description
<<accurate−fact−attributes>>	Operation	indicates that this attribute is an accurate measure
<<holistic−fact−attributes>>	Operation	indicates that this attribute is an holistic measure

We give an example for <<accurate−fact−attributes>> which is an extension for UML operation in this table:

<STEREOTYPE> <NAME>accurate−fact−attribute< /NAME>
<DESCRIPTION>extension of UML operation< /DESCRIPTION>
<BASECLASSES> <BASECLASS>UMLOperation< /BASECLASS>
< /BASECLASSES> <RELATEDTAGDEFINITIONSET>accurate-fact-
attribute< /RELATEDTAGDEFINITIONSET>
< /STEREOTYPE>

In Fig. 1, we propose the delivery TDW model with Unified TDW Profile.

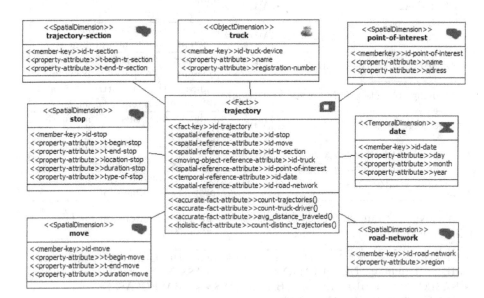

Fig. 1. Delivery TDW model with our profile

5   Conclusion

In this paper, we proposed a unified representation for TDW conceptual model. This model is able to propose a unified way for modeling facts, dimensions, and members while taking into account requirements for TDW models. Through a motivating example for delivery system, we prove the validity of our model. Finally, we realized a unified TDW UML profile. As future work, we suggest also a general model of ontology allowing the construction of mobility data warehouse regardless the case of study involving mobile objects.

References

1. Arfaoui, N., Akaichi, J.: Modeling herd trajectory data warehouse. International Journal of Engineering Trends and Technology (2011)
2. Braz, F.J.: Trajectory data warehouses: Proposal of design and application to exploit data. In: GeoInfo, pp. 61–72 (2007)
3. de C. Leal, B., de Macêdo, J.A.F., Times, V.C., Casanova, M.A., Vidal, V.M.P., de Carvalho, M.T.M.: From conceptual modeling to logical representation of trajectories in dbms-or and dw systems. JIDM 2(3), 463–478 (2011)
4. Campora, S., de Macedo, J.A.F., Spinsanti, L.: St-toolkit: A framework for trajectory data warehousing. AGILE 2(3), 18–22 (2011)
5. Güting, R.H., Schneider, M.: Moving Objects Databases. Morgan Kaufmann (2005)
6. Hägerstrand, T.: What about people in Regional Science. Papers in Regional Science 24, 6–21 (1970)
7. Hornsby, K., Egenhofer, M.J.: Modeling moving objects over multiple granularities. Ann. Math. Artif. Intell. 36(1-2), 177–194 (2002)
8. Marketos, G., Theodoridis, Y.: Mobility data warehousing and mining. In: Rigaux, P., Senellart, P. (eds.) VLDB PhD Workshop. VLDB Endowment (2009), http://dblp.uni-trier.de/db/conf/vldb/vldb2009phd.html#MarketosT09
9. Moreno, F., Arango, F.: A conceptual trajectory multidimensional model: An application to public transportation. DYNA 78(166) (2011)
10. Orlando, S., Orsini, R., Raffaetà, A., Roncato, A., Silvestri, C.: Trajectory data warehouses: Design and implementation issues. JCSE 1(2), 211–232 (2007)
11. Oueslati, W., Akaichi, J.: Mobile information collectors trajectory data warehouse design. International Journal of Managing Information Technology (IJMIT) 2(3) (2010)
12. Parent, C., Spaccapietra, S., Zimányi, E.: Spatio-Temporal Conceptual Models: Data Structures + Space + Time. In: Proceedings. of the 7th Symposium on Advances in Geographic Information Systems, ACM GIS, pp. 26–33. ACM Press, Kansas City (November 1999), http://cs.ulb.ac.be/publications/P-99-01.pdf
13. Renso, C., Spaccapietra, S., Zimnyi, E. (eds.): Mobility Data: Modeling, Management, and Understanding. Cambridge Press (2010)
14. Schneider, M.: A general model for the design of data warehouses. International Journal of Production Economics 112(1), 309–325 (2008)
15. Spaccapietra, S., Parent, C., Damiani, M.L., de Macedo, J.A., Porto, F., Vangenot, C.: A conceptual view on trajectories. Data and Knowledge Engineering 65(1), 126–146 (2008)
16. Theriault, M., Claramunt, C., Seguin, A.M., Villeneuve, P.: Temporal gis and statistical modelling of personal lifelines. In: 9th Spatial Data Handling symposium, pp. 9–12. Springer (2002)
17. Tryfona, N., Price, R., Jensen, C.S.: Chapter 3: Conceptual models for spatio-temporal applications. In: Sellis, T.K., Koubarakis, M., Frank, A., Grumbach, S., Güting, R.H., Jensen, C., Lorentzos, N.A., Manolopoulos, Y., Nardelli, E., Pernici, B., Theodoulidis, B., Tryfona, N., Schek, H.-J., Scholl, M.O. (eds.) Spatio-Temporal Databases. LNCS, vol. 2520, pp. 79–116. Springer, Heidelberg (2003)

Data Quality Issues Concerning Statistical Data Gathering Supported by Big Data Technology

Jacek Maślankowski

University of Gdańsk, Department of Business Informatics, Poland
jacek@ug.edu.pl

Abstract. The aim of the paper is to show the data quality issues concerning statistical data gathering supported by Big Data technology. An example of statistical data gathering on job offers was used. This example allowed comparing data quality issues in two different methods of data gathering: traditional statistical surveys vs. Big Data technology. The case study shows that there are lots of barriers related to data quality when using Big Data technology. These barriers were identified and described in the paper. The important part of the article is the list of issues that must be tackled to improve the data quality in the repositories that comes from Big Data technology. The proposed solution gives an opportunity to integrate it with existing systems in organization, such as the data warehouse.

Keywords: Big Data, unstructured data, data quality.

1 Introduction

In recent years Big Data has been treated as an opportunity for enhancing information management in various organizations. For most of companies, Big Data is related to the technology that support collecting data from unstructured data sources, mostly from websites [5]. Although filtering web data to feed databases is not new [17], Big Data gave an opportunity to have a platform that will filter the data in very efficient way, especially by using Apache Hadoop platform for parallel processing the data. That's why Big Data is mostly defined as three V's, which are volume, velocity and variety [24] and additional V's, especially in terms of data quality, such as validity, veracity and value. Big Data can be supported by a large number of tools used for embedded data analytics and statistics analyses [15]. To provide reliable data analyses it is necessary to use various tools to manage Big Data algorithms [14]. But the key issue is to ensure that the data quality is high and data sources are relevant as well as reliable.

There are several issues concerning Big Data that researchers should tackle with, such as ethical aspects of Big Data [20] or inconsistencies in Big Data analysis [28]. Another issue is that Big Data implementation may be followed by organizational changes necessary for a company to take full benefits created by data [2]. Implementing a Big Data system in a company may reflect in the

S. Kozielski et al. (Eds.): BDAS 2014, CCIS 424, pp. 92–101, 2014.

necessity of managing changes effectively, including such areas like leadership, talent management, technology, decision making and company culture [19].

The use of Big Data technology is not only related to business data but also to other types of data. For instance researchers may analyze historical data in more efficient way – by scanning lots of historical papers and using Big Data algorithms to analyze all the information included in them [11].

Recently, Big Data technology is also used to retrieve and process statistical data in large repositories, such as mobile call records (MCR) and unstructured data. In this paper the focus is on the statistical data retrieving from unstructured sources, such as websites and internet portals containing job offers.

The goal of this article is to show the way of using Big Data to collect statistical data, capturing all aspects of data quality. The proposed solution gives an opportunity to integrate it with existing systems in organization such as the data warehouse because the results of analysis are usually stored in a structured database. The thesis of this article is as follow: the data gathered using Big Data technology is much more vulnerable to statistical standard errors than using traditional data sources. In the paper there is a comparison of collecting statistical data in traditional way versus by using big data technology.

The paper has been divided into 6 parts. In the first part there is an introduction with a short review of the literature on Big Data. The second part of the paper shows the prerequisites to use Big Data to collect statistical data. In the third part there is a descriptive analysis of the data quality issues concerning Big Data solutions. The fourth part shows typical problems with data related to Big Data. The fifth part shows the comparison between collecting statistical data in traditional way and by using big data technology. In this paper the focus is on the data quality issues concerning two different types of statistical data sources. The last part shows conclusion.

2   Feeding the Big Data Repository with Statistical Data

The failing response rates of the statistical surveys in last several decades [9] had a big impact on the necessity of finding new ways to collect statistical data. One of the opportunity is to use Big Data technology to filter web data that comes from websites to find useful information on various statistical topics and areas. However it has been demonstrated in scientific papers that typical software architecture makes it easier to collect and store data than to analyze it [12]. Therefore the goal is to increase the data quality by avoiding data noise or redundancy.

Big Data is mostly related to unstructured data analysis. It is the result of the fact that more than 80 per cent of all potentially useful information is unstructured data [5]. However Big Data technology combines all of the internal as well as external data that comes from multiple sources both structured and unstructured [4]. Therefore the main challenge is to make a Big Data repository flexible, secure and efficient [27]. Thus it is necessary to select data sources properly. One of the possibilities to accomplish this challenge is to find relations between web sources based on link analysis [25].

Another benefit of using Big Data to collect statistical data is that this technology can offload data warehouses that are usually very expensive to maintain [22]. However Big Data technology can be used as one of the sources to feed data warehouse [18].

The successful plan of Big Data implementation will focus on matching investment priorities with business strategy, balancing speed, cost and acceptance as well as focusing on frontline engagement and capabilities [1].

3   Data Quality and Big Data

The quality is very often defined based on ISO 8402:1986 with further changes. Following this document, a quality is a set of attributes and properties of product or service that decides about fulfilling declared or default requirements [3].

Big Data technology as a tool to retrieve statistical data may be regarded by experts as a low quality data source [7]. Based on GIGO (Garbage In-Garbage Out) rule, using low quality data source may result in low quality output tables. In this part of the paper there is an analysis of possible methods to improve the quality of the data that comes from various unstructured data sources supported by Big Data tools.

There is no unified definition of the data quality. However most of the definitions of data quality are often referred to data reliability and fulfilling user requirements [16]. A good data quality is needed to make high quality decisions [13]. It is especially important in a culture of evidence-based decision making [21].

It is obvious that managing large amount of data creates new challenges in the context of data quality. Poor management of data leads to "data silos" where data are redundantly stored, managed and processed. Therefore poor data quality is the major challenge in the companies of today [10]; [16].

However problems with data quality in databases is not new. Surveys conducted in the past also showed that more than half of large companies had problems with data quality [26]. Data quality issues are not only important during data gathering in traditional databases but are also very relevant even in the process on collecting data based on sensors signals [23].

To increase the data quality in traditional statistical survey a good practice is to include don't know answer in particular questions. It has been proved in the context of brand image measurement. Respondents who are unfamiliar with the product category should not be filtered out [6]. The similar situation would be if the keyword-matching algorithm in Big Data is unable to identify the value of the particular variable based on the analyzed content.

Data quality in literature was described even in up to 179 attributes of the data [26]. However the data quality in Big Data solutions can be related to data characteristics such as: consistency, reliability, accuracy, validity, completeness, reasonableness and timeliness. The traditional data quality dimensions with importance level related to Big Data technology has been shown in Tab. 1.

The importance level in Tab. 1 has been assessed by analysing the repository containing statistical data based on the case study used in this paper. Firstly,

Table 1. The data quality dimensions and categories (Source: Own elaboration based on: [16])

Category	Dimension	Big Data importance level
Core (intrinsic)	Accuracy, Objectivity, Relevancy, Reputation	High, rather difficult to achieve
Contextual	Value added, Relevancy, Timeliness, Completeness, Appropriate amount of data	High, easy to achieve
Representational	Interpretability, Ease of understanding, Representational consistency, Brief representation	Medium, easy to achieve
Access	Accessibility, Access security	Medium, easy to achieve

the core category is related to accuracy, objectivity, relevancy and reputation of data. It is very important to achieve a good quality data in this category, however when filtering data from unstructured data sources there is a large amount of data noise. As a result the estimates of the number of particular occurrences in the data may not be accurate, what is highly plausible.

The second category is contextual. It is obvious that amount of data in the Big Data sources is large, especially in the context of unstructured data sources. However some dimensions such as value added and timeliness are rather easy to achieve. The major problem is its completeness. Please note that typical statistical social survey is based on a statistical sampling. Therefore there is a need to estimate missing values. As the statistical data are stored in time series, it is necessary to ensure that the data has proper time stamp. This time stamp is retrieved usually from the day of the message and it is very likely that it can be wrongly identified.

The third category – representational is always referred to its interpretability, ease of understanding and consistency. It is rather easy to achieve because in this paper the focus is on the data that is filtered from websites to produce the results that are strictly defined.

The last category – access is related to the data issues that are not important to statistical data gathering in the example presented in this paper. However this issue is very relevant when processing large amount of personal data, such as mobile call records (MCR), used for instance to analyse a daily mobility of population.

All aspects above led to the conclusion that Big Data technology needs various mechanisms of data quality improvement depending on the type of processed data. In general use of Big Data technology the most important issues of data quality are its usefulness, accurateness and relevancy.

4   Data Quality Barriers in Big Data

As written in introduction, the example used in this paper presumes working with website data to gather statistical information. There are two essential steps in the suggested data quality process for statistical data retrieving from the websites.

First is to use the data pattern to make a structured data that can be put in the traditional relational database, for instance a staging area. By using proper algorithm it is possible to make first filtering which will not accept the data that are below the particular level of quality. The level of the quality is assessed by the keyword-matching algorithm.

The second step is to feed the target repository which could be a data warehouse repository with variable dimension used to identify different business metadata. This process was described more detailed in [18].

The following presumptions were identified when working with Big Data, concerning the websites as the data sources containing relevant information:

1. There is a noise in the data.
2. Data is not clean.
3. There are lots of sources that should be prioritize.

Due to this fact several issues concerning data quality must be regarded. During working with the framework for gathering unstructured information, the possibility of occurrences of different problems was listed in Tab. 2.

Table 2. Most important issues concerning Big Data quality (Source: Own elaboration based on: [16])

Type of data	Occurrence in Big Data
Wrong data	High
Noisy data	High
Irrelevant data	High
Inadequate data	High
Hard data	Low
Redundant data	High
Right data	Medium
Reach data	Medium

As shown in Tab. 2 the most likely problems with data gathering supported by Big Data technology is to cope with wrong, noisy, irrelevant, inadequate and redundant data.

5   Case Study

To assess the possibility of retrieving high quality statistical data it is necessary to compare the typical survey with the one that use Big Data technology. The goal was to identify the possibility of using web portals on job offers to collect the data on job vacancies. The job vacancy rate is used in statistics based on surveys conducted by official statistics offices. The goal of this measure is to match supply and demand on labor market. This indicator is used by European Commission and European Central Bank to monitor an evolution in labor market on national (country) and European level.

The typical job offer portal has ca. 20 thousand job offers. A short overview of such portals is shown in Tab. 3.

Table 3. Overview of selected web portals with job offers

Name of the website	Number of job offers (in thous.) as of December 2013	Country
jobboerse.arbeitsagentur.de	673	Germany
thelocal.de/jobs	17	Germany
www.leforem.be	12	Belgium
www.jobs.ie	5	Ireland
pracuj.pl	21	Poland
praca.wp.pl	21	Poland
jobpilot	20+	International (selected countries)

Based on [8] statistical data on job offers is occupied from institutions with at least one employee but some European Union countries collect data just from enterprises having 10+ employees. The typical survey is addressed to 1-10 thousand of units that are obliged to send data on job offers, which is enough in terms of statistical sampling.

The question is whether it is possible to assess the benefits from Big Data technology use to collect data on job offers instead of collecting them in traditional way. Several issues must be regarded concerning the differences between statistical survey and the one based on Big Data technology. Main differences are presented in Tab. 4.

As presented in Tab. 4 the number of observations in typical traditional survey is limited. This is the result of the fact that respondents (units surveyed) are usually selected based on sampling algorithm. Comparing to Big Data technology, the number of observations cannot be easily increased. Using Big Data technology gives an opportunity to increase the number of observations by identifying and processing new data sources.

Typical statistical survey conducted in traditional way is prepared by statisticians who are using sampling method to find the best representation for surveyed topic. The consequence of such survey planning is high quality data with small standard errors. Different situation is when using Big Data algorithm to feed the database on surveyed topic. The data quality in Big Data is the consequence of the quality of data source which cannot be treated as a highly reliable data

Table 4. Overview of selected web portals with job offers

Attribute	Traditional survey	Supported by Big Data
Number of observations	limited	unlimited
Data quality	high	low due to redundancy, problems with matching keyword algorithms
Duplicates	no	possible
Duration between data collecting and results publishing	long (usually up to 3 months after conducting the survey)	short
Cost	high	low

when the data source is unstructured. However the reliability of the data source is not the only aspect of data quality in Big Data. There is also an issue relating to keyword-matching algorithm from the data source, which should be changed to fit the content of the source. Another issue that decrease the data quality of Big Data is duplicated information in various sources. In described case, it is very possible to have the same job offer published on different websites: job offer portals as well as website of the company that offer particular job. The key issue is to identify whether there are more than one job offer for particular workplace or this is the same job offer. Therefore it is necessary to prioritize the data sources, as it was written in previous parts of the paper to eliminate problems related to ambiguousness.

It is obvious that it takes long time to publish the results of the traditional survey. It is needed to apply algorithms to generalize the results and to test threshold values and standard errors. It is necessary because in many cases, the time period between data collecting and its publishing is usually up to three months. However it can be much longer depending on the type of the survey. Using Big Data technology this period can be reduced. However the most time is spent on preparing keyword-matching algorithms.

Another very important aspect of comparing traditional surveys with supported by Big Data technology is cost. As it was written in this paper, traditional processing the data, for instance in data warehouse, is much more expensive than using Big Data technology. That is due to the fact that collecting data in traditional way needs to prepare the survey and implement it by printing on paper or preparing electronic application to gather the data from respondents. When Big Data technology is used to collect the data, the main challenge is to prepare matching-keyword algorithms to identify the relationship between context of the website and surveyed topic.

To conclude, the following sequence should be used to increase the data quality in Big Data:

1. Understand consumer preferences regarding the data quality.
2. Ensure that data source is reliable and is strictly related to the topic of the survey.
3. Check if the matching-keyword algorithms used to explore the website are eligible to be used on the particular website – try to test keywords for its ambiguousness.
4. Ensure that there are no duplicates of the data between websites, otherwise the duplicates must be eliminated by data profiling with HiveQL or Pig Latin tools.
5. Do not filter the data if there is no certainty that the data source is relevant and reliable.
6. Collect the results in traditional database and use the same statistical methods to count standard errors and threshold values.

Technical issues of storing such information in traditional data warehouse was shown in [18].

6   Conclusion

The paper shows that there is a strong potential in Big Data technology as a supporting tool in retrieving statistical data from various sources. However there are lots of barriers related to data quality when using Big Data technology. Those barriers were identified and described in the paper. Big Data technology was treated in this paper as a tool to filter large amounts of unstructured data that is available on websites.

The main conclusion from the paper is that there is a necessity to define strict rules to increase the data quality in repositories that are supported by Big Data tools for information processing. Although the definition of the data quality is the same as in typical database, especially by taking three dimensions of the data: usefulness, accurateness and relevancy, it is necessary to apply a special rules written in fifth part of this paper. Improving the data quality is the next step after understanding user or consumer preferences concerning the data.

As it was identified in this paper, the data quality of Big Data is the result of the reliability of the data sources used in data processing. The second aspect of the quality of Big Data is the algorithm used to match the relationship between surveyed topic and the website used as a source.

The thesis written in the introduction of the paper was reconfirmed. It means that data quality of the information retrieved by Big Data technology is rather low and typical problems with Big Data are predictable. Therefore it must be accepted that unstructured information used to produce information will not be as reliable as data gathered using traditional survey. However Big Data can easily enhance the knowledge about specific topic much faster than by using traditional survey.

References

1. Biesdorf, S., Court, D., Willmott, P.: Big data: What's your plan? McKinsey Quarterly, 40–51 (2013)
2. Brown, B., Court, D., Willmott, P.: Mobilizing your c-suite for big-data analytics. McKinsey Quarterly, 76–87 (2013)
3. Central Statistical Office of Poland: Central statistical office of poland notes, http://www.stat.gov.pl/gus/5466_PLK_HTML.htm (accessed December 1, 2013)
4. Church, A.H., Dutta, S.: The promise of big data for od: Old wine in new bottles or the next generation of data-driven methods for change? OD Practitioner 45, 23–31 (2013)
5. Das, T.K., Kumar, P.: Big data analytics: A framework for unstructured data analysis. International Journal of Engineering Science & Technology 5, 153–156 (2013)
6. Dolnicar, S., Grun, B.: Including Don't know answer options in brand image surveys improves data quality. International Journal of Market Research 55, 2–14 (2013)
7. Durand, M.: Can big data deliver on its promise? OECD Observer,17 (2012)
8. Eurostat: Eurostat notes, http://epp.eurostat.ec.europa.eu/cache/ITY_SDDS/en/jvs_esms.htm (accessed December 12, 2013)
9. Hansen, J., Smith, S.: The impact of two-stage highly interesting questions on completion rates and data quality in online marketing research. International Journal of Market Research 54, 241–260 (2012)
10. Haug, A., Arlbjorn, J., Zachariassen, F., Schlichter, J.: Master data quality barriers: an empirical investigation. Industrial Management & Data Systems 113, 234–249 (2013)
11. Hoffmann, L.: Looking back at big data. Communications of the ACM 56, 21–23 (2013)
12. Jacobs, A.: The pathologies of big data. Communications of the ACM 52, 36–44 (2009)
13. Karr, A., Sanil, A., Banks, D.: Data quality: A statistical perspective. Statistical Methodology, 137–173 (2006)
14. Kumar, A., Niu, F., Re, C.: Hazy: Making it easier to build and maintain big-data analytics. Communications of the ACM 56, 40–49 (2013)
15. Louridas, P., Ebert, C.: Embedded analytics and statistics for big data. IEEE Software 30, 33–39 (2013)
16. Mandal, P.: Data quality in statistical process control. Total Quality Management & Business Excellence 15, 89–103 (2004)
17. Maślankowski, J.: The evolution of the data warehouse systems in recent years. Journal of Management and Finance 11, 42–54 (2013)
18. Maślankowski, J.: The integration of web-based information and the structured data in data warehousing. In: Wrycza, S. (ed.) SIGSAND/PLAIS 2013. LNBIP, vol. 161, pp. 66–75. Springer, Heidelberg (2013)
19. McAffee, A., Brynjolfsson, E.: Big data: The management revolution. Harvard Business Review, 61–68 (2012)
20. Nunan, D., Di Domenico, M.: Market research and the ethics of big data. International Journal of Market Research 55, 2–13 (2013)

21. Ross, J., Beath, C.M., Quaadgras, A.: You May Not Need Big Data After All. Harvard Business Review, 90–91 (2013)
22. Schroeder, J.: Big data, big business and the future of enterprise computing. NetworkWorld Asia 10, 17 (2013)
23. Sidi, F., Mohamed, K., Jabar, M., Ishak, I., Ibrahim, H., Mustapha, A.: A review of current trend on data management and quality in data communication. Australian Journal of Basic & Applied Sciences 7, 755–760 (2013)
24. Stonebraker, M.: What does 'big data' mean? Communications of the ACM 56, 10 (2013)
25. Vaughan, L., Yang, R.: Web data as academic and business quality estimates: A comparison of three data sources. Journal of the American Society for Information Science & Technology 63, 1960–1972 (2012)
26. Wang, R., Strong, D.: Beyond accuracy: What data quality means to data consumers. Journal of Management Information Systems 12, 5–33 (1996)
27. Yiu, D.: 5 storage system challenges in the big data era. NetworkWorld Asia 10, 26 (2013)
28. Zhang, D.: Granularities and inconsistencies in big data analysis. International Journal of Software Engineering & Knowledge Engineering 23, 887–893 (2013)

Reasoning with Projection in Multimodular Description Logics Knowledge Bases

Krzysztof Goczyła, Aleksander Waloszek, and Wojciech Waloszek

Gdańsk University of Technology, Department of Software Engineering,
ul. Gabriela Narutowicza 11/12, 80-233 Gdańsk, Poland
{kris,alwal,wowal}@eti.pg.gda.pl

Abstract. We present an approach to reasoning with projection, i.e. reasoning in which it is possible to focus on a selected part of knowledge (by neglecting some non-interesting fragments). Projection is most useful for modular knowledge bases in which only parts of knowledge have to be exchanged or imported to other modules. In this paper we present an optimized method of reasoning over results of projection. The tests indicate that the method can drastically reduce the time of inferencing, enabling our reasoner to efficiently work with modular knowledge bases embracing several dozens of modules, even in the presence of cyclic interrelationships between them.

Keywords: modular knowledge base, description logics, reasoning.

1 Introduction

Ontologies are a convenient tool for creating standardized object-oriented descriptions of selected domains of interest [10]. Such a standardized description can prove itself useful in integrating data from various information systems and processing them with respect to various perspectives and requirements.

Ontologies can draw knowledge from each other. In fact this reuse process is often considered essential for ontology creation and design (simply because that without it every ontologist would be forced to describe various areas of interest and to repeat the work of others). In the influential NeOn project a proposal of networked ontologies [8] has been formulated. In this vision ontologies are related to each other with links specifying the kind of dependency. The nature of the specification could vary, however it is worth noting that in this project a proposal of algebraic operations for ontologies has been formulated.

Algebraic description of interrelationships between ontologies is attractive, because a user is provided with a flexible apparatus for specifying a variety of relationships. Within our team we created a proposal of so-called tarset algebra for ontology modules. Not delving into details, we can say that with the algebra a user can extract interesting part of a third party ontology and include it in their own work.

One of the most important operations of tarset algebra that allows the user to perform this task is projection. Projection allows to focus on the needed

S. Kozielski et al. (Eds.): BDAS 2014, CCIS 424, pp. 102–111, 2014.

information while excluding the spurious details. (For example we might want to import some information about human beings from a biological ontology, but without its full classification as a vertebrate etc. Concepts like *Vertebrates* can thus be projected out from the imported part of the ontology.) Projection was also very extensively used by us in modeling knowledge bases consisting of mutually dependent modules (so that a knowledge base itself begins to resemble a network of ontologies).

In this paper we present an optimized method of reasoning with ontological modules connected with algebraic relationships. We introduced this method in CongloS reasoner handling tarset algebra expression, though the method itself is more general and can be also introduced in other types of inference engines. Due to introduction of the method we gained a major improvement of reasoning efficiency, especially in knowledge bases consisting of more than 10 modules, connected with cyclic dependencies.

The rest of the paper is organized as follows: in section 2 we elaborate upon the role of projection in general, and in organizing modular knowledge bases. In section 3 we describe the devised method. In section 4 we show results of experiments performed in order to assess the performance gain. Section 5 summarizes the paper.

Throughout the paper we assume familiarity with basics of Descripton Logics (DL). For a general introduction to DL we refer the Reader to [1].

2   Motivation

At the heart of combing ontology modules lie two general, and to some extent contradictory, needs: the need to maintain dependency of modules, and the need to reuse the knowledge contained in other modules. An elegant formulation of this phenomenon was ·proposed in [3] in the form of two rules: locality and compatibility. Not delving into details we may express them in a simplified form: Let us assume that we perform inferences from the perspective of one selected module (the target). Locality means that we reason over the domain of the target, taking into consideration all the conclusions drawn from the sentences contained in the target; while compatibility means that the other modules are for us a source of new conclusions for terms specified in the target, and the new conclusions should not contradict the original contents of the target.

The rules of locality and compatibility have been reflected in many systems, one of the most notable being Distributed Description Logic (DDL) [2]. DDL has been developed specifically for integrating knowledge from different ontologies. Specifics of integration between two ontologies are described by a set of bridge rules. A bridge rule specifies a direction of knowledge flow (always unidirectional), and the relationship between terms (usually concepts) from the two ontologies.

As an example of use of DDL consider two ontologies $O_1 = \{Animal \sqsubseteq \top, Wolf \sqsubseteq \top\}$ and $O_2 = \{Canid \sqsubseteq Mammal, Mammal \sqsubseteq Vertebrate\}$. The first ontology simply introduces the concepts *Animal* and *Wolf* not establishing any relationship between them. In the second ontology a simple hierarchy of concepts is defined. Introduction of bridge rules of the following form (prefixes preceding concepts indicate the original ontology of the concept): $\{2 : Canid \overset{\sqsupseteq}{\rightarrow} 1 : Wolf, 2 : Vertebrate \overset{\sqsubseteq}{\rightarrow} 1 : Animal\}$ will cause the transfer from O_2 to O_1 (according to the compatibility rule) of the conclusion that $Wolf \sqsubseteq Animal$. Very important is that the conclusion affects only concepts from O_1, concepts like $Mammals$ or $Vertebrates$ and relationships between them remain hidden from O_1.

A similar principle is maintained within *Structural Interpretation Model* (SIM [4]). SIM is a method of organizing multimodular knowledge bases proposed by our group. Modules in a SIM base form a hierarchy. Not delving into details of construction of a SIM base (we refer the interested Reader to [4]), we may say that a characteristic feature of SIM is that the level of details of describing knowledge increases down the hierarchy, i.e. the lower modules contain more specific concepts, while the higher modules more general ones. The flow of conclusions is determined by the structure itself: the conclusions, in a generalized form, are transferred up the hierarchy, and then, after the translation, are distributed to all lower branches. In the exemplary knowledge base shown in Fig. 1, information about $Mary$ being a soprano (from M_5) would be translated into more general knowledge that $Mary$ is a woman (in M_4), which in turn results in the conclusion that $Mary$ is a mother in the module M_6. Once again all the intermediate steps, and the concepts like $Soprano$, remain hidden from the perspective of the module M_6.

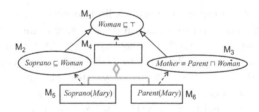

Fig. 1. An exemplary SIM knowledge base

A tool that describes this processes in a more generalized way is the algebra of tarsets [6]. Our algebraic framework is general enough to capture the most important part of DDL (see [5]) and to express interrelationships within a SIM base [7].

Tarset algebra introduces a set of operations on tarset universe. The elements of the universe are pairs $(\mathbf{S}, \mathbf{W})$, where $\mathbf{S}$ is a *signature* (or a *vocabulary*, i.e. a set of names), and $\mathbf{W}$ is a set of Tarski style interpretations of the vocabulary $\mathbf{S}$ (tarset models). By $\mathbf{S}(t)$, where t is a tarset, we understand its signature, and by $\mathbf{W}(t)$ its set of models.

The set of operations introduced by the algebra embraces union, intersection, selection and projection. Each operation is defined in the terms of their effect on either signature or models, or both. Intersection ($\cap$) is one of the conceptually simplest operation, defined as follows:

$$t \cup t' = (\mathbf{S}(t) \cup \mathbf{S}(t'), \mathbf{W}(t) \cap \mathbf{W}(t')) \tag{1}$$

As it can be seen, intersection reduces the set of possible models: effectively all the conclusions from the two tarsets being intersected can be drawn from the intersection.

The operation that enable us to hide concepts from other modules is projection. Projection is defined in the following way:

$$\pi_{\mathbf{S}}(t) = (\mathbf{S}, \mathbf{W}(t)|\mathbf{S}) \tag{2}$$

where by $\mathcal{I}|\mathbf{S}$ we understand an interpretation $\mathcal{I}$ projected to the signature $\mathbf{S}$ so that all the terms in $\mathbf{S}$ preserve their way of being interpreted: $\mathcal{I}|\mathbf{S} = \{\mathcal{J} : \Delta^{\mathcal{J}} = \Delta^{\mathcal{I}} \wedge \forall X \in \mathbf{X} : X^{\mathcal{J}} = X^{\mathcal{I}}\}$.

The two aforementioned operations: intersection and projection, allow us to describe an interestingly broad range of problems. Let us assume rather straightforward notational conventions: by $\mathbf{S}(S)$, where S is a set of sentences, we denote the signature of S (a set of all terms used in S) and by $T(S)$ we denote the tarset $(\mathbf{S}(S), \{\mathcal{I} : \mathcal{I} \models S\})$.

Using this convention, the result of the first example (the result of combining two DDL ontologies) can easily be described as $t_1' = \pi_{\mathbf{S}(O_1)}(T(O_1) \cap T(O_2) \cap T(\{Wolf \sqsubseteq Canid, Vertebrate \sqsubseteq Animal\}))$. The concepts from O_2 are hidden from a user due to projection, however the conclusion about every $Wolf$ being an $Animal$ perseveres.

A similar approach can be taken to describe the SIM base from Fig. 1. In this case it is better to assume that the relationships between modules are defined as inequalities ($t_1 \leq t_2$ means that $\mathbf{W}(t_1) \leq \mathbf{W}(t_2)$), and that modules M_i, $i \in [1..6]$ are represented as *tarset variables* (i.e. as variables whose values are tarsets). These inequalities, called *couplers*, maintain the modular structure of the base, and have to be satisfied also after every update to the base.

Interrelationships between modules M_4, M_5, M_6 can thus be expressed with the following couplers:

$$M_5 \leq \pi_{\{Soprano, Woman, Mary\}}(M_4 \cap M_2) \tag{3}$$

$$M_4 \leq \pi_{\{Woman, Mary\}}(M_5 \cap M_1) \tag{4}$$

$$M_6 \leq \pi_{\{Parent, Mother, Woman, Mary\}}(M_4 \cap M_3) \tag{5}$$

$$M_4 \leq \pi_{\{Woman, Mary\}}(M_6 \cap M_1) \tag{6}$$

Note that the projection assures that locality of modules is preserved. Only conclusions influencing their domain of interest are transferred.

As it has been shown in this section, the notion of projection is very important for maintaining locality and compatibility. In the next section we elaborate on the issue of reasoning with projection, especially in the presence of cyclic relationships between modules, like those expressed by couplers (3)-(6).

3   The Algorithm

3.1   The Problem

In this section we more closely describe the problems of reasoning with projection within tarset algebra framework. The currently utilized methods employ Description Logics sentences (axioms and assertions) to represent contents of modules. With this assumption, the only way to consider in the target context (the one representing the point of view from which we reason) conclusions from other contexts is by transferring sentences from one representation to another. Such a transfer is conducted basically by executing operations of tarset algebra described by couplers defined between modules of a knowledge base.

The presence of cyclic relationships between modules may pose a serious problem: if M_1 and M_2 are connected in a cyclic way, they require transfer of the knowledge—in the form of sentences—both: from M_1 to M_2 and from M_1 to M_2. Consequently, very often redundant knowledge is transferred to the target module representation, resulting in vast growth of the number of sentences that have to be considered during reasoning.

To describe this problem in more details we have to recall the definition of sentential representability of tarsets [6].Given the set of dummy names $\mathbf{D}$, a tarset t is $\mathcal{L}$-(2)-$\mathbf{D}$-$representable$ iff there exists a (finite) set of sets of sentences $\mathcal{S} = \{\mathcal{O}_i\}_{i\in[1..k]}, k \in N$, being its $\mathcal{L}$-(2)-$\mathbf{D}$ $representation$ (denoted $t \sim \mathcal{S}$), which means that all the conclusions that can be drawn from t can also be drawn from $\mathcal{S}$ and, conversely, all the conclusions that can be drawn from $\mathcal{S}$ and do not concern terms from $\mathbf{D}$ can also be drawn from t.

For example, the contents of modules of standard SIM knowledge bases can be easily shown to be $\mathcal{L}$-(2)-$\mathbf{D}$ representable, since (1) the user initially fills the modules with sentences, and (2) the SIM couplers consist only of two tarset algebra operations: intersection, and projection[1], and both of them preserve $\mathcal{L}$-(2)-$\mathbf{D}$ representability. An algorithm for calculation of $\mathcal{L}$-(2)-$\mathbf{D}$ representation of the results of these operations is sketched below:

1. If $t_1 \sim \mathcal{S}_1$, $\mathcal{S}_1 = \{\mathcal{O}_{1:i}\}_{i\in[1..k]}$ and $t_2 \sim \mathcal{S}_2$, $\mathcal{S}_2 = \{\mathcal{O}_{2:j}\}_{j\in[1..m]}$, then $t_1 \cap t_2 \sim \{\mathcal{O}_{1:i} \cup \mathcal{O}_{2:j} : i \in [1..k], j \in [1..m]\}$.
2. If $t \sim \mathcal{S}$, $\mathcal{S} = \{\mathcal{O}_i\}_{i\in[1..k]}$, then $\pi_{\mathbf{S}}(t) \sim \{\gamma_{t,\mathbf{S},\mathbf{D}}(\mathcal{O}_i) : i \in [1..k]\}$; where by $\gamma_{t,\mathbf{S},\mathbf{D}}$ we understand a function renaming terms from t not included in $\mathbf{S}$ to terms from $\mathbf{D}$.

In the case when no non-standard couplers have been defined by the user, the above formulas can be significantly simplified by the observation that all the representations may contain only a single set of sentences (representation calculation rules):

1. If $t_1 \sim \{\mathcal{O}_1\}$ and $t_2 \sim \{\mathcal{O}_2\}$, then $t_1 \cap t_2 \sim \{\mathcal{O}_1 \cup \mathcal{O}_2\}$.
2. If $t \sim \{\mathcal{O}\}$, then $\pi_{\mathbf{S}}(t) \sim \{\gamma_{t,\mathbf{S},\mathbf{D}}(\mathcal{O})\}$; $\gamma_{t,\mathbf{S},\mathbf{D}}$ is defined as above.

---

[1] In reality, also Rename operation is used but it does not influence the argument.

It is worth noticing that the projection is simply executed by "hiding" the unwanted terms through assigning them names from the dummy set **D**. This simplified approach can be a source of problems in the presence of cyclic relationships between modules of a knowledge base, as shown in the following example.

Let us consider a knowledge base K with four modules from Fig. 2. The initial contents of modules is: $M_1 := \{C \sqsubseteq \top\}, M_2 := \{D \sqsubseteq C\}, M_3 := \{\}, M_4 := \{D(a)\}$. Moreover, assume that the set of dummy names **D** consists of infinite (but countable) number of names of the form X_i (for concepts) and x_i (for individuals), $i \in N$.

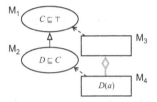

Fig. 2. An exemplary knowledge base K

The interesting part of the base are relationships between modules M_3 and M_4, which can be described with the following couplers:

$$M_3 \leq \pi_{\{C,d\}}(M_4 \cap M_1) \tag{7}$$

$$M_4 \leq \pi_{\{C,D,a\}}(M_3 \cap M_1) \tag{8}$$

The naive algorithm consists in "triggering" each of the couplers in turn by recalculating the sentential representation of the module in accordance with the representation calculation rules. For instance, triggering the coupler (7) would result in assigning the new representation M_3' of the module M_3 by calculating M_3' as $M_3 \cap \pi_{\{C,a\}}(M_4 \cap M_1)$.

Execution of this naive algorithm for K would result in infinite growth of representations of modules M_3 and M_4. Below we present results of some first iterations of the algorithm for these modules (for the sake of readability we denote their representation in successive iterations with subsequent primes):

1. $M_4' \sim \{\{D(a), D \sqsubseteq C\}\}$
2. $M_3' \sim \{\{C \sqsubseteq \top, X_1(a), X_1 \sqsubseteq C\}\}$
3. $M_4'' \sim \{\{D(a), D \sqsubseteq C, X_1(a), X_1 \sqsubseteq C\}\}$
4. $M_3'' \sim \{\{C \sqsubseteq \top, X_1(a), X_1 \sqsubseteq C, X_2(a), X_2 \sqsubseteq C\}\}$

3.2   The Solution

As shown in the previous subsection, the proper approach to the problem of reasoning with projection is necessary in order to perform efficient reasoning from tarset knowledge bases with cyclic couplers.

To solve this problem we developed a reduction algorithm which removes redundant subsets of sentences containing dummy terms. The reduction takes place after triggering a coupler, so the reduction algorithm is generally not dependent on the set of tarset algebra operations (and can thus be used for all kinds of tarset knowledge bases, not only SIM ones).

The reduction of redundant sentences can be easily explained by the observation that dummy terms can be interpreted as "some examples" of concepts, individuals, etc. So, for instance, the sentences $\{X_1(a), X_1 \sqsubseteq C\}$ can be read as "there exists some concept being subsumed by C, and a is a member of the concept." Such an approach allows us to come to a conclusion that a is also a member of C.

However, in the effect of repeated projection such subsets of sentences multiply within the representation. In the 4th step of the naive algorithm, there appears a representation $M_3'' \sim \{\{C \sqsubseteq \top, X_1(a), X_1 \sqsubseteq C, X_2(a), X_2 \sqsubseteq C\}\}$. The sentences $X_2(a)$ and $X_2 \sqsubseteq C$ are clearly superfluous, as they convey the same knowledge as the subset $\{X_1(a), X_1 \sqsubseteq C\}$.

Such superfluous subsets of sentences can be detected due to the fact that there exists a substitution of dummy variables which can be used to convert them into sentences already present in the representation. In the case of M_3'' the substitution is $X_2 \to X_1$, after which the representation assumes the form $\{\{C \sqsubseteq \top, X_1(a), X_1 \sqsubseteq C, X_1(a), X_1 \sqsubseteq C\}\} = \{\{C \sqsubseteq \top, X_1(a), X_1 \sqsubseteq C\}\}$.

One, however, has to be careful with substitutions. For example, in the set $\{C(x_1), C(x_2), x_1 \neq x_2, x_2 \neq x_1\}$ there seems that with use of $x_2 \to x_1$ we can remove superfluous sentences. However the result of the substitution $\{C(x_1), x_1 \neq x_1\}$ clearly does not convey the original knowledge.

The adverse effect described above can be easily avoided by the assumption that in the process of reduction we may only remove sentences from the original set S. This guarantees the correctness of the algorithm. The new set of sentences S' is a subset of S. Thus every model $\mathcal{I}$ of S is also a model of S'. Conversely, every model $\mathcal{J}$ of S' can easily be extended to a model of S by assigning the substituted terms exactly the same interpretation as their substitutions. Consequently, though the sets of models for S and S' are different, they give the same results of projection to the signature from which dummy terms are removed.

In accordance with the above discussion the reduction algorithm was divided into two major stages. In the first stage potentially redundant subsets are identified. In the second stage the superfluous sentences are removed.

To identify the redundant subsets along with the needed substitutions, we used a procedure for calculating graph simulation [9]. A simulation relation is defined for a labeled directed graph G understood as a quadruple (V, E, L, l), where V is a set of vertices, E set of edges, L set of labels, and l is a function which assigns each vertex from V to a label from L. A relation $\leq \in V \times V$ is a simulation iff for every $u \neq v$ it follows that (1) $l(u) = l(v)$ and (2) for all vertices u' such that $(u, u') \in E$ there exists a vertex v' such that $(v, v') \in E$ and $u' \leq v'$.

In order to use the simulation calculation algorithm for identifying redundant subsets of sentences, the $\mathcal{L}$-(2)-**D** representation is transformed into a tri-partite graph, where labels are strings of characters. The first partition consists of vertices representing axioms. Only axioms of exactly the same grammatical structure (e.g. $C \sqcap D \sqsubseteq E$ and $E \sqcap F \sqsubseteq D$) are assigned the same labels. Vertices in the second partition represent possible reassignments, i.e. i-th position in the grammar structure of each axiom. Their label is the concatenation of the label of the axiom and the ordinal number of the position. The third partition embraces vertices representing actual terms used in module representation, and are assigned one of the labels: "concept", "individual", or "role". Edges in the graph are set as follows: we connect bi-directly a vertex for every axiom with vertices for each of its assignments and, also bi-directly a vertex for every assignment (see Fig. 3 for the illustration of the process).

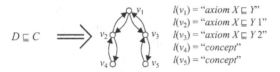

$D \sqsubseteq C$

$l(v_1) = $ "axiom $X \sqsubseteq Y$"
$l(v_2) = $ "axiom $X \sqsubseteq Y\,1$"
$l(v_3) = $ "axiom $X \sqsubseteq Y\,2$"
$l(v_4) = $ "concept"
$l(v_5) = $ "concept"

Fig. 3. An illustration of converting a sentence to a graph. Vertex v_4 represents the concept D, v_5 the concept C, and v_1 the sentence $D \sqsubseteq C$.

Such manner of constructing the graph allows us to almost directly use the simulation calculation algorithm to identify similar sets of axioms and terms. The resulting simulation would embrace the axioms of the same grammatical structure which hold similar terms in the same positions. Only the terms which occur in the same positions in similar axioms are considered similar (so the simulation between terms directly determines possible substitutions).

In the second stage of the algorithm, the resulting simulation is being analyzed, and the subsets of axioms that can be reduced to other subsets by substituting terms to those of similar terms are calculated. Only valid substitution (i.e. those of dummy terms) are taken into account. A simple heuristic is used here, according to which more frequently used terms are substituted with similar terms in the first place (in a greedy fashion). After these substitutions the set of axioms is reduced, by removal of the axioms that are already present in the representation.

In the example with the base K application of the reduction algorithm would result in modifying the 4th step of "triggering" couplers. The representation $\{\{D(a), D \sqsubseteq C, X_1(a), X_1 \sqsubseteq C\}\}$ of M_4'' would be reduced to $\{\{D(a), D \sqsubseteq C\}\}$ on the basis of similarity between X_1 and D. X_1 is identified as a term used in exactly the same positions of similar sentences as D and is therefore substituted with D, which results in the reduction. Further consequence of this reduction is reaching the fixpoint in the 4th step and termination of the process of "triggering" couplers.

4 Evaluation

In order to assess the impact of the sentence-set reduction, we performed a set of experiments with modular knowledge bases developed with use of CongloS system within the SYNAT project.

Since with use of a naive algorithm it was no possible to reach the fixed point (and the end of the algorithms), a special course of action had to be undertaken for the comparison. Namely, in the first step the optimized version has been executed. During the execution it was recorded which couplers and in which order were "triggered". Then the same order was applied to the naive version of the algorithm.

Three knowledge bases of different sizes have been selected for the test. The first base, "Kowalski", is very similar to the one shown in Fig. 1 and it contains 6 modules. The second base, "Simple Faculties", is slightly larger, and it describes functions of people at different faculties. The base consists of 10 modules. The third, largest, knowledge base "Faculties" is an extended version of the second one and contains 16 modules. All of the bases are SIM bases, which means that interrelationships between modules are circular.

Tests were performed with use of Pentium i5 3.0 GHz processor with 8GB RAM. Each test consisted in performing default inferences for Protégé (class inferences like subsumption, disjointness and equivalence, property inferences like domains and inverses, and individual inferences like types) on all of the modules. Every test has been performed 10 times to avoid potential influence of other processes. The results of the tests are gathered in Tab. 1.

Table 1. Evaluation of the reduction algorithm for various knowledge bases (KBs)

	Execution time with reduction [ms]	Naive algorithm execution time [ms]
"Kowalski" KB	111	62
"Simple Faculties" KB	1659	10624
"Faculties" KB	2663	> 30 minutes

As it can be seen from the table, for a small knowledge base the overhead of calculation simulation was larger than the potential growth of efficiency, as the naive algorithm proved itself to be faster in this simplest case. However, for the base with 10 modules the increase in speed was very significant (the algorithm with reduction performed over 6 times faster than the naive one). It can be observed that the gain grows with the size of the knowledge base, as for 16-modules' "Faculties" the improved algorithm achieved an acceptable reasoning time of approx. 2.5 seconds, while naive algorithm could not complete the inference tasks within 30 minutes (we terminated the tests after 30 minutes wait).

5   Summary

In the article we presented an improved algorithm for reasoning from modular knowledge bases with projection and in the presence of cyclic relationships between modules. As shown in the evaluation results, the algorithm allowed for significant reduction of reasoning time. The improved algorithm has been applied in CongloS system (available from *http://www.conglos.org*), which allows a user to work with modular knowledge bases consisting of dozens of modules.

The results presented in this article may also be generalized for use with all reasoners exploiting the sentential representation of ontological modules. Moreover, as the results of projection can be interpreted as sentences concerning "any concepts" or "any individual", the presented method of reasoning may be perceived as an improvement of expressiveness of DL ontologies.

Acknowledgments. This work was partially supported by the Polish National Centre for Research and Development (NCBiR) within the strategic scientific research and experimental development program: "SYNAT — Interdisciplinary System for Interactive Scientific and Scientific-Technical Information" under Grant No. SP/I/1/77065/10.

References

1. Baader, F., Calvanese, D., McGuiness, D.L., Nardi, D., Patel-Schneider, P.F. (eds.): Cambridge University Press, 2nd edn. (2007)
2. Borgida, A., Serafini, L.: Distributed description logics: Assimilating information from peer sources. In: Spaccapietra, S., March, S., Aberer, K. (eds.) Journal on Data Semantics I. LNCS, vol. 2800, pp. 153–184. Springer, Heidelberg (2003)
3. Ghidini, C., Giunchiglia, F.: Local model semantics, or contextual reasoning = locality + compatibility. Artificial Intelligence 127(2), 221–259 (2001)
4. Goczyła, K., Waloszek, A., Waloszek, W.: Contextualization of a dl knowledge base. In: Proc. DL 2007, Brixen/Bressanone, Italy (2007)
5. Goczyła, K., Waloszek, A., Waloszek, W.: S-modules - approach to capture semantics of modularized dl knowledge bases. In: Proc. of KEOD, pp. 117–122 (2009)
6. Goczyła, K., Waloszek, A., Waloszek, W.: A semantic algebra for modularized description logics knowledge bases. In: Proc. DL 2009, Oxford, United Kindgom (2009)
7. Goczyła, K., Waloszek, A., Waloszek, W., Zawadzka, T.: Owl api-based architectural framework for contextual knowledge bases. In: Bembenik, R., Skonieczny, L., Rybiński, H., Niezgódka, M. (eds.) Intelligent Tools for Building a Scientific Information Platform: Advanced Architectures and Solutions. Springer (2013)
8. Gomez-Perez, A., Suárez-Figueroa, M.C.: Scenarios for building ontology networks within the neon methodology. In: K-CAP 2009 (2009)
9. Henzinger, M.R., Henzinger, T.A., Kopke, P.W.: Computing simulations on finite and infinite graphs. In: Proc. of the 36th Annual Symposium on Foundations of Computer Science, pp. 453–462 (1995)
10. Staab, S., Studer, R. (eds.): Handbook on Ontologies, 2nd edn. Springer (2010)

Contextualizing a Knowledge Base
by Approximation - A Case Study

Krzysztof Goczyła, Aleksander Waloszek, and Wojciech Waloszek

Gdańsk University of Technology, Department of Software Engineering,
ul. Gabriela Narutowicza 11/12, 80-233 Gdańsk, Poland
{kris,alwal,wowal}@eti.pg.gda.pl

Abstract. Modular knowledge bases give their users opportunity to
store and access knowledge at different levels of generality. In this paper
we present how to organize a modular knowledge bases organized into
contexts in which a user can express their knowledge in much simplified
way, yet without losing its precision. The work is centered around the
notion of approximation - i.e. reducing the arity of predicates used. The
presentation is based on a case-study conducted with CongloS system
within a large-scale national project.

Keywords: modular knowledge base, description logics, reasoning.

1 Introduction

1.1 Foreword

The impact of context on the meaning of sentences is the phenomenon that has
been studied for many years. At the beginning it was the area of interest for
linguists and cognitive psychologists only. For some time, it has also become
an object of interest for knowledge engineers. Recently, due to the initiative of
Semantic Web, work on this phenomenon has been remarkably intensified.

The main aspect on which researchers on the contextual phenomena concentrate their efforts, is ontology merging. Creating ontologies is very costly, so it
is sensible to reuse them as often as possible. But if we do not intend to reuse
an ontology directly (as our needs are different than the needs of its authors)
we have to merge two points of view. This task results in a lot of problems and
there are many approaches to solve them.

In our work we focus on a quite different aspect of the phenomenon of context.
We are convinced that contexts should be integral parts of every created ontology. There are many arguments justifying such a convincement. The first group
of them are widely known arguments supporting generally ontology modularization. According to these arguments, it is much easier to proceed with smaller
portions of knowledge. But beside of them we find also many arguments that
are centered around the essentials of the process of knowledge base design and
exploitation.

S. Kozielski et al. (Eds.): BDAS 2014, CCIS 424, pp. 112–123, 2014.
© Springer International Publishing Switzerland 2014

One of the problems with designing a knowledge base is the choice of proper level of abstraction, i.e. approximation level on which the domain is pre-sented to the user. A standard approach here is to conceptually use a relation (or a predicate) for describing some class of phenomena (we are referring here to broad range of phenomena, it may be an event, but also like a fact of holding some position by some person etc.). Desire of being precise urges us to increase the arity of relation by including in it subsequent details like time, place, and other circumstances (so that we are not simply saying that someone is a *Dean*, but we include the information about the faculty, the time of their election, term of office, etc.). On the other hand, such decision leads to complication of querying the knowledge base, and of appending new knowledge to it, and also reduces the efficiency of reasoning.

The problem is even more apparent in Description Logics (DL), theoretic basis of OWL (recently the most extensively used standard for defining ontologies in the Internet. DL supports unary predicates (called concepts), and binary ones (called roles). Predicates of higher arity needs to be simulated here by reification. Therefore, our choice here is either to abandon precision and choose natural DL expressions, (like $Dean(x)$), or to introduce new individuals representing single instances of phenomena of interest and assigning the new individual y all the needed attributes (like $DeanOffice(y), hasPerson(y, x), hasDate(y, 2004-05-13), hasTermOfOffice(y, 4)$, etc.). As the second approach needs binding the new individual with many other with roles, we call it *role-based approach* in contract to the first one, called *concept-based.*

Using contextual methods of designing ontologies let us use the concept-based approach without losing the precision offered by the role-centric one. This can be done by using multiple contexts within one knowledge base, each context represents additional circumstances of phenomena being described. A designer can manipulate contexts, adjusting the level of approximation. With this in mind, we consider the ability of approximation as a fundamental feature of every contextual method. We illustrate it by contextualization of an exemplary non-contextual ontology conducted with use of our own contextual framework called CongloS (see http://conglos.org).

The structure of the paper is as follows. In section 2 we describe what is the role-centric approach using the cDnS ontology as an advanced example. Section 3 describes the object of contextualization – SYNAT Ontology. The main part of the paper – section 4 – contains description of the way of conceptualization and presents results. Section 5 concludes these results and section 6 summarizes the paper.

2   c.DnS as an Extreme Example of Role-Centric Approach

This section explains what is the role-centric approach and what are its advantages and drawbacks in comparison with the concept-centric approach.

As an illustration we present *c.DnS* ontology, which is an extreme example of the role-centric approach. It was described in [3]. The abbreviation means

Constructive Descriptions and Situations (according to a philosophical stance called constructivism, reality is described as a mental structure depending on a context).

The authors propose a system containing very few classes describing the most general objects of a domain of interest (e.g. *Entity*, *SocialAgent*, *Situation*, *Collection*, *InformationObject*, *Time*), and also some meta-classes (e.g. *Description*, *Concept*) for objects that represent elements of a description of a domain, not a domain itself.

Every knowledge base is understood as a *c.DnS* relation, i.e. a set of tuples, each containing eight elements:

$$c.DnS(d, s, c^*, e^*, a^*, k^*, i^*, t^*) \rightarrow$$
$$D(d) \land S(s) \land C(c^*) \land E(e^*) \land A(a^*) \land K(k^*) \land I(i^*) \land T(t^*)$$

The asterisk means that a given variable is an ordered list of values of this same type.

The meaning of a tuple is as follows: a social agent $a(A = SocialAgent)$ as a member of knowledge communities $k^*(K = Collection)$ perceives a situation $s(S = Situation)$ and describes it using a description $d(D = Description)$ and assigning entities $e^*(E = Entity)$ concepts $c^*(C = Concept)$. Information objects $i^*(I = InformationObject)$ are supposed to express the description d. Time intervals $t^*(T = Time)$ play two roles: firstly, they are temporal attributes for s, informing when the situation occurred; secondly, they describe a time interval when the description was carried out.

An example of a *c.DnS* tuple is:

$c.DnS(KnowledgeOfPreviousCases\#1, KillingSituation\#1,$
    $\{Precedent, Killer, Tool, HypotheticalIntention\},$
    $\{Event\#1, PhysicalAgent\#1, Tool\#1, Plan\#1\},$
    $Detective\#1, InvestigationTeam\#1, CriminalCode\#1,$
    $\{TimeOfEvent\#1, TimeOfInterpretation\#1\})$

The real power of cDnS ontology is *redescription*. Redescription allows us to describe this same situation from other points of view. For example, while the above tuple describes a situation of intentional murder, during an investigation the interpretation of the same event may be changed to self-defense:

$c.DnS(KnowledgeOfPreviousCases\#1, KillingSituation\#1,$
    $\{Precedent, Killer, Tool, HypotheticalIntention\},$
    $\{Event\#1, PhysicalAgent\#2, Tool\#1, SelfDefence\#1\},$
    $Detective\#2, InvestigationTeam\#1, CriminalCode\#1,$
    $\{TimeOfEvent\#1, TimeOfInterpretation\#1\})$

The constructive stance is some specific set of solutions directed towards multi-aspectual description of different aspects of reality within a non-contextual ontology. Such a non-contextual ontology must combine attributes of high-level ontology and of operational knowledge base, so it must be able to be expanded by facts in order to make reasoning over them possible. According to Gruber [7], the former should be characterized by the weakest possible ontology commitment, while the latter should strengthen the commitment, making it possible to particularize meaning of terms to one selected context.

Another requirement to satisfy is the monotonicity rule. This requirement prevents introduction to an ontology sentences that are in contradiction with its current contents. For example, it is infeasible to state that "John Doe is a rector" and then "John Doe is not a rector" to model a situation when John Doe ends his term of office and is not elected a rector for the next period. This is the reason why a non-contextual ontology cannot constrain the meaning of the concept Rector to "current rector".

Consequently, descriptions of various phenomena within such bases have to contain large amounts of additional contextual information. The knowledge base based on c.DnS is able to store information that is described very precisely. But the cost of it is cessation of designing hierarchic taxonomies of classes towards graph solutions.

Assigning an individual to an extension of a concept is in accordance with specific character of DL languages. The grammar of these languages bases on a kind of descriptions, what therefore enables us to formulate sentences similar to natural language utterances. Expressiveness of these descriptions supports ability of terminological reasoning according to the open world assumption (OWA). In case of the role-centric approach this power is being replaced by the power of reasoning similar to the one used in logical programming based on Horn rules-reasoning from facts based on the close world assumption (CWA). Such a kind of reasoning has *extensional character* rather than *intensional one*.

The concept-centric approach gives us much better ability to implement intensional reasoning. Describing things by concepts, i.e. unary predicates, allows us to utilize all advantages of DL languages.

3   SYNAT Ontology as a Case Study for Contextualization

In this section we present SYNAT Ontology (see [11]). It was created in the course of work on the SYNAT project using Protégé (http://protege.stanford.edu/). It is formulated in DL language with expressivity $\mathcal{ALCHOIQ(D)}$ and contains 472 classes and 296 properties.

The ontology has many features that make it a remarkable experimental material for contextualization. Firstly, it describes a very broad domain of interest which embraces different but somehow related problem sub-domains. Examples of the sub-domains are various aspects of institutions and their structure, activities, effects of activities, events, work of people, their career paths, geographical locations, etc. Diversity of the domains alone suggests the possibility of distinguishing some subsystems, which, while being autonomous, are connected with others by various relationships. Secondly, SYNAT Ontology bears many characteristic features of role-centric approach (as described above).

The main purpose of SYNAT Ontology is to gather information about objects and events concerning academic and scientific society. It is internally divided into five parts centered around five generic concepts: *Agent, InformationResource, Event, Project* and *Characteristic*.

These five concepts are roots of separated trees of taxonomies. First four concepts embrace entities existing in the reality. The last one is the essence of the ontology, typical for the constructive approach. It is the root for the richest tree of classes defining reified properties of different kinds and arities for all the other classes of individuals. The best example is characteristic of information resources gathered under the class *InformationResourceCharacteristic*. While the class *InformationResource* describes only physical carriers, its characteristic contains information about the content. First, there is a group of concepts describing content types, e.g. Catalogue, Guide, Encyclopedia, Norm, Patent, News, Report, etc. Then there are concepts for periodical resources, for describing structure of documents, for access conditions, or a very rich group of notions concerning web resources.

Another example is characteristic of persons. The main concept for this branch of conceptualization is *PersonCharacteristic*. It contains basic personal data, like names and addresses, as well as all data connected with the career in science and education. Of course, the organization of the data relies on reification of relations (see Fig. 1).

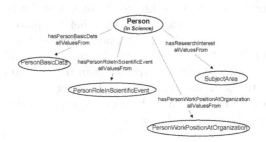

Fig. 1. Properties describing persons (from [11])

The ontology is divided into four modules, however this division is not realized as modularization in its common sense, i.e. as a way allowing to reason from smaller parts of knowledge. The purpose of the division is to separate information taken from two external ontologies, GIO (Geographic Information Objects, see [9]) and SCPL (Science Classification elaborated by the Polish government), and to adapt them to the needs of designed structure of concepts. This design decision allows a user to import the original ontologies without harming existing classification.

The SYNAT Ontology is also ready to reuse some other globally known ontologies (available under "free" licenses, like CC and BSD), like for example FOAF [2], VIVO [8], SWRC [10] or BIBO [1]. This is realized by relating local terms to the terms defined by those ontologies with appropriate axioms (the ontologies are imported by the SYNAT Ontology).

4 Contextualization of SYNAT Ontology

In this section we present how contextual approach helps to solve the contradiction between the two approaches described above: the role-centric and the concept-centric, taking as an example a fragment of the contextualized version of SYNAT Ontology. Beforehand (in Sec. 4.1) we briefly describe the technical platform used during the contextualization process.

4.1 Preliminaries

The method we have used to contextualize SYNAT Ontology is SIM. It was proposed in 2007 in [4]. According to the SIM method TBox and ABox of an ontology constitute two structures organized in different manners. Division of TBox is based on a relation very similar to OWL *import*. Contextualized TBox $T = (\{T_i\}_{i \in I}, \trianglelefteq)$ is a set of context types $\{T_i\}$ (partial TBoxes) connected with inheritance relation $\trianglelefteq$ on a set of indexes I. $\trianglelefteq$ is a poset with a minimal element m, T_m being a top context type.

A contextualized TBox cannot have a model by itself (or the model has lesser meaning). What determines the structure of the interpretation is a division of ABox. Contextualized ABox is defined as $A = (\{A_j\}_{j \in J}, inst, \ll)$ and consists of context instances $\{A_j\}$ (partial ABoxes), function $inst \colon J \to I$ (called instantiation function) assigning each partial ABox to its partial TBox and the aggregation relation $\ll$ established on a set of indexes J; $\ll$ is a poset with a minimal element n, A_n being a top context instance, $inst(n) = m$. Each partial ABox has its local interpretation defined as $\mathcal{I}_j = (\Delta^{\mathcal{I}_j}, \bullet^{\mathcal{I}_j})$, $\bullet^{\mathcal{I}_j}$ assigns elements of the domain to concepts and roles defined in $T_i \colon inst(j) = i$ and its ancestors. Contextualized interpretation $\mathcal{I} = (\{\mathcal{I}_j\}_{j \in J}, \ll)$ is a set of local interpretations connected with the relation $\ll$ (analogical as in the case of ABoxes). The flow of conclusions is assured by aggregation conformance rules obligatory for all models:

1. $\bigcup_{k \in \{k \colon j\} \ll k\}} \Delta^{\mathcal{I}_k} \subseteq \Delta^{\mathcal{I}_j}$,
2. $\bigcup_{k \in \{k \colon j\} \ll k\}} C^{\mathcal{I}_k} \subseteq C^{\mathcal{I}_j}$,
3. $\bigcup_{k \in \{k \colon j\} \ll k\}} R^{\mathcal{I}_k} \subseteq R^{\mathcal{I}_j}$,
4. $a^{\mathcal{I}_k} = a^{\mathcal{I}_j}$ for $j \ll k$.

$C^{\mathcal{I}}$ is the interpretation of a given concept C, $R^{\mathcal{I}}$ is the interpretation of a given role R and $a^{\mathcal{I}}$ is the interpretation of a given individual in $\Delta^{\mathcal{I}}$.

Generally, every SIM knowledge base is a special case of a tarset knowledge base (see [5]), and the nodes of the structure may be perceived as *tarsets*, and the relationships as *couplers*. The transformation is straightforward. There are only two tarset types predefined: the first one for context types and the second one for context instances. And, correspondingly, there are three coupler types: the first one for expressing inheritance, the second one for instantiation, and the third one for aggregation relationships.

The tool we used to contextualization of SYNAT Ontology is called Con-glos (see [6] and http://conglos.org). This tool allows a user to create tarset knowledge bases accordingly to the SIM method. Technically it is a set of plug-ins for Protégé – the most popular editor of ontologies in the OWL standard (http://protege.stanford.edu/). These plug-ins enrich the interface of the editor by adding elements necessary from the point of view of the SIM method and provide the system with mechanisms for management of the structure of the knowledge base being edited. Expressing SIM in terms of tarset algebra gave us possibility to extend the SIM method with user defined couplers, which increases expressivity of created structures. A user is also offered an inference engine capable of reasoning from contextual knowledge bases.

4.2 Process of Contextualization

In this section we present the contextualization, in the form of SIM knowledge base, of a fragment of the domain of interest covered by SYNAT Ontology. Within this fragment we set ourselves a goal to examine how to design different levels of approximation allowing to exploit the concept-centric approach.

As the discussed fragment of the domain we picked the one embracing publishing of articles, their editors, and employment of persons in institutions. The original contents of the fragment is illustrated in Fig. 2.

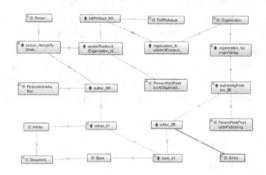

Fig. 2. A fragment of an exemplary ABox provided with SYNAT Ontology

For the sake of this presentation we chose a fragment which allows for answering to a query about employees of Institute of Computer Science who published articles in Springer-Verlag (in the further part of the paper this query is denoted as Q). The picture in Fig. 2 was generated from OntoGraf plug-in to Protégé editor.

The graph structure is typical for role-centric approach. It gives an ontology user considerable flexibility of specifying complex contextual information, however at the cost of complication of the structures that need to be inserted to the base. Ground objects (*Agents* and *InformationResources*) in the picture are *person_henrykRybinski*, *organization_InstituteOfComputerScience*, *organization_SpringerVerlag*, *article_*01, *book_*01. The remaining individuals are reifications of characteristics of the objects and specify relationships between them. For instance individual *author_HR* describes the authorship of the *article_*01 by *person_henrykRybinski* (and can be associated with attributes one would like to relate to this authorship – e.g. its date). The cost of this flexibility is, among others, complexity of queries: the query *Q* is in fact a query about instances of the following concept:

$\exists hasPersonWorkPositionAtOrganization.\exists hasRoleAtOrganization.$
$\exists holdsPersonRoleAtOrganization.\{organization\_InstituteOfComputerScience\}$
$\sqcap$
$\exists hasPersonAuthorship.\exists isAuthorOfArticle.\exists isIncludedIn.$
$\exists Inverse(isEditorOf).\exists Inverse(hasPersonRoleInPublishing).$
$\exists hasWorkPositionInPublishingOrganization.\{organization\_SpringerVerlag\}$

In the effect we obtained the modular structure of the knowledge base depicted in Fig. 3. In the figure context types are denoted with light ovals, and context instances with rectangles (tringle-ended lines between them represent inheritance, diamond-ended aggregation, and dashed instantiation). The names of the context types are prefixed with *C-* and should be interpreted as referring to the contents of their context instances, e.g. the name *C-Universities* means that each instance of this context type represents a group of universities; similarly, the name *C-University* indicates that each context instance of this type represents a single university. The names of context instances are prefixed with *I-* and may carry additional information, e.g. about which university is represented by this instance.

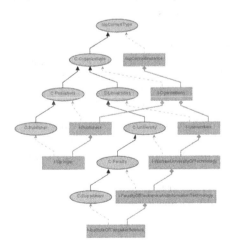

Fig. 3. Structure of the contextualized SYNAT Ontology fragment

As it can be seen in Fig. 3, the hierarchy of context types is divided into two major branches. At the top of this tree there is the context type *C-Organizations*. In this context only very general vocabulary is introduced, namely concepts: *Person*, *Employee* (subsumed by *Person*), and *Product*. (It is worth noting, that the vocabulary does not concern organizations themselves – they are represented not as individuals but as modules – other entities important from the point of view of an organization.) The vocabulary is generally taken from SYNAT Ontology, though sometimes with some major changes in meaning, e.g. *Employee* in SYNAT Ontology denotes a social role, while in the contextualized version it relates to a person.

The left branch concerns publishers: in the context type *C-Publishers* we introduce a set of new terms connected with publishing, among others concepts: *Book*, *Article* (both being subsumed by *Product*), *Editor* (subsumed by *Employee*) and roles: *hasAuthor*, *hasEditor*, *includesArticle*. (The same remark as before also applies to roles: their meaning is similar, though not identical, to that in SYNAT Ontology).

The context type *C-Publisher* is intended to embrace context instances representing single publishers. The most important concepts introduced there are *LocalArticle* (subsumed by *Article*), *LocalAuthor* (subsumed by *Author*), *LocalEmployee* (subsumed by *Employee*), *LocalEditor* (subsumed by *LocalEmployee* and *Editor*), denoting respectively articles published by the specific publisher, their authors, and employees and editors of the specific publisher. Defining these concepts allows a user to state within a single sentence that an entity is bound somehow with the publisher. Moreover, these concepts are defined in such a place in the hierarchy that the knowledge about individuals being their instances does not flow between context instances representing single publishers.

The right branch of the context types tree is organized analogously. The only difference is that also some aspects of organizational structure are reflected in the tree: as a consequence, here a university is understood as both a single university and a group of faculties it consists of (an analogous relation takes place between a faculty and its institutes).

At the top level of the branch, in the context type *C-Universities*, several concepts are introduced. For our discussion an important concept is *Professor*, subsumed by *Employee*. At the level of *C-University* new concepts are introduced: *UniversityEmployee* (subsumed by *Employee*) and *UniversityProfessor* (subsumed by *UniversityEmployee* and *Professor*). The concepts are analogous to concepts *LocalEmployee* and *LocalEditor* defined for publishers, and have their counterparts in the lower levels of the hierarchy: *C-Faculty* introduces *FacultyEmployee* and *FacultyProfessor*, and *C-Institute* introduces *InstituteEmployee* and *InstituteProfessor*.

In order to reflect the ABox fragment from Fig. 2 in this contextual structure, following assertions should be formulated in the context instance *I-Springer*:

LocalAuthor(person _henrykRybinski)
LocalArticle(article _01)
hasAuthor(article _01, person _henrykRybinski)
LocalBook(book _01)
includesArticle(book _01, article _01)
LocalEditor(person _ZbigniewRas)
hasEditor(person _ZbigniewRas)

Additionally, in the context instance *I-InstituteOfComputerScience*, the single assertion has to be formulated: *InstituteProfessor(person _henrykRybinski)*.

It is worth noting that there are no individuals reifying relations of higher arity in the ontology. All objects are instances of natural concepts and roles that describe relations between these objects.

This information suffices to answer query Q, which in the contextualized knowledge base has to be issued in two steps: first, one has to ask tarset-algebraic query for the module *I-InstituteOfComputerScience* ⊓ *I-Springer*, and then ask the resulting module about instances of concept *InstituteEmployee* ⊓ *LocalAuthor*.

To sum up, the case study showed that it is possible to create a contextualized knowledge base in SIM model with the assumption of concept-centric approach which conforms to the specified requirements. This approach may lead to substantial simplification of inserting new knowledge and querying a knowledge base.

The study was also an opportunity for observing a worrisome effect that may be an obstacle with use of the presented methods in more complex engineering solutions, namely the necessity of redefining *Professor* at each level of context hierarchy. This drawback we plan to alleviate by introducing a new kind of coupler.

5   Conclusions

The introduced structure was prepared strictly in accordance with assumed requirements about queries. Extension of the reauirements may lead to rendering the structure infeasible. In such a situation, a user of the knowledge base would find himself lacking of some important information, including perhaps implicit contextual parameters intended by a designer of the base.

Naturally, lack of this knowledge is simply a consequence of an incompatibility between specified requirements and user expectations, and as such might be neglected. Nonetheless, the ability to adjust the base to changing user needs is an important feature influencing the operability of the base. During the course of contextualization we identified the kind of changes in requirements that cause difficulties with the adjustment. Specifically, these are the changes that force to assign some attributes to the fragments of the original domain that have been modeled as context instances (or more generally, elements of the structure of the base) rather than individual objects.

An example of such a situation may be an addition of a new competency question about articles published by publishers from Berlin. It requires us to extend stored information by geographical locations of publishers, however there seems to be no natural way of doing this.

One of the possibilities here is to introduce a new context type *C-LocationHolders* containing a concept like *GeographicalLocation* and to inherit *C-Publisher* from this context type (see Fig. 4). A user may then assign each publisher a location by specifying an assertion like *GeographicalLocation(Berlin)* in *I-Springer*. However effective, this solution may easily lead to perplexity, as it mixes different levels of descriptions.

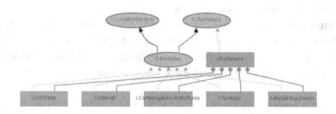

Fig. 4. A fragment of contextualized SYNAT Ontology after addition of a new context type

Analysis of this example leads to the conclusion that the most natural course of action is to assign attributes directly to publishers, that is directly to context instances. It in turn shows the direction of refining the description of contextual structure towards a manner similar to describing individual objects of the basic domain of interest. In the case of SIM method such a refinement might consist in treating context types in an analogous way to concepts and context instances to individuals. As a result we could gain a possibility of flexible adjustments of the knowledge base structure to answer for variable user needs, without necessity of utilizing unnatural constructs.

6 Summary

The conducted case study showed that with use of modular contextual design it is possible to maintain concept-based approach without the loss of precision of knowledge description. Moreover it allows for maintaining different levels of approximation within one knowledge base. As a result we have obtained a base which is extendible and relatively easy to be augmented and queried.

The effect was achieved due to use of multiple contexts and assigning each context a set of implicit contextual parameters. The process of querying a knowledge base became two-staged: first the user establish interesting context, and then asks the query.

Deeper analysis of the case shows ways of improving our methods. The necessity of redefining concepts (like $Professor$) at different levels of context hierarchy can be removed by introducing a new kind of coupler, while the cumbersomeness of extending the base by new knowledge (not covered by the initial version of requirements) may be alleviated by introducing a possibility of binding between an individual and a context; the binding can then be used in queries to direct them to context representing specific individuals (like, e.g. German publishers).

Acknowledgements. This work was partially supported by the Polish National Centre for Research and Devel-opment (NCBiR) under Grant No. SP/I/1/77065/ 10 within the strategic scientific research and experimental development program: "SYNAT – Interdisciplinary System for Interactive Scientific and Scientific-Technical Information".

References

1. Bibo ontology, `http://bibotools.googlecode.com/svn/` `bibo-ontology/trunk/doc/index.html`
2. Foaf vocabulary specification 0.99, `http://xmlns.com/foaf/spec/`
3. Gangemi, A., Lehmann, J., Catenacci, C.: Norms and plans as unification criteria for social collectives. In: Proc. of Dagstuhl Sem., vol. 07122, pp. 48–87 (2007)
4. Goczyła, K., Waloszek, A., Waloszek, W.: Contextualization of a dl knowledge base. In: Proc. DL 2007, Brixen/Bressanone, Italy (2007)
5. Goczyła, K., Waloszek, A., Waloszek, W.: A semantic algebra for modularized description logics knowledge bases. In: Proc. DL 2009, Oxford, United Kindgom (2009)
6. Goczyła, K., Waloszek, A., Waloszek, W., Zawadzka, T.: Owl api-based architectural framework for contextual knowledge bases. In: Bembenik, R., Skonieczny, L., Rybiński, H., Niezgódka, M. (eds.) Intelligent Tools for Building a Scientific Information Platform: Advanced Architectures and Solutions. Springer (2013)
7. Gruber, T.: Towards principles for the design of ontologies used for knowledge sharing. International Journal of Human-Computer Studies 43, 907–928 (1995)
8. Krafft, D.B., Cappadona, N.A., Caruso, B., Corson-Rikert, J., Devare, M., Lowe, B.J., VIVO Collaboration : Vivo: Enabling national networking of scientists. In: Web-Sci 2010: Extending the Frontiers of Society On-Line (2010)
9. Paliouras, G., Spyropoulos, C.D., Tsatsaronis, G. (eds.): Multimedia Information Extraction. LNCS, vol. 6050. Springer, Heidelberg (2011)
10. Sure, Y., Bloehdorn, S., Haase, P., Jens Hartmann, J., Oberle, D.: The swrc ontology - semantic web for research communities. In: Bento, C., Cardoso, A., Dias, G. (eds.) EPIA 2005. LNCS (LNAI), vol. 3808, pp. 218–231. Springer, Heidelberg (2005)
11. Wróblewska, A., Podsiadły-Marczykowska, T., Bembenik, R., Rybiński, H., Protaziuk, G.: SYNAT system ontology: Design patterns applied to modeling of scientific community, preliminary model evaluation. In: Bembenik, R., Skonieczny, Ł., Rybiński, H., Kryszkiewicz, M., Niezgódka, M. (eds.) Intelligent Tools for Building a Scientific Information. SCI, vol. 467, pp. 323–340. Springer, Heidelberg (2013)

SMAQ – A Semantic Model
for Analytical Queries

Teresa Zawadzka

Gdańsk University of Technology, Department of Software Engineering,
ul. Gabriela Narutowicza 11/12, 80-233 Gdańsk, Poland
tegra@eti.pg.gda.pl

Abstract. While the Self-Service Business Intelligence (BI) becomes an important part of organizational BI solutions there is a great need for new tools allowing to construct ad-hoc queries by users with various responsibilities and skills. The paper presents a Semantic Model for Analytical Queries – SMAQ allowing to construct queries by users familiar with business events and terms, but being unaware of database or data warehouse concepts and query languages such as MDX or SQL. The key idea of SMAQ is to provide a reference model that allows building user interfaces, even those based on natural language, independently on data models and query languages.

Keywords: Business Intelligence, data warehouse, query language, ontology.

1 Introduction

Analytical results of Business Intelligence activities have great influence on the decisionmaking process at many different management levels across the organization. Diversity of analytical tools is strongly connected with responsibilities and skills of BI users. The standard classification of consumer communities for BI tools [6] en-compasses power users, business users, casual users, data aggregators or information providers, operational analytics users, extended enterprise users and IT users. However, the actual problems of Self-Service BI, which is in the spotlight of this paper, concerns mostly power users and business users. Power users conduct "sophisticated analysis" using a BI semantic layer and issuing ad-hoc queries to prepare analytical reports for business users. On the other hand, business users, basing on analytical reports prepared by power users as well as their own ad-hoc queries, make business decisions. The idea of Self-Service BI is to loosen cooperation of these two groups of users with IT users. To meet this objective analytical tools should provide an interface allowing to formulate ad-hoc queries in business terms, independently of models used in semantic layers and query languages like MDX or SQL.

In this article author presents the Semantic Model for Analytical Queries – SMAQ based on 7W Framework [3]. The 7W: who, what, when, where, how

S. Kozielski et al. (Eds.): BDAS 2014, CCIS 424, pp. 124–138, 2014.

many, why and how is used to discover and model data requirements in the process of data warehouse design. Following the BEAM (Business Event Analysis & Modeling) creators [3] "The 7W questions you[1] ask to discover event details, mirror the questions that stakeholders will ask themselves when they define queries and reports." SMAQ defines a query model, which allows to formulate queries in the form of 7W. As this kind of queries is understandable to stakeholders in the process of gathering requirements for data warehouse design, the authors believe that it is also understandable in the process of querying a BI solution. Basing on this assumption the architectural framework and query model has been elaborated.

2   Architectural Framework

Introducing the framework that would allow a user to issue SMAQ queries, i.e. queries conformed to the SMAQ reference model, we want it to be independent of various semantic layers defined by vendors. Semantic layer is defined as a layer providing translation of the underlying database structures into business user oriented terms and constructs [9]. The vendors provide new layers (e.g. Microsoft introduced Tabular Model and PowerPivot in MS SQL Server 2012 [10]) to ease end users to create Business Intelligence solutions and analyzing data. The presented framework is not intended to replace semantic layer as it doesn't define translations between external data sources and business terms and constructs. It is designed as a set of tools allowing to issue queries to BI solutions independently of vendors, their solutions, data structures and query languages. The framework is presented in Figure 1.

Within the framework five levels are defined:

1. Various Semantic Layers – the first level that defines variety of semantic layers provided by vendors, e.g. BISM (Business Intelligence Semantic Model) [10] by Microsoft, BOSL (BusinessObjects Semantic Layer) [2] by SAP, or CEIM (Common Enterprise Information Model) [13] by Oracle.
2. Unified Model for Semantic Layers – the second level intended to unify access to Business Intelligence Tools by providing an ontology for multidimensional model, defined in OWL (Web Ontology Language [20]).
3. Query Model – the third level that defines reference model for typical BI queries. The model defines in SMAQ ontology, in the form of 7W queries, the semantics and structure of the most popular types of BI queries, like comparison or rating.
4. Various interfaces – the fourth level that allows for specifying user interfaces basing on the SMAQ ontology.
5. User – the fifth level, which represents power and business users of BI solutions.

Additionally, in Figure 1 mappings between particular levels of framework are depicted. They are represented as arrows with Roman numerals. Firstly,

---

[1] A data warehouse developer.

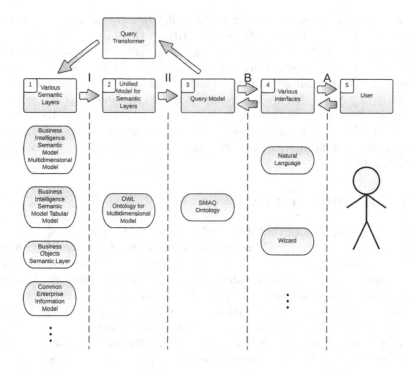

Fig. 1. Architectural framework for SMAQ

mappings between specified Semantic Layer and an OWL ontology for multidimensional model are specified (I). These mappings allow for transforming notions defined in semantic layers, like measures, facts, dimensions, hierarchies and others into OWL concepts, properties and their members. Mappings between Unified Model for Semantic Layer and Query Model (II) allow for projecting notions used to define queries in SMAQ ontology into concepts, properties and their members defined in the ontology for multidimensional model. Thanks to these two kinds of mappings (I and II) it is possible to express query in the form of 7W and formally model it in terms of the SMAQ ontology. This in consequence allows for transforming SMAQ queries into MDX statements expressed in XMLA [8], which is supported by most vendors. The last two arrows illustrate usage of user interface (A) and algorithms used to build SMAQ queries by various interfaces (B).

3   Unified Model for Semantic Layer

The Unified Model for Semantic Layer has been defined in the form of an ontology. The ontology, defined as "a formal specification of a shared conceptualization" [4] allows for defining a formal model independently of vendors, in a standardized language and in the form recommended by Semantic Web Initiative. In literature there is a discussion about usefulness of ontologies in the process of designing data warehouses. In [14] there are defined situations in which ontologies may help designers in development of data warehouses – among others, the following issues are mentioned:

- requirement analysis for multidimensional design should be complemented with formalized business model, managed as an ontology, for better understanding business issues by stakeholders,
- enrichment of multidimensional model with new data extracted from external ontologies like WordNet,
- introducing the formal description of units, magnitudes and scales for measures,
- using various classifications of measures with respect to possible mathematical operations carried on them,
- introducing to OLAP analysis inference based on logics and not only on calculating aggregations or data mining algorithms,
- matching visualization information with multidimensional model.

More profound analysis of works published in this area can be found in the Related Works section of the paper. In the context of selection the best formalism to define a unified model the ontology seems to be the most appropriate. The issues listed above are important with respect to building interfaces based on business terms rather than on data models.

The whole model consists of two ontologies. The first one is the Multidimensional Model Terminology that is independent of the particular multidimensional model and encompasses these aspects of standard multidimensional model which are used in query expressions. The second one is the Model-Specific Ontology, that is built according to mappings I (see Figure 1), imports Multidimensional Model Terminology and adds dimension elements.

3.1 Multidimensional Model Terminology

The UML diagram of Multidimensional Model Terminology is depicted in Figure 2. The class *Multidimensional Model* defines a multidimensional model, containing exactly one fact (class *Fact*), its measures (class *Measure*) and dimensions (class *Dimension*). Each fact is contained in exactly one multidimensional model. Facts corresponding to many-to-many relationships but not generating new measures are not required to be modeled. Otherwise, a new multidimensional model is created and some dimensions become conformed. Conformed dimensions can be elements of more than one multidimensional model. One dimension e.g. *Product* contains several category dimensions (class *CategoryDimension*) e.g. *product name* or *product type*. Additionally, for each dimension there may have been one or more hierarchies (class *HierarchyDimension*) defined. Each hierarchy consists of two or more levels (class *DimensionLevel* having a category dimension and a *hasParent* role).

The Multidimensional Model Terminology has been implemented as an OWL-DL (DL – description logics) [1] ontology in $\mathcal{ALCHIQ}$ dialect. All depicted in

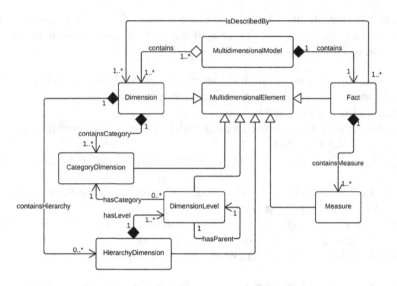

Fig. 2. UML model for Multidimensional Model Terminology

Figure 2 classes were implemented as disjoint concepts. Roles were implemented as object properties. Roles *containsHierarchy*, *containsCategory* and *contains-Measure* are implemented as subproperties of *contains* object property. Appropriate cardinality constraints are defined with the use of subsumption axioms. To simplify the notion of those axioms as an addition an inverse roles were introduced for each role whose subject cardinality is different from *0. . . ** (for example inverse role of *containsCategory* is *isCategoryFor*). Transitiveness of contains role has been intentionally omitted without any loss of accuracy of the model. Having transitive role or role defined as a chain of roles together with cardinality restriction put on this role causes ontology to be undecidable [5]. Thus only cardinality constraints were implemented. Furthermore, in order to validate the model consistency the terminology addresses two more issues:

- the first one is how to assign dimension levels connected together with *has-Parent* role to the same hierarchy,
- the second one is how to express that dimension levels being part of one hierarchy and the hierarchy itself must be assigned to the same dimension.

The most obvious way to resolve these issues would be to define appropriate chains of roles. However, having in mind the impact of introducing chain of roles for ontology decidability, a different approach must be applied. And so the solution for the first issue introduces two additional concepts being subsumes of the *DimensionLevel* concept: the *TopDimensionLevel* concept for the first level of the hierarchy and the *NotTopDimensionLevel* for the others. The individuals being instances of the *TopDimensionLevel* concept are not allowed to define *hasParent* object property assertion, for which allows the *NotTopDimension-Level* concept. However the *TopDimensionLevel* concept is stated as a domain for *isLevelFor* object property – in contrary to the *NotTopDimensionLevel* concept. The solution for the second issue demands adding axioms and assertions depending on the particular multidimensional model, thus it is described in the next point.

3.2   Model-Specific Ontology

The Model-Specific Ontology imports Multidimensional Model Terminology and adds axioms and assertions specific for particular model. This ontology is expressed in $\mathcal{ALCHOIQ}$ DL dialect, due to the fact that axioms using isets are added. On the basis of the standard multidimensional model concerning simplified book sale process, depicted in Tab. 1, the set of rules for creating new axioms and assertions is presented.

Table 1. The multidimensional model for a book sale process

Fact definition:
The sale of a book title at the given day by the given seller within a single transaction.
Measures:
Number of items sold
Price
Number of transactions

Dimensions	Category Dimensions	Hierarchy Dimensions
Transaction	Transaction number	
Book	ISBN	
	Genre	
	Title	
	Price range	
Seller	Personal Identification Number	Education hierarchy
	Name and surname	1. Education
	Age range	2. Age range
	Education	3. NameAndSurname
Date of purchase	Date	Date hierarchy
	Year	1. Year
	Month	2. Month
	Day	3. Day
	Holiday	

– An individual for the model is added:

$$Multidimensional Model(iModel)$$

– An individual for the fact is added and assigned to the model:

$$Fact(iSale)$$
$$contains(iModel, iSale)$$

- Individuals for measures *iPrice*, *iNumberOfTransactions* and *iNumberOf-ItemsSold* are added and assigned to the fact:

$$Measure(iPrice)$$
$$containsMeasure(iSale, iPrice)$$

and analogically for other measures.
- Individuals for the dimensions are added and assigned to the model and fact:

$$Dimension(iTransaction)$$
$$contains(iModel, iTransaction)$$
$$describes(iTransaction, iSale)$$

and analogically for other dimensions: *iBook*, *iSeller* and *iDateOfPurchase*.
- Individuals for category dimensions are added and assigned to appropriate dimensions:

$$CategoryDimension(iTransactionNumber)$$
$$containsCategory(iTransaction, iTransactionNumber)$$

and analogically for other category dimensions: *iISBN*, *iGenre*, *iTitle*, and *iPriceRange* for *iBook* dimension; *iPersonalIdentificationNumber*, *iName-AndSurname*, *iAgeRange* and *iEducation* for *iSeller* dimension; *iDate*, *iYear*, *iMonth*, *iDay* and *iHoliday* for *iDateOfPurchase* dimension.
- For each dimension a concept defining its category dimensions is created. Additionally, all those concepts are disjoint with each other and each individual being an instance of category dimension concept must be an instance of exactly one *CategoryDimension* subclass:

$$CategoryDimensionForBook \equiv \exists isCategoryFor.\{iBook\} \tag{1}$$
$$CategoryDimensionForSeller \equiv \exists isCategoryFor.\{iSeller\}$$
$$CategoryDimensionForBook \sqcap CategoryDimensionForSeller$$
$$\equiv Nothing$$

- Individuals for hierarchy dimensions are added and assigned to appropriate dimensions:

$$HierarchyDimension(iEducationHierarchy)$$
$$containsHierarchy(iSeller, iEducationHierarchy)$$
$$HierarchyDimension(iDateHierarchy)$$
$$containsHierarchy(iDate, iDateHierarchy)$$

- Analogically as for category dimensions for each dimension a concept defining its hierarchy dimensions is created:

$$HierarchyDimensionForDate \equiv \tag{2}$$
$$\exists\ isHierarchyFor.\{iDateOfPurchase\}$$
$$HierarchyDimensionForSeller \equiv$$
$$\exists\ isHierarchyFor.\{iSeller\}$$
$$HierarchyDimensionForDate \sqcap HierarchyDimensionForSeller$$
$$\equiv Nothing$$
$$HierarchyDimensionForDate \sqcup HierarchyDimensionForSeller$$
$$\equiv HierarchyDimension$$

- For each hierarchy dimension individuals for top and other levels are created, object assertions for *hasParent* role are added and top level individuals are assigned to appropriate hierarchies. The following assertions concern *iEducationHierarchy*. Analogical assertions are stated for *iDateHierarchy*:

$$TopLevelDimension(iEducationHierarchyLevel1)$$
$$hasCategory(iEducationHierarchyLevel1, iEducation)$$
$$isLevelFor(iEducationHierarchyLevel1, iEducationHierarchy)$$
$$NotTopLevelDimension(iEducationHierarchyLevel2)$$
$$hasCategory(iEducationHierarchyLevel2, iAgeRange)$$
$$hasParent(iEducationHierachyLevel2, iEducationHierarchyLevel1)$$
$$NotTopLevelDimension(iEducationHierarchyLevel3)$$
$$hasCategory(iEducationHierarchyLevel3, iNameAndSurname)$$
$$hasParent(iEducationHierarchyLevel3, iEducationHierarchyLevel2)$$

- Analogically as for hierarchies and category dimensions for each dimension a concept defining dimension levels for its hierarchies is created. The following example shows added axioms for education hierarchy of a seller:

$$NotTopDimensionLevelForEducationHierarchy \equiv \tag{3}$$
$$\exists hasParent.(\exists hasCategory.CategoryDimensionForSeller)$$
$$TopDimensionLevelForEducationHierarchy \equiv$$
$$\exists isLevelFor.HierarchyDimensionForSeller \sqcap$$
$$\exists isParent.NotTopDimensionLevelForEducationHierarchy$$
$$DimensionLevelForEducationHierarchy \equiv$$
$$NotTopDimensionLevelForEducationHierarchy \sqcup$$
$$TopDimensionLevelForEducationHierarchy$$
$$DimensionLevelForEducationHierarchy \equiv$$
$$\exists hasCategory.CategoryDimensionForSeller$$

Introducing new axioms (*1, 2* and *3*) allows to ensure that all levels of the specified hierarchy are assigned (by category dimension) to the same dimension as hierarchy whose levels they constitute. Otherwise the ontology becomes inconsistent due to one of the two: (1) dimension category will be assigned to more than one dimension, which is forbidden due to number restrictions defined in Multidimensional Model Terminology, or (2) top level dimension individual will be a dimension level for more than one hierarchy, which is also forbidden.

The last step is to enrich Model-Specific Ontology with dimension elements. The ontology is not enriched with measure values due to the fact that they are not needed to formulate a query. Dimension elements are added to *CategoryDimensionElement* and then the instances of *CategoryDimensionValue* are created. Each instance of *CategoryDimensionValue* is assigned an individual of *CategoryDimension* via *hasName* role and an instance of *CategoryDimesnionElement* via *hasValue* role.

4   SMAQ Ontology

The last (but not least) issue to discuss is the reference model for formulating queries. The most important advantage of MDX language is its expressiveness, but in consequence it is pretty difficult to use. In contrary, the presented SMAQ model is intended to be a base for building various types of interfaces – easy to understand but more expressive than a simple pivot table. Thus the model contains the set of ontologies importing Model-Specific Ontology. Each of those ontologies describes a particular type of query. The most popular types of analytical queries are comparison queries, statistical queries and scoring queries.

4.1   SMAQ Simple Query Model

The base model for SMAQ is Simple Query Model. It provides the functionality of pivot table and is further extended with other query models. The set of ontologies for SMAQ model is not closed – the model can be further extended with new types of queries. The UML model for the Simple Query Model ontology is depicted in Figure 3 and implemented as OWL ontology in analogical dialect as Model-Specific Ontology.

The essence of the model are seven concepts corresponding to *7W* framework: *HowMany*, **When**, **What**, **Why**, **Where**, **Who**, *How*. *HowMany* class represents values which are to be calculated. The rest six classes represent reference information (class *Description*) for which those values should be calculated. Consider for example a query that should calculate number of items sold at given dates. The *iNumberOfItemsSold* is a value to be calculated and the *iDateOfPurchase* is a reference date (which is an instance of *Where* concept – a subclass of *Description*). However, the meaning of date of purchase must be refined – it must be clarified which dates are in the field of interest. To do that the *WhenAttribute* is taken into account. The ontology defines also *WhatAttribute*, *WhyAttribute*, *WhereAttriubute*, *WhoAttribute* and *HowAttribute* concepts whose role is exactly the same as role of *WhenAttribute* concept, i.e. to define which reference

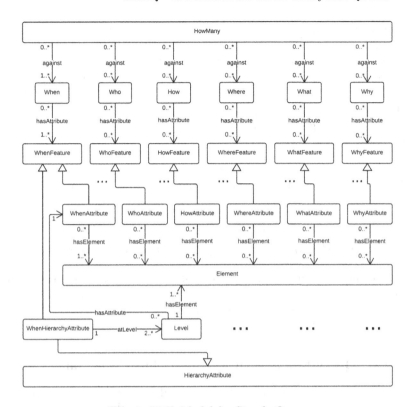

Fig. 3. UML Model for Simple Query

information will be used to specify calculation. So in our exemplary model the year (*iYear*), the month (*iMonth*), the day (*iDay*), the date (*iDate*) and the holiday (*iHoliday*) are instances of *WhenAttribute*. For the question: "Calculate number of sold items in 2001" the *iYear* attribute is chosen. In respect to ontology it means that *hasAttribute* object property assertion is defined: *hasAttribute(iDateOfPurchase, iYear)*. Moreover, the query demands defining which years are taken into account. In our question the year of 2001 is considered and thus the 2001 is an instance of *Element* class. The *Element* class contains all possible values for all attributes, but only 2001 is an object of *hasElement* role for *iYear* individual. What is more, as it can be seen in the UML diagram, instead of simple attributes like year it is also possible to choose the whole hierarchy (concepts *WhenHierarchyAttribute, WhatHierarchyAttribute, . . .*). With the use of *Level* concept it is also possible to define elements of hierarchies. Hierarchical and simple attributes are grouped into so-called features – concepts such as *WhenFeature, WhatFeature*, and so on.

In the example in Figure 3 there are individuals defined in Model-Specific Ontology. In order to provide proper task execution time, before defining values to calculate and reference information, the individuals must be assigned to appropriate concepts defined for Simple Query Model. This assignment is carried out against the mappings II (see Figure 1) in the form of OWL axioms:

1. *Description* ⊑ *Dimension* (each description is a dimension),
2. *Attribute* ⊑ *CategoryDimension* ⊓ *HierarchyDimension* (each attribute is a category dimension or hierarchy dimension),
3. *HowMany* ⊑ *Measure* (each *how many* value is a measure),
4. *HierarchyAttribute* ⊑ *HierarchyDimension* (each hierarchy attribute is a hierarchy dimension),
5. *Elements* ⊑ *CategoryDimensionElement* (each hierarchy element is a category dimension element),
6. *Level* ⊑ *DimensionLevel* (each level is a dimension level).

Additionally, for each of 6W attributes the subsumption axioms in the form of:

$$HowAttribute \sqsubseteq \exists isCategoryFor.How$$

$$HowHierarchyAttribute \sqsubseteq \exists isHierarchyFor.How$$

are added. It allows for automatic assignment of category and hierarchy dimensions to 6W attributes. The last step is to assign appropriate dimensions to 6W descriptions. In the book sale example that is *What(iBook)*, *How(iTransaction)* and *When(iDateOfPurchase)*.

4.2 SMAQ Query Model with Ad-Hoc Defined HowMany Values

This model defines types of queries that demand defining new values to calculate. The exemplary of those could be: (A) "Calculate the average price of the book within particular transactions," (B) "Calculate the difference between the number of sold items in 2010 and 2011 year," or (C) "Calculate percentage ratio of the number of transactions completed by young people among all sellers with bachelor degree." To allow formulating such queries, subclasses of *HowMany* class are defined. In Figure 4 new concepts are depicted.

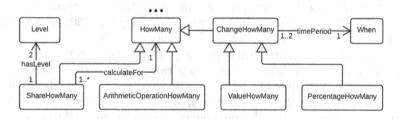

Fig. 4. UML Model for Query with ad-hoc defined HowMany values

The new *ArithmeticalOperationHowMany* concept is introduced to describe values that are to be calculated by an arithmetical operation – as in the exemplary A query. The average price of the book is defined as *iPrice / iNumberOf-ItemsSold*. The current version of SMAQ model supports operations defined as a textual data type property, however it is planed to integrate some formal language like MathML [19] in future versions. Another new concept *ChangeHowMany* defines the set of values to be calculated as the changes between some periods of time (while preserving the natural order of time). The *ChangeHowMany* concept is a domain for the following two roles: (1) *calculateFor*, which defines the change of interest (an instance of *HowMany* concept) and (2) *timePeriod*, which defines the time periods of interest (an instance of *WhenAttribute* concept with elements of interest assigned). Now, as the change can be calculated either as a percentage or as a value, the new subclasses of *ChangeHowMany* are introduced: *PercentageHowMany* and *ValueHowMany* concepts. So, for (B) query an instance of *ValueHowMany* (*howManyForBQuery*) is defined and the object property assertion *timePeriod(howManyForBQuery, iYear)* is added. Then the elements of *iYear* are chosen, as it is described in Simple Query Model i.e. via *hasElement* role.

The next new concept *ShareHowMany* defines the share of particular value between two subsequent levels of hierarchy. Analogically as for *ChangeHowMany*, the *calculateFor* role is defined. Moreover, the two object property assertions *shareFor* indicating consequent hierarchy levels (class *Level*) of the same hierarchy are to be defined. Also the percentage and value variant of the *ShereHowMany* measure is defined. For (C) query the *iNumberOfTransaction* is calculated for two hierarchy levels *iAgeRange* and *iEducation* with respectively "young" and "bachelor" values. The very important fact is that this model can be further extended with new types of defined *HowMany* values. It can be done easily by defining new subclasses of *HowMany* concept.

4.3 SMAQ Scoring Query Model

The last presented model is Scoring Query Model, which is depicted in Figure 5. The new concept *Score* is defined. There are two types of score: best and worst, modeled as subclasses of *Score* concept (*BestScore* and *WorstScore*). For each score the role *hasCalculate* is defined and its object defines which value is taken into account when the score is calculated. Moreover the datatype property *numberOfElementsInTheScore* is defined to limit the number of results returned. An example of such query is "Give me top ten bestsellers in 2013". The ontological representation of this query would consist of the following assertions:

$$BestScore(iBestsellers)$$

$$numberOfElementsInTheScore(iBestsellers, 10)$$

$$hasCalculate(iBestsellers, iNumberOfItemsSold)$$

$$against(iNumberOfItemsSold, iDateOfPurchase)$$

$$hasAttribute(iDateOfPurchase, iYear)$$

$$hasElement(iYear, 2013)$$

$$against(iNumberOfItemsSold, iBook)$$

$$hasAttribute(iBook, iISBN)$$

$$hasElement(iISBN, \ldots)$$

where ... stands for all ISBNs stored.

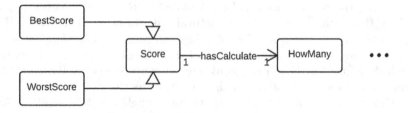

Fig. 5. UML Model for Scoring Query

5   Related Works

The presented model concerns two wide scientific areas: building interfaces controlled by ontologies and utilization of ontologies in the process of designing Business Intelligence solutions. In the first area much work has been devoted to building Graphical User Interface referencing domain ontology [17,7,18]. This field of scientific research is loosely connected with the subject of the article as the SMAQ Model does not specify the graphical presentation of formulated query in any way. Much more interesting would be research results concerning designing structure of a query on the basis of ontologies. However, to the best of our knowledge there is no work done that addresses this subject. In the field of utilization ontologies in designing Business Intelligence solutions: a profound analysis of this subject is described in [14]. The common part of this work and the current version of SMAQ Model is multidimensional model ontology. There are some papers concerning this subject [11,21,12], however, following the authors of [20], either the ontology structure is implicit [11,21], or some important aspects of dimension model are omitted [12]. The only ontology that was considered to be used in SMAQ model is the multidimensional ontology published in [15]. This ontology is based on UML model of multidimensional model presented in [16] and has been the starting point in the process of designing Multidimensional Model Terminology. However, the SMAQ Multidimensional Model Terminology is conceptually simpler than the one mentioned above, e.g. it defines separate concepts for dimensions, its attributes and hierarchies.

6   Conclusion and Future Work

The presented SMAQ Model is the first step towards providing Self-Service solution for formulating ad-hoc queries in BI solutions in a business process oriented way and not data models oriented. The contribution of this paper is the entire idea of the SMAQ framework with described ontologies implemented and validated. What is more the exemplary analytical queries have been implemented over the presented framework and successfully validated. Currently, the key issue is to develop the prototype of user interface utilizing the SMAQ Model.

References

1. Baader, F., Calvanese, D., McGuinness, D.L., Nardi, D., Patel-Schneider, P.F.: The Description Logic Handbook: Theory, Implementation, Applications. Cambridge University Press (2003)
2. Brogden, J., Sinkwitz, H., Holden, M., Marks, D., Orthous, G.: SAP BusinessObjects Web Intelligence: The Comprehensive Guide, 2nd edn. Galileo Press (2012)
3. Corr, L., Stagnitto, J.: Agile Data Warehouse Design Collaborative Dimensional Modeling from Whiteboard to Star Schema. DecisionOne Press (2013)
4. Gruber, T.R.: A Translation Approach to Portable Ontology Specification. Knowledge Acquisition 5(2), 199–220 (1993)
5. Horrocks, I., Kutz, O., Sattler, U.: The even more irresistible SROIQ. In: Proc. of the 10th Int. Conf. on Principles of Knowledge Representation and Reasoning (KR 2006), pp. 57–67. AAAI Press (2006)
6. Loshin, D.: Business Intelligence The Savvy Manager's Guide, 2nd edn. Morgan Kaufmann (2013)
7. Luo, S.Y., Wang, Y.L., Guo, J.: Research on Ontology-Based Usable User Interface Layout Approach. In: Computer Sciences and Convergence Information Technology (ICCIT), vol. 1, pp. 234–238 (2009)
8. Microsoft Corporation and Hyperion Solutions Corporation: XML for Analysis Specification, version 1.1 (November 2002),
   http://xml.coverpages.org/xmlaV11-20021120.doc
9. Mundy, J.: Making sense of the semantic layer, http://www.kimballgroup.com/2013/08/05/design-tip-158-making-sense-of-the-semantic-layer
10. Myers, P.: Introducing the BI Semantic Model in Microsoft SQL Server (October 2012), http://msdn.microsoft.com/en-us/library/jj735264.aspx
11. Nebot, V., Berlanga, R., Pérez, J.M., Aramburu, M.J., Pedersen, T.B.: Multidimensional integrated ontologies: A framework for designing semantic data warehouses. In: Spaccapietra, S., Zimányi, E., Song, I.-Y. (eds.) Journal on Data Semantics XIII. LNCS, vol. 5530, pp. 1–36. Springer, Heidelberg (2009)
12. Niemi, T., Niinimäki, M.: Ontologies and summarizability in OLAP. In: Proceedings of SAC 2010. pp. 1349–1353 (March 2010)
13. Oracle Corp.: Oracle Business Intelligence Foundation Suite (2013),
    http://www.oracle.com/technetwork/middleware/bi/bi-foundation-suite-wp-215243.pdf?ssSourceSiteId=ocomen
14. Pardillo, J., Mazón, J.: Using Ontologies for the Design of Data Warehouses. International Journal of Database Management Systems (IJDMS) 3(2), 73–87 (2011)
15. Prat, N., Akoka, J., Comyn-Wattiau, I.: Transforming multidimensional models into OWL-DL ontologies. In: Proceedings of the 2012 Sixth International Conference on Research Challenges in Information Science (RCiS), pp. 1–12. IEEE Computer Society Press (2012)
16. Prat, N., Comyn-Wattiau, I., Akoka, J.: Combining objects with rules to represent aggregation knowledge in data warehouse and OLAP systems. Data and Knowledge Engineering 70, 732–752 (2011)
17. Shahzad, S.K., Granitzer, M., Helic, D.: Ontological Model Driven GUI Development: User Interface Ontology Approach. In: Computer Sciences and Convergence Information Technology (ICCIT), pp. 214–218 (2011)
18. Tudorache, T., Noy, N.F., Falconer, S.M., Musen, M.A.: A knowledge base driven user interface for collaborative ontology development. In: Proceedings of the 16th International Conference on Intelligent User Interfaces, pp. 411–414 (February 2011)

19. W3C Recommendation: Mathematical Markup Language (MathML) Version 3.0 (October 2010), http://www.w3.org/TR/MathML3/
20. W3C Recommendation: OWL.: 2 Web Ontology Language (December 2012), http://www.w3.org/TR/owl-overview
21. Xie, G.T., Yang, Y., Liu, S., Qiu, Z., Pan, Y., Zhou, X.: EIAW: Towards a business-friendly data warehouse using semantic web technologies. In: Aberer, K., Choi, K.-S., Noy, N., Allemang, D., Lee, K.-I., Nixon, L.J.B., Golbeck, J., Mika, P., Maynard, D., Mizoguchi, R., Schreiber, G., Cudré-Mauroux, P. (eds.) ASWC 2007 and ISWC 2007. LNCS, vol. 4825, pp. 857–870. Springer, Heidelberg (2007)

An Ontology-Enabled Approach
for Modelling Business Processes

Thi-Hoa-Hue Nguyen and Nhan Le-Thanh

WIMMICS - The I3S Laboratory - CNRS - INRIA
University of Nice Sophia Antipolis
Sophia Antipolis, France
nguyenth@i3s.unice.fr, Nhan.LE-THANH@unice.fr

Abstract. Coloured Petri Nets (CPNs) have formal semantics and can describe any type of workflow system, behavioral and syntax wise simultaneously. They are widely studied and successfully applied in modelling of workflows and workflow systems. There is an inherent problem regarding business processes modelled with CPNs sharing and subsequently their reuse need to be considered. The Semantic Web technologies, such as ontologies, with their characteristics demonstrate that they can play an important role in this scenario. In this paper, we propose an ontological approach for representing business models in a meta-knowledge base. Firstly, the CPN ontology is defined to represent CPNs with OWL DL. Secondly, we introduce four basic types of manipulation operations on process models used to develop and modify business workflow patterns. To the best of our knowledge, representing business process definitions and business workflow patterns as knowledge based upon ontologies is a novel approach.

Keywords: Business Process, CPN, Manipulation Operation, OWL DL Ontology, Representing.

1 Introduction

To date, software systems that automate business processes have been becoming more and more available and advanced. According to [16], process models, which are first designed during built-time phase on the basis of design requirements, are then automated by software systems during run-time. Therefore, grasping the requirements properly and then transforming them without losing any information into a semantically rich specification play an important role in supporting business process management. However, the existing practice of modelling business processes is mostly manual and is thus vulnerable to human error, resulting in a considerable number of failed projects. Consequently, it is desirable to develop an alternative approach, which ensures high quality and semantically rich business process definitions.

On one hand, Coloured Petri Nets (CPNs) [12] have been developed into a full-fledged language for the design, specification, simulation, validation and

S. Kozielski et al. (Eds.): BDAS 2014, CCIS 424, pp. 139–147, 2014.
© Springer International Publishing Switzerland 2014

implementation of large software systems. CPNs are a well-proven language suitable for modelling of workflows or work processes [10]. Although CPNs are widely studied and successfully applied in modelling of workflows and workflow systems, the lack of semantic representation of CPN components can make business processes difficult to interoperate, share and reuse.

On the other hand, an ontology with its components, which can provide machine-readable definitions of concepts, plays a pivotal role in representing semantically rich business process definitions. Once business process semantics are machine-processable, IT experts can easily develop their appropriate software systems from business process definitions.

Our objective, outlined in Fig. 1, is to represent Control flow-based Business Workflow Patterns (CBWPs) in a meta-knowledge base which intend to make them easy to be shared and reused among process-implementing software components. We focus on proposing an ontological model to represent Coloured Petri Nets (CPNs) with OWL DL. We first define a meta-knowledge base for CBWPs management. We then introduce manipulation operations on process models for the purpose of developing and modifying CBWPs. Our ongoing work is developing a graphical interface to design and simulate CBWPs. To the best of our knowledge, this is a novel approach for representing business process definitions and patterns as knowledge based upon ontologies.

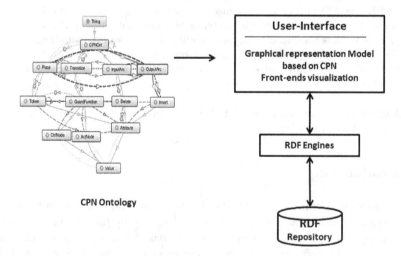

Fig. 1. An overall approach for representing Control flow-based Business Workflow Patterns (CBWPs) in knowledge base

The rest of this paper is structured as follows: In section 2, we recall the main notions of CPNs and ontologies. In section 3, we introduce a novel ontology for CPNs. We then elucidate the realization of our ontology in section 4. In section 5, we introduce four basic types of manipulation operations to support the creation and modification of CBWPs. Finally, section 6 concludes the paper with an outlook on the future research.

2   Foundations

In this section, the main notions of CPNs and ontologies are introduced. On this basis, we will present an ontology for business processes in the upcoming section.

2.1   Coloured Petri Nets

Coloured Petri Nets (CPNs) are extended from Petri nets with colour, time and expressions attached to arcs and transitions. A CPN, which is a directed bipartite, consists of *places* (drawn as ellipses) and *transitions* (drawn as rectangles) connected by *directed arcs* (drawn as arrows). Each place holds a set of markers called *tokens*. The number of tokens in a place can vary over time. Each token can carry both a data value called its colour and a time-stamp. The type of tokens in a place is the same type as the type of the place.

Since transitions may consume and produce tokens, it is necessary to use *arc expressions* to determine the input-output relations. An incoming arc indicates that tokens may be removed by the transition from the corresponding place while an outgoing arc indicates that tokens may be added by the transition.

In addition, CPNs allow us to structure the descriptions in a hierarchical way. Large models may be divided into sub-models, also known as sub-processes or modules, in order to get a layered hierarchical description. As a result, sub-models can be reused. For more details on CPNs, please refer to [12].

The Workflow Management Coalition (WfMC) [21] identified four routing constructs including *sequential, parallel, conditional* and *iteration*. We use five building blocks, i.e., *Sequence, And − split, And − join, Xor − split, Xor − join* to model these types of routing. Consequently, a routing construct might contain control nodes (transitions in the building blocks) and activity nodes (other transitions). An example of a business process modelled with CPNs is shown in Fig. 2. This process model comprises five activity nodes T_1, T_3, T_4, T_6 and T_8 connected by three control nodes T_2, T_5, T_7.

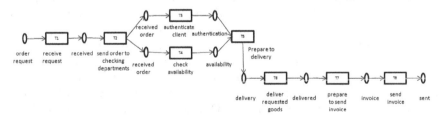

Fig. 2. An example of an order management process

2.2   Ontologies

Ontology definition languages such as RDFS[1], OIL[2] or OWL[3] can be used to define ontologies. Here, we focus on the Web Ontology Language (OWL), a W3C

---

[1] http://www.w3.org/2001/sw/wiki/RDFS
[2] http://www.w3.org/TR/daml+oil-reference
[3] http://www.w3.org/2004/OWL/

Recommendation, which is "a family of knowledge representation languages for authoring ontologies" [2]. OWL ontologies can be categorised into three sub-languages OWL Lite, OWL DL, and OWL Full.

OWL DL, which stands for OWL Description Logic, is equivalent to Description Logic $\mathcal{SHOIN}(\mathcal{D})$. OWL DL supports all OWL language constructs with restrictions (e.g., type separation), provides maximum expressiveness while always keeping computational completeness and decidability. Therefore, we choose OWL DL language for our work with the aim of taking advantage of off-the-shelf reasoning technologies. For more details on OWL DL, please refer to [20].

It is important to underline that an ontology consists of several different components including classes (concepts), individuals (instances), attributes, relations and axioms, etc. Therefore, by using OWL DL language to formulate an ontology, we get an OWL DL ontology that contains a set of OWL DL (class, individual, data range, object property, data type property) identifiers, and a set of axioms used to represent the ontology structure and the ontology instances.

3   An Ontology for Business Processes Modelled with Coloured Petri Nets

3.1   Represention of Coloured Petri Net with OWL DL Ontology

In this subsection, we define semantic metadata for business processes modelled with CPNs. The main purpose is to facilitate business process models easy to be shared and reused among process-implementing software components. We continue our work at [13, 14] to develop the CPN ontology. Each element of CPNs is concisely translated into a corresponding OWL concept. Fig. 3 depicts the core concepts of our CPN ontology. In the next step, we will describe the main constructs of the ontology modelled with OWL DL.

The CPN ontology comprises the concepts: **CPNOnt** defined for all possible CPNs; **Place** defined for all places; **Transition** defined for all transitions; **InputArc** defined for all directed arcs from places to transitions; **OutputArc** defined for all directed arcs from transitions to places; **Token** defined for all tokens inside places (We consider the case of one place containing no more than one token at one time); **GuardFunction** defined for all transition expressions; **CtrlNode** defined for occurrence condition in control nodes; **ActNode** defined for occurrence activity in activity nodes, **Delete** and **Insert** defined for all expressions in input arcs and output arcs, respectively; **Attribute** defined for all attributes of individuals); **Value** defined for all subsets of $I_1 \times I_2 \times \ldots \times I_n$ where I_i is a set of individuals.

Properties between the concepts in the CPN ontology are also indicated in Fig. 3. For example, a class *Place* has two properties *hasMarking* and *connectsTrans*. Consequently, the concept *Place* can be glossed as 'The class *Place* is defined as the intersection of: (i) any class having at least one property *connectsTrans* whose value is equal to the class *Transition* and; (ii) any class having one property *hasMarking* whose value is restricted to the class *Token*'.

$CPNOnt \equiv \geq 1hasTrans.Transition \sqcap \geq 1hasPlace.Place$
  $\sqcap \geq 1hasArc.(InputArc \sqcup OutputArc)$
$Place \equiv connectsTrans.Transition \sqcap = 1hasMarking.Token$
$Transition \equiv connectsPlace.Place \sqcap = 1hasGuardFunction.GuardFunction$
$InputArc \equiv \geq 1hasExpresion.Delete \sqcap \exists hasPlace.Place$
$OutputArc \equiv \geq 1hasExpression.Insert \sqcap \exists hasTrans.Transition$
$Delete \equiv \forall hasAttribute.Attribute$
$Insert \equiv \exists hasAttribute.Attribute$
$GuardFunction \equiv \geq 1hasAttribute.Attribute \sqcap = 1hasActivity.ActNode$
  $\sqcup = 1hasControl.CtrlNode$
$Token \equiv \geq 1hasAttribute.Attribute$
$Attribute \equiv \geq 1valueAtt.Value$
$CtrlNode \equiv \leq 1valueAtt.Value$
$ActNode \equiv = 1valueAtt.Value$
$Value \equiv valueRef.Value$

Fig. 3. CPN ontology expressed in a description logic

3.2 Realization

We rely on OWL DL and use Protégé[4], an OWL editor, to create our CPN ontology. We here describe some axioms created for the CPN ontology.

It is necessary to note that two OWL classes, $owl : Thing$ and $owl : Nothing$, are particularly predefined. The class extension of $owl : Thing$ is employed to denote the set of all individuals. The class extension of $owl : Nothing$ is the empty set. As a result, each user-defined class is absolutely a subclass of $owl : Thing$.

- The class *Place* comprises two properties *connectsTrans* and *hasMarking*. The class axiom is created as follows:
  $Class(Place\ complete\ restriction(connectsTrans\ hasValue(Transition))$
    $restriction(hasMarking\ allValuesFrom(Token)$
    $qualifiedCardinality(1)));$
- The class *Place* is a sub-concept of the class *CPNOnt*. The class axiom is created as follows:
  $SubClassOf(Place\ CPNOnt);$
- The classes *Place* and *Transition* are mutual disjoint. The class axiom is created as follows:
  $DisjointClasses(Place\ Transition);$
- The domain of the property *connectionsTrans* is a union of the class *Place* with the class *InputArc*. The range of this property is a union of the class *Transition* with the class *OutputArc*. In addition, the properties *connects–Trans* and *connectsPlace* are inverse properties. We create the property axiom for *connecsTrans* as follows:
  $ObjectProperty(connectsTrans\ domain(unionOf(Place\ InputArc))$
    $range(unionOf(Transition\ OutputArc))\ inverseOf(connectsPlace));$

---

[4] http://protege.stanford.edu/

- The class *Attribute* contains at least one *Value*. We create the class axiom for *Attribute* as follows:
  EquivalentClasses(Attribute restriction(valueAtt allValuesFrom(Value)
  minQualifiedCardinality(1)));

After presenting some axioms created for *Classes* and *Properties* of our ontology, we now introduce the modelling of *Individuals* being the third OWL element. Individuals or instances are generated based on the modeller and depend on the modelling objective. For example, Fig. 4 shows the mapping of the transition *receive request* named T_1, which is depicted in Fig. 2, to the classes and properties of the CPN ontology.

```
<Transition rdf:ID="T_1">
 <hasGuardFunction rdf:resource="#Guard_Request"/>
 <connectsPlace rdf:resource="#Received"/>
</Transition>
```

Fig. 4. Mapping Individuals to Classes and Properties of the CPN ontology

We have been introducing the CPN ontology represented in OWL DL. In order to develop or modify CBWPs (CPN models), in the next section, we will present manipulation operations on their elements. We also introduce the corresponding manipulation statements written in the SPARQL language[5] to store concrete CBWPs (CPN models) in RDF[6].

4   Manipulation Operations on Basic Business Process Models

For the purpose of modelling basis business processes with CPNs, the following basic types of manipulation operations on its elements are required:

1. Inserting new elements (i.e., places, transitions or arcs, etc.) into a process model.
2. Deleting existing elements from a process model.
3. Updating existing elements for adapting to a process model.
4. Editing the order of existing elements in a process model.

More complex operations can be developed based upon these basic operations. For example, merging two separate CBWPs, which represent two process models into one, can be considered as inserting all places, transitions and arcs from one pattern to the other. Besides, one new arc is also inserted in order to link these CBWPs.

We next define the basic manipulation statements by the corresponding pseudocodes. With the aim of storing CBWPs in RDF, we introduce the SPARQL statements being suitable to the manipulation statements.

---

[5] http://www.w3.org/TR/sparql11-query/

[6] http://www.w3.org/TR/2004/REC-rdf-concepts-20040210/

- Inserting new elements into a process model.
  INSERT ELEMENT $\{e_1, e_2, \ldots, e_n\}$ *INTO PROCESS wf*
  $[WHERE\ cond_1, cond_2, \ldots, cond_m]$; $(n \geq 1, m \geq 1)$
  This statement means that elements $e_1, e_2, \ldots, e_n$, each of which has been created, are inserted into a process model named wf. Conditions $cond_1$, $cond_2$, ..., $cond_m$ in the *WHERE* clause (if any) specify how to insert these new elements in the process model wf.
  The INSERT DATA statement or the INSERT WHERE statement in the SPARQL query language fits for inserting new elements into the RDF file format. As an example, in Fig. 5, a new place, which contains a token and is connected to a transition, is inserted into a process model[7].

```
INSERT DATA{
 myWF:NameOfPlace a CPNOntology:Place;
   CPNOntology:hasMarking myWF:NameOfToken.
 myWF:NameOfWF CPNOntology:hasPlace myWF:NameOfPlace.}
```

Fig. 5. An example of the INSERT DATA statement

- Deleting existing elements from a process model.
  DELETE ELEMENT $\{e_1, e_2, \ldots, e_n\}$ *FROM PROCESS wf*; $(n \geq 1)$
  This statement means that existing elements $e_1, e_2, \ldots, e_n$ are completely deleted from a process model named wf.
  The DELETE DATA statement or the DELETE WHERE statement in the SPARQL query language fits for deleting existing elements from the RDF file format.
- Updating existing elements for adapting to a process model.
  UPDATE ELEMENT $\{e_1, e_2, \ldots, e_n\}$ *ON PROCESS wf*
  $[WHERE\ cond_1, cond_2, \ldots, cond_m]$; $(n \geq 1, m \geq 1)$
  This statement means that elements $e_1, e_2, \ldots, e_n$ in a process model named wf, each of which has been created, are updated. Conditions $cond_1, cond_2, \ldots,$ $cond_m$ in the *WHERE* clause (if any) specify how to update these elements in the process model wf.
  In this case, some statements in the SPARQL query language can be used, such as the INSERT DATA statement, the INSERT WHERE statement or the DELETE INSERT WHERE statement.
- Editing the order of existing elements in a process model.
  MODIFY PROCESS wf
  $WHERE\ cond_1, cond_2, \ldots, cond_n$
  $REPLACE\ condR_1, condR_2, \ldots, condR_m$; $(n \geq 1, m \geq 1)$
  This statement is used to edit ordering relationships in a process model. No element inserted, deleted or updated in the model.
  The DELETE INSERT DATA statement is used to edit the order of existing elements in the RDF file format.

---

[7] Two prefixes are assumed as:
  $PREFIX\ CPNOntology:\ \ <http://myOnt.tutorial.org/2013\#>$
  $PREFIX\ myWF:\ \ <http://myOnt.tutorial.org/2013-instances\#>$

5 Related Work

As of today, many workflow modelling languages have been proposed, some of which have become widely accepted, used and replacing others. Most of them are based upon textual programming languages or graphical notations. For example, the Business Process Execution Language (BPEL) [5] is a standard way of orchestrating Web service execution in a business domain; the Business Process Model and Notation (BPMN) [15] is the de-facto standard for business process modelling; the Yet Another Workflow Language (YAWL) [4] is a domain specific language based on rigorous analysis of workflow patterns [3]; Petri Nets [1] and its extensions are well applied process modelling and analysis, etc. However, they use different concepts and different terminologies. This makes them difficult to inter-operate, share or reuse different workflow systems. As a result, it is a need for the semantics of concepts and their relationships.

We know that the ontology-based approach for modelling business process is not a new idea. There are some works made efforts to build business workflow ontologies, such as [7–9, 11, 17–19, 22], etc., to support (semi-)automatic system collaboration, provide machine-readable definitions of concepts and interpretable format. In [11], the authors defined an ontology for Petri Nets based business process description. Their ontology aims to facilitate the semantic interconnectivity of semantic business processes allowing semantical information exchange. Our CPN ontology is very close to the one proposed by [11], however there are some differences. Our work focuses on representing process models in a meta-knowledge base, which is defined based upon the CPN ontology, in order to share and reuse them.

6 Conclusion

In this paper, we have introduced an ontological approach for modelling business processes. We defined the CPN ontology to represent CPNs with OWL DL. Firstly, each element of CPNs was translated into a corresponding OWL concept. Secondly, some of the axioms created for *Classes* and *Properties* in the CPN ontology were presented. The third OWL element, *Individuals* was also considered. As a result, the combination between CPNs and ontologies provides not only semantically rich business process definitions but also machine-processable ones.

In order to model business processes, four basic types of manipulation operations on the elements of process models are required. Therefore, we introduced the basic manipulation statements written in the pseudocode. The SPARQL statements, which correspond to those statements, were indicated to store CB-WPs in the RDF file format. In spite of lacking the graphics, layout and GUI descriptions, the RDF files afterwards could be sent to other process-implementing systems to share and reuse them.

For validating CBWPs, we are planning to develop a run-time environment, which relies on the CORESE [6] semantic engine answering SPAQRL queries asked against an RDF knowledge base.

Acknowledgments. This research is conducted within the UCN@Sophia Labex.

References

1. Petri net, http://en.wikipedia.org/wiki/Petri_net
2. Web ontology language, http://en.wikipedia.org/wiki/Web_Ontology_Language
3. The workflow patterns home page, http://www.workflowpatterns.com/
4. Yawl: Yet another workflow language, http://www.yawlfoundation.org/
5. Andrews, T., Curbera, F., Dholakia, H., et al.: Business process execution language for web services version 1.1 (May 2003), http://msdn.microsoft.com/en-us/library/ee251594v=bts.10.aspx
6. Corby, O., et al.: Corese/kgram, https://wimmics.inria.fr/corese
7. Gašević, D., Devedžić, V.: Reusing petri nets through the semantic web. In: Bussler, C.J., Davies, J., Fensel, D., Studer, R. (eds.) ESWS 2004. LNCS, vol. 3053, pp. 284–298. Springer, Heidelberg (2004)
8. Gasevic, D., Devedzic, V.: Interoperable petri net models via ontology. Int. J. Web Eng. Technol. 3(4), 374–396 (2007)
9. Hepp, M., Roman, D.: An ontology framework for semantic business process management. Wirtschaftsinformatik (1), 423–440 (2007)
10. Jørgensen, J.B., Lassen, K.B., van der Aalst, W.M.P.: From task descriptions via colored petri nets towards an implementation of a new electronic patient record workflow system. STTT 10(1), 15–28 (2008)
11. Koschmider, A., Oberweis, A.: Ontology based business process description. In: EMOI-INTEROP, pp. 321–333. Springer (2005)
12. Kristensen, L.M., Christensen, S., Jensen, K.: The practitioner's guide to coloured petri nets. STTT 2(2), 98–132 (1998)
13. Nguyen, T.H.H., Le-Thanh, N.: Representation of coloured workflow nets with owl dl ontoloty. In: Second International Workshop "Rencontres Scientifiques UNS-UD" (RUNSUD 2013), pp. 29–41 (2013)
14. Nguyen, T.H.H., Le-Thanh, N.: Representation of rdf-oriented composition with owl dl ontology. In: Web Intelligence/IAT Workshops, pp. 147–150 (2013)
15. OMG: Business process model and notation, v2.0., http://www.bpmn.org/
16. OMG: Workflow management facility specification, v1.2 (2000), http://www.workflowpatterns.com/documentation/documents/00-05-02.pdf
17. Salimifard, K., Wright, M.: Petri net-based modelling of workflow systems: An overview. European Journal of Operational Research 134(3), 664–676 (2001)
18. Sebastian, A., Noy, N.F., Tudorache, T., Musen, M.A.: A generic ontology for collaborative ontology-development workflows. In: Gangemi, A., Euzenat, J. (eds.) EKAW 2008. LNCS (LNAI), vol. 5268, pp. 318–328. Springer, Heidelberg (2008)
19. Sebastian, A., Tudorache, T., Noy, N.F., Musen, M.A.: Customizable workflow support for collaborative ontology development. In: 4th International Workshop on Semantic Web Enabled Software Engineering (SWESE) at ISWC 2008 (2008)
20. W3C: Owl web ontology language reference. W3C Recommendation (2004), http://www.w3.org/TR/owl-ref/
21. WFMC: Workflow management coalition terminology and glossary (wfmc-tc-1011), document number wfmc-tc-1011. Tech. rep. (1999)
22. Zhang, F., Ma, Z.M., Ribaric, S.: Representation of petri net with owl dl ontology. In: FSKD, pp. 1396–1400 (2011)

Rule-Based Algorithm Transforming OWL Ontology Into Relational Database

Teresa Podsiadły-Marczykowska[1], Tomasz Gambin[2], and Rafał Zawiślak[3]

[1] Nalecz Institute of Biocybernetics and Bioengineering, Warsaw, Poland
tpodsiadly@ibib.waw.pl
[2] Institute of Computer Science, Warsaw University of Technology, Warsaw, Poland
tgambin@ii.pw.edu.pl
[3] Institute of Automatics, Lodz University of Technology, Lodz, Poland
Rafal.Zawislak@p.lodz.pl

Abstract. In the last decade, an increasing number of biomedical, scientific and business domains have made significant effort to standardize their terminology and organize domain knowledge using ontologies, for example in order to facilitate data exchange. It is a tempting idea to model domain knowledge using ontology and then use relational database to provide persistent information storage, search and retrieval utilizing transaction management, security and integrity control. For this it is necessary to transform an ontology into relational database. The goal of the paper is to present rule-based algorithm transforming OWL ontology into relational database. The algorithm not only transforms ontology structure (TBox) into database schema, ontology instances (ABox) into RDB instances, but also automatically creates user interface for database content access and exploration. The Authors tested and applied algorithm to convert FSTPPDO (the model representing knowledge concerning physiological models of mass transport phenomenon) into Transport Model Coefficients, a database containing the results of peritoneal dialyze research.

Keywords: OWL transforming, OWL, Ontology, Relational Database.

1 Introduction

Ontologies are considered as one of the most popular solutions in knowledge representation, currently they also offer an expressive formal language allowing to express, share and integrate knowledge. Unfortunately experience has shown that ontologies implemented as OWL constructors, representing complex domain knowledge remain difficult and uneasy to utilize in software projects. Main difficulties are related to two problems: user interface construction and query performance of native triple-stores[1]. Knowledge-rich semantic models result in

---

[1] Native triple-stores exploit the RDF data model to efficiently store and access the RDF data. To enumerate a few implemented form scratch: Sesame [http://www.openrdf.org/], OWLIM [http://www.ontotext.com/owlim], Allegro-Graph [http://www.franz.com/agraph/allegrograph/]

S. Kozielski et al. (Eds.): BDAS 2014, CCIS 424, pp. 148–159, 2014.
© Springer International Publishing Switzerland 2014

complex, deeply nested software interface. Unfortunately, not only for user but even for programmer, it is also not intuitive and awkward. Whats more, contemporary standard database engines, largely outperform querying and security mechanisms offered by ontologies or by native triple-store. Building knowledge-rich applications requires efficient, persistent storage of ontology structure and instances, thus the synthesis of possibilities offered by ontologies in knowledge representation with standard every-day software engineering practice of relational databases (RDB).

The goal of the paper is to present an ontology structure preserving algorithm transforming OWL DL ontology into RDB. The methodology is applied to build relational Database of Transport Models Coefficients using Ontology for Fluid and Solute Transport Processes in Peritoneal Dialysis (FSTPPD Ontology) as input ontology [12]. The paper is structured as follows: section 2 gives background knowledge on ontologies and relational models in the aim to analyze their relationships and reasons for similarities and differences; section 3 summarizes related research on OWL ontologies to RDB transformation algorithms and presents our approach; section 4 presents algorithm implementation, and example of its work.

2 Ontologies and Relational Models, Background Knowledge, Similarities and Differences

2.1 Ontology Basics - Definition, Language and Modeling Primitives

According to Gruber [10] ontology is a formal, explicit specification of a shared conceptualization. Formal means that ontology is machine readable and process-able, explicit specification stipulate the necessity of explicit and unambiguous definitions of the concepts described by ontology, shared refers to the fact that ontological model is meant to contain consensual knowledge, and finally conceptualization de-notes an abstract model of a phenomenon existing in the world created by identification of relevant concepts and relations existing between them. Ontologies are considered as computational models allowing for shared knowledge representation using different paradigms [8], construction of unambiguous concepts definitions using formal apparatus, and for automated reasoning. The last feature of ontologies, capability of inferring additional knowledge using increases their modeling power.

Ontology Web Language (OWL)recommended by the W3C Web Ontology Working Group, Description Logics [4] is currently regarded as an effective and expressive knowledge modeling language serving to define and instantiate Web ontolgies. OWL is based on RDF/RDF-S (Resource Description Framework and RDF Schema), it exists in three increasingly expressive sublanguages, OWL Lite, OWL DL and OWL Full. OWL Lite provides support for class hierarchies construction and simple constraints, while OWL Full offers no guarantees concerning effective reason-ing. OWL DL is the most frequently used OWL dialect because includes all OWL language constructs without losing computational completeness and decidability. Domain knowledge in ontologies is mainly expressed using five basic types of modeling primitives (ontology components), which should be transformed into RDB components: classes (concepts), properties (relations), instances (individuals) and axioms (assertions).

In OWL DL classes represent sets of objects with common properties and are organized in taxonomy. In OWL classes can be: named (defined by name), enumerated (defined by listing all their members) or complex. Complex classes may be defined: by specifying restrictions on sets of their properties, as complement of another class or by using standard set of operators i.e. intersection, union and complement. It is possible to define Boolean combinations of classes. Additionally in OWL classes can be declared as: disjoint with other classes (without common instances), primitive (unsuitable for reasoning) or defined i.e. enabling reasoning.

In OWL DL properties relate a class to another class (object property) or between class and a datatype value (data type property), both of them organized in taxonomy. Properties are characterized by their domain, range and algebraic characteristics (transitivity, symmetry, functionality and inverse functionality). Properties can be restricted using quantifiers (universal or existential), one can also restrict the number [2] of values for specified property using so called cardinality restrictions. Finally instances represent class occurrences and axiom sentences that are always true in domain of interest. Those above described modeling primitives have to be transformed into main elements of the most popular relational model proposed by Codd [6].

2.2   Ontologies Versus RDB, Similarities and Differences

Ontologies and data models both represent domain knowledge but are showing some intrinsic differences [1,14] namely application dependencies, knowledge coverage and expressive power. While ontologies should be generic and task-independent, RDB are conceived for specific tasks of data storage and querying. Compared to RDB, ontologies have more expressive power for information and knowledge representing. Another crucial distinction between both paradigms results from different ways of interpreting data according to Open World Assumption (OWA) in ontologies and Closed World Assumption (CWA) in RDB. OWA presumes that missing information cant be treated as false, while CWA states that anything not known to be true must be false. Taking into account the real meaning of the missing information requires precise assessment of the possible question meaning according to user needs and intensions. The differences between OWA and CWA are deepened by the use of Unique Name Assumption[3] (UNA) in RDB and lack of UNA [4] in ontologies. To sum up, OWA is suitable for describing knowledge in a precise and extensible way, CWA is more appropriate for constraining and validating data. Both paradigm have their strong points,

---

[2] Minimal, maximal and exact number of possible values.

[3] The unique names assumption presumes that all individual names refer to distinct objects.

[4] One individual may have more than one name.

for example the use of CWA provides high RDB performance. To use them both simultaneously it is necessary to indentify the areas of computer system in which OWA and CWA should be used in the first place, then design and implement some technique to perform the tasks. The paper [9] proposes solution enabling the implementation of both data interpretation paradigm in a knowledge-based system.

Despite the differences in general assumptions, the idea of correspondences between ontologies and RDB and simple, general mapping rules between them was proposed in 1998 by Tim Berners-Lee [5] and can be summarized as follows:

- class to table: an ontological concept/class corresponds to a table,
- property to column: an ontological property corresponds to a column of a table,
- class instance to row of a table correspondence.

The comparison of general modeling assumptions of ontologies and RDB systems is given in Tab. 1, more detailed comparisons of ontology and RDB constructs is given in Tab. 2. Authors have used these comparisons as basis for the formulation of transformation rules that in turn form the basis of ontology to RDB transformation algorithm.

Table 1. Ontologies and RDB Schema comparison general key assumptions

	ONTOLOGY	**RDB Schema**
Core Purpose	Formal, reusable knowledge representation and reasoning	Efficient data storage and querying
Primary Focus	Focus on **semantics**, shared understanding, individuals are optional	Focus on **data**
World Description Assumption *Missing Information* *Individual naming*	OWA Open World Assumption Missing information treated as unknown No UNA individuals may have more than one name	CWA Closed World Assumption Missing information treated as false UNA - Unique Name Assumption, each individual has a single, unique name
Notation *Notation Syntax* *Notation Semantics*	Recommended by W3C XML-based syntax, Strong focus on formal semantics	ER diagrams, no standard serialization Minimal focus on formal semantics

Table 2. Correspondences between components of OWL ontology and RDB, the basis for transformation rules

	OWL-DL Ontology	Relational Database
Basic Constructs	Classes Object properties Data type properties Instances Restrictions	Tables Foreign Keys Table Columns Tables Rows Constraints
Inheritance, Taxonomy	**Supported** (relation between classes and properties)	**Not Supported**
Binary Relationships	**Supported** (object properties)	**Supported** (foreign keys)
Bidirectional Binary Relationships	**Supported** (inverse object properties)	**Not Supported**
Ternary and Higher Degree Relationships	**Supported** (N-ary Relation Patterns)	**Supported indirectly** by reification (i.e. reduction to N binary relationships)
Data Types	**Supported indirectly** (OWL doesnt define directly any data types, but it makes use of built-in XML data types)	**Supported**
Cardinalities	**Supported**	**Very Limited Support** (Unique and Not Null Constraints only 0 and 1 card.)
Value Restrictions	**Supported** Value restrictions are used for defining meaning and checking model consistency	**Supported** (through CHECK Constraints) Constraints used for integrity
Anonymous Concepts	**Supported** (anonymous classes)	**Not Supported**
Identity	**Supported** (all instances have an implicit unique identifier, that doesnt depend on any property value)	**Supported** (through primary keys)

3   Description of Ontology into RDB Rule-Based Transformation Algorithm

Transformation of ontologies to relational databases is based on a set of transformation rules (five rules for ontology classes and their hierarchy, five for data type properties and three for object properties) specifying how to map constructs of the ontological model to RDB. It is assumed that source ontology is consistent and modular. The algorithm pareses OWL file (source), using rules to transform ontology and generates DDL scripts (target) containing database descriptions with included domain ontology restrictions. Detailed mapping rules are presented below. It should be noted, that because of new OWL2 functionality with respect to OWL-DL the described transformation rules are not sufficient for OWL2 ontologies.

3.1   Related Research

Interdisciplinary research on relationships between ontologies resulted in approaches which can be grouped into two main cases. In the first approach there are no differ-ences between ontology classes and instances, they are both stored using a native triple-store approach, i.e. in subject-predicate-object tables. This design is suitable for accessing and storing ontology specified content and is optimal for reasoning on the knowledge base [16]. Yet, it is quite inefficient for accessing and querying individuals following traditional database techniques [11]. What is more domain knowledge stored in class taxonomy is lost. The second approach stores ontology concepts and instances in RDB schema created using specific transformation preserving ontology constructs. The second approach does not lose information and unites capabilities of ontologies and RDB systems. There are several described in the literature type two approaches to transformation of ontologies into RDB [2,3,7,15]. However, all these approaches suffer from one or more of the following problems:

- all above cited approaches [2,3,7,15] dont tackle the problem of user interface (UI) automatic generation,
- they ignore capturing additional semantics represented in ontologies as class restrictions [3,7],
- they are not implemented [2,7,15], or the described implementation is not accessible [3].

In an attempt to overcome these problems, an algorithm allowing the transformation of ontology into a RDB along with automatic generation of UI has been created, implemented and tested. The described algorithm proposes solutions of above mentioned unresolved problems, taking advantage of the described approaches. It is assumed, that ontology is expressed in OWL-DL [13], consistent, and that RDB is defined in SQL DDL. The transformation algorithm does not contain any mechanisms allowing for the implementation of OWA data interpretation in the database, all properties restrictions are assumed to be expressed using closure axiom.

3.2   Transformation Rules for Ontology Classes, Class Instances and Subsumption Relations

During the first step of transformation the ontology classes are transformed into RDB tables according to the following five mapping rules:

- **Class rule 1** - every named ontology class with specified restrictions or assigned instance maps to an RDB table (in order to keep the overall number of tables as small as possible only tables with specified restrictions or having instances are mapped to RDB tables),
- **Class rule 2** - RDB table is named with the name of the class,
- **Class rule 3** - an instance of the class maps to a tuple (row) in a corresponding table,
- **Class rule 4** - RDB table is assigned a primary key to ensure identity of concept mapping (primary key is build automatically from the table name and suffix string ID),
- **Class rule 5** - subsumption relations between a class and its subclass and between a class and its super-class in the ontology are represented in RDB as a foreign key.

3.3 Transformation Rules for Ontology Data Type Properties

Data type properties are relations between instances of OWL classes and XML Schema data types. The algorithm, transforming ontology data type properties into relational database columns of tables, searches and parses all data type properties in OWL file. When all data type properties are parsed, the process of data type properties transformation is finished. According to property *rdfs:domain* value algorithm finds database table and creates data column with the name of the property. If data type property is functional, single value in the column field is allowed according to property *rdfs:range* value. If data type property is non-functional, value in column filed is restraint to the specified by property *rdfs:range* value selection list. To sum up, in the process of transforming domain ontology into relational database, data type-properties are transformed into RDB data columns according to the following five rules:

- **Data type property rule 1** - data type property in the ontology is transformed into RDB column, the table is determined using rdfs:domain property value, the col-umn data type is determined using rdfs:range property value,
- **Data type property rule 2** - if data type property is functional, single value in the column field is allowed according to property rdfs:range value,
- **Data type property rule 3** - if data type property is non-functional, value in column filed is restraint to the specified by property rdfs:range value selection list,
- **Data type property rule 4** - a value restriction on a data type property maps to a CHECK constraint on the corresponding column,
- **Data type property rule 5** - an enumerated data type maps to a CHECK constraint with enumeration.

3.4 Transformation Rules for Ontology Object Properties

Object property in an ontology is a relation between instances of two classes. Fully defined object property has the domain and range specified. Object property can be defined to be a specialization (sub-property) of an existing property.

When OWL classes are mapped to tables non-functional object-properties are transformed into RDB foreign keys.

One of the problems during the design of the algorithm was the existence of non-functional object properties that correspond to many-to-many relationships in relational database, which are realized by the use of joining tables. The field type *list:reference* available in Web2py framework allowed to successfully simulate many-to-many relationship without introducing joining tables. Transformation of ontology object properties is performed according to the following three mapping rules:

- **Object property rule 1** - object properties between ontology classes are mapped to RDB foreign keys or joining tables. Object property domain and range point to tables joined by foreign key,
- **Object property rule 2** - if object property is functional, or restricted to one instance (using *hasValue* restriction), the relation is created using one to one foreign key association,
- **Object property rule 3** - if object property is non-functional, the many to many relation is created using Web2py field type *list:reference*.

3.5   Rule-Based Transformation Algorithm Steps

The ontology to RDB transformation algorithm is composed of two main tracks, presented schematically in Fig. 1. Track number one is transforming ontology constructs into RDB schema and populate it with ontology instances, while during the track number two user interface UI to the created database is automatically generated. Algorithm track number one is divided into following six stages:

- **Stage 1** - comprises the conversion of the OWL-DL class and its children into the unordered list of database tables. This part of the algorithm is named Generate Tables and is described in more details in 3.6,
- **Stage 2** - during second stage, tables order is determined taking into account the information about classes subsumption relation from stage 1. The stage output is the ordered list of tables definitions,
- **Stage 3** - during third stage a DLL query for each ontology class is generated,
- **Stage 4** - during fourth stage RDB tables are created using Web2py framework,
- **Stage 5** - during fifth stage RDB tables are populated with ontology data,,
- **Stage 6** - during sixth stage UI for RDB is created using metadata created in algorithm track number 2.

Second algorithm track (see Fig. 1) prepares the metadata for automatic UI generation. The output of the second algorithm track serves as input in the last, sixth stage of the track one. Track two is divided into the following two stages:

- **Stage 1** - during this stage the metadata for creating tree structure of the ontology is created; ontology tree structure is guiding the process of browsing the database content,
- **Stage 2** - during the second stage the metadata for creating insert controllers is generated.

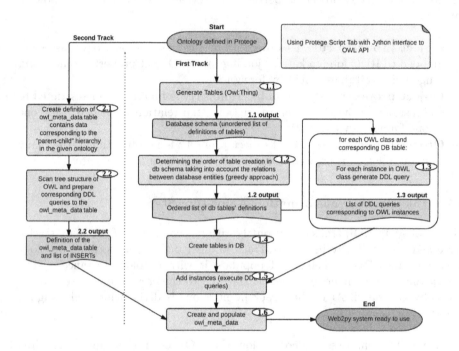

Fig. 1. Two tracks of ontology to RDB transformation algorithm

3.6 GenerateTables Algorithm - Short Description

The general goal of the first part in the first track of ontology to RDB transformation algorithm is the conversion of the given OWL-DL class and its children into the ordered list/set of database tables. Single named OWL class (with specified restrictions or assigned instances) corresponds to single RDB table. OWL properties are transformed either into standard DB fields in case of data type properties, or into relations to other tables in case of object properties. The *GenerateTables* algorithm is composed of 16 steps. The steps can be divided into two types, the first type step performs and action, the second one is checking the conditions for future actions. Fig. 2 presents the schema of Generate Tables part of the whole algorithm.

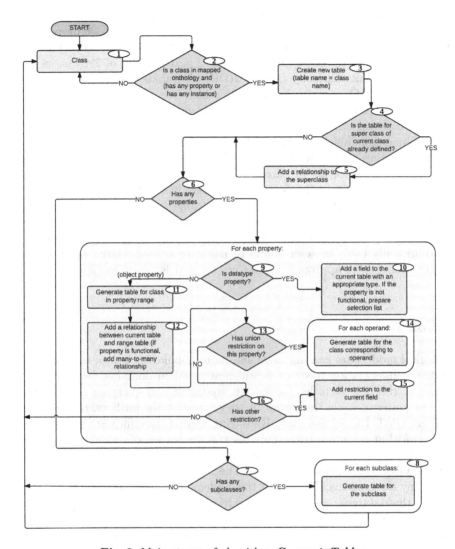

Fig. 2. Main stages of algorithm *Generate Tables*

4   Algorithm Implementation and Testing

The algorithm has been implemented using Web2py, a free open source full-stack framework for rapid development of fast, scalable, secure and portable database-driven Web-based applications, written and programmable in Python. Web2py follows the Model View Controller design pattern. The main advantage of this framework is a clear separation of a model (data), logic and presentation layers. Moreover, Web2py implements own Database Abstraction Layer, which makes the code independent from selected database system (MySQL, SQLlite, Postgres, ORACLE and several of other database management systems are supported).

The algorithm has been tested using two ontologies from Bioportal: *AminoAcid* ontology (46 classes 6 properties) and *BioPax* ontology (68 class, 96 properties). For final testing FSTPPD ontology [1], created by the authors, has been used. The model has been developed using Protege 3.5. It consists of 299 classes and 112 properties. Automatically created RDB - Database of Transport Models Coefficients (DTMC) contains in total 240 tables.

User interface for Web-based DTMC is divided into two main parts (panels). Left panel contains original OWL-DL class hierarchy presented in the form of a dynamically expandable tree, in which nodes correspond to DB tables defined by transformation algorithm. Tree view of the ontology enables user to browse RDB tables by exploration of parent-child class hierarchy defined originally in OWL-DL. The reconstruction of ontology class hierarchy uses the metadata created in track 2 of the stage 1 of the transformation algorithm. Right panel contains records browser enabling user to explore the corresponding database. After clicking on the particular node user can browse or edit the selected table, additionally table browser allows to navigate among related records from different tables (e.g., appropriate links are created for each foreign key in the table).

5   Conclusions

Ontological models are gaining more and more popularity as a perspective way to enhance Information Systems in different problem domains, but databases still ensure the best facilities for storing, updating and querying the information. The paper presents the algorithm, developed by the authors, transforming consistent OWL-DL ontology into RDB. But what is more important, the result of the algorithm execution is not only the transformation of ontology knowledge base (ontology structure and instances) into RDB schema and content, but also generation of database User Interface. The transformation algorithm does not contain any mechanisms allowing for the implementation of OWA data interpretation in the database, all properties restrictions are assumed to be expressed using closure axiom.

Effectiveness of the algorithm has been preliminary confirmed using two ontologies from BioPortal: BioPax Ontology and AminoAcid Ontology. Final test of the algorithm was carried out on the ontology developed by the authors - FSTPPD Ontology, modeling complex physiological knowledge concerning solute and solution transport in peritoneal dialysis. Correctness of obtained Web-based database structure has been approved by nephrologists domain experts. Automatically generated UI was assessed as intuitive, once the user familiarized him/herself with the structure of the ontology, this last task was rated as cumbersome.

Continuation of work should at first concentrate on analysis of the optimality of the created database structure. As it is obvious, that complex knowledge structure will result in large databases, the second research goal is the construction of semantic, using natural language, database query interface.

Acknowledgements. The study is cofounded by the European Union from resources of the European Social Fund. Project PO KL Information technologies: Research and their interdisciplinary applications", Agreement UDA-POKL.04.01. 01-00-051/10-00.

References

1. Abiteboul, S., Hull, R., Vianu, V.: Foundations of databases, vol. 8. Addison-Wesley, Reading (1995)
2. Astrova, I., Kalja, A.: Storing owl ontologies in sql3 object-relational databases. In: AIC 2008: Proceedings of the 8th Conference on Applied Informatics and Communications, pp. 99–103 (2008)
3. Astrova, I., Kalja, A., Korda, N.: Automatic transformation of owl ontologies to sql relational databases. In: IADIS European Conf. Data Mining (MCCSIS), pp. 5–7 (2007)
4. Baader, F.: The description logic handbook: theory, implementation, and applications. Cambridge University Press (2003)
5. Berners-Lee, T.: Relational databases on the semantic web (2013)
6. Codd, E.F.: A relational model of data for large shared data banks. Communications of the ACM 13(6), 377–387 (1970)
7. Gali, A., Chen, C.X., Claypool, K.T., Uceda-Sosa, R.: From ontology to relational databases. In: Wang, S., Tanaka, K., Zhou, S., Ling, T.-W., Guan, J., Yang, D.-q., Grandi, F., Mangina, E.E., Song, I.-Y., Mayr, H.C. (eds.) ER Workshops 2004. LNCS, vol. 3289, pp. 278–289. Springer, Heidelberg (2004)
8. Goczyła, K., Grabowska, T., Waloszek, W., Zawadzki, M.: Cartographic approach to knowledge representation and management in kasea. In: International Workshop on Description Logics (2005)
9. Goczyla, K., Grabowska, T., Waloszek, W., Zawadzki, M.: Designing world closures for knowledge-based system engineering (2005)
10. Gruber, T.R.: A translation approach to portable ontology specifications. Knowledge Acquisition 5(2), 199–220 (1993)
11. Lee, J., Goodwin, R.: Ontology management for large-scale enterprise systems. Electronic Commerce Research and Applications 5(1), 2–15 (2006)
12. Linholm, B., Podsiadly-Marczykowska, T., Stachowska-Pietka, J., Galach, M., Debowska, M., Waniewski, J.: Ontology for fluid and solute transport processes in peritoneal dialysis (pd). J. Am. Soc. Nephrol. 23(2) (2012)
13. McGuinness, D.L., Van Harmelen, F., et al.: Owl web ontology language overview. W3C recommendation 10(2004-03), 10 (2004)
14. Spyns, P., Meersman, R., Jarrar, M.: Data modelling versus ontology engineering. ACM SIGMod Record 31(4), 12–17 (2002)
15. Vysniauskas, E., Nemuraite, L.: Transforming ontology representation from owl to relational database. Information Technology and Control 35(3A), 333–343 (2006)
16. Wilkinson, K., Sayers, C., Kuno, H.A., Reynolds, D., et al.: Efficient rdf storage and retrieval in jena2. In: SWDB, vol. 3, pp. 131–150 (2003)

Grouping Multiple RDF Graphs
in the Collections

Dominik Tomaszuk[1] and Henryk Rybiński[2]

[1] Faculty of Mathematics and Informatics,
University of Bialystok, Poland
dtomaszuk@ii.uwb.edu.pl
[2] Institute of Computer Science,
Warsaw University of Technology, Poland
h.rybinski@ii.pw.edu.pl

Abstract. This paper defines a document-oriented Resource Description Framework (RDF) graph store. It proposes collections for grouping multiple graphs. We define a lightweight representation of graphs which emphasizes legibility and brevity. We also present an implementation of our system, an algorithm of mapping to a pure RDF model and algorithms of generation and normalization. Our proposal supports knowledge metrics for RDF graphs.

Keywords: Semantic Web, graph store, document-oriented database, serialization, provenance, metric, semistructural data.

1 Introduction

Knowledge representation deals with how knowledge is represented, in the case at hand based on Semantic Web standards including RDF [9] and OWL [1], while knowledge storing is the way in which the knowledge is retained in a computer: RDF, one of the foundations of Linked Data and the Semantic Web at large, is used for knowledge representation on the Web. The tools supporting processing and storage of RDF appeared in the beginning of the 21st century, but they have a number of drawbacks and limitations:

- No mechanisms to store access and subgraph selection in compliance with the Linked Data principles.
- No possibility of grouping graphs, which make graph provenance and other related metrics hard if not impossible to realise.
- Problems with capacity related to data processing and access to data, resulting from no normalisation of structures; the existing proposals do not allow for generating optimal structures, needed in certain use cases.

The current solutions do not offer complete storage access mechanisms. The RDF graph store in connection with such proposals as semi-structured documents together with serialisation means a complete solution of problems related

S. Kozielski et al. (Eds.): BDAS 2014, CCIS 424, pp. 160–169, 2014.

to knowledge storing and processing in a Linked Data environment, resulting in an improvement of the possibilities to access it in the graph store.

In this paper is presented an approach of grouping multiple RDF graphs in the collections, which providing various metrics. Our proposal extends RDF graphs to values, which can symbolize metrics such as temporal, uncertainty and trust. We also introduce implementation and algorithms of generating and normalization for this approach. Our proposal allows to store knowledge metrics near RDF graphs. Moreover, we propose document serialisation, which can contain additional metadata about stored RDF triples.

2   Collections in RDF Graph Store

In this section we introduce a document-oriented graph store with collections and document serialization for the graph store.

2.1   Collections and Graph Store

In this subsection we propose a document-oriented graph store that is not bound to any predefined database types. Instead, it is close to the RDF data, so that no predefined structure is needed. The graph store can be thought of as a store including containers so called *data collections* or simply *collections*. A data collection is similar to a relation from relational databases. A collection is represented by a graph, provenance and list of metrics. These collections include multiple *documents* and documents store serialized RDF statements. The concept of a *document* is a central element of the graph store. The documents consist of RDF data. For the sake of generality in our considerations, we define here a document as an ordered set of keys with associated values, which can be one of several different datatypes.

Hence, a collection can be seen as a group of RDF triples (representing documents). A *collection* is a tuple $C = \langle r, [v_1, v_2, \ldots, v_i], G \rangle$, where:

1. $r \in \mathcal{I}$ is the provenance of a graph, which can be interpreted as IRI,
2. $[v_1, v_2, \ldots, v_i]$ is a list of metrics ($v \in \mathcal{L}$), which can be interpreted as temporal [13], uncertainty [18] and/or trust metrics [20],
3. G is an RDF graph.

A provenance provides information about a graph's origin, such as who created it, when it was modifed, or how it was created. It is used for building representations of entities, involved in producing a piece of data. Special metrics provide information about RDF graph characteristics.

A *document-oriented graph store* is $GS_D = \{C_1, C_2, \ldots, C_i\}$, where every C_i is a collection, $i \geq 1$.

2.2   Document Serialization

In this subsection we introduce a concept of *RDF in JSON Document* (in the following sections denoted by RDFJD) and their serialization. Serialization is the process of converting a data structure into a format that can be stored and transmitted across the web and reconstructed later in the same or another computer environment. We define a document as a resource that serves as the container of semistructural data. One of the semistructural data formats is JavaScript Object Notation (JSON) [8], which is a syntax designed for human-readable data interchange and easy for machines to generate. It uses both simple datatypes, such as number, string or boolean and composite data types, such as array and object.

We propose serialization based on JSON, which is equivalent to the RDF model. The proposal is a lightweight textual syntax that can easily be modified by humans, servers and clients. The advantage of this syntax is that it can easily be transformed from other syntaxes. Another benefit of serializing RDF graphs in JSON is that there are many software libraries and built-in functions, which support the serialization.

The difference between regular JSON and RDFJD is that the above RDFJD object uniquely identifies itself on the World Wide Web and can be used, without introducing ambiguity across the Web Service using a document-oriented graph store.

The proposed structure can be modeled as a set of an abstract data structure with two operations:

1. $\mathcal{U} = get(C, \mathcal{Y})$ – returning a list of objects $\mathcal{U}$, where C is a collection, $\mathcal{Y}$ is a key,
2. $set(C, \mathcal{Y}, \mathcal{U})$ – causes a key $\mathcal{Y}$ and a list of objects $\mathcal{U}$ to be stored at a collection C.

We propose two types of RDFJD documents:

1. **directive document**, which expresses the context of statement documents,
2. **statement document**, which expresses RDF statements.

A *directive document* is associated with a collection and implements the knowledge metrics, provenance, and defines the short-hand names that are used throughout an RDFJD statement document. The directive document is a metadata package of a collection. This document should be unique. All the possible keys in a directive document are presented in Tab. 1. The list of metrics and provenance keys should impose a unique key constraint.

In the Listing 1 we present a directive document. The RDFJD document contains fields which define the provenance (`http://example.org/g1`) and trust (`0.9`) of the collection. It also defines a `foaf` prefix as an abbreviation for `http://xmlns.com/foaf/0.1/`.

```
1  {
2      "_prov": "http://example.org/g1",
3      "_metric": [0.9],
4      "foaf": "http://xmlns.com/foaf/0.1/"}
```

Listing 1. Directive document

Table 1. RDFJD directive document keys

Key	Description
`_metric`	a predefined value of collection metric
`_prov`	a predefined value of collection provenance
prefix ID	abbreviating IRIs

A *statement document* is the main part, which stores RDF statements with extensions. A statement document uses subject-centric syntax, and it represents one or more properties of a subject. Often these documents occur more than once in the context of collection. They implement the subject as predefined keys, predicates as keys and objects as values. Plain literals with a language tag and typed literals are supported by special predefined keys. All the possible keys in a statement document are presented in Tab. 2.

In the Listing 2 we present a statement document. Key `foaf.name` is expand to value from a directive document (see Listing 1). The RDFJD document contains fields which define RDF statements:

1. triple 1: `http://example/voc#me`, `rdf:type`, `http://example/voc#Teacher`
2. triple 2: `http://example/voc#me`, `http://xmlns.com/foaf/0.1/name`, John Smith

```
1  {
2    "_subject": "http://example/voc#me",
3    "_type": {"_value": "http://example/voc#Teacher"},
4    "foaf.name": {"_value": "John Smith"}
5  }
```

Listing 2. Statement document

Table 2. RDFJD statement document keys

Key	Description
`_subject`	Used to identify subject that are being described
`_type`	Used to set the datatype of a subject
predicate key	Used to describe object
Possible values of predicate key	
`_value`	Used to specify the data that is associated with a particular predicate
`_lang`	Used to specify the native language for a particular object
`_datatype`	Used to specify the datatype for a particular object

3 Generating Algorithms

In this section we propose algorithms for serialization, normalization, and mapping into named graph model.

3.1   Serialization and Normalization

Algorithm 2 shows the process of generating RDFJD statement document. The algorithm creates triples. The algorithm takes into account the simple literals without a language tag, simple literals with a language tag and typed literals.

There is the possibility that the same subject could occur in different RD-FJD statement documents (e.g. because of the insertion of new statements). To improve the speed of data retrieval operations on a subject-centric statement there is the necessity to merge two or more statement documents with the same subject. Algorithm 1 presents the process of merging RDFJD documents. After this action an index may be applied to the *subject*.

> **input** : set of statement document SD
> **output**: statement document SD_M
> 1 SDt ← sort(SD);
> 2 **foreach** $s \in SDt$ **do**
> 3 | **if** *equal(current(), next())* **then**
> 4 | | merge(current(), next());

Algorithm 1. Merging statement documents

3.2   Mapping into Named Graph Model

In this subsection the mapping from our approach to the named graph model [6] is presented. A collections $C = (r, [], G)$ is equivalent to named graph $ng = (n, G)$, where $n \in \mathcal{I}$ is name of graph G. To case where $C = (r, [v_1, v_2, \ldots, v_i], G)$ we proposed to use `value` object property defined in [20], which allows to include metric values. Algorithm 3 presents the process of transformation, which uses named graphs.

The RDF graph store can also be mapped to an RDF dataset. Following [14], RDF dataset DS consists of one graph, called the default graph, which does not have a name, and zero or more named graphs, each identified by IRI. We assume that $\mathcal{NG} = \{(u_1, G_1), (u_2, G_2), \ldots, (u_n, G_n)\}$ is a set of named graphs, where all IRI references are disjoint. An *RDF dataset* is $DS = \{G, \mathcal{NG}\}$, where G is called default graph and $\mathcal{NG}$ is a set of named graphs. If in $GS_D = \{C_1, C_2, \ldots, C_i\}$, $C_1 = (\varnothing, \varnothing, G)$ then DS is equivalent to GS_D. Otherwise, we suggest to map from C_i to (u_{i+1}, G_{i+1}) and use Algorithm 3. It is also possible to use RDF reification with the same metric in all statements, but this solution is much more verbose than our proposal.

4   Implementation and Experiments

In this section we present the implementation and experiments of our approach. We used NoSQL database MongoDB[1] as the development platform. The testbed

---

[1] http://www.mongodb.org/

```
input  : set of RDF triples T
output: set of statement documents SD
```
1 create root object;
2 **foreach** $t \in T$ **do**
3   get subject s from t;
4   insert s into "_subject" key;
5   get predicate p from t;
6   get object o from t;
7   **if** *equal(p, "rdf:type")* **then**
8   create "_type" key;
9   insert o into "_type" key;
10   **else**
11   add prefix(p) to directive document;
12   create key abbreviation(p);
13   **if** *o is literal without a language tag* **then**
14   insert o into abbreviation(p) key;
15   **else if** *o is literal with a language tag* **then**
16   create "_value" key in abbreviation(p) key;
17   insert o into "_value" key;
18   get language lg from o;
19   create "_language" key in abbreviation(p) key;
20   insert lg into "_language" key;
21   **else**
22   create "_value" key in abbreviation(p) key;
23   insert o into "_value" key;
24   get datatype dt from o;
25   create "_datatype" key in abbreviation(p) key;
26   insert dt into "_datatype" key;

Algorithm 2. Generating statement document

```
input  : collection C
output: named graph NG, default graph G
```
1 get r from C;
2 get v from C;
3 create NG with r as a name;
4 create default graph G;
5 **foreach** $q \in C$ **do**
6   get triple t from triple with metric;
7   insert t into NG;
8   get metric m;
9   **if** *equal(m, $\varnothing$)* **then**
10   insert (r, "value", v) into G;
11   **else**
12   insert (r, "value", m) into G;

Algorithm 3. Mapping to Named Graphs

consists of the following three parts: query engine (applying matching Application Programming Interface), resources stored in collections (a part of the RDF graph store), and Representational State Transfer (REST) [11] client. The main part of the prototype is the matching API, which maps Hypertext Transfer Protocol (HTTP) request methods to object-oriented imperative query language.

Now we present load tests, which we performed on the Berlin SPARQL Benchmark [3]. We also discuss the results of these tests. The load experiment measures the time required to load on the testbed and Virtuoso Open-Source Edition 6.1, which is the leading graph store supporting the biggest Linked Data knowledge base DBpedia[2].

In Fig. 1a. we show loading of normalized RDFJD serialization into our testbed and RDF into Virtuoso. This plot shows that loading triples into Virtuoso is much faster than loading statements into testbed. For the loading 40000 statements Virtuoso is up to 60 times faster. The testbed times are nearly quadratic to the number of quads and the coefficient of determination $R^2 \approx 0.99$. The Virtuoso times are nearly linear to the number of triples and the coefficient of determination $R^2 \approx 0.98$.

Taking into consideration that the times of textual RDFJD are nearly quadratic, we propose binary representation of RDFJD. The design goals for it emphasized performance. In particular, it is designed to be smaller and faster than textual version and it is fully compatible. Compared to textual RDFJD, binary RDFJD is designed to be efficient both in storage space and scan-speed. Our proposal represents data types in little-endian format. Large elements are prefixed with a length field to facilitate scanning. In Fig. 1b. we show the loading of binary normalized RDFJD serialization into the testbed. This plot shows that loading statements into the testbed is much faster than loading statements into Virtuoso. The load times of the testbed with binary serialization are approximately 2.4 times faster than the load times of Virtuoso. At 40000 statements the loading of binary RDFJD into the testbed is up to 10 times faster than the loading of RDF into Virtuoso.

5   Related Work

While the pure RDF does not allow referring to whole RDF graphs, named graphs introduced in [6] provide the means to group a set of statements in a graph. This approach may be sufficient for RDF graph stores only with provenance metrics. Unfortunately, it is not satisfactory for other metrics. Schenk et al. propose Networked Graphs [16]. It allows a user to define RDF graphs by using a SPARQL CONSTRUCT clause and a named graph model. Unfortunately, this approach may be insufficient for RDF graph stores, which do not support SPARQL queries or named graphs. Shaw et al. [17] propose vSPARQL which allows to define virtual graphs and use recursive subqueries to iterate over paths of arbitrary lengths. It also extends SPARQL by allowing to create new entities based upon the data encoded in existing datasets.

---

[2] http://dbpedia.org/

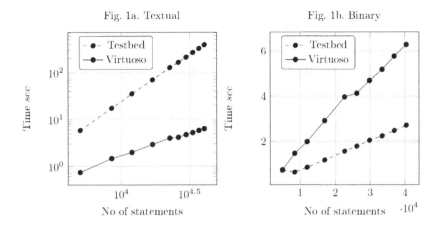

Fig. 1. Load test: Testbed comparison to Virtuoso

On the other hand there are RDF serializations [12,7,2,15,5,4]. RDF/XML
[12] is XML compatible syntax, which nodes and predicates must be represented
in the names of elements, names of attributes, contents of elements or values of
attributes. RDF/XML may not be fully described by such schemes as DTD or
XML Schema. Another disadvantage of this syntax is its incapability of encoding
all legal RDF graphs. It not handle named graphs, while Triples In XML (TriX)
[6] serialisation does. TriX used XML syntax as well but it is not compatible
with [12]. Another proposal refers to Terse RDF Triple Language (Turtle) [15]
is simplification and subset of [2]. This solution offers textual syntax that makes
it possible to record RDF graphs in a completely compact form. The drawbacks
of this proposal include the fact that it is not capable of handling named graphs
and its possibility to represent RDF triples in an unnormalised form. N-Triples
[5] and N-Quards [4] are also a textual format of RDF serialisation. It is based
on Turtle. Unfortunately, there are sign restrictions imposed on older version
by US-ASCII standard and it does not handle named graphs. There are also
serializations based on the JSON syntax [21,19]

Foregoing serializations are supported by various graph stores. One of them is
Virtuoso [10]. It is a row-wise transaction oriented database. It is re-targeted as
an RDF store and inference. It is also revised to column-wise compressed storage
and vectored execution.

6   Conclusions

The problem of how to group RDF triples and support metrics in these groups
has produced many proposals. We assume that RDF, being more functional,
should provide a method to set the metrics and provenance at the graph level.

We have produced a simple and thought-out proposal for grouping multiple
RDF graphs in collections. We propose how our approach can be used in combination with various metrics. We believe that our idea is an interesting approach,

because it can be transformed to the pure RDF and named graphs models. More importantly, we have provided algorithms for the generation and normalization of these semistructural data. Our approach extends the classical case of RDF with collections. The implementation shows its great potential.

We believe that our approach offers a flexible way to represent RDF data, we acknowledge, however, that there are areas that are subject to future investigation, such as replication of collections, versioning and access control.

References

1. Bao, J., Kendall, E.F., McGuinness, D.L., Patel-Schneider, P.F.: Owl 2 web ontology language quick reference guide. Tech. rep., World Wide Web Consortium, 2nd edn. (2013)
2. Berners-Lee, T., Connolly, D.: Notation3 (n3): A readable rdf syntax. Tech. rep., World Wide Web Consortium (2008)
3. Bizer, C., Schultz, A.: Benchmarking the performance of storage systems that expose sparql endpoints. World Wide Web Internet and Web Information Systems (2008)
4. Carothers, G.: Rdf 1.1 n-quads. Tech. rep., World Wide Web Consortium (2014)
5. Carothers, G., Seaborne, A.: Rdf 1.1 n-triples. Tech. rep., World Wide Web Consortium (2014)
6. Carroll, J.J., Bizer, C., Hayes, P., Stickler, P.: Named graphs, provenance and trust. In: Proceedings of the 14th International Conference on World Wide Web, pp. 613–622. ACM (2005)
7. Carroll, J.J., Stickler, P.: Rdf triples in xml. In: Proceedings of the 13th International World Wide Web Conference on Alternate Track Papers & Posters, pp. 412–413. ACM (2004)
8. Crockford, D.: The application/json media type for javascript object notation (json). Tech. rep., Internet Engineering Task Force (2006)
9. Cyganiak, R., Wood, D., Lanthaler, M.: Rdf 1.1 concepts and abstract syntax. Tech. rep., World Wide Web Consortium (2014)
10. Erling, O., Mikhailov, I.: RDF support in the virtuoso DBMS. In: Pellegrini, T., Auer, S., Tochtermann, K., Schaffert, S. (eds.) Networked Knowledge - Networked Media. SCI, vol. 221, pp. 7–24. Springer, Heidelberg (2009)
11. Fielding, R.T.: Architectural styles and the design of network-based software architectures. Ph.D. thesis, University of California (2000)
12. Gandon, F., Schreiber, G.: Rdf 1.1 xml syntax. Tech. rep., World Wide Web Consortium (2014)
13. Gutierrez, C., Hurtado, C.A., Vaisman, A.: Introducing time into rdf. IEEE Transactions on Knowledge and Data Engineering 19(2), 207–218 (2007)
14. Harris, S., Seaborne, A.: Sparql 1.1 query language. Tech. rep., World Wide Web Consortium (2012)
15. Prud'hommeaux, E., Carothers, G.: Rdf 1.1 turtle. Tech. rep., World Wide Web Consortium (2014)
16. Schenk, S., Staab, S.: Networked graphs: a declarative mechanism for sparql rules, sparql views and rdf data integration on the web. In: Proceedings of the 17th International Conference on World Wide Web, pp. 585–594. ACM (2008)
17. Shaw, M., Detwiler, L.T., Noy, N., Brinkley, J., Suciu, D.: vsparql: A view definition language for the semantic web. Journal of Biomedical Informatics 44(1), 102–117 (2011)

18. Straccia, U.: A minimal deductive system for general fuzzy RDF. In: Polleres, A., Swift, T. (eds.) RR 2009. LNCS, vol. 5837, pp. 166–181. Springer, Heidelberg (2009)
19. Tomaszuk, D.: Named graphs in rdf/json serialization. In: Zeszyty Naukowe Politechniki Gdańskiej, pp. 273–278 (2011)
20. Tomaszuk, D., Pąk, K., Rybiński, H.: Trust in RDF graphs. In: Morzy, T., Härder, T., Wrembel, R. (eds.) Advances in Databases and Information Systems. AISC, vol. 186, pp. 273–283. Springer, Heidelberg (2013)
21. World Wide Web Consortium: Flat triples approach to RDF graphs in JSON (2010)

Optimization of Approximate Decision Rules Relative to Coverage

Beata Zielosko

Institute of Computer Science, University of Silesia
39, Będzińska St., 41-200 Sosnowiec, Poland
beata.zielosko@us.edu.pl

Abstract. We present a modification of the dynamic programming algorithm. The aims of the paper are: (i) study of the coverage of decision rules, and (ii) study of the size of a directed acyclic graph (the number of nodes and edges) for a proposed algorithm. The paper contains experimental results with decision tables from UCI Machine Learning Repository and comparison with results for the dynamic programming algorithm.

Keywords: decision rules, optimization relative to coverage, dynamic programming.

1 Introduction

Decision rules are used in many areas connected with knowledge representation and data mining [8, 10, 11]. Approximate decision rules have usually smaller number of attributes, so they are better from the point of view of understanding. Classifiers based on approximate decision rules have often better accuracy than the classifiers based on exact decision rules. Therefore, recent years particular attention has been devoted to the study of approximate decision rules [4, 7, 11].

In this paper, we study a modification of a dynamic programming algorithm for construction and optimization of decision rules relative to coverage. The rule coverage is a measure that allows to discover major patterns in the data. Construction and optimization of decision rules relative to coverage can be considered as important task for knowledge representation and knowledge discovery.

There are different approaches for construction of decision rules, for example, Boolean reasoning [11, 12], different kinds of greedy algorithms [9, 11], separate and conquer approach [7, 8], dynamic programming approach [2–4]. Also, there are different rule quality measures that are used for induction or classification processes [5, 13].

The paper is a continuation of research connected with a problem of scalability for decision rule optimization relative to coverage based on dynamic programming approach [14, 15]. We try to find a heuristic, modification of a dynamic programming algorithm that allows us to find the values of coverage of decision rules close to optimal ones.

S. Kozielski et al. (Eds.): BDAS 2014, CCIS 424, pp. 170–179, 2014.

There are two aims for a proposed algorithm: (i) study the coverage of approximate rules and comparison with the coverage of decision rules constructed by the dynamic programming algorithm [4], (ii) study the size of a directed acyclic graph (the number of nodes and edges) and comparison with the size of a directed acyclic graph constructed by the dynamic programming algorithm.

To work with approximate decision rules, we use an uncertainty measure $R(T)$ that is the number of unordered pairs of rows with different decisions in the decision table T. Then we consider β-decision rules that localize rows in subtables of T with uncertainty at most β.

We based on dynamic programming algorithm to construct approximate decision rules. For a given decision table T a directed acyclic graph $\Delta_\beta(T)$ is constructed. Nodes of this graph are subtables of a decision table T described by descriptors (pairs attribute = value). We finish the partition of a subtable when its uncertainty is at most β. This parameter helps us to control computational complexity and makes the algorithm aplicable to more complex problems. In comparison with algorithm presented in [4], subtables of the graph $\Delta_\beta(T)$ are constructed for one attribute with the minimum number of values, and for the rest of attributes from T - the most frequent value of each attribute (value of an attribute attached to the maximum number of rows) is chosen. So, the size of the graph $\Delta_\beta(T)$ is smaller than the size of the graph constructed by the dynamic programming algorithm[4]. This fact is important from the point of view of scalability. Based on the graph $\Delta_\beta(T)$ we can describe sets of approximate decision rules for rows of table T. Then, based on a procedure of optimization of the graph $\Delta_\beta(T)$ relative to coverage we can find for each row r of T a β-decision rule with the maximum coverage.

The paper consists of six sections. Section 2 contains main notions connected with a decision table and approximate decision rules. In section 3, proposed algorithm for construction of a directed acyclic graph is presented. Section 4 contains a description of a procedure of optimization relative to coverage. Section 5 contains experimental results with decision tables from UCI Machine Learning Repository, and section 6 - conclusions.

2   Main Notions

In this section, we present notions corresponding to decision tables and decision rules.

A *decision table* T is a rectangular table with n columns labeled with conditional attributes $f_1, \ldots, f_n$. Rows of this table are filled with nonnegative integers that are interpreted as values of conditional attributes. Rows of T are pairwise different and each row is labeled with a nonnegative integer that is interpreted as a value of a decision attribute.

We denote by $N(T)$ the number of rows in table T. By $R(T)$ we denote the number of unordered pairs of rows with different decisions. We will interpret this value as *uncertainty* of the table T.

The table T is called *degenerate* if T is empty or all rows of T are labeled with the same decision, in this case, $R(T) = 0$.

A minimum decision value that is attached to the maximum number of rows in T is called *the most common decision for T*.

A table obtained from T by the removal of some rows is called a *subtable* of the table T. Let T be nonempty, $f_{i_1}, \ldots, f_{im} \in \{f_1, \ldots, f_n\}$ and $a_1, \ldots, a_m$ be nonnegative integers. We denote by $T(f_{i_1}, a_1) \ldots (f_{i_m}, a_m)$ the subtable of the table T that contains only rows that have numbers $a_1, \ldots, a_m$ at the intersection with columns $f_{i_1}, \ldots, f_{i_m}$. Such nonempty subtables (including the table T) are called *separable subtables* of T.

We will say that an attribute $f_i \in \{f_1, \ldots, f_n\}$ is *not constant* on T if it has at least two different values. For the attribute that is not constant on T we can find *the most frequent value*. It is an attribute's value attached to the maximum number of rows in T. If there are two or more such values then we choose the most frequent value for which exists the most common decision.

We denote by $E(T)$ a set of attributes from $\{f_1, \ldots, f_n\}$ that are not constant on T. The set $E(T)$ contains one attribute with the minimum number of values and attributes with the most frequent value. For any $f_i \in E(T)$, we denote by $E(T, f_i)$ a set of values of the attribute f_i in T. Note, that if $f_i \in E(T)$ is the attribute with the most frequent value then $E(T, f_i)$ contains only one element.

The expression

$$f_{i_1} = a_1 \wedge \ldots \wedge f_{i_m} = a_m \to d \tag{1}$$

is called a *decision rule over T* if $f_{i_1}, \ldots, f_{i_m} \in \{f_1, \ldots, f_n\}$, and $a_1, \ldots, a_m, d$ are nonnegative integers. It is possible that $m = 0$. In this case (1) is equal to the rule

$$\to d. \tag{2}$$

Let $r = (b_1, \ldots, b_n)$ be a row of T. We will say that the rule (1) is *realizable for r*, if $a_1 = b_{i_1}, \ldots, a_m = b_{i_m}$. If $m = 0$ then the rule (2) is realizable for any row from T.

Let β be a nonnegative real number. We will say that the rule (1) is *β-true for T* if d is the most common decision for $T' = T(f_{i_1}, a_1) \ldots (f_{i_m}, a_m)$ and $R(T') \leq \beta$. If $m = 0$ then the rule (2) is β-true for T if d is the most common decision for T and $R(T) \leq \beta$.

If the rule (1) is β-true for T and realizable for r, we will say that (1) is a *β-decision rule for T and r*. Note that if $\beta = 0$ we have an exact decision rule for T and r.

Let τ be a decision rule over T and τ be equal to (1). The *coverage* of τ is the number of rows in T for which τ is realizable and which are labeled with the decision d. We denote it by $c(\tau)$. If $m = 0$ then $c(\tau)$ is equal to the number of rows in T that are labeled with decision d.

The main aim of the paper is to present a heuristics that allows to find, for β-decision rule for T and r, value of coverage that is close to optimal one obtined using dynamic programming algorithm for opimization of β-decision rules [4].

3   Algorithm for Directed Acyclic Graph Construction

In this section, we present an algorithm that construct, for a given decision table T, a *directed acyclic graph* $\Delta_\beta(T)$. Based on this graph we can describe set of decision rules for T and each row r of T. Nodes of the graph are separable subtables of the table T. During each step, the algorithm processes one node and marks it with the symbol *. At the first step, the algorithm constructs a graph containing a single node T that is not marked with *.

Let us assume that the algorithm has already performed p steps. We describe now the step $(p+1)$. If all nodes are marked with the symbol * as processed, the algorithm finishes its work and presents the resulting graph as $\Delta_\beta(T)$. Otherwise, choose a node (table) Θ, that has not been processed yet. If $R(\Theta) \leq \beta$, then mark Θ with the symbol * and go to the step $(p+2)$. If $R(\Theta) > \beta$ then for each attribute $f_i \in E(\Theta)$, draw a bundle of edges from the node Θ if f_i is the attribute with the minimum number of values. If f_i is the attribute with the most frequent value draw one edge from the node Θ. Let f_i be the attribute with the minimum number of values and $E(\Theta, f_i) = \{b_1, \ldots, b_t\}$. Then draw t edges from Θ and label them with pairs $(f_i, b_1) \ldots (f_i, b_t)$ respectively. These edges enter to nodes $\Theta(f_i, b_1), \ldots, \Theta(f_i, b_t)$. For the rest of attributes from $E(\Theta)$ draw one edge, for each attribute, from the node Θ and label it with pair (f_i, b_1), where b_1 is the most frequent value of the attribute f_i. This edge enters to a node $\Theta(f_i, b_1)$. If some of nodes $\Theta(f_i, b_1), \ldots, \Theta(f_i, b_t)$ are absent in the graph then add these nodes to the graph. We label each row r of Θ with the set of attributes $E_{\Delta_\beta(T)}(\Theta, r) \subseteq E(\Theta)$. Mark the node Θ with the symbol * and proceed to the step $(p+2)$.

The graph $\Delta_\beta(T)$ is a directed acyclic graph. A node of this graph will be called *terminal* if there are no edges leaving this node. Note that a node Θ of $\Delta_\beta(T)$ is terminal if and only if $R(\Theta) \leq \beta$.

In the next section, we will describe a procedure of optimization of the graph $\Delta_\beta(T)$ relative to the coverage. As a result we will obtain a graph G with the same sets of nodes and edges as in $\Delta_\beta(T)$. The only difference is that any row r of each nondegenerate table Θ from G is labeled with a set of attributes $E_G(\Theta, r) \subseteq E_{\Delta_\beta(T)}(\Theta, r)$. It is possible also that $G = \Delta_\beta(T)$.

Now, for each node Θ of G and for each row r of Θ we describe a set of decision rules $Rul_G(\Theta, r)$. We will move from terminal nodes of G to the node T.

Let Θ be a terminal node of G and d be the most common decision for Θ. Then $Rul_G(\Theta, r) = \{\to d\}$.

Let now Θ be a nonterminal node of G such that for each child Θ' of Θ and for each row r' of Θ' the set of rules $Rul_G(\Theta', r')$ is already defined. Let $r = (b_1, \ldots, b_n)$ be a row of Θ. For any $f_i \in E_G(\Theta, r)$, we define the set of rules $Rul_G(\Theta, r, f_i)$ as follows:

$$Rul_G(\Theta, r, f_i) = \{f_i = b_i \wedge \gamma \to s : \gamma \to s \in Rul_G(\Theta(f_i, b_i), r)\}.$$

Then $Rul_G(\Theta, r) = \bigcup_{f_i \in E_G(\Theta, r)} Rul_G(\Theta, r, f_i)$.

To illustrate the presented algorithm we consider a simple decision table T depicted on the top of Fig. 1. We set $\beta = 2$, so during the construction of the graph $\Delta_2(T)$ we stop the partitioning of a subtable Θ of T if $R(\Theta) \leq 2$. We denote $G = \Delta_2(T)$.

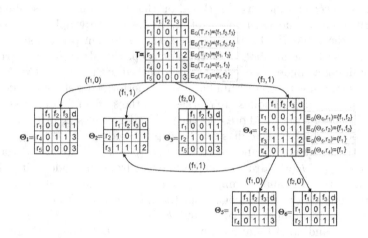

Fig. 1. Directed acyclic graph for decision table T

Now, for each node Θ of the graph G and for each row r of Θ we describe the set $Rul_G(\Theta, r)$. We will move from terminal nodes of G to the node T. Terminal nodes of the graph G are Θ_1, Θ_2, Θ_3, Θ_5, and Θ_6. For these nodes, $Rul_G(\Theta_1, r_1) = Rul_G(\Theta_1, r_4) = Rul_G(\Theta_1, r_5) = \{\rightarrow 3\}$; $Rul_G(\Theta_2, r_2) = Rul_G(\Theta_2, r_3) = \{\rightarrow 1\}$; $Rul_G(\Theta_3, r_1) = Rul_G(\Theta_3, r_2) = Rul_G(\Theta_3, r_5) = \{\rightarrow 1\}$; $Rul_G(\Theta_5, r_1) = Rul_G(\Theta_5, r_4) = \{\rightarrow 1\}$; $Rul_G(\Theta_6, r_1) = Rul_G(\Theta_6, r_2) = \{\rightarrow 1\}$. Now, we can describe the sets of rules attached to rows of nonterminal node Θ_4. For this subtable children (subtables Θ_2, Θ_5 and Θ_6) are already treated, and we have:
$Rul_G(\Theta_4, r_1) = \{f_1 = 0 \rightarrow 1, f_2 = 0 \rightarrow 1\}$; $Rul_G(\Theta_4, r_2) = \{f_1 = 1 \rightarrow 1, f_2 = 0 \rightarrow 1\}$; $Rul_G(\Theta_4, r_3) = \{f_1 = 1 \rightarrow 1\}$; $Rul_G(\Theta_4, r_4) = \{f_1 = 0 \rightarrow 1\}$.
Finally, we can describe the sets of rules attached to rows of T:
$Rul_G(T, r_1) = \{f_1 = 0 \rightarrow 3, f_2 = 0 \rightarrow 1, f_3 = 1 \wedge f_1 = 0 \rightarrow 1, f_3 = 1 \wedge f_2 = 0 \rightarrow 1\}$;
$Rul_G(T, r_2) = \{f_1 = 1 \rightarrow 1, f_2 = 0 \rightarrow 1, f_3 = 1 \wedge f_2 = 0 \rightarrow 1, f_3 = 1 \wedge f_1 = 1 \rightarrow 1\}$;
$Rul_G(T, r_3) = \{f_1 = 1 \rightarrow 1, f_3 = 1 \wedge f_1 = 1 \rightarrow 1\}$;
$Rul_G(T, r_4) = \{f_1 = 0 \rightarrow 3, f_3 = 1 \wedge f_1 = 0 \rightarrow 1\}$;
$Rul_G(T, r_5) = \{f_1 = 0 \rightarrow 3, f_2 = 0 \rightarrow 1\}$.

4   Procedure of Optimization Relative to Coverage

In this section, we present a procedure of optimization of the graph G relative to the coverage c. For each node Θ in the graph G, this procedure assigns to each row r of Θ the set $Rul_G^c(\Theta, r)$ of decision rules with the maximum coverage from $Rul_G(\Theta, r)$ and the number $Opt_G^c(\Theta, r)$ – the maximum coverage of a decision rule from $Rul_G(\Theta, r)$.

We will move from the terminal nodes of the graph G to the node T. We will assign to each row r of each table Θ the number $Opt_G^c(\Theta, r)$ and we will change the set $E_G(\Theta, r)$ attached to the row r in the nonterminal node Θ of G. We denote the obtained graph by G^c.

Let Θ be a terminal node of G and d be the most common decision for Θ. Then we assign to each row r of Θ the number $Opt_G^c(\Theta, r)$ that is equal to the number of rows in Θ that are labeled with the decision d.

Let Θ be a nonterminal node of G and all children of Θ have already been treated. Let $r = (b_1, \ldots, b_n)$ be a row of Θ. We assign the number

$$Opt_G^c(\Theta, r) = \max\{Opt_G^c(\Theta(f_i, b_i), r) : f_i \in E_G(\Theta, r)\}$$

to the row r in the table Θ and we set

$$E_{G^c}(\Theta, r) = \{f_i : f_i \in E_G(\Theta, r), Opt_G^c(\Theta(f_i, b_i), r) = Opt_G^c(\Theta, r)\}.$$

Figure 2 presents the directed acyclic graph G^c obtained from the graph G (see Fig. 1) by the procedure of optimization relative to the coverage.

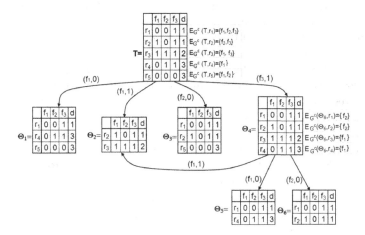

Fig. 2. Graph G^c

Using the graph G^c we can describe for each row r_i, $i = 1, \ldots, 5$, of the table T the set $Rul_G^c(T, r_i)$ of decision rules for T and r_i with the maximum coverage. We will give also the value $Opt_G^c(T, r_i)$ which is equal to the maximum coverage of decision rule for T and r_i. This value was obtained during the procedure of optimization of the graph G relative to the coverage. We have

$Rul_G(T, r_1) = \{f_1 = 0 \to 3, f_2 = 0 \to 1, f_3 = 1 \wedge f_2 = 0 \to 1\}$, $Opt_G^c(T, r_1) = 2$;

$Rul_G(T_0, r_2) = \{f_2 = 0 \to 1, f_3 = 1 \wedge f_2 = 0 \to 1,\}$, $Opt_G^c(T, r_2) = 2$;

$Rul_G(T_0, r_3) = \{f_1 = 1 \to 1, f_3 = 1 \wedge f_1 = 1 \to 1\}$, $Opt_G^c(T, r_3) = 1$;

$Rul_G(T_0, r_4) = \{f_1 = 0 \to 3\}$, $Opt_G^c(T, r_4) = 2$;

$Rul_G(T_0, r_5) = \{f_1 = 0 \to 3, f_2 = 0 \to 1\}$, $Opt_G^c(T, r_5) = 2$.

5   Experimental Results

We made experiments on decision tables from UCI Machine Learning Repository [6] using system Dagger [1] created in King Abdullah University of Science and Technology. Some decision tables contain conditional attributes that take unique value for each row. Such attributes were removed. In some tables there were equal rows with, possibly, different decisions. In this case each group of identical rows was replaced with a single row from the group with the most common decision for this group. In some tables there were missing values. Each such value was replaced with the most common value of the corresponding attribute.

Let T be one of these decision tables. We consider, for this table, values of β from the set $B(T) = \{R(T) \times 0.15, R(T) \times 0.25, R(T) \times 0.35\}$. Let $\beta \in B(T)$.

Tab. 1 presents the average values of the maximum coverage of β-decision rules. Column *rows* contains the number of rows in T, column *attr* - the number of attributes in T. For each row r of T, we find the maximum coverage of a β-decision rule for T and r. After that, we find for rows of T the average coverage of rules with the maximum coverage - one for each row. The results can be found in a column *avg*. To make comparison with the average coverage of β-decision rules constructed by the dynamic programming algorithm [4] we made experiments and present results in a column *avg-dp*. The last column *rel diff* presents a relative difference that is equal to $(Opt\_Coverage - Coverage)/Opt\_Coverage$, where *Coverage* denotes the average coverage of β-decision rules constructed by the proposed algorithm, *Opt_Coverage* denotes the average coverage of β-decision rules constructed by the dynamic programming algorithm. The last row in Tab. 1 presents the average value of the relative difference for considered decision tables.

Based on the presesnted results we can see that the average relative difference is decreasing when the value of β is increasing. The biggest relative difference exists, for $\beta = R(T) \times 0.15$, for "monks-3-train" - 25%. For $\beta = R(T) \times 0.35$, the relative difference is equal to 0 for ten data sets. For data sets "lymphography", "soybean-small", "teeth" and "zoo-data", for $\beta \in B(T)$, the relative difference is equal to 0.

Tab. 2 presents a size of the directed acyclic graph, i.e., number of nodes (column *nd*) and number of edges (column *edg*) in the graph constructed by the proposed algorithm and dynamic programming algorithm (columns *nd-dp* and *edg-dp* respectively).

Tab. 3 presents comparison of the number of nodes (column *nd diff*) and number of edges (column *edg diff*) of the directed acyclic graph. Values of these columns are equal to the number of nodes/edges in the directed acyclic graph constructed by the dynamic programming algorithm divided by the number of nodes/edges in the directed acyclic graph constructed by the proposed algorithm.

Presented results show that the size of the directed acyclic graph constructed by the proposed algorithm is smaller than the size of the directed acyclic graph constructed by the dynamic programming algorithm. In particular, for the data sets "lymphography", "soybean-small", "teeth" and "zoo-data", the results of the average coverage are the same (see Tab. 1) but there exists a difference

Table 1. Average coverage of β-decision rules

Decision table	rows	attr	$\beta = R(T) \times 0.15$			$\beta = R(T) \times 0.25$			$\beta = R(T) \times 0.35$		
			avg	avg-dp	rel diff	avg	avg-dp	rel diff	avg	avg-dp	rel diff
adult-stretch	16	4	6.75	7	0.04	7.5	7.5	0	7.5	7.5	0
balance-scale	625	4	88.48	92.31	0.04	88.48	92.31	0.04	88.48	92.31	0.04
breast-cancer	266	9	81.2	81.84	0.01	114.9	117.39	0.02	124.35	125.08	0.01
cars	1728	6	486.54	499.26	0.03	486.54	499.26	0.03	486.54	499.26	0.03
house-votes	279	16	148.76	150.33	0.01	156.69	157.54	0.01	160.15	160.55	0
lymphography	148	18	56.33	56.35	0	57.94	57.94	0	59.06	59.06	0
monks-1-test	432	6	77	81	0.05	108	108	0	108	108	0
monks-1-train	124	6	25.12	28.01	0.1	28.36	30.22	0.06	34.4	34.4	0
monks-2-test	432	6	95.3	100.74	0.05	135.11	136.19	0.01	146.5	146.5	0
monks-2-train	169	6	34.43	37.39	0.08	41.91	49.83	0.16	52.54	54.23	0.03
monks-3-test	432	6	97.11	116	0.16	120	120	0	120	120	0
monks-3-train	122	6	24.07	32.15	0.25	27.75	33.52	0.17	33	35.04	0.06
nursery	12960	8	2511.42	3066.1	0.18	2705.29	3066.1	0.12	2917.33	3066.1	0.05
soybean-small	47	35	15.64	15.64	0	17	17	0	17	17	0
teeth	23	8	1	1	0	1	1	0	1	1	0
zoo-data	59	16	16.85	16.86	0	17.83	17.83	0	18.85	18.85	0
average					0.06			0.04			0.01

Table 2. Size of the directed acyclic graph

Decision table	$\beta = R(T) \times 0.15$				$\beta = R(T) \times 0.25$				$\beta = R(T) \times 0.35$			
	nd	edg	nd-dp	edg-dp	nd	edg	nd-dp	edg-dp	nd	edg	nd-dp	edg-dp
adult-stretch	36	37	72	108	16	17	52	76	16	17	52	76
balance-scale	23	22	117	140	9	8	21	20	9	8	21	20
breast-cancer	305	494	1927	3630	147	211	844	1234	72	88	390	499
cars	56	60	505	736	23	22	179	248	23	22	118	129
house-votes	11950	26070	27268	58594	3764	6560	8808	14178	1479	2243	3573	4908
lymphography	4610	14739	12885	54927	2379	6827	6556	23600	1281	3054	3170	9596
monks-1-test	72	112	441	726	30	43	137	203	24	25	74	77
monks-1-train	80	108	294	437	30	37	116	147	27	31	84	91
monks-2-test	74	103	411	667	45	55	163	268	24	25	122	161
monks-2-train	91	132	423	706	48	61	161	241	30	37	101	119
monks-3-test	62	76	363	507	29	37	125	161	24	25	94	105
monks-3-train	79	98	264	376	37	43	138	174	31	37	93	105
nursery	103	115	583	833	47	51	249	293	28	27	78	77
soybean-small	1993	8512	3220	24015	1098	3296	2121	8879	585	1335	1211	3458
teeth	92	245	126	644	63	137	112	362	47	84	91	219
zoo-data	3013	15722	4159	33941	1975	7372	3096	15915	1240	3453	2041	7381

relative to the number of nodes (more than one time) and relative to the number of edges (more than two times).

Tab. 4 presents the average coverage of exact decision rules for the proposed algorithm (column *avg*), for the algorithm presented in [15] (column *avg-mod1*), and for the dynamic programming algorithm (column *avg-dp*). The last column contains the relative difference of the average coverage of decision rules between proposed algorithm and the dynamic programming algorithm. Presented results show also that the proposed algorithm is better than algorithm presented in [15] because the average values of the maximum coverage of exact decision rules are greater and we don't have rows without rules.

Table 3. Comparison of the size of the directed acyclic graph

Decision table	$\beta = R(T) \times 0.15$		$\beta = R(T) \times 0.25$		$\beta = R(T) \times 0.35$	
	nd diff	edg diff	nd diff	edg diff	nd diff	edg diff
adult-stretch	2.00	2.92	3.25	4.47	3.25	4.47
balance-scale	5.09	6.36	2.33	2.50	2.33	2.50
breast-cancer	6.32	7.35	5.74	5.85	5.42	5.67
cars	9.02	12.27	7.78	11.27	5.13	5.86
house-votes	2.28	2.25	2.34	2.16	2.42	2.19
lymphography	2.80	3.73	2.76	3.46	2.47	3.14
monks-1-test	6.13	6.48	4.57	4.72	3.08	3.08
monks-1-train	3.68	4.05	3.87	3.97	3.11	2.94
monks-2-test	5.55	6.48	3.62	4.87	5.08	6.44
monks-2-train	4.65	5.35	3.35	3.95	3.37	3.22
monks-3-test	5.85	6.67	4.31	4.35	3.92	4.20
monks-3-train	3.34	3.84	3.73	4.05	3.00	2.84
nursery	5.66	7.24	5.30	5.75	2.79	2.85
soybean-small	1.62	2.82	1.93	2.69	2.07	2.59
teeth	1.37	2.63	1.78	2.64	1.94	2.61
zoo-data	1.38	2.16	1.57	2.16	1.65	2.14
average	4.17	5.16	3.64	4.30	3.19	3.55

Table 4. Comparison of the coverage for exact decision rules

Decision table	$\beta = R(T) \times 0.0$			
	avg-mod1	avg	avg-dp	rel diff
adult-stretch	6.00	6.25	7.00	0.11
balance-scale	0.51	3.07	4.21	0.27
breast-cancer	4.22	6.15	9.53	0.35
cars	323.00	325.58	332.76	0.02
house-votes	73.03	73.08	73.52	0.01
lymphography	20.52	20.69	21.54	0.04
monks-1-test	31.00	33.50	45.00	0.26
monks-1-train	5.77	6.70	13.45	0.50
monks-2-test	11.26	12.11	12.36	0.02
monks-2-train	3.55	4.32	6.38	0.32
monks-3-test	36.33	37.28	56.00	0.33
monks-3-train	6.21	9.03	12.20	0.26
nursery	1476.71	1483.58	1531.04	0.03
soybean-small	12.53	12.53	12.53	0.00
teeth	0.70	1.00	1.00	0.00
zoo	11.07	11.07	11.07	0.00
average				0.16

6   Conclusions

We presented a modification of the dynamic programming algorithm for optimization of β-decision rules relative to the coverage. Experimental results show that the size of the directed acyclic graph constructed by the proposed algorithm is smaller than the size of the directed acyclic graph constructed by the dynamic programming algorithm, and in the case of edges, the difference is at least two times. The average coverage of β-decision rules constructed by the proposed algorithm is equal to optimal values for four data sets, for $\beta \in B(T)$. In comparison with the algorithm presented in [15], proposed algorithm, usually, gives greater values of the average coverage of exact decision rules and each row has assigned at least one rule. In the future works, we will use another uncertainty measure, e.g., difference between number of rows in decision table T and number of rows with the most common decision for T [3].

Acknowledgements. The author wishes to thanks the anonymous reviewers for useful comments.

References

1. Alkhalid, A., Amin, T., Chikalov, I., Hussain, S., Moshkov, M., Zielosko, B.: Dagger: A tool for analysis and optimization of decision trees and rules. In: Computational Informatics, Social Factors and New Information Technologies: Hypermedia Perspectives and Avant-Garde Experiences in the Era of Communicability Expansion, pp. 29–39. Blue Herons (2011)
2. Alsolami, F., Chikalov, I., Moshkov, M., Zielosko, B.: Optimization of inhibitory decision rules relative to length. Studia Informatica 33(2A(105)), 395–406 (2012)
3. Amin, T., Chikalov, I., Moshkov, M., Zielosko, B.: Dynamic programming approach for partial decision rule optimization. Fundam. Inform. 119(3-4), 233–248 (2012)
4. Amin, T., Chikalov, I., Moshkov, M., Zielosko, B.: Dynamic programming approach to optimization of approximate decision rules. Inf. Sci. 221, 403–418 (2013)
5. An, A., Cercone, N.: Rule quality measures improve the accuracy of rule induction: An experimental approach. In: Raś, Z.W., Ohsuga, S. (eds.) ISMIS 2000. LNCS (LNAI), vol. 1932, pp. 119–129. Springer, Heidelberg (2000)
6. Asuncion, A., Newman, D.J.: UCI Machine Learning Repository (2007), http://www.ics.uci.edu/~mlearn/
7. Błaszczyński, J., Słowiński, R., Szeląg, M.: Sequential covering rule induction algorithm for variable consistency rough set approaches. Inf. Sci. 181(5), 987–1002 (2011)
8. Dembczyński, K., Kotłowski, W., Słowiński, R.: Ender: a statistical framework for boosting decision rules. Data Min. Knowl. Discov. 21(1), 52–90 (2010)
9. Moshkov, M., Piliszczuk, M., Zielosko, B.: Partial Covers, Reducts and Decision Rules in Rough Sets - Theory and Applications. SCI, vol. 145. Springer, Heidelberg (2008)
10. Moshkov, M., Zielosko, B.: Combinatorial Machine Learning - A Rough Set Approach. SCI, vol. 360. Springer, Heidelberg (2011)
11. Nguyen, H.S.: Approximate boolean reasoning: Foundations and applications in data mining. In: Peters, J.F., Skowron, A. (eds.) Transactions on Rough Sets V. LNCS, vol. 4100, pp. 334–506. Springer, Heidelberg (2006)
12. Pawlak, Z., Skowron, A.: Rough sets and boolean reasoning. Inf. Sci. 177(1), 41–73 (2007)
13. Sikora, M., Wróbel, Ł.: Data-driven adaptive selection of rule quality measures for improving rule induction and filtration algorithms. Int. J. General Systems 42(6), 594–613 (2013)
14. Zielosko, B.: Coverage of decision rules. In: Decision Support Systems, pp. 183–192, University of Silesia (2013)
15. Zielosko, B.: Coverage of exact decision rules. Studia Informatica 34(2A(111)), 251–262 (2013)

Nondeterministic Decision Rules
in Rule-Based Classifier

Piotr Paszek and Barbara Marszał-Paszek

Institute of Computer Science, University of Silesia
Będzińska 39, 41-200 Sosnowiec, Poland
{paszek,bpaszek}@us.edu.pl

Abstract. In the paper is discussed the truncated nondeterministic rules and their role in an evaluation of classification model. The nondeterministic rules are created as the result of shorting deterministic rules in accordance with the principle of minimum description length (MDL). As deterministic rules in database we treat the full objects description in a meaning of descriptors conjunction. The nondeterministic rules are calculated in polynomial time by using greedy strategy.

The classification model is composed in two steps process. In the first step deterministic and nondeterministic rules are constructed. Next these rules are used for classifier evaluation. The evaluation results are compared with classifiers only based on deterministic rules creating by different algorithms. The experiments shows that such nondeterministic rules could be treat as an extra knowledge about data. This knowledge is able to improve the classification quality. It should be pointed out that classification process requires tuning some of their parameters relative to analyzed data.

Keywords: classification, decision tables, nondeterministic decision rules, rough sets, rule-based classifier.

1 Introduction

Over the years many methods based on rule induction and rule-based classification systems were developed (see, e.g. different versions of Michalski's AQ system [11], Holta's 1R system [9], systems based on rough sets [8,1,2,13,20], and many others [17,19]).

In this paper, we show that exist possibility for improving the rule-based classification systems by using nondeterministic decision rules. We discuss a method for rule inducing based on searching for strong rules for a union of a few relevant decision classes – nondeterministic decision rules. Because shortening the deterministic rules creates these rules, so they are called truncated nondeterministic rules.

In the paper, the following classification problem is considered: for a given decision table T [15] and a new object v generate a value of the decision attribute on v using values of conditional attributes on v.

S. Kozielski et al. (Eds.): BDAS 2014, CCIS 424, pp. 180–190, 2014.

In [18] Skowron and Suraj are shown that there exist information systems $S = (U, A)$ [15], where U is a finite set of objects and A is a finite set of attributes, such that the set U can't be described by deterministic rules. In [12] Moshkov et al. are shown that for any information system, the set can be described by inhibitory rules. Inhibitory rules [6] are a special case of nondeterministic rules. These results inspired us to use the nondeterministic rules in a classification process [10].

The application of (truncated) nondeterministic rules in construction of rule-based classifiers is presented in the paper. Also the results of experiments are included. It shows that by combining the rule-based classifiers based on deterministic decision rules with nondeterministic decision rules that have sufficiently large support [3] – truncated nondeterministic decision rules – it is possible to improve the classification quality and reduce the classification error.

The paper consists of six sections. In section 2, we recall the notions of a decision table and deterministic and nondeterministic decision rules. In sections 3 and 4 we present a greedy algorithm for nondeterministic decision rule construction and main steps in construction of classifiers enhanced by nondeterministic rules. In section 5 the results of the experiments with real-life data from the UCI Machine Learning Repository [7] are discussed. Section 6 contains short conclusions.

2   Basic Notations

When we think about rule-based classifier, the decision table is considered.

A *decision table*, is a triple $T = (U, A, d)$, where $U = \{u_1, \ldots, u_n\}$ is a finite nonempty set of *objects*, $A = \{a_1, \ldots, a_m\}$ is a finite nonempty set of *conditional attributes* (functions defined on U), and d is the *decision attribute* (function defined on U).

We assume that for each $u_i \in U$ and each $a_j \in A$ the value $a_j(u_i)$ belong to $V_{a_j}(T)$ and the value $d(u_i)$ belong to $V_d(T)$, where $V_d(T)$ denotes the set of values of the decision attribute d on objects from U.

2.1   Deterministic Decision Rules

In general, the *deterministic decision rule* in T has the following form:

$$(a_{j_1} \in V_1) \wedge \ldots \wedge (a_{j_k} \in V_k) \rightarrow (d = v),$$

where $a_{j_1}, \ldots, a_{j_k} \in A$, $V_j \subseteq V_{a_j}$, for $j \in \{1, \ldots, k\}$ and $v \in V_d(T)$. The predecessor of this rule is a conjunction of generalized descriptors and the successor of this rule is a descriptor.

Rule-based classifiers are composed of rules in the shape of Horn Clauses mostly:

$$(a_{j_1} = b_1) \wedge \ldots \wedge (a_{j_k} = b_k) \rightarrow (d = v)$$

where $k > 0$, $a_{j_1}, \ldots, a_{j_k} \in A$, $b_1, \ldots, b_k \in V_A(T)$, $v \in V_d(T)$ and numbers $j_1, \ldots, j_k$ are pairwise different. The predecessor of this rule (conditional part) is a conjunction of descriptors.

2.2 Nondeterministic Decision Rules

In this paper, we also consider nondeterministic decision rules. A *nondetermin-istic decision rule* in a given decision table T is of the form:

$$(a_{j_1} \in V_1) \land \ldots \land (a_{j_k} \in V_k) \to d = (c_1 \lor \ldots \lor c_s), \tag{1}$$

where $a_{j_1}, \ldots, a_{j_k} \in A$, $V_j \subseteq V_{a_j}$, for $j \in \{1, \ldots, k\}$, numbers $j_1, \ldots, j_k$ are pairwise different, and $\emptyset \neq \{c_1, \ldots, c_s\} \subseteq V_d(T)$.

Let us introduce some notation about (nondeterministic) rules.

If r is the nondeterministic rule of the form (1) then by $lh(r)$ we denote its left hand side, i.e., the formula $(a_{j_1} \in V_1) \land \ldots \land (a_{j_k} \in V_k)$, and by $rh(r)$ its right hand side, i.e., the formula $d = (c_1 \lor \ldots \lor c_s)$.

By $\|lh(r)\|_T$ (or $\|lh(r)\|$, for short) we denote all objects from U satisfying $lh(r)$ [5]. To measure the quality of such rules we use coefficients called the support and the confidence [3]. If r is a nondeterministic rule then the support of this rule in the decision table T is defined by

$$supp(r) = \frac{|\ \|lh(r)\| \cap \|rh(r)\|\ |}{|U|},$$

and the confidence of r in T is defined by

$$conf(r) = \frac{|\ \|lh(r)\| \cap \|rh(r)\|\ |}{|\ \|lh(r)\|\ |}.$$

We also use a normalized support of r in T defined by

$$norm\_supp(r) = \frac{supp(r)}{\sqrt{|V(r)|}},$$

where $V(r) \subseteq V_d(T)$ is a decision values set from right hand side of the rule $(rh(r))$.

2.3 Truncated Nondeterministic Rules

Now we can define a parameterized set of truncated nondeterministic decision rules. This type of nondeterministic rules appears as a result of shortening rules according to the principle MDL (Minimum Description Length) [16].

This parameterized set $Rules_{nd}(\alpha, k)$ is defined as the subset of nondeterministic rules r (over attributes in T) such that:

1. On the left hand sides of such rules are only conditions of the form $a = v$, where $a \in A, v \in V_a$ (descriptor);
2. $conf(r) \geq \alpha$, where $\alpha \in [0.5, 1]$ is a threshold;
3. $|V(r)| \leq k < |V_d(T)|$, where k is a threshold used as an upper bound on the number of decision values on the right hand sides of rules.

Hence, the *truncated nondeterministic decision rules* are of the form:

$$(a_{j_1} = b_1) \wedge \ldots \wedge (a_{j_k} = b_k) \rightarrow d = (c_1 \vee \ldots \vee c_s), \tag{2}$$

where $a_{j_1}, \ldots, a_{j_k} \in A$, for $j \in \{1, \ldots, k\}$, $b_j \in V_{b_j}(T)$, numbers $j_1, \ldots, j_k$ are pairwise different, and $\emptyset \neq \{c_1, \ldots, c_s\} \subseteq V_d(T)$.

The algorithm presented in section 3 is searching for truncated nondeterministic rules with sufficiently large support and relatively small (in comparison to the set of all possible decisions), sets of decisions defined by the right hand sides of such rules for the decision table T.

3 Algorithm for Nondeterministic Decision Rule Construction

TruNDeR (*Tru*ncated *N*ondeterministic *De*cision *R*ules) is the name of the algorithm that constructs truncated nondeterministic decision rules for decision table T with parameters $\alpha \in [0.5, 1]$ and k.

The algorithm consists of two main steps. In the first step, the deterministic decision rules are constructed for a given decision table T. In the second step of the algorithm, the set of nondeterministic rules was obtained by shortening (truncation) deterministic rules.

There are many algorithms for the set of deterministic rules construction. Therefore, the choice of the algorithm, for the set of deterministic rules construction, has the influences on algorithm computational complexity and also on the quality of the constructed classifier.

In the work [14], we showed that in the Trunder algorithm, we could replace the minimal rules (some kind of deterministic rules, which are created using the exhaustive algorithm from RSES [2]) with the complete rules (rows of decision table) without any loss of the classification quality classifier using nondeterministic rules. This remark is very important for the reason of reduction computational complexity. Therefore, algorithm Trunder, for truncated nondeterministic rules construction, has polynomial computational complexity, which depends on number of objects and number of attributes in the decision table.

The main idea of the algorithm Trunder, is shortening (deterministic) rules according to the principle Minimum Description Length [16].

Trunder algorithm reduces the rules by removing a conditional attribute that maximizes the normalized support of the rule. In shorting process we obtain nondeterministic rules because more objects from a decision table fits to the shortened conditional part of the rule. Thus objects could belong to different decision classes. This algorithm uses a greedy strategy for choosing decision classes (decision part of the rule). The algorithm selects decision classes which are the mostly occur among the objects.

Algorithm 1 contains pseudo-code of the algorithm Trunder.

Tab. 1 contains data and nondeterministic rules that were created by the algorithm Trunder for that data.

Algorithm 1. *TruNDeR* – greedy algorithm for truncated nondeterministic
decision rule construction

Input: T – decision table, $\alpha \in [0.5, 1]$, k – upper bound on the number of decision
values;
Output: $Rules_{nd}(\alpha, k)$ – a set of nondeterministic decision rules for T.

$R_{nd} \leftarrow \emptyset$;
for all $r \in T$ **do**
   $\{r : L \to (d = v); \; L = D_1 \wedge \ldots \wedge D_m; \; v \in V_d; \}$
   $STOP \leftarrow false$;
   $\lambda_L \leftarrow norm\_supp(L)$;
   repeat
      for all condition attributes from r **do**
         $L^i = D_1 \wedge \ldots \wedge D_{i-1} \wedge D_{i+1} \wedge \ldots \wedge D_m$;
         $\{L^i$ is obtained by dropping i-th attribute from the left hand side of rule $r\}$
         $\theta = \{v \in V_d : \exists_{x \in U_{L^i}} d(x) = v\}$;
         Sorting in decreasing order θ;
         $\theta_i \subset \theta : conf(L^i \to (d = \theta_i)) = \frac{|||L^i|| \cap ||\theta_i|||}{|||L^i|||} \geq \alpha$; $\{\theta_i$ greedy selection$\}$
         if $|\theta_i| \leq k$ **then**
            $\lambda_{L^i} \leftarrow norm\_supp(L^i \to \theta_i)$
         else
            $\lambda_{L^i} \leftarrow 0$;
         end if
      end for
      $\lambda_{max}^i \leftarrow argmax\{\lambda_{L^i}\}$;
      if $\lambda_{max}^i \geq \lambda_L$ **then**
         $L \leftarrow L^i$; $\lambda_L \leftarrow \lambda_{max}^i$; $\{r_{nd} : L \to (d = \theta_i); \; \lambda_L\}$
      else
         $STOP \leftarrow true$;
      end if
   until STOP
   $R_{nd} \leftarrow R_{nd} \cup \{r_{nd}\}$;
end for
return R_{nd};

---

4 Classifiers

In this section, we present an application of nondeterministic rules for the classification process.

We constructed two kinds of classifiers. First group of classifiers uses only the
set of deterministic decision rules, generated by using *RSESlib* library (Rough
Set Exploration System library) [4], and standard voting procedure to resolve
conflicts between rules – *DR* classifiers. Second group of classifiers uses deterministic decision rules (created in *RSES*) and truncated nondeterministic rules
(created in *TruNDeR* algorithm) – *TNDR* classifiers.

TNDR classifier is constructed in the following way:

Table 1. Sample data (a) and nondeterministic rules (b) generated by algorithm *Trunder* for these data

<table>
<tr><th colspan="5">(a)</th><th>(b)</th></tr>
<tr><th>a1 a2 a3 a4</th><th>cl</th><th></th></tr>
<tr><td>0 1 0 1</td><td>2</td><td></td></tr>
<tr><td>0 1 1 1</td><td>1</td><td>$Rules_{nd}(\alpha = 1.0, k = 2)$</td></tr>
<tr><td>0 1 1 0</td><td>2</td><td>(a3=1) → (cl=1[3]) ∨ (cl=2[1])</td></tr>
<tr><td>1 0 0 1</td><td>1</td><td>(a1=0) ∧ (a3=0) → (cl=2[1]) ∨ (cl=0[2])</td></tr>
<tr><td>0 0 0 1</td><td>0</td><td>(a2=0) → (cl=1[2]) ∨ (cl=0[1])</td></tr>
<tr><td>1 1 1 0</td><td>1</td><td>(a1=0) ∧ (a4=0) → (cl=0[1]) ∨ (cl=2[1])</td></tr>
<tr><td>0 1 0 0</td><td>0</td><td>(a2=1) ∧ (a4=1) → (cl=1[1]) ∨ (cl=2[1])</td></tr>
<tr><td>1 0 1 1</td><td>1</td><td></td></tr>
</table>

- A set of deterministic rules, using the selected algorithm from *RSES* is created;
- A set of truncated nondeterministic decision rules, using the *TruNDeR* algorithm is created too;
- Next, the collections of deterministic and nondeterministic rules are combined.

It is normal that, there are conflicts between rules during combining process. To resolve these conflicts we used modified standard voting procedure. Modification of standard voting procedure for the nondeterministic rules involves consideration of the support for all values of decision attribute occurring in the nondeterministic rule.

For each new object, during classification process using the classifier of the second type, we obtain (predicted) a single decision value using modified standard voting procedure.

For the construction of the classifier, the first and second type, we have used four different types of deterministic rules, which can be obtained in the RSES [2]. We used the following algorithms from RSES [4] to receive nondeterministic rules:

- Covering algorithm;
- Genetic algorithm (rules are created using genetic algorithm, with a certain number of reducts);
- LEM2 algorithm [8];
- Exhaustive algorithm (minimal rules).

Hence, we have received eight classifiers. Four classifiers using deterministic rules (*DR* - classifier) and four classifiers using deterministic and nondeterministic rules (*TNDR* - classifiers).

5   Experiments

The experiments on decision tables from the UCI Machine Learning Repository [7] using proposed *DR* and *TNDR* classification algorithms were performed.

The following data sets were selected for the experiments: Balance Scale, Ecoli, Iris, Postoperative, Primary Tumor and Zoo.

Table 2. Classification accuracy for classifiers DR and $TNDR$ for data set Balance scale

Classification factor	DR	Rules					
		DR + TNDR, α					
		1.0	0.9	0.8	0.7	0.6	0.5
Covering algorithm							
cover	60.0	94.8	100	100	100	100	100
acc	68.67	70.46	80.4	82.8	84.0	84.4	84.8
acc × cover	41.20	66.80	80.40	82.80	84.00	84.40	**84.80**
Genetic algorithm							
cover	97.6	99.2	100	100	100	100	100
acc	77.0	61.7	78.8	78.0	78.8	80.4	79.2
acc × cover	75.15	61.21	78.80	78.00	78.80	**80.40**	79.20
LEM2 algorithm							
cover	51.2	93.2	100	100	100	100	100
acc	78.1	71.2	80.8	83.6	84.8	85.6	85.2
acc × cover	39.99	66.36	80.80	83.60	84.80	**85.60**	85.20
Exhaustive algorithm							
cover	97.6	99.2	100	100	100	100	100
acc	79.5	69.0	80.8	83.6	84.8	86.8	85.6
acc × cover	77.59	68.45	80.80	83.60	84.80	**86.80**	85.60

Table 3. Classification accuracy for classifiers DR and $TNDR$ for data set Ecoli

Classification factor	DR	Rules					
		DR + TNDR, α					
		1.0	0.9	0.8	0.7	0.6	0.5
Covering algorithm							
cover	80.00	100	100	100	100	100	100
acc	56.48	22.96	20.74	23.70	26.67	21.48	21.48
acc × cover	**45.19**	22.96	20.74	23.70	26.67	21.48	21.48
Genetic algorithm							
cover	94.07	100	100	100	100	100	100
acc	51.97	51.85	53.33	53.33	52.59	52.59	52.59
acc × cover	48.89	51.85	**53.33**	**53.33**	52.59	52.59	52.59
LEM2 algorithm							
cover	38.52	100	100	100	100	100	100
acc	53.85	6.67	17.78	28.15	22.22	37.78	40.74
acc × cover	20.74	6.67	17.78	28.15	22.22	37.78	**40.74**
Exhaustive algorithm							
cover	94.07	100	100	100	100	100	100
acc	51.97	51.85	51.85	53.33	54.07	52.59	52.59
acc × cover	48.89	51.85	51.85	53.33	**54.07**	52.59	52.59

Table 4. Classification accuracy for classifiers DR and $TNDR$ for data set Iris

Classification	DR	Rules					
		DR + TNDR, α					
factor		1.0	0.9	0.8	0.7	0.6	0.5
Covering algorithm							
cover	96.67	100	100	100	100	100	100
acc	87.93	93.33	93.33	93.33	93.33	93.33	91.67
acc × cover	85.00	**93.33**	**93.33**	**93.33**	**93.33**	**93.33**	91.67
Genetic algorithm							
cover	98.33	100	100	100	100	100	100
acc	91.53	96.67	96.67	96.67	96.67	96.67	91.67
acc × cover	90.00	**96.67**	**96.67**	**96.67**	**96.67**	**96.67**	91.67
LEM2 algorithm							
cover	68.33	91.67	91.67	91.67	91.67	91.67	91.67
acc	95.12	94.55	94.55	94.55	94.55	94.55	92.73
acc × cover	65.00	**86.67**	**86.67**	**86.67**	**86.67**	**86.67**	85.00
Exhaustive algorithm							
cover	98.33	100	100	100	100	100	100
acc	91.53	96.67	96.67	96.67	96.67	96.67	91.67
acc × cover	90.00	**96.67**	**96.67**	**96.67**	**96.67**	**96.67**	91.67

In evaluation of the accuracy of classification algorithms on a decision tables (i.e., the percentage of correctly classified objects) the train and test validation method was used. The data set was split at the ratio of 60 – 40. On testing sets the arithmetic mean of accuracy and the coverage factor were calculated.

For each considered data table and for $TNDR$ classification algorithm, we used different values of parameter α.

Tab. 2 contains the results of our experiments for Balance Scale data set. For all four rule-based classifiers, by adding nondeterministic rules to the deterministic rules, we increased the quality of classification.

Tab. 3 contains the results of our experiments for Ecoli data set. For three of the four rule-based classifiers (genetic, Lem2, exhaustive algorithm), by adding nondeterministic rules to the deterministic rules, we increased the quality of classification. For the covering algorithm, we received the highest classification quality for the classifier DR, using deterministic rules.

For Iris data set (Tab. 4) and Post Operative data set (Tab. 5), for all four rule-based classifiers, by adding nondeterministic rules to the deterministic rules, we increased the quality of classification.

Tab. 6 contains the results of our experiments for Primary Tumor data set. For all four rule-based classifiers, by adding nondeterministic rules to the deterministic rules, we increased the quality of classification.

For Zoo decision table (Tab. 7), for two rule-based classifiers (covering, Lem2 rules) increased the quality of the classification after the addition of nondeterministic rules to deterministic rules. For the remaining two rule-based classifiers (genetic, exhaustive rules), the quality of the classification is not changed by adding nondeterministic rules to deterministic rules.

Table 5. Classification accuracy for classifiers DR and $TNDR$ for data set Post operative

Classification factor	DR	Rules					
		DR + TNDR, α					
		1.0	0.9	0.8	0.7	0.6	0.5
Covering algorithm							
cover	2.78	97.22	97.22	97.22	97.22	97.22	97.22
acc	100	31.43	22.86	22.86	25.71	77.14	77.14
acc × cover	2.78	30.56	22.22	22.22	25.00	**75.00**	**75.00**
Genetic algorithm							
cover	100	100	100	100	100	100	100
acc	44.44	36.11	33.33	33.33	47.22	69.44	69.44
acc × cover	44.44	36.11	33.33	33.33	47.22	**69.44**	**69.44**
LEM2 algorithm							
cover	47.22	100	100	100	100	100	100
acc	52.94	41.67	36.11	36.11	47.22	75.00	75.00
acc × cover	25.00	41.67	36.11	36.11	47.22	**75.00**	**75.00**
Exhaustive algorithm							
cover	100	100	100	100	100	100	100
acc	41.67	36.11	36.11	36.11	47.22	69.44	69.44
acc × cover	41.67	36.11	36.11	36.11	47.22	**69.44**	**69.44**

Table 6. Classification accuracy for classifiers DR and $TNDR$ for data set Primary tumor

Classification factor	DR	Rules					
		DR + TNDR, α					
		1.0	0.9	0.8	0.7	0.6	0.5
Covering algorithm							
cover	8.82	100	100	100	100	100	100
acc	66.67	40.44	54.41	54.41	54.41	55.15	55.15
acc × cover	5.88	40.44	54.41	54.41	54.41	**55.15**	**55.15**
Genetic algorithm							
cover	100	100	100	100	100	100	100
acc	52.94	49.26	51.47	51.47	51.47	55.15	55.15
acc × cover	52.94	49.26	51.47	51.47	51.47	**55.15**	**55.15**
LEM2 algorithm							
cover	47.79	100	100	100	100	100	100
acc	56.92	41.18	52.94	52.94	52.94	54.41	54.41
acc × cover	27.21	41.18	52.94	52.94	52.94	**54.41**	**54.41**
Exhaustive algorithm							
cover	100	100	100	100	100	100	100
acc	51.47	50.00	56.62	56.62	56.62	55.15	55.15
acc × cover	51.47	50.00	**56.62**	**56.62**	**56.62**	55.15	55.15

Table 7. Classification accuracy for classifiers DR and $TNDR$ for data set Zoo

Classification factor	DR	DR + TNDR, α					
		1.0	0.9	0.8	0.7	0.6	0.5
Covering algorithm							
cover	46.34	100	100	100	100	100	100
acc	100	51.22	51.22	46.34	46.34	46.34	46.34
acc × cover	46.34	**51.22**	**51.22**	46.34	46.34	46.34	46.34
Genetic algorithm							
cover	100	100	100	100	100	100	100
acc	97.56	97.56	97.56	97.56	97.56	97.56	97.56
acc × cover	**97.56**	**97.56**	97.56	97.56	97.56	97.56	97.56
LEM2 algorithm							
cover	82.93	100	100	100	100	100	100
acc	100	95.12	82.93	90.24	82.93	82.93	82.93
acc × cover	82.93	**95.12**	82.93	90.24	82.93	82.93	82.93
Exhaustive algorithm							
cover	100	100	100	100	100	100	100
acc	95.12	95.12	95.12	95.12	95.12	95.12	95.12
acc × cover	**95.12**	**95.12**	95.12	95.12	95.12	95.12	95.12

6   Conclusions

In this paper we presented the algorithm for creating truncated nondeterministic rules, which has a polynomial computational complexity due to the number of attributes and the number of objects in a decision table.

The truncated nondeterministic rules are created as the result of shorting deterministic rules in accordance with the principle of the minimum description length. As deterministic rules in decision table we treat the full objects description in a meaning of descriptors conjunction.

The truncated nondeterministic rules and deterministic rules we use to build rule-based classifiers - $TNDR$. We have combined these sets of rules into a single set of rules and we used modified standard voting procedure to resolve conflicts between rules. We compared $TNDR$ classifiers with classifiers based only on deterministic rules DR classifiers. We used four different algorithms, from $RSES$ program, to create deterministic rules: Covering algorithm, Genetic algorithm, LEM2 algorithm and the Exhaustive algorithm (minimal rules). We conducted experiments on data from the UCI ML Repository.

For all four types of deterministic rules, it was confirmed (experimental), that by adding nondeterministic rules to the deterministic rules, we increased the quality of classification. It should be pointed out that classification process requires tuning some of their parameters (α – rules confidence) for each data table. This means that the parameter α should be tuned for each data set.

References

1. Rosetta, http://www.lcb.uu.se/tools/rosetta/
2. Rough Set Exploration System, http://logic.mimuw.edu.pl/~rses/
3. Agrawal, R., Imieliński, T., Swami, A.: Mining associations rules between sets of items in massive databases. In: Buneman, P., Jajodia, S. (eds.) Proc. of the ACM-SIGMOD 1993 International Conference on Management of Data, Washington, D.C., pp. 207–216 (1993)
4. Bazan, J., Szczuka, M.S., Wojna, A., Wojnarski, M.: On the evolution of rough set exploration system. In: Tsumoto, S., Słowiński, R., Komorowski, J., Grzymała-Busse, J.W. (eds.) RSCTC 2004. LNCS (LNAI), vol. 3066, pp. 592–601. Springer, Heidelberg (2004)
5. Delimata, P., Marszał-Paszek, B., Moshkov, M., Paszek, P., Skowron, A., Suraj, Z.: Comparison of some classification algorithms based on deterministic and non-deterministic decision rules. In: Peters, J.F., Skowron, A., Słowiński, R., Lingras, P., Miao, D., Tsumoto, S. (eds.) Transactions on Rough Sets XII. LNCS, vol. 6190, pp. 90–105. Springer, Heidelberg (2010)
6. Delimata, P., Moshkov, M., Skowron, A., Suraj, Z.: Inhibitory rules in data analysis: A rough set approach. SCI, vol. 163 (2009)
7. Frank, A., Asuncion, A.: UCI machine learning repository (2010), http://archive.ics.uci.edu/ml
8. Grzymala-Busse, J.: Lers - a data mining system. In: Maimon, O., Rokach, L. (eds.) Data Mining and Knowledge Discovery Handbook, pp. 1347–1351. Springer, US (2005)
9. Holte, R.: Very simple classification rules perform well on most commonly used datasets. Mach. Learn. 11(1), 63–90 (1993)
10. Marszał-Paszek, B., Paszek, P.: Minimal templates and knowledge discovery. In: Kryszkiewicz, M., Peters, J.F., Rybiński, H., Skowron, A. (eds.) RSEISP 2007. LNCS (LNAI), vol. 4585, pp. 411–416. Springer, Heidelberg (2007)
11. Michalski, R.: http://www.mli.gmu.edu/michalski/
12. Moshkov, M., Skowron, A., Suraj, Z.: Maximal consistent extensions of information systems relative to their theories. Inf. Sci. 178(12), 2600–2620 (2008)
13. Nguyen, H.S.: Scalable classification method based on rough sets. In: Alpigini, J.J., Peters, J.F., Skowron, A., Zhong, N. (eds.) RSCTC 2002. LNCS (LNAI), vol. 2475, pp. 433–440. Springer, Heidelberg (2002)
14. Paszek, P., Marszał-Paszek, B.: Nondeterministic decision rules in classification process. In: Herrero, P., Panetto, H., Meersman, R., Dillon, T. (eds.) OTM 2012 Workshops. LNCS, vol. 7567, pp. 485–494. Springer, Heidelberg (2012)
15. Pawlak, Z.: Rough Sets. Theoretical Aspects of Reasoning about Data. Data. Kluwer Academic Publishers, Dordrecht (1991)
16. Rissanen, J.: Modeling By Shortest Data Description. Automatica 14, 465–471 (1978)
17. Simiński, R., Nowak-Brzezińska, A., Jach, T., Xięski, T.: Towards a practical approach to discover internal dependencies in rule-based knowledge bases. In: Yao, J., Ramanna, S., Wang, G., Suraj, Z. (eds.) RSKT 2011. LNCS, vol. 6954, pp. 232–237. Springer, Heidelberg (2011)
18. Skowron, A., Suraj, Z.: Rough sets and concurrency. Bulletin of the Polish Academy of Sciences 41, 237–254 (1993)
19. Triantaphyllou, E., Felici, G.: Data Mining and Knowledge Discovery Approaches Based on Rule Induction Techniques. Springer Science and Business Media (2006)
20. Tsumoto, S.: Modelling medical diagnostic rules based on rough sets. In: Polkowski, L., Skowron, A. (eds.) RSCTC 1998. LNCS (LNAI), vol. 1424, pp. 475–482. Springer, Heidelberg (1998)

Extraction of Rules Dependencies
for Optimization
of Backward Inference Algorithm

Roman Simiński

University of Silesia, Institute of Computer Science
ul. Będzińska 29, 41-200 Sosnowiec, Poland
roman.siminski@us.edu.pl

Abstract. This work presents the modification of backward inference algorithm for rule knowledge bases. Proposed algorithm extracts information of internal rules dependencies and performs only promising recursive calls. Optimization relies on reducing the number of rules searched for each run of inference and reducing the number of unnecessary recursive calls. We assume that the rule knowledge base itself contains enough information, which allow to improve the efficiency of the classic algorithms of the inference and we propose the decision units conception as tool for extracting and modeling such information. The first part of the work briefly presents backward inference algorithms in its classical version, next part of the work describes the decision units conception, then the utilization of decision units in optimization of inference algorithm is described and the modified versions of algorithm are presented. The preliminary evaluation of modified versions of algorithm finish presented work.

Keywords: knowledge base, inference, decision units.

1 Introduction

The rules are probably the most popular form of representing knowledge in the intelligent information systems. Regardless of the development of different knowledge representations – semantic networks, object-oriented representations and frame systems, probabilistic methods of knowledge representation and processing (Bayesian networks for example) [2,6] – the rule representation was still popular. Recent years have brought a renaissance of rules representation for knowledge bases, currently, the rules are considered as standard result form of data mining methods, for example, decision rules bind the values of conditional attributes with decision attribute, describing in this way the relation between attributes in the decision tables.

Rules are important and useful material for constructing knowledge bases for different types of decision support systems. Such systems solve problems using the knowledge stored in the knowledge base and inference methods. Two main methods of inference are used, forward and backward inference. Forward

S. Kozielski et al. (Eds.): BDAS 2014, CCIS 424, pp. 191–200, 2014.

inference is a top-down method which takes facts as they become available and attempts to draw conclusions which lead to extending fact base. Backward inference is a bottom-up procedure which starts with goals and queries the fact base about information which may satisfy the conditions contained in the rules. Forward inference is data-driven, it may do lots of work which is irrelevant to the query. Backward inference is goal-driven, appropriate for problem-solving. Methods of inference dedicated to the rule-bases have a long descent and are fairly well known. There were no significant changes in inference since RETE algorithm [5] and works described in [9] and [1]. All well known modifications were focused on the specific proposals for changes in rule knowledge representation, without changing the general approach to inference.

In this work we will focus on backward inference. We plan to use the rule based classifier within the system security software. The classifier will evaluate in real time the programs abnormal and suspicious behaviors using goal-driven algorithm. Due the real time nature of classifier any efficiency improvement is important and necessary. Backward inference algorithm is a recursive, recursion is used to confirm sub-goals of inference, which are the conditions of rules that are not facts currently. Unfortunately, the number of unnecessary recursive may be very high for large rule-bases. Any unnecessary recursive call takes time and slows the process of confirming the main goal of inference. In this work we present the modification of backward inference algorithm for rule knowledge bases. Proposed algorithm extracts information of internal rules dependencies and performs only promising recursive calls. Optimization relies on reducing the number of rules searched for each run of inference and reducing the number of unnecessary recursive calls. We assume that the rule knowledge base itself contains enough information, which allow to improve the efficiency of the classic algorithms of the inference and we propose the decision units conception as tool for extracting and modeling such information.

The first part of the work briefly presents inference algorithms in their classical versions. The next part of the work describes the decision units conception, then the utilization of decision units in optimization of inference algorithms is described and the modified versions of these algorithms are presented. The preliminary evaluation of modified versions of algorithms finish presented work.

2   Inference Algorithms in Classical Versions

The structure of a rule-based system contains mainly a fact base, called the working memory of the system where the known facts at different moments are stored, a rule base which contains the rules used to infer new facts, and an inference engine which selects some applicable rules to infer new facts. When rules are examined by the inference engine, new fact are added to fact base if the current content of fact base satisfies the conditions in the rules. Both types of inference – forward and backward – reveal their weakness when they operate on large rule-bases. Backward inference occurs at query time and starts from particular query. It attempts to prove mechanisms for how the original query would be true in terms of something else and hopefully you can chain these conditional proofs back until you eventually hit something which already is true in your knowledge base and stop the backward search. Backward inference algorithm is a recursive, recursion is used to confirm sub-goals of inference, which are the conditions of rules that are not facts currently. Unfortunately, the number of unnecessary recursive may be very high for large rule-bases. Any unnecessary recursive call takes time and slows the process of confirming the main goal of inference.

The classical versions of the algorithms have been repeatedly published, for example in [2,6], for this reason they are not reproduced in this work more detailed. In this section we present only simple example of the backward inference. A very simple rules set is shown on Fig. 1. We use rules containing literals in the form of the attribute-value pair, more formal specification of the literals representation will be presented in the next section.

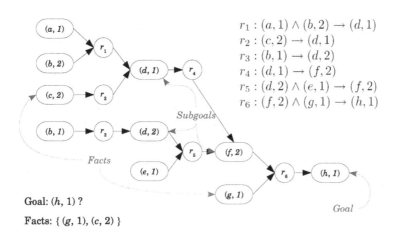

$$r_1 : (a, 1) \wedge (b, 2) \rightarrow (d, 1)$$
$$r_2 : (c, 2) \rightarrow (d, 1)$$
$$r_3 : (b, 1) \rightarrow (d, 2)$$
$$r_4 : (d, 1) \rightarrow (f, 2)$$
$$r_5 : (d, 2) \wedge (e, 1) \rightarrow (f, 2)$$
$$r_6 : (f, 2) \wedge (g, 1) \rightarrow (h, 1)$$

Goal: $(h, 1)$?

Facts: $\{ (g, 1), (c, 2) \}$

Fig. 1. An example knowledge base – a structure for backward inference

According to the model shown by Fig. 1, the main goal of the inference described by literal $(h, 1)$ does not belong to the fact set. The algorithm selects the rule set containing rules, which posses conclusion relevant to the goal $(h, 1)$. In our example this is $\{r_6\}$, now it is necessary to confirm the truth of the each condition of the selected rule, in our example r_6. First each condition is checked against current fact set. In our example second condition $(g, 1)$ of rule r_6 is a fact, but first condition $(f, 2)$ is not in the fact set. A backward inference algorithm will try to confirm whether $(f, 2)$ is a fact. Now $(f, 2)$ becomes a new subgoal of inference and backward inference algorithm runs itself recursively for it (inference will start backwards for $(f, 2)$ as goal recursively). If recursive function call result is equal *true*, all conditions of rule r_6 are true. Recursive call must be repeated for condition $(d, 1)$ of rule r_4. Finally, goal $(h, 1)$ is confirmed by rule chain: $r_2 \rightarrow r_4 \rightarrow r_6$, which use the input facts $\{(g, 1), (c, 2)\}$ and the new facts $\{(f, 2), (d, 1)\}$, confirmed by recursive function calls. The classic version of the algorithm doesn't known whether the call is promising – ie, it is unknown whether there is a rule that fits the new goal of inference. Indeed, in real cases very often there is no such rule and the recursive call is unnecessary. The aim of this work is to define a modified algorithm, the proposed algorithms allow to reduce the number of unnecessary recursive call in the case of backward inference algorithm, owing to selected properties of the decision units model of the rule-base.

3   Decision Units

Decision units originally came into existence as a tool for global and local rule knowledge base verification. This approach is devoted to knowledge bases, in which there are rules that probably create deep inference path. Decision units can be considered as a global model of dependencies occurring in knowledge base. It allows to *retrieve the decision model*, hidden in, potentially numerous, set of rules. Decision units conception assumes, that we use backward chaining inference, so naturally they can be used to optimize this kind of inference.

Decision units conception was presented in our previous work e.g. [8]. For this reason we present only short description of issues relevant to the main goal of this work – the optimization of inference processes in rule-based knowledge bases. The decision units idea allow us to divide a set of rules into subsets – we group all rules with the same conclusion attribute, hence single decision unit contains the set of rules with the same attribute in the conclusion part. Decision units conception assumes, that knowledge base contains rules with subgoals – which means that many literals appearing in the conclusion parts of the rules also appearing in the conditional parts of the rules. Those connections between conclusion and conditional literals are typically hidden in rule knowledge base and only during inference we discover how sometimes deep are inference chains. However in many cases rule knowledge base are "flat" – sometimes numerous set of rules contains only one, this same attribute in the conclusion of the rules. This situation is typical for rule bases generated during data mining process. Such knowledge base generate only one decision unit, containing all rules. In such situations decision units approach can be of course useful in limited scope, but many of the decision units interesting properties can not be applied.

3.1   Basic Notations

We shall introduce conception of elementary decision units dedicated for a rule base containing the *Horn* clause rules, where literals are coded using *attribute-value* pairs. Let $\Re$ is the rule knowledge base $\Re = \{r_1, r_2, \ldots r_m\}$ containing m rules, where each rule looks as follows: $r_i : l_{i1} \wedge l_{i2} \wedge \ldots \wedge l_{in} \to c_i$, where: l_{ij} – the j-th conditional literal of r_i rule, c_i – the conclusion literal of r_i rule. Let A is a non-empty finite set of conditional and decision (conclusion) attributes and for every $a_i \in A$ the set V_{ai} is called the value set of a_i, respectively $A = \{a_1, a_2, \ldots a_i, \ldots a_n\}$ and $V_{a_i} = \{v_1^{a_i}, v_2^{a_i}, \ldots, v_k^{a_i}\}$. As noted earlier each attribute in $a \in A$ may be conditional and/or decision attribute – a conclusion of particular rule r_i can be a condition in other rule r_j, it means that rule r_i and r_j are connected and it is possible that inference chain occurs. Common attribute in conclusion and conditions of two different rules indicates connection between rules – this remark will be important in the further part of this chapter.

Let literals of the rules from $\Re$ becomes (a, v_i^a) where $a \in A$ and $v_i^a \in V_a$ and let notations (a, v_i^a) and $a = v_i^a$ are equivalent. For clarity of presentation we will sometimes use small letter $a, b, c, \ldots$ for attributes and $1, 2, \ldots$ for values of attributes. The conception of decision units will be illustrated with an example knowledge base that contains six rules, presented in Fig. 1. We will also consider also for each rule r four functions: $conclAttr(r)$ – a conclusion attribute of rule r is the value of this function, $concl(r)$ – a conclusion (attribute-value pair) of rule r is value of this function, $condAttribs(r)$ – a set of attributes from conditional part of rule r is value of this function, $cond(r)$ – a set of attribute-value pairs from conditional part of rule r is value of this function.

An decision unit D is defined as a triple $D = (R, I, O)$. Set R is called *decision rules set* of D, contains those rules from $\Re$, which have this same attribute in the conclusion part of the rule, defined as follows (for $i, j = 1, 2, \ldots card(\Re)$ an $i \neq j$): $R = \{r \in \Re : \forall r_i, r_j \in \Re \; conclAttrib(r_i) = conclAttrib(r_j)\}$. Set I is called *input entries set* of D, contains attribute-value pairs appearing in the conditional part of the each rule $r \in R$, defined as follows: $I = \{(a, v^a) : \exists r \in R \; (a, v^a) \in cond(r)\}$. Set O is called *output entries set* of D, contains attribute-value pairs appearing in the conclusion part of the each rule $r \in R$, defined as follows: $O = \{(a, v^a) : \exists r \in R \; (a, v^a) = concl(r)\}$. Decision unit D contains the set of rules R, each rule $r \in R$ contains the same attribute in the literal appearing in the conclusion part – as shown by Fig. 2. All rules grouped within a decision unit take part in an inference process confirming the aim described by attribute, which appears in the conclusion part of each rule. All pairs attribute-value appearing in the conditional part of each rule are called decision unit *input entries*, while all pairs appearing in the conclusion part of each set rule R are called decision *unit output entries*.

3.2   The Connections among Decision Units

The modified algorithm proposed in this work extracts information of internal rules dependencies and performs only promising recursive calls. Such information are extracted from decision units. Analyzing of relations between decision units requires comparisons of input and output entries. Important role in the modification of inference algorithms plays the information obtained from the set of *connected input entries* and the set of *connected output entries*, defined for each decision unit. The set of connected input entries $IC \subseteq I$ contains attribute-value pairs appearing also in at least one rule's $(r \in \Re)$ conclusion: $IC = \{(a, v^a) \in I : \exists r \in \Re \; (a, v^a) = concl(r)\}$. The set of connected output entries $OC \subseteq O$ contains attribute-value pairs appearing also in at least one rule's $(r \in \Re)$ condition: $OC = \{(a, v^a) \in O : \exists r \in \Re \; (a, v^a) \in cond(r)\}$. The

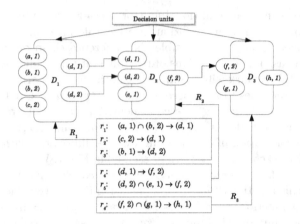

Fig. 2. Decision units – rules subsets

Fig. 3 shows the interpretations of sets IC and OC, those sets will be specially useful for modification of backward inference algorithms.

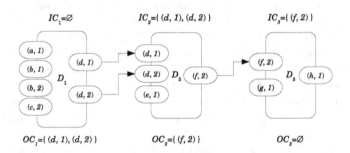

Fig. 3. Connected input and output entries

4   Modified Backward Inference Algorithm

Each attributive rule knowledge base can be presented in the form of decision units net[1]. Modification of classical inference algorithm will be based on particular information obtained from decision units net. This representation can be created with linear time complexity once. Recreation of decision units information is required when the content of rule base changes. The modification of the backward inference algorithm consists on the initiation only promising recursive calls of classical backward algorithm. The decision units net provides information which allow a preliminary assessment of whether the call could potentially

---

[1] Decision units are dedicated for rules with literals in the form of attribute-value pairs. It is possible to adapt decision units specification for some other representations of literals.

confirm the nominated subgoal of inference. According to the model shown by Fig. 4 for confirmation of the goal of the inference described by literal $(h, 1)$ it is necessary to confirm the truth of the premise $(f, 2)$. Because it isn't in the fact set, a classic backward inference algorithm will try to determine that $(f, 2)$ is a fact – now it is a new subgoal of inference and backward inference algorithm runs itself recursively for it. The classic version of the algorithm doesn't known whether the call is promising – ie, it is unknown whether there is a rule that fits the new goal of inference.

In our proposed modification, the recursive call is made only if the subgoal of inference is the connected input entry in the considered decision unit – then there is a possibility of the confirmation of subgoal through the recursive call of algorithm. In the case of deep inference, where there are the big number of subgoals for confirmation, the concept of verifying the reasonableness of recursive calls can significantly improve the inference algorithm. In the example shown in Fig. 4 $(d, 1)$ and $(d, 2)$ will be the subgoals of inference. They are also appearing as the connected input entries of decision unit D_2, therefore the next recursive calls of backward inference for subgoals $(d, 1)$ and $(d, 2)$ are promising.

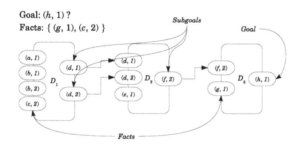

Fig. 4. Backward inference in the decision units net

Below we present the backward inference algorithm, which use the information from decision units net described above. For each decision unit $d \in DS$ we will use following functions:

- $R(d)$ – the set of *rules* R of decision unit d is the result of the function,
- $IC(d)$ – the set of *connected input entries* IC of decision unit d is the result of the function, $IC(d) \subseteq I(d)$;

Input data for forward inference algorithm: $\mathcal{D}$ – the set of decision units: $\mathcal{D} = \{D_1, D_2, \ldots, D_{dn}\}$, F – the set of facts: $F = \{f_1, f_2, \ldots, f_{fn}\}$, g – the goal of inference.

The output data of the algorithm: F – the set of facts with potential new facts obtained through inference, function result as boolean value, *true* if the goal g is in the set of facts: $g \in F$, *false* otherwise.

The algorithm uses the working variables: d – the working decision unit, $A \subseteq R(d)$ – the set of rules already activated, $r \in R(d)$ – the rule currently considered;

truePremise – boolean variable, *true* if currently considered premise of rule r is true, *false* otherwise, w – currently considered premise's condition of rule r.

function *backwardInferenceDU*($\mathcal{D}$, g, **var** F) : **bool**
begin
  if $g \in F$ **do**
    return true
  else
    $A \leftarrow \emptyset$
    select $d \in \mathcal{D}$ **where** $g \in O(d)$
    truePremise $\leftarrow$ **false**
    while $\neg truePremise \wedge \{R(d) - A\} \neq \emptyset$ **do**
      select $r \in \{R(d) - A\}$ **using current selection strategy**
      forall $w \in cond(r)$ **do**
        truePremise $\leftarrow$ ($w \in F$)
        if $\neg truePremise \wedge w \in IC(d)$ **then**
          truePremise $\leftarrow$ *backwardInferenceDU*($\mathcal{D}$, w, F)
          if $\neg truePremise$ **then**
            truePremise $\leftarrow$ *environmentConfirmsFact*(w)
            if $\neg truePremise$ **then**
              break
            endif
          endif
        endif
      endfor
      if $\neg truePremise$ **then**
        $A = A \cup \{r\}$
      endif
    endwhile
  endif
  if *truePremise* **then**
    $F = F \cup \{g\}$
  endif
  return *truePremise*
end

The introduced algorithm realizes the inference for the single goal of backward inference g. This algorithm can be performed repeatedly in the case when there is many goals (when there is the set of goals G similarly to the forward inference) or when the usage of variables in the literals appearing in the condition and conclusions of the rules causes the necessity of backtracking. This version of the algorithm will not be introduced due to the limited frames of the this study and such algorithms will be proposed in future work.

Proposed modification of backward inference algorithm takes into account information from decision units to reduce the number of unnecessary recursive calls and the number of rules searched for each run of inference. Only promising decision units are selected for further processing (**select** $d \in D$ **where** $g \in O(d)$), and only selected subset of the whole rules ($R(d) - A$) is processed in each iteration, finally only promising recursive calls are made ($w \in IC(d)$). In each iteration the set $R(d)$ simply contains proper rules set matching to the goal currently considered. In fact, it completely eliminates the searching for rules with conclusion matching to the inference goal – it is necessary in the classical version of backward inference algorithm. It is not necessary to search within the whole set of rules $\Re$, this information is simply stored in the decision-units and does not have to be generated.

5   Evaluation of the Algorithm and Conclusions

The complexity of decision units building process is $O(n)$ – we need scan rules list only once. We may think about decision units as a *goal-oriented indexing structure* for rule base. Additional memory occupation for data structures isn't very high too. For n rules and m decision units we need approximately $is * n + ps * m$ bytes of additional memory for data structures (is – size of unsigned short integer, ps – size of pointer). Therefore for $n = 1000$ rules, $m = 100$ decision units we need approximately $2.5KB$. Precise amount of memory depends on used programming language and designated system platform. Therefore for $n = 1000$ rules, $m = 100$ decision units, pure ANSI C, $is = 2$, $ps = 4$ we need approximately $2.5KB$. The number of decision units m depends on the number of different attribute-value pairs appearing in the conclusion part of the rules k: $1 <= m <= n$. Proposed algorithm has been tested on artificial knowledge bases prepared for tests and on two[2] real-world knowledge bases:

1. Knowledeg base *media.kb*:
    - the number of rules: 135, the number of decision units: 2, the number of rules in the units 1: 21, 2: 114,
    - the number searched rules, optimistic case: 16% of rules against classical algorithm, the number searched rules, pessimistic case: 85% of rules against classical algorithm.
2. Knowledeg base *credit.kb*:
    - the number of rules: 171, the number of decision units: 8, the number of rules in the units 1: 6, 2: 4, 3: 3, 4: 4, 5: 109, 6: 30, 7: 12, 8: 3.
    - the number searched rules, optimistic case: 2% of rules against classical algorithm, the number searched rules, pessimistic case: 74% of rules against classical algorithm.

Decision units are not originally built to optimize the inference. They was introduced as a tool for verifying rule knowledge bases and then as a tool for knowledge base modeling [8]. The working hypothesis of this study was that the decision units can improve the efficiency of classical algorithms of inference, through the use of selected properties of the decision units model. The proposed modification of the backward inference algorithm allow us to search only few percentages of all rules during the inference process. Proposed algorithm extracts information of internal rules dependencies and performs only promising recursive

---

[2] It is very difficult to find a knowledge base with deep chains of inference, typical knowledge bases from repositories are unfortunately flat.

calls. We introduce decision units as the source of the such information. Decision units also allow us to reduce the number of rules searched for each run of inference. Only promising decision units are selected for further processing and only selected subset of the whole rules is processed in each iteration, finally only promising recursive calls are made. The decision units are not only a structure for improving the efficiency of inference. The issues presented in this paper are a fragment of the broader work [3,7,4], including rule knowledge base modeling, static and dynamic verification and nontrivial problem of visualization and visual manipulation of the contents of such knowledge bases.

Acknowledgments. This work was partially supported by the Polish National Science Center grant $2011/03/D/ST6/03027$.

References

1. Chandru, V., Hooker, J.: Optimization Methods for Logical Inference. John Wiley & Sons (2011)
2. Luger, G.: Artificial Intelligence. Addison Wesley, England (2000)
3. Nowak, A., Siminski, R., Wakulicz-Deja, A.: Two-way optimizations of inference for rule knowledge bases. In: Proceedings of International Conference CS&P 2008, Concurrency, Specification and Programming, pp. 398–409 (2009)
4. Nowak-Brzezińska, A., Simiński, R.: Knowledge mining approach for optimization of inference processes in rule knowledge bases. In: Herrero, P., Panetto, H., Meersman, R., Dillon, T. (eds.) OTM 2012 Workshops. LNCS, vol. 7567, pp. 534–537. Springer, Heidelberg (2012),
   http://dx.doi.org/10.1007/978-3-642-33618-8_70
5. Online information: Reasoning About RETE (2001), http://www.haley.com
6. Russell, S., Norvig, P.: Artificial Intelligence: A Modern Approach, 2nd edn. Prentice Hall (2003)
7. Simiński, R., Nowak-Brzezińska, A., Jach, T., Xięski, T.: Towards a practical approach to discover internal dependencies in rule-based knowledge bases. In: Yao, J., Ramanna, S., Wang, G., Suraj, Z. (eds.) RSKT 2011. LNCS, vol. 6954, pp. 232–237. Springer, Heidelberg (2011),
   http://dx.doi.org/10.1007/978-3-642-24425-4_32
8. Siminski, R., Wakulicz-Deja, A.: Application of decision units in knowledge engineering. In: Tsumoto, S., Słowiński, R., Komorowski, J., Grzymała-Busse, J.W. (eds.) RSCTC 2004. LNCS (LNAI), vol. 3066, pp. 721–726. Springer, Heidelberg (2004),
   http://dx.doi.org/10.1007/978-3-540-25929-9_91
9. Smith, D.E., Genesereth, M.R., Ginsberg, M.L.: Controlling recursive inference. Artificial Intelligence 30(3), 343–389 (1986),
   http://www.sciencedirect.com/science/article/pii/0004370286900032

The Incompleteness Factor Method as a Support of Inference in Decision Support Systems

Agnieszka Nowak-Brzezińska and Tomasz Jach

University of Silesia, Institute of Computer Science
ul. Będzińska 29, 41-200 Sosnowiec, Poland
{agnieszka.nowak,tomasz.jach}@us.edu.pl

Abstract. The authors propose the incompleteness factor (IF) method to improve the effectiveness of browsing in knowledge bases with missing data. The paper explains the whole method, which is based on certainty factors and cluster analysis. The experiments' results conducted to obtain optimal parameters for the algorithm are presented. The evaluation is made by using recall, precision, F-measure and other factors.

Keywords: rule knowledge bases, inference, cluster analysis, data mining.

1 Introduction

Decision support systems (DSS) are a vital part of modern computer science. In order to properly support the users, those systems have to be properly optimised. The classical DSS [20] system consists of the knowledge base (usually the flat set of "IF ... THEN..." clauses) and inference algorithms which give new knowledge after the inference process. One of the most time-consuming task is the process of searching for the relevant rules[1]. Those are to be activated in order to provide the user with previously not known knowledge.

The way to improve the inference process is to change the structure of knowledge base from a flat to a hierarchical one using the cluster analysis. The proposed approach was presented in previous works by the authors [9,14,19,8,13,7,12,18]. The scope of this paper is to summarize previous works and provide additional computational experiments to verify the optimal parameters for the clustering algorithm applied to the set of rules. All the experiments presented here is novel and not presented before. The authors have chosen to present the impact of various clustering parameters on the validity of clusters. As it was shown in previous experiments [12,13], the better quality of clusters grants the better overall quality of DSS.

The change of the structure of the knowledge base allows to optimise the search for the relevant rules by using the method of the most promising path [13,7] along with the modifications proposed by the authors in previous works

---

[1] Rules, which have all the premises satisfied by the facts set.

S. Kozielski et al. (Eds.): BDAS 2014, CCIS 424, pp. 201–210, 2014.
© Springer International Publishing Switzerland 2014

[14,19]. Furthermore, when there is uncertainty and vagueness involved, the clustering process can provide the information on the not fully, but most relevant rules from the knowledge base, where the classical inference would lead to no results. The method shown in this paper proposes the activation of rules, which have not all the premises satisfied by the facts set. The new knowledge provided by these rules will be marked as not fully certain by using the method of the incompleteness factor (IF) which was introduced in previous works [14,19].

It is quite often, when the deployed decision support systems suffers from the incomplete input data needed to make the inference process possible. The proposed method of IF factor can lead to an improvement in this matter and is the main goal achieved by the works presented in authors' previous papers [14,19]. The structure of the clusters being the new form of representation of the rules set base will provide a fast way to look up the relevant rule or cluster within the hierarchical structure, even when the facts set does not consist of the sufficient number of facts to find a rule which is completely satisfied by known facts. By having the facts set, the browsing process commences by comparing the facts vector with the representants of the clusters and then choosing the most relevant one. The browsing takes place until the most relevant cluster (or rule) is found. If the cluster (rule) is fully satisfied by the facts set – it is activated with the incompleteness factor of 1 and new knowledge is used to further increase the knowledge by inference. However, if the found rule (cluster) is not fully satisfied by the facts set – it is analysed and activated when the uncertainty brought by it is below given threshold. The conclusion provided by it is added to the facts set, but with the decreased value of IF.

The proposed solution is valid for all the types of data – in particular both categorical and numeric. It is also possible to apply it to knowledge bases of various complexity: either consisting of a large number of rules or of a large number of premises within rules.

The existing approaches utilising the rough set theories, Bayesian networks and others appear to be insufficient to the problem stated in this paper. The modified cluster algorithms stated by the authors along with the IF value lead to the improvement of the inference processes in DSSs.

2   Inference with Incomplete Data

Inference is a process leading to proving (or not) the conclusions provided that the premises are true. There are mainly three types if inference: forward inference (sometimes called data driven), backward inference (goal driven) and mixed inference. The proposed system at it's current state is designed to support the first type. When the fact's set is not sufficient to activate even one rule, the user is faced with the deadlock situation, when no new knowledge can be obtained. The proposed solution provides the method to obtain new, potentially useful knowledge with the computed degree of belief by using the incompleteness factor (IF) method.

Forward inference algorithm [6] is relatively easy. By having the knowledge base and facts set all the rules are analysed if all the premises of a particular rule are satisfied by the current facts set. The order of analysing and activating the rules is given by the inference strategies which were discussed in authors' previous works [14,19]. If the relevant rule is found, it is activated and it's conclusion is added to the facts set. The process is repeated until either all the rules are activated, there are no more rule which can be activated or the hypothesis provided at the beginning of the algorithm is proven.

It is commonly thought, that when there are more than 100 decision rules in knowledge base, the search for the relevant rules can consume more than 90% of the time which application is working[3].

During the first years of decision support systems development, the necessity of effective and fast matching of the rules was spotted. The literature gives mainly four algorithms which lead to improvement of this process.

The first one, RETE algorithm, was proposed by Forgy in 1979 [4]. It builds the network of nodes, which each and every one except the root matches the pattern (or the set of attribute and value pairs) occurring in the premises part of a rule from knowledge base. Each path from root to the leaf is associated with a particular rule. Each node provides the list of facts which match the pattern stored within. By visiting the nodes and marking them as satisfied the inference is made. When the whole path of a particular rule is visited, the rule can be activated and the coclusion of it can be added to the facts set, potentially providing more visited paths.

The building of RETE network is a very time consuming task, therefore TREAT algorithm [11] was developed. The main improvement consists of ignoring some of the intermediate nodes decreasing the memory complexity of the algorithm. The same principle applies to GATOR algorithm [5], where the number of paths coming to intermediate nodes is decreased.

Another optimisation of the inference algorithms is the LEAPS [1] algorithm. The improvement of effectiveness is obtained by adding the extra data structure of pointers to the rules which are possible to activate in a given iteration. After adding new facts, the algorithm decides whether to resume the previously interrupted process or to start all over again.

None of these algorithms can deal with the incompleteness of the facts set. Both RETE and TREAT will not be able to provide any results when there is no rule which has all the premises satisfied. The additional time taken by building RETE network will be consumed without any proper reason. Furthermore, the network is very inefficient when the user request to know which facts should be added to be able to activate at least one rule. The same goes for LEAPS, where rules which have most premises set are pushed away out of scope of the algorithm. As previously, user is not provided with the information about which premisses should be set in order to activated at least one rule.

On the other hand, the proposed approach also modifies the structure of knowledge base. It uses the cluster analysis to cluster similar rules. The hierarchical structure (dendrogram) allows fast browsing through the knowledge base, as well as the ability to find rules which are the most relevant given current facts set.

3 CLSearch - The Proposed Solution

Proposed solution is implemented in CLSearch system based on SMART system by Salton [15]. It was used to cluster textual documents by using the similarity between them. As well as in the proposed solution, the browsing speed was vastly improved because of using the representatives of the clusters and the hierarchical structure of clusters.

The proposed approach uses the *AHC* algorithm [10] to create hierarchical structure of clusters (dendrogram) in a place of a flat structure of knowledge base. The drawbacks, which include extra time needed to create such a structure, are compensated by the ability to browse the rules set much quicker and to give the user partial results, when there is no exact rule to be activated.

After creating the cluster, the most promising path algorithm is executed to find the most relevant rule (or cluster of rules). On each step of the algorithm, the facts set Q is compared to the representants of left and right sub-tree of the currently analysed node which is shown on fig. 1. The path which has greater value of similarity is chosen and the process commences until reaching the desired level of depth[2].

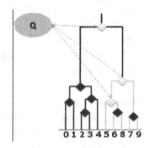

Fig. 1. Search process using the hierarchical structure of knowledge base

Having the most relevant cluster found, the rules $r_1 \ldots r_n$ belonging to it are activated if their *IF* value is above given threshold. Of course, not every premise of each rule is satisfied, that is why the IF of every rule is computed:

$$IF(r_i) = \frac{\sum IF(f_j)}{card(D_i)}; f_j \in (F \cap D_i) \tag{1}$$

where $IF(r_i)$ being the incompleteness factor of i-th rule, $IF(f_j)$ – incompleteness factor of j-th fact being the member of both the fact's set F and the set of premises of a i-th rule D_i.

The facts originally submitted by the user have the value of $IF = 1$. After activating rules with lower *IF*s, their conclusions dec_i are added to a facts set

---

[2] From the level of clusters, through smaller cluster, up to the leaves level, meaning single rule from the knowledge base.

with the computed value of IF. The members of fact set are now considered as threes:

$$f_i =< a_i, v_{a_i}, IF(f_i) > IF(f_i) \in [0 \dots 1], f_i \in F \tag{2}$$

where $IF(f_i)$ is a IF factor of fact f_i, a_i is the attribute and v_i is the value of i-th fact respectively. The authors use the term descriptor $d_j = (a_j, v_{a_j})$ to further note the attribute-value pair.

The IF factor is inspired by CF factor proposed in MYCIN [17], however IF utilises even simpler way to indicate the incompleteness of knowledge. It is also possible to apply it to fine-tune the DSS in order to obtain either potentially useful new knowledge but with some vagueness or the knowledge that is fully backed up by the facts set.

The f_{sim} function is given to properly choose the sub-tree based on the similarity between the facts set and the representatives of the left $Tree[L]$ and right $Tree[R]$ sub-tree. The function was presented by authors as follows:

Descriptor Path. given as:

$$f_{sim_d}(k, l) = card(d_k \cap d_l) \tag{3}$$

where d_l and d_k are the attribute-value pairs set of clusters l and k respectively.

Attribute Path. which can be computed as:

$$f_{sim_a}(k, l) = card(a_k \cap a_l) \tag{4}$$

where a_k and a_l are the attribute sets of k and l cluster respectively.

Hybrid Method. where both the above stated factors are combined using the weighting coefficients B_1 and B_2:

$$f_{sim_h}(k, l) = card(d_k \cap d_l) \cdot B_1 + card(a_k \cap a_l) \cdot B_2. \tag{5}$$

It is worth mentioning, that in the process of experiments the authors firstly chose the values B_1 and B_2 independent. After conduction several experiments, the authors chose to scale those coefficients so $B_1 + B_2 = 1$.

The parameters of the AHC algorithm which were optimised in these experiments were:

1. The method of computing the similarity matrix (*simpleSimilarity* – SS and *weightedSimilarity* – WS) presented in detail in [13,7].
2. The cluster joining criteria: Simple Linkage (SL), Complete Linkage (CL) and Average Linkage (AL) discussed in detail in [12,18].
3. The modification of the most promising path: Attribute Path (AP), Descriptor Path (DP), Hybrid Path with $B_1 = 0, 25$ (HPB=0,25), Hybrid Path with $B_1 = 0, 75$ (HPB=0,75).
   The full algorithm is given as follows:

Algorithm 1. The inference based on an *IF* method with DSS with incomplete knowledge

---

Data: $U = \{r_1 \ldots r_n\}$ – flat rules set; Fact set Q with *IF* values
Result: Updated facts set
begin
    /* Clustering stage   */
    Establish clustering algorithm parameters;
    Compute the similarity matrix given the distance measure;
    while *All the rules are not in one cluster* **do**
        $R_1, R_2 :=$ Two most similar rule (clusters);
        Join R_1 and R_2 into one cluster using cluster joining criterion;
        Update the similarity matrix;
    end
    repeat
        /* Browsing the rules stage   */
        $W :=$ Root of the dendrogram ;
        while *Desired depth is not reached* **do**
            $s_1 = f_{sim}(Q, Tree[L])$;
            $s_2 = f_{sim}(Q, Tree[R])$;
            if $s_1 >= s_2$ **then**
                | $W := Tree[L]$;
            else
                | $W := Tree[R]$;
            end
            /* Inference stage   */
            Compute *IF* factors of every rule belonging to cluster W;
            Activate rule R from cluster W according to a given strategy;
            Update the facts set;
        end
    until *There are no rules to be activated*;
end

---

4   The Experiments

The authors conducted experiments using the Machine Learning Repository Databasese. The chosen data sets were the base for the generation of minimal rules using LEM2 algorithm implemented in RSES system [2]. Those rules were considered input to the solution presented in this paper. One of the rules from knowledge base was randomly chosen. It's premises and conclusion were added to the facts set. The algorithm was commenced and cluster validation parameters were computed.

To easily present the cluster validity measures used here, we propose the following marks similar to the ones used in error analysis[16]:

The Tab. 1 should be understood in a following way: after commencing the clustering process, the a, b, c, d coefficients are given numerical values depending

Table 1. Error coefficient measuring

	Relevant class	Irrelevant class
Relevant cluster	a	b
Irrelevant cluster	c	d

on their labels. E. g. a is the number of objects belonging to one particular cluster matching the one defined before the clustering process. The defined cluster is assumed to be consisting only from relevant objects, that is objects which have all the premisses and conclusions satisfied by the facts set.

Now we can define the cluster quality coefficients used in this paper:

Precision. is a ratio of relevant found objects to all found objects. In the presented case, the relevant objects are rules which have all the premisses satisfied by the facts set.

Recall. is a ratio of relevant found objects to all relevant objects in the system. The relevant objects (rules) are defined in the same way as previously.

F-measure. is the combination of recall and precision with the weighting factor β:

$$F = \frac{(1 + \beta) \cdot PRECISION \cdot RECALL}{\beta \cdot PRECISION + RECALL}. \tag{6}$$

The most widely used F-measure uses $\beta = 1$ and so it is used in the context of this paper.

Rand Statistics. gives the ratio of good matches to all matches:

$$RAND = \frac{d + a}{a + b + c + d}. \tag{7}$$

Jaccard Factor. is similar to Rand Statistics, but not takes into account objects belonging to different classes matched to different clusters:

$$JACCARD = \frac{a}{a + b + c}. \tag{8}$$

Γ Hubert Statistics. gives boost to a proper matching (both proper group to matching cluster and improper group to non matching cluster):

$$\Gamma = \frac{(a + b + c + d) \cdot a - (a + b) \cdot (a + c)}{\sqrt{(a + b) \cdot (a + c) \cdot (c + d) \cdot (b + d)}}. \tag{9}$$

The experiments were performed in such a way, that the result of the system was the cluster consisting of about 10 rules. After the clustering process, the IF value of the cluster was computed and the information about finding the rule which premisses had been chosen at the beginning of the experiment was presented.

The first experiment was performed using the Lymphography database, where the second one was conducted using the Spect dataset. The results are shown in Tab. 2 and Tab. 3 respectively.

Table 2. Experiment on Lymphography database

Sim. Matrix	Cluster join	Path	Recall	Prec.	F1 score	Rand	Jaccard	Hubert	IF	Success?
SS	SL	AP	0,75	0,33	0,46	0,07	0,3	0,48	0,73	YES
SS	SL	DP	0,75	0,33	0,46	0,07	0,3	0,48	0,73	YES
SS	SL	HPB=0,25	0,75	0,33	0,46	0,07	0,3	0,48	0,73	YES
SS	SL	HPB=0,75	0,75	0,33	0,46	0,07	0,3	0,48	0,73	YES
SS	CL	AP	0,5	0,17	0,25	0,09	0,14	0,25	0,11	NO
SS	CL	DP	1	0,44	0,62	0,07	0,44	0,65	0,68	YES
SS	CL	HPB=0,25	1	0,44	0,62	0,07	0,44	0,65	0,68	YES
SS	CL	HPB=0,75	1	0,44	0,62	0,07	0,44	0,65	0,68	YES
SS	AL	AP	1	0,44	0,62	0,07	0,44	0,65	0,7	YES
SS	AL	DP	1	0,44	0,62	0,07	0,44	0,65	0,7	YES
SS	AL	HPB=0,25	1	0,44	0,62	0,07	0,44	0,65	0,7	YES
SS	AL	HPB=0,75	1	0,44	0,62	0,07	0,44	0,65	0,7	YES
WS	SL	AP	0,75	0,43	0,55	0,06	0,38	0,55	0,44	NO
WS	SL	DP	0,75	0,43	0,55	0,06	0,38	0,55	0,44	NO
WS	SL	HPB=0,25	0,75	0,43	0,55	0,06	0,38	0,55	0,44	NO
WS	SL	HPB=0,75	0,75	0,43	0,55	0,06	0,38	0,55	0,44	NO
WS	CL	AP	1	0,57	0,73	0,06	0,57	0,75	0,74	YES
WS	CL	DP	1	0,57	0,73	0,06	0,57	0,75	0,74	YES
WS	CL	HPB=0,25	1	0,57	0,73	0,06	0,57	0,75	0,74	YES
WS	CL	HPB=0,75	1	0,57	0,73	0,06	0,57	0,75	0,74	YES
WS	AL	AP	0,25	0,15	0,18	0,05	0,1	0,15	0,3	NO
WS	AL	DP	1	0,5	0,67	0,06	0,5	0,7	0,7	YES
WS	AL	HPB=0,25	1	0,5	0,67	0,06	0,5	0,7	0,7	YES
WS	AL	HPB=0,75	1	0,5	0,67	0,06	0,5	0,7	0,7	YES

Table 3. Experiment on Spect database

Sim. Matrix	Cluster join	Path	Recall	Prec.	F1 score	Rand	Jaccard	Hubert	IF	Success?
SS	SL	AP	0,75	0,25	0,375	0,06	0,23	0,41	0,43	NO
SS	SL	DP	0,75	0,25	0,375	0,06	0,23	0,41	0,43	NO
SS	SL	HPB=0,25	0,75	0,25	0,375	0,06	0,23	0,41	0,43	NO
SS	SL	HPB=0,75	0,75	0,25	0,375	0,06	0,23	0,41	0,43	NO
SS	CL	AP	0,06	0,25	0,09	0,09	0,04	0,07	0,02	NO
SS	CL	DP	0,5	0,18	0,27	0,06	0,15	0,28	0,26	NO
SS	CL	HPB=0,25	0,06	0,25	0,09	0,09	0,04	0,07	0,02	NO
SS	CL	HPB=0,75	0,5	0,18	0,27	0,06	0,15	0,28	0,26	NO
SS	AL	AP	0,25	0,14	0,18	0,03	0,1	0,16	0,18	NO
SS	AL	DP	0,5	0,13	0,2	0,08	0,11	0,22	0,43	NO
SS	AL	HPB=0,25	0,25	0,14	0,18	0,03	0,1	0,16	0,18	NO
SS	AL	HPB=0,75	0,25	0,14	0,18	0,03	0,1	0,16	0,18	NO
WS	SL	AP	0,25	0,08	0,13	0,06	0,07	0,11	0,02	NO
WS	SL	DP	0,5	0,22	0,31	0,05	0,18	0,31	0,43	YES
WS	SL	HPB=0,25	0,25	0,08	0,13	0,06	0,07	0,11	0,02	NO
WS	SL	HPB=0,75	0,5	0,22	0,31	0,05	0,18	0,31	0,43	YES
WS	CL	AP	0,75	0,43	0,55	0,04	0,38	0,56	0,46	YES
WS	CL	DP	0,75	0,43	0,55	0,04	0,38	0,56	0,46	YES
WS	CL	HPB=0,25	0,75	0,43	0,55	0,04	0,38	0,56	0,46	YES
WS	CL	HPB=0,75	0,75	0,43	0,55	0,04	0,38	0,56	0,46	YES
WS	AL	AP	0,75	0,43	0,55	0,04	0,38	0,56	0,46	YES
WS	AL	DP	0,75	0,43	0,55	0,04	0,38	0,56	0,46	YES
WS	AL	HPB=0,25	0,75	0,43	0,55	0,04	0,38	0,56	0,46	YES
WS	AL	HPB=0,75	0,75	0,43	0,55	0,04	0,38	0,56	0,46	YES

5   Conclusions

The results of the experiments confirmed the previously drawn conclusion regarding the usefulness of the proposed solution. The IF method is consistent with other quality factors. By using the proposed algorithm with optimal parameters we were able to find relevant clusters in both experiments. As it was stated in previous research, *weightedSimilarity* combined with complete linkage and the modification of the most promising path with the boost for common descriptors (HPB=0,75) provided optimal results.

The Lymphography database experiments achieved better results. This fact is connected with an overall better quality of rules generated for this database by RSES. The rules were more distinguishable than in Spect database. The results for the Spect database were often suboptimal when the algorithm has found similar, but not exactly the same results. This observation will become the base for further works.

Among all the modifications of the most promising path algorithm, the attribute path method is significantly worst. It was anticipated by the authors before the experiments were conducted. The most significant clustering parameter seems to be the cluster joining criterion. During the experiments, the impact generated by this parameter, was the greatest.

The success ratio on Lymphography experiment was above 80%, where on Spect – only 38%. However, the optimal set of parameters provided success regardless of the test. The hybrid method of the most promising path has a slight advantage when $B = 0,75$. It is also natural, when comparing the attribute and descriptor path method.

The final conclusion drawn from the experiments is that the *simpleSimilarity* is completely unusable when dealing with the clusters of rules.

The proposed solution was implemented in a PhD thesis of one of the authors and is one of the major parts of a project *"Exploration of rule knowledge bases"* founded by the Polish National Science Centre.

Acknowledgments. This work is a part of the project *"Exploration of rule knowledge bases"* founded by the Polish National Science Centre (2011/03/D/ST6/03027).

References

1. Batory, D.: The LEAPS Algorithms (1995),
   http://reports-archive.adm.cs.cmu.edu/anon/1995/CMU-CS-95-113.pdf
2. Bazan, J., Szczuka, M.: RSES and RSESlib - A collection of tools for rough set computations. In: Ziarko, W.P., Yao, Y. (eds.) RSCTC 2000. LNCS (LNAI), vol. 2005, pp. 106–113. Springer, Heidelberg (2001)
3. Forgy, C.L.: On the efficient implementation of production systems. Ph.D. thesis, Carnegie-Mellon University (1979)
4. Forgy, C.L.: Rete: A fast algorithm for the many pattern/many object pattern match problem. Artificial Intelligence (1981)

5. Hanson, E., Hasan, M.S.: Gator: An optimized discrimination network for active database rule condition testing. Tech. rep. (1993)

6. Ignizio, J.P.: An introduction to Expert Systems. McGraw-Hill (1991)

7. Jach, T.: Wnioskowanie w systemach z wiedzą niepechhuą. Systemy wspomagania decyzji. Wydawnictwo Uniwersytetu Śląskiego (2011)

8. Jach, T.: Wybrane aspekty wnioskowania w systemach z wiedzą niepełną. Systemy wspomagania decyzji. Wydawnictwo Uniwersytetu Śląskiego (2012)

9. Jach, T.: Metody wyznaczania współczynnika niepełności wiedzy w systemach z wiedzą niepełną. Systemy wspomagania decyzji. Wydawnictwo Uniwersytetu Śląskiego (2013)

10. Kaufman, L., Rousseeuw, P.J.: Finding Groups in Data: An Introduction to Cluster Analysis. Wiley (1990)

11. Miranker, D.P.: Treat: A better match algorithm for ai production systems. Tech. rep., Department of Computer Sciences, University of Texas at Austin (1987)

12. Nowak-Brzezińska, A., Jach, T.: Wnioskowanie w systemach z wiedzą niepełną. Studia Informatica, Zeszyty Naukowe Politechniki Śląskiej (2011)

13. Nowak-Brzezińska, A., Jach, T.: Wybrane aspekty wnioskowania w systemach z wiedzą niepełną. Studia Informatica, Zeszyty Naukowe Politechniki Śląskiej (2012)

14. Nowak-Brzezińska, A., Jach, T.: Metoda współczynników niepełności wiedzy w systemach wspomagania decyzji. Studia Informatica, Zeszyty Naukowe Politechniki śląskiej (2013)

15. Salton, G.: Automatic information organization and retreival. McGraw-Hill (1975)

16. Sheskin, D.: Handbook of Parametric and Nonparametric Statistical Procedures. CRC Press (2004)

17. Swinburne, R.G.: An introduction to confirmation theory. Methuen (1973)

18. Wakulicz-Deja, A., Nowak-Brzezińska, A., Jach, T.: Inference processes in decision support systems with incomplete knowledge. In: Yao, J., Ramanna, S., Wang, G., Suraj, Z. (eds.) RSKT 2011. LNCS, vol. 6954, pp. 616–625. Springer, Heidelberg (2011)

19. Nowak-Brzezińska, A., Jach, T., Wakulicz-Deja, A.: Inference processes using incomplete knowledge in decision support systems – chosen aspects. In: Yao, J., Yang, Y., Słowiński, R., Greco, S., Li, H., Mitra, S., Polkowski, L. (eds.) RSCTC 2012. LNCS, vol. 7413, pp. 150–155. Springer, Heidelberg (2012)

20. Wakulicz-Deja, A., Nowak-Brzezińska, A., Simiński, R.: Sztuczna Inteligencja - systemy ekspertowe. Instytut Informatyki UŚl., wydanie elektroniczne (2009)

Policy Clusters: Government's Agenda Across Policies and Time

Hossein Rahmani[1,2] and Christine Arnold[1]

[1] Faculty of Arts and Social Sciences, Maastricht University, The Netherlands
[2] Dept. of Knowledge Engineering, Maastricht University, The Netherlands
{hossein.rahmani,c.arnold}@maastrichtuniversity.nl

Abstract. In the last decade, Machine Learning research has developed several data analysis algorithms for real-world problems. On the other hand, analyzing the attention governments allocate to different policy areas is important since it helps us to understand the extent to which the limited resources of governments are focused or diversified. We classify the previous studies on government agenda representation into Individual and Total approaches. While the Individual approaches focus on one policy area at a time and traces the extent of attention each one received, the Total approaches propose aggregated data analysis methods to represent the government agenda considering all the policy areas. In this paper, we use hierarchical clustering to propose an intermediate type of policy analysis called "Policy Cluster" which considers the relationships among different policy areas. For the evaluation, we built and analysed the Policy Clusters for the Irish government covering the time period 1945 to 2012. Comparing to previous Individual and Total approaches, the proposed intermediate approach reduces the search space in which we are looking for informative patterns by 57% and the results of our analysis represent the political agenda in more modular and informative way, taking into account intra-relationships of policies.

Keywords: Policy Clustering, Digging into Legislative Documents, Government Representation.

1 Introduction

Today, a historically unprecedented volume of data is available in the public domain with the potential of becoming useful for researchers. More than at any other time before, political parties and governments are making data available such as speeches, legislative bills and acts. However, as the size of available data increases, the need for sophisticated methods for web-harvesting and data analysis simultaneously grows. Accordingly, in recent years, data analysis methods have become more and more of interest for political scientists. The collaboration among political and computer scientists forms an interdisciplinary field of study called "computational political science" at the interface of political science and computer science, promoting an exchange of ideas in both directions.

Analyzing the attention governments allocate to different policy areas is an important research questions for political scientist. One view in political science is that governments turn their attention to policy domains if there is a need to address pressing

S. Kozielski et al. (Eds.): BDAS 2014, CCIS 424, pp. 211–221, 2014.

problems. Attention in this view is limited by the cognitive constraints of policy makers that is their ability to process complex information and reach meaningful conclusions. Furthermore, the possibility to allocate attention to new policy areas is limited by the extent to which political institutions can respond to new challenges. A second view in political science emphasizes the importance of the preferences of the public in shaping and guiding policy change. A central tenet of representative democracy is that the provisions of public policy are related to the wishes of the citizens and policy changes are linked to changing preferences of citizens [14,3,16,20,18]. At the same time, many political commentators believe that contemporary democratic institutions in Europe are facing a severe challenge, not least due to the scope of present economic turbulence. With the imperatives of financial markets and the constraints placed on governments by policy commitments made on the supranational level with the deepening of European integration, national decision-makers may well find themselves less able to take policy positions in line with public preferences and to change national public policies in response to the demands of their citizens [19]. Therefore, measuring the degree of attention governments allocate to different policy areas is a first stept towards building a model of political representation.

The aim of this paper is to contribute to the ongoing discussions about the quality of government agenda representation in Ireland by utilizing a large dataset of legislative documents covering more than 60 years. The work discussed in this paper has been developed as part of a larger project, The Policy Votes Project, which is funded by the Dutch Political Science Association, with Christine Arnold, Mark Franklin, and Christopher Wlezien as the Principle Investigators. From this project we will use the data collected and apply advanced data mining techniques to study the changes in the clustering of policies and the extent to which EU legislation might be changing national legislation. We applied hierarchical clustering to national legislation considering the time periods before and after joining EU, respectively. Analyzing the two sets of Policy Clusters indicate that policy profile of Irish legislation changes in a way that brings the Irish government closer to the market-making logic of the EU with strong emphasis on freedom of movement, freedom of goods and service, freedom of finance.

The paper is structured in five sections. We briefly review the previous methods and categorize them into Individual and Total approaches in section 2. In section 3, we consider policies intra-relationships and discuss the details of our proposed Policy Cluster approach. Section 4 includes the empirical result of applying techniques implementing previous approaches and result of applying our Policy Cluster approach on the same dataset. Additionally, we analyse the effect of joining EU on Policy Clusters policies by considering the time periods before and after joining EU, respectively. The conclusions and future works are discussed in section 5.

2   Background

Government agenda represnetation has been widely studied by the previous methods [7,2,1,12,9]. We categorize the previous methods into Individual and Total approaches. The Individual approaches analyze policy areas seperately, tracing the extent of attention each one received [2,1,9] and the Total approaches propose aggregated data analysis methods to represent the government agenda considering all the policy areas [7,2,1,12,9]. In this section, we briefly review these approaches and discuss what techniques are typically used.

2.1 Individual Approaches: "Trend Analysis" and "Attention Allocation Analysis"

The most direct way of implementing Individual approach is to extract the total or relative government attention to each policy p through the time period $[t_1...t_2][9,2,1]$. Trend analysis uses scatter diagram in which the X axis shows time t_i and the Y axis is government's total allocated attention for policy area p. Attention Allocation Analysis uses stacked-area graph in which the total area of the graph represents the government total attention allocation and the region for each policy represents the proportion of the government's attention for that policy.

2.2 Total Approaches: "Entropy Analysis'"and "Agenda Stability Analysis"

Among the techniques implementing Total Approaches, Entropy analysis and Agenda Stability analysis have the most prominent roles. Entropy analysis measures the attention diversity of government towards high-level policy areas using Shannons H information entropy ($H = (-1)\sum_{i=1}^{n} RGA(p_i)ln(RGA(p_i))$) [8,7,1]. In our case, $RGA(p_i)$ shows the relative government's allocated attention for each policy p_i.

Agenda Stability Analysis determines whether elections and shifts of government changes policy agenda of governments [17]. The issue overlap measure proposed by Sigelman et al., [17] is one way of calculating Agenda Stability (AS). They first, caluclate the relative allocated attention for each policy area and then they apply Formula $AS_t = 100 - (\sum_{i=1}^{n} |RGA_t - RGA_{t-1}|)/2$ to calculate Agenda Stability. RGA_t and RGA_{t-1} are the relative government allocated attention to particular policy area at time t and $t - 1$, respectively.

3 Policy Cluster Approach

When one considers one policy area at a time each policy domain is treated independently of the others. The assumption that is made in studies that treat policy areas separately is that the policy domains are independent and have no effect on each other. This family of methods are mainly affected by the way the concrete pre-defined policy hierarchies such as Eurovoc [4] are defined. Policy hierarchies define main policies as concrete independent concepts with no relationships among them. But the policy-making process does not fit too well with this assumption. Decision-makers are faced with a range of problems that need to be addressed and attention allocated to one problem might reduce the attention to other problems. At the same time, some policy areas tend to be going together. If the set of policies discovered to be associated to each other, then, attention allocation for one policy area increases a need for attention allocation for its policy associators. In this study, we assume that policies are no longer independent and there might be some hidden relationships among them. We propose Policy Clusters as one possible way of discovering the hidden relationships among policies.

3.1 Use of Machine Learning Approach

To discover Policy Clusters automatically, we make benefit from the Machine Learning approaches. Machine Learning is a subfield of Artificial Intelligence in which the main goal is to learn knowledge through experience. Based on the problem definition and the type of training data (whether it is labeled or unlabeled), we focus on two high level main machine learning tasks: *Supervised learning*[10] and *Unsupervised learning*[5]. An output of supervised learner is a classifier that has the ability to predict the correct label for any valid input data while an unsupervised learner tries to infer hidden structure among unlabeled data.

In this paper, our input data is unlabeled and we do not access to any training set, therefore we should apply Unsupervised learning (Clustering) to discover hidden relationships among policy areas. We define a Policy Clustering as the task of grouping set of policies in a way that policies in the same group (called cluster) are more similar to each other comparing to those in other clusters. The outcome of policy clustering depends directly on a definition of similarity measure among the policies. Since the policy-making is a process that unfolds over time, therefore this is necessary that the proposed similarity measure has a time perspective. Here, we propose a hybrid similarity measure $HybridSim(p_i, p_j)$ (Formula 1) which considers not only the exact value of a government's attention to different policy areas but also the attention trends of government for those policies over time.

$$HybridSim(p_i, p_j) = \alpha * Sim_{EAV}(p_i, p_j) + \beta * Sim_{PT}(p_i, p_j) \tag{1}$$

In Formula 1, Sim_{EAV} and Sim_{PT} calculate the similarity value between policies p_i and p_j with respect to Exact Attention Value and Policy Trend behaviors, respectively. α and β are configurations parameters that determine the preference bias of end users toward either of exact attention value or policy trend metrics. To calculate the policies similarities according to exact attention value, first, we normalize the policy attention values and then, we apply Manhattan Similarity measure [11] ($ManhattanSim(p_i, p_j) = 1 - \frac{\sum_{y_k=t_1}^{y_k=t_2} |GA_{p_i}(y_k) - GA_{p_j}(y_k)|}{t_2 - t_1 + 1}$. $GA_{p_i}(y_k)$ shows the Government's Attention to policy p_i in $year = y_k$. We use Pearson Correlation ($corel(p_i, p_j) = \frac{cov(GA_{p_i}, GA_{p_j})}{\sqrt{var(GA_{p_i})}\sqrt{var(GA_{p_j})}}$) to calculate the similarity measure among policies with respect to policies trends.

Considering the proposed hybrid similarity measure, we have similarity value for each policy pair (p_i, p_j) rather than the description vector for each policy area. So, the only applicable clustering approach is hierarchical clustering [13]. We used "Average Linkage Strategy" as a linkage criteria in the clustering process. The difficulties regarding determining the cluster count and the initial assignment of data to clusters in some clustering algorithms do not exist anymore in this clustering algorithm.

4 Empirical Results

In this paper, we analyse the legislative documents of Irish Government for time period 1945 to 2012. We harvest the legislative documents from [6] and then, we predict 6

Table 1. High-level policies of EuroVoc concept hierarchy

Index	Category	Index	Category	Index	Category
1	LAW	2	TRADE	3	BUSINESS AND COMPETITION
4	AGRI-FOODSTUFFS	5	INDUSTRY	6	PRODUCTION, TECHNOLOGY AND RESEARCH
7	SOCIAL QUESTIONS	8	FINANCE	9	EDUCATION AND COMMUNICATIONS
10	POLITICS	11	TRANSPORT	12	INTERNATIONAL RELATIONS
13	GEOGRAPHY	14	ENERGY	15	EMPLOYMENT AND WORKING CONDITIONS
16	SCIENCE	17	ENVIRONMENT	18	AGRICULTURE, FORESTRY AND FISHERIES
19	EUROPEAN COMMUNITIES	20	ECONOMICS	21	INTERNATIONAL ORGANISATIONS

EuroVoc [4] policy areas using JEX [15] classifier. JEX is a multi-label classification software that learns from manually labelled data to automatically assign EuroVoc policies to new documents. EuroVoc policies are in hierarchical order. In this paper, we focus only on the most general level of EuroVoc policies which are shown in Tab. 1. In the following subsections, we analyse the result of applying Individual and Total approaches on Irish legislative documents and we compare these results with our proposed Policy Cluster approach.

4.1   Individual Approach

Fig. 1 shows Trend Analysis of Ireland considering three policies "Science", "Trade" and "Industry" in time period 1945 to 2012. According to Fig. 1, we could simply conclude that Trade policy seems more important to Irish government comparing to industry and science. Detailed interpretation of attention intonation for all 21 high-level policy areas is out of scope of this paper and needs comprehensive separate study.

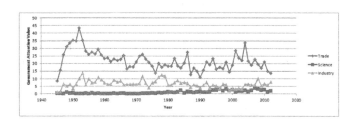

Fig. 1. Ireland attention allocation for three policies Science, Trade and Industry in time period 1945 to 2012

Fig. 2 shows Attention Allocation Analysis of Irish government in time period 1945 to 2012 considering all high-level Eurovoc categories shown in Tab. 1. The impressions one gets from Fig. 2 are 1-The rough dominance of "Agri-FoodStuffs" and "Trade" in time period 1945 to 1960. 2-The rough dominance of "Transport" in time period 1960 to 1990. 3-The rough dominance of "Agriculture, Forestry and Fisheries" in time period 1990 to 2012.

4.2   Total Approach

Fig. 3 shows the Entropy Analysis of Irish government in time period 1945 to 2012. The entropy values of Irish government become minimum and maximum in 2005 and 2008

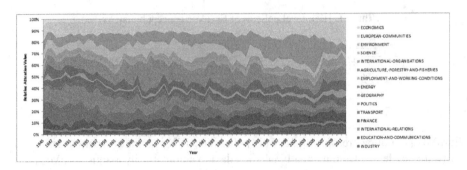

Fig. 2. Attention Allocation Analysis of Irish government in time period 1945 to 2012 considering all high-level Eurovoc categories shown in Tab. 1

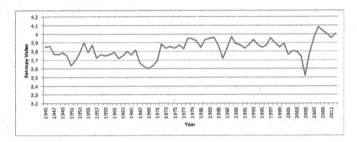

Fig. 3. Entropy Analysis of Irish governments in time period 1945 to 2012. The entropy values become minimum and maximum in 2005 and 2008, respectively.

respectively. These entropy values are in line with our Attention Allocation Analysis shown in Fig. 2. In $year = 2005$, Irish government's attention is more biased towards "Agriculture, Forestry and Fisheries" and this results in minimum entropy value. While in $year = 2008$, Irish government's attention is more evenly distributed among different policies and results in maximum entropy value for this year.

Fig. 4 shows the Agenda Stability Analysis for Irish government in time period 1945 to 2012.

4.3   Policy Cluster Approach

Using the Policy Cluster approach discussed in section 3, Fig. 5, shows the Policy Clusters of Irish government considering time period 1945 to 2012. The 9 extracted Policy Clusters are: 1-{trade, agri-foodstuffs, industry}, 2-{geography, law, international relations, environment, politics, business and comptetition}, 3-{european communities, social questions, education and communications, finance}, 4-{economics}, 5-{science, production, technology and research}, 6-{international organizations}, 7-{energy} , 8-{employment and working conditions, transport}, 9-{agriculture forestry and fisheries}. These 9 Policy Clusters provide better representation of the attention allocation of Irish government. The clusters visually provide a tool which allows us to see which policy domains go together. We can clearly observe distinct clusters, such as transport

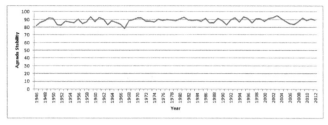

Fig. 4. Agenda Stability Analysis of Irish governments in time period 1945 to 2012. The Stability values become minimum and maximum in 1967 and 2003, respectively.

and employment and working conditions in one cluster, and transport, employment and working conditions, production, and science in a an other cluster. A third cluster can be seen in economics, finance, education and communications, social questions, and European Communities. And a fourth cluster can be seen in politics, environment, international relations, law and geography. A final cluster can be seen in Agri-Foodstuff and Trade. With these clusters and their associations, we can reduce the input space from 21 to 9 (57%Reduction). Furthermore, the clusters tells us more about the policy priorities of the government, since we can see the associations between different policy domains. These associations, have information and are more insightful than simply, for instance adding the number of legislations in each policy domain seperately.

4.4   Effect of Joining European Union on Policy Clusters

Joining the European Union has been seen by political scientists as an event which can be expected to change the attention allocation of governments and subsequently the legislative profile of a governments. In this section, we analyse the effect of joining the EU on Policy Clusters policies by considering the time periods before and after joining EU, respectively. Comparing the two sets of Policy Clusters provide some intuition about the effect of joining EU on Policy Clusters.

Ireland became a member of the EU in 1973. In order to capture any possible effect of this event, we build Policy Clusters before and after 1973. The results are shown in Fig. 6(a) and 6(b).

The first impression from comparing the Irish attentiion allocation before joining the EU and after is is that after joining EU, the attention allocation strategy of the Irish government become more even across the different policies. This is in congruous with the Irish entropy analysis discussed above, which shows that the average entropy values increases after joining the EU. The second impression from comparing Fig. 6(a) and 6(b) is that before joining EU, the attention allocation of the Irish government was more diversified. On the third level of clustering we have three key clusters, each with a rich sub-division of clusters. After joining the EU, Irish legislation clusters in two main branches: one representing the core legislation that continues to matter on the national level, such as finance, social questions, education, etc and a one representing the new dimension of policies that matter on the EU level, such as agri-foodstuff, trade, industry and employment and working conditions. Clearly, the attention the Irish government

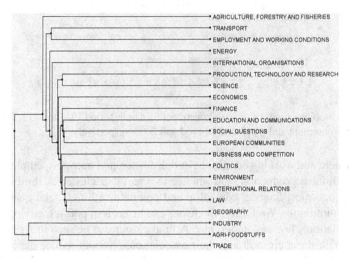

Fig. 5. Policy Clusters of Irish government considering the time period 1945 to 2012

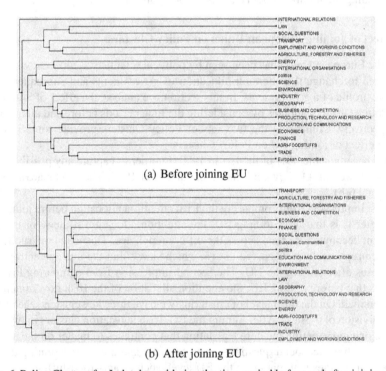

(a) Before joining EU

(b) After joining EU

Fig. 6. Policy Clusters for Ireland considering the time period before and after joining EU

allocates to all policy areas together has changed since joining the EU. Before joining the EU the different policy areas were clustering in a larger number of distinct clusters. We can observe three distinct second-tier nodes. After joinging the EU most policies now cluster in two groups. One cluster which covers policy domains such as "Business and Competition", "Economics", and "Finance", but also interestingly enough policy domains such as "Social Questions", "European Communities", and "Environment". This cluster fits with our understanding of a European Union which is seen as being strong in market-making policies and in which Social policies go together with questions of freedom of movement and mobility rights.

A third interesting cluster which emerges in Irish legislation after having joined the EU is a cluster of policies such as "Agri-Foodstuff", "Trade", and "Industry" and "Employment and Working Conditions." Again, this cluster shows a changing policy profile of Irish legislation which brings the Irish government closer to the market-making logic of the EU and strong emphasis on freedom of movement, freedom of goods and service, freedom of finance. Already in the snapshot captured in this figure can we get a sense of the changing profile of the legislative output of the Irish government since joining the EU. This changing profice we would not be able to see if we would only examine each policy domain seperately.

5   Conclusions

To our knowledge, our proposed Policy Cluster approach is the first study that considers the hidden relationships among different policy areas, over time and by taking into account the volume of legislation developed at the level of the European Union. Most of the previous methods analyze the policy areas in one of the two extreme Individual or Total approaches. In individual approaches, they focus on one policy area at a time and traces the extent of attention each one received and in Total approaches they represent the overall agenda with the aggregate value considering all the policy areas. We believe that the previous methods suffer from neglecting the existence of informative hidden relationships among different policies. To discover the informative policy intra-relationships, we proposed Policy Cluster as a intermediate, more nuanced, type of policy analysis approach. To detect Policy Clusters, first, we introduced a hybrid similarity measure which capable of not only considering the governments policies attention values but also the attention trends of governments for those policies as well. Second, according to the proposed similarity measure, we found hierarchical clustering as the most suitable Machine Learning approach to discover the hidden relationships among different policy areas. We built Policy Clusters for Ireland and we concluded that Policy Clusters approach reduces the search space in which we are looking for a informative patterns and represents the political agenda in more modular and informative way, comparing to previous Individual and Total approaches. We also analysed the effect of joining the EU on Policy Clusters policies by considering the time periods before and after joining EU, respectively. Comparing the two sets of Policy Clusters indicate that

1-Attention allocation strategy of the Irish government become more even across the different policies. 2-Policy profile of Irish legislation changes in a way that brings the

Irish government closer to the market-making logic of the EU with strong emphasis on freedom of movement, freedom of goods and service, freedom of finance.

In general, the results of our proposed Policy Clusters approach are in line with previous Individual and Total approaches with capacity of discovering new complementary knowledge which has been not considered before by those approaches. As a future work, we could apply Policy Cluster approach to national legislation of other members of European Union and discover the direction and magnitude of EU's effect on national legislation of each member.

References

1. Alexandrova, P., Carammia, M., Timmermans, A.: Policy punctuations and issue diversity on the European council agenda. Policy Studies Journal 40 (2012),
   http://onlinelibrary.wiley.com/doi/10.1111/j.1541-0072.2011.00434.x/pdf
2. Breeman, G., Lowery, D., Poppelaars, C., Resodihardjo, S.L., Timmermans, A., de Vries, J.: Political Attention in a Coalition System: Analysing Queen's Speeches in the Netherlands 1945-2007. Acta Politica 44(1), 1–27+ (2009),
   http://dx.doi.org/10.1057/ap.2008.16
3. Dahl, R.A.: Polyarchy, Participation and Opposition. Yale University Press, New Haven (1971)
4. Eurovoc: The eu's multilingual thesaurus (June 2013), http://eurovoc.europa.eu/
5. Grira, N., Crucianu, M., Boujemaa, N.: Unsupervised and semi-supervised clustering: a brief survey. In: A Review of Machine Learning Techniques for Processing Multimedia Content, Report of the MUSCLE European Network of Excellence (FP6) (2004)
6. Ireland (June 2013), http://www.irishstatutebook.ie/statutory.html
7. Jennings, W., Bevan, S., Timmermans, A., Breeman, G., Brouard, S., Chaqués-Bonafont, L., Green-Pedersen, C., John, P., Mortensen, P.B., Palau, A.M.: Effects of the core functions of government on the diversity of executive agendas. Comparative Political Studies 44(8), 1001–1030 (2011), http://eprints.soton.ac.uk/336602/
8. John, P., Jennings, W.: Punctuations and turning points in British politics: the policy agenda of the queen's speech, 1940-2005. British Journal of Political Science 40(3), 561–586 (2010), http://eprints.soton.ac.uk/336583/
9. Jones, B.D., Baumgartner, F.R.: Representation and Agenda Setting. Policy Studies Journal 32(1), 1–24 (2004), http://dx.doi.org/10.1111/j.0190-292x.2004.00050.x
10. Kotsiantis, S.B.: Supervised machine learning: A review of classification techniques. In: Proceedings of the 2007 Conference on Emerging Artificial Intelligence Applications in Computer Engineering: Real Word AI Systems with Applications in eHealth, HCI, Information Retrieval and Pervasive Technologies, pp. 3–24. IOS Press, Amsterdam (2007), http://dl.acm.org/citation.cfm?id=1566770.1566773
11. Manhattan: Manhattan distance - catalog of similarity measures for ontology mappings wiki atom (August 2013), http://simeon.wikia.com/wiki/Manhattan_distance
12. Mortensen, P.B., Green-Pedersen, C., Breeman, G., Chaqués-Bonafont, L., Jennings, W., John, P., Palau, A.M., Timmermans, A.: Comparing government agendas: executive speeches in the Netherlands, United Kingdom and Denmark. Comparative Political Studies 44(8), 973–1000 (2011), http://eprints.soton.ac.uk/336603/
13. Murtagh, F., Contreras, P.: Methods of hierarchical clustering. CoRR abs/1105.0121 (2011)
14. Pitkin, H.F.: The Concept of Representation. University of California Press, Berkeley (1967)

15. Pouliquen, B., Steinberger, R., Ignat, C.: Automatic annotation of multilingual text collections with a conceptual thesaurus. In: Proceedings of the Workshop Ontologies and Information Extraction at EUROLAN 2003, pp. 9–28 (2003)

16. Powell, B.G.: Elections as instruments of democracy: majoritarian and proportional visions. Yale Univeristy Press, New Haven (2000)

17. Sigelman, L., Buell, E.H.: Avoidance or engagement? issue convergence in U.S. presidential campaigns, 1960-2000. American Journal of Political Science 48(4), 650–661 (2004), http://dx.doi.org/10.1111/j.0092-5853.2004.00093.x

18. Soroka, S.N., Wlezien, C.: Degrees of Democracy: Politics, Public Opinion, and Policy. Cambridge University Press, Cambridge (2010)

19. Streeck, W., Mertens, D.: An index of fiscal democracy. MPIfG Working Paper 10/3, Max Planck Institute for the Study of Societies (2010),   •
http://ideas.repec.org/p/zbw/mpifgw/103.html

20. Thomassen, J.J., Schmitt, H.: Policy representation. European Journal of Political Research 32(2), 165–184 (2003)

A Novel Clustering Approach:
Simple Swarm Clustering

Seyed Ghasem RazaviZadegan[1] and Seyed Mohammad RazaviZadegan[2]

[1] Departmentof Software, South Pars Gas Complex (SPGC), Asalouyeh, Iran
rzv.ksm@gmail.com
[2] Department of Health Information Management, School of Health Management
and Information Sciences, Tehran University of Medical Sciences, Tehran, Iran
Razavi_sm@yahoo.com

Abstract. Clustering categorizes data into meaningful groups without
any prior knowledge. This paper presents a novel swarm-base clustering algorithm inspired from flock movement. Many algorithms solve the
problem by optimizing a cost function but ours clusters data by applying
one rule on data agent movements. We demonstrated that not only this
simple rule is sufficient but completely effective in accurately dividing
the data into natural clusters. It is a good model of how simply nature
solves complex problems. Unlike some algorithms, this one does not need
number of desired cluster in advance and discovers it by itself correctly.
Eight data sets were used to compare the algorithm with five well-known
algorithms. K-means and k-harmonic fail to find none-Gaussian clusters
and two other swarm-base algorithms suffer severely from performance
but our algorithm works successfully in both cases. The result confirms
the superiority of our method.

Keywords: clustering, swarm intelligence, flock algorithm.

1 Introduction

Clustering analysis gives intuition about a mass of objects by dividing them into
groups called clusters, so that objects in a cluster are more similar to each other
than to ones in other clusters [1]. Grouping objects into meaningful categories
is a common scheme of learning and understanding. In fact, by organizing objects and concepts into meaningful groups the underlying hidden structures and
patterns are revealed. So we can handle the complexity caused by variety.

A wide range of disciplines and theories have employed many techniques to
tackle the problem of clustering including fuzzy set [2], statistic [3], neural network [4], graph theory [5] and swarm intelligence [6]. Despite these techniques,
clustering algorithms can be categorized into two groups: hierarchical and partitioning. Hierarchical algorithms organize the objects into a tree. Leafs are single
data object and interior nodes are the subsets created by combining the most
similar subsets represented by descendant nodes. Partitioning algorithms divide
objects into partitions with no intersection.

S. Kozielski et al. (Eds.): BDAS 2014, CCIS 424, pp. 222–237, 2014.
© Springer International Publishing Switzerland 2014

We devised a partitioning algorithm base on the swarm intelligence which is a new branch of science. It has been inspired from biology of ants, termites and bees social life in their colony and the study of movement of flock of bird, school of fish and herd of land. The key point is that every individual has a simple behavior but when they collaborate a complex behavior emerge on the society level. This complicity and intelligence come from the individual interactions. A good analogy to swarm system is fractal. By repeating a simple pattern in different ways, an extraordinary complex shape can be built.

There are many clustering algorithms that used swarm intelligence ideas. Some of them use swarm-base optimization algorithm like PSO to minimize a cost function [7]. They usually need the number of cluster in advance to formulate the cost function. In fact, they use an optimization mechanism that is swarm-base and it can be whatever else e.g. genetic algorithm, simulated annealing, etc. There are another class of swarm-base algorithms that try to solve the problem in the real problem space e.g. ASM [8], DSC [9], artificial bees clustering [10]. Many of these algorithms suffer from performance and have problem with large data sets or do not produce high quality clusters.

Our algorithm belongs to the second class. It just defines one simple rule on the movement of data agent according to the movement of its similar neighbors. It is the only interaction of the data agent with its neighbors and environment. Simulation of the swarm driving by this simple rule proves that not only the rule is sufficient but also effective in dividing the data objects into its natural cluster accurately and within an acceptable time span. Another advantage is that the algorithm does not need the number of clusters as the parameter and, in all cases, it discovers them correctly.

LF-base algorithms usually need a post-processing step for merging different clusters that have small intra-cluster dissimilarity because they actually belong to the same cluster but formed in different region of the swarm plane [11]. In our algorithm, all the objects that belong to a same cluster coalesce in a single region of swarm space and therefore, do not need any post-processing step in order to merge the clusters. The algorithm is not sensitive to the initial condition and always produces the same result in different runs. The way that an agent behaves (moves) is not exact and, of course, different in every run but the way that the whole swarm behaves is exact enough to produce right clusters. It means that the algorithm is stable and tolerates individual bad behavior.

2 SSC Algorithm

At the following we describe the motivation and inspiration for the algorithm and then formally state the algorithm with some example.

2.1 Algorithm Inspiration

There were two sources of the inspiration for the proposed algorithm: First, Reynolds [12] ideas for simulating flying bird and second the food forging behavior of multi-species bird in some tropical regions.

For the first time Reynolds established a set of rules in order to simulate a flock of bird flying in spite of obstacles like columns, trees, mountains etc. depending on the environment for an animation program. The purpose was to keep the birds in the flock and avoid them to disperse. Controlling the movement of each individual bird, one by one, is a complicated task and needs a lot of computations. Instead Reynolds drew up three movement codes for each bird movement according to the movement of its neighbors. Figuratively, each animated bird looks at the surrounding birds and decides what direction to fly. Therefore, the decision is made according to local information and each individual is responsible for its movement and staying in the flock. Three Reynolds movement codes are:

1. A bird must match its velocity with nearby flock mates.
2. A bird must keep away from striking nearby flock mates.
3. A bird must stay close to nearby flock mates.

The second source of inspiration came from observing some different flock species of insectivorous bird mingle together in a place over an abundant number of insects in some tropical regions. Then, in the middle of total chaos, different flock become separated and flew in a direction at the end of the fest. How is it possible? No birds have a global view at the flock that is flying in, and there is no coordinating bird. Every bird decides on the direction of the flight according to its local perception. No birds have the idea what direction its flock is going to fly in advance but surprisingly the way that each individual behaves lead to reaching some hidden consensus that no individual is aware of. It exists in the hidden wisdom of the flock. In other words, there must be some rules among birds like what keep a bird in flock described by Reynolds but this time lead different flock species become separated.

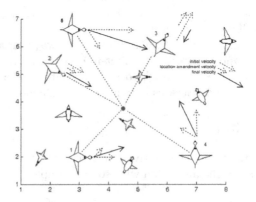

Fig. 1. Similar looking mates streer to the center of their location

We have discovered and proved by simulation that this kind of collective swarm behavior can be justified by only one rule:

- Every bird looks at its local similar neighbor mates and steer to the center of their location.

The situation is illustrated in Fig. 1.

The key point to the success of the algorithm is the way that we define similar mate at the neighborhood. The concept of similar neighborhood consists of two competing parameter: 1) the intrinsic similarity between the two birds and 2) the distance of the observing bird from the one that actually is being observed. Please note that they are two somehow similar but distinct concepts: first how much similar two bird are and second how much a bird perceives another bird similar. The first concept is something per-determined and fixed but at the second one, the distance makes an important role and it is not fixed as the bird fly and its distance changes relative to the others. By competing parameter we mean that the farther a bird is, the less similar it looks to the observer. it is crystal clear that the more similar a bird is, it looks more similar to the observer. Otherwise stated, if a similar bird fly farther it will look less similar. So, in order to formulate the second concept we have to consider the distance as well as the intrinsic similarity between a pair of bird.

2.2   Algorithm Formalism

Suppose we have a set of data object $Data = \{o_1, o_2, , o_n\}$ and a set of agents that is called boids $B = \{b_1, b_2, , b_n\}$ so that each data object can be assigned to a boid. The set $Pos = \{pos_1, pos_2, , pos_n\}$ and $Velo = \{velo_1, velo_2, , velo_n\}$ are the position and the velocity of boids correspondingly in 3-dimensional space. Two boids are similar if they carry similar data object. Our goal is to define the local interactions between individual in order to affect their movement so that they separate into sub-swarms of similar mates.

At the following, we formalized the concept of similar-looking and then explained the algorithm formally.

"Similar-looking" mate: The idea of similar-looking was formulated by the srd measure (Eq 1). The greater its value, the more similar two boids look like.

$$srd_{ij} = d_{max} \times s_{ij} - fr_{ij} \times d_{ij} \tag{1}$$

There is the description of constituents' parts of srd formula below:

- d_{max}, d_{ij}
  $pos_i = (x_i, y_i, z_i)$ is the position of boid b_i in the swarming space that all three axes are restricted to x_{max}. The Euclidean distance between two boids i and j, d_{ij} , and the maximum possible, d_{max}, are (Eq 2 and Eq 3)

$$d_{ij} = \sqrt[2]{(x_j - x_i)^2 + (y_j - y_i)^2 + (z_j - z_i)^2} \tag{2}$$

$$d_{max} = \sqrt{3}x_{max} \tag{3}$$

$-$ s_{ij}

s_{ij} is the similarity between data object o_i and o_j. It is a number between 0 and 1: 0 for sheer dissimilarity and 1 for identity. Similarity is problem-specific and could be based on Correlation, Euclidean, Cosine distance etc.

$-$ r_{ij}

$Rank_{ij}$ is the order of o_j relative to o_i in term of their similarity. For example $Rank_{ij} = r$, says that o_j is the rth most similar object to o_i.

You should note that, in contrast to S that is a symmetric relation ($s_{ij} = s_{ji}$), $Rank$ is not ($Rank_{ij} \neq Rank_{ji}$). Because Objects i and j are not potentially at the same rank of each other. $Rank_{ij}$ varies from 1 to n but it is mapped to interval $(0, 1]$ by function f .(Eq 5)

$$fr = f(Rank) : I \rightarrow (0, 1] \tag{4}$$

$$f_i = (i/n) \tag{5}$$

We apply the function introduced by Eq 5 in order to map the Rank in the same interval and scale as the similarity s_{ij}.

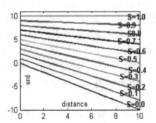

Fig. 2. Srd changes versus distance

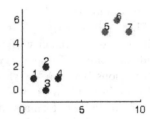

Fig. 3. Example dataset has two clusters

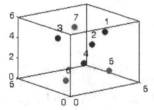

Fig. 4. Dataset objects assigned to boids

Table 1. Example dataset

Label	x	y
1	1	1
2	2	2
3	2	0
4	3	1
5	7	5

To comprehend the concept, imagine we have a dataset with two clusters and seven objects, as an example. (Table. 1 and Fig. 3)

The matrix S (Table. 3) is similarity between data objects based on Euclidian distance. You see that point 1, 2, 3 and 4 are more similar because of thier vicinity, the same as the points 5, 6 and 7. Using similarity matrix, we can compute rank matrix (Table. 2). For example $rank_{5,6} = 2$ means 6 is the second most similar data object to 5 (of course the first most similar is itself). Despite S matrix, rank matrix is not symmetric: $rank_{4,6} = 6$ but $rank_{6,4} = 4$. Please note that the data objects are assigned to boids and then randomly scatter over a 3-daimentional swarming space (Fig. 4). $d = [d_{ij}]$ contains the distance of every boid to the other. Using similarity, rank and distance matrix, srd measure can be computed easily. A closer look at the srd measure reveals that by default two boids "look similar" as much as a magnitude of the value of s_{ij} (the first term of srd equation ($d_{max} \times s_{ij}$)). The first term is the fixed part of the equation because d_{max} and s_{ij} do not change in the course of the algorithm. What changes is the second term ($-fr_{ij} \times d_{ij}$). fr_{ij} is greater for the more dissimilar biods and it is obvious that d_{ij} is larger for farther object. Therefore, the second term imposes a serious penalty on far and dissimilar boids.

In order to grasp different aspect of srd measure, we can divide all the boids in the swarming space to four categories in respect to the boid b_i: the boids that are "far and similar", "far and dissimilar", "near and similar" and finally the ones that are "near and dissimilar". This kind of categorization is not exact nor maybe true because of the fuzzy nature of similar, dissimilar, far and near concepts but it does help to comprehend how srd measure works as the similarity and distance of boid change.

Fig. 2 demonstrates how srd measure varies in respond to the changes of similarity and distance. Note that each line shows dependence of srd to distance for the fixed value of similarity. It is easy to notice that as the similarity increases the slope of the lines decrease. It means that the more the similarity, the less the distance makes a difference in the value of srd. The value of srd is less dependent to distance for the similar boid. In the other world, it is irrelevant that a similar boid is near or far. Therefore, "far and similar" and "near and similar" boids are treated the same by srd and both map to high values. Remember that boids with greater value of srd are considered as neighbors by the observing boid. And it is the neighbors that determine what direction a boid move. For the dissimilar boids the distance really matter and they have low values of srd because high line slope.

Algorithm: Now that we have described "similar-looking" concept, we are able to explain the algorithm by defining some notations. pos_i^j and s_i^j are the position (Eq 6) and similarity (Eq 7) of jth "similar looking" boid to b_i according to srd.

$$pos_i^j = pos_{\hat{c}} \quad if \quad srd_{i\hat{c}} = j^{th} \max_{c=1...n} (srd_{ic}) \tag{6}$$

$$s_i^j = s_{i\hat{c}} \tag{7}$$

Table 2. Objects rank based on their similarity

rank	1	2	3	4	5	6	7
1	1	2	3	4	5	6	7
2	2	1	4	3	5	6	7
3	2	4	1	3	5	6	7
4	4	2	3	1	5	6	7
5	7	5	6	4	1	2	3
6	7	5	6	4	2	1	3
7	7	5	6	4	3	2	1

Table 3. Object similarity

S	1	2	3	4	5	6	7
1	1	0.8	0.8	0.8	0.2	0	0
2	0.8	1	0.8	0.8	0.3	0.2	0.1
3	0.8	0.8	1	0.8	0.2	0.1	0
4	0.8	0.8	0.8	1	0.4	0.2	0.2
5	0.2	0.3	0.2	0.4	1	0.8	0.8
6	0	0.2	0.1	0.2	0.8	1	0.8
7	0	0.1	0	0.2	0.8	0.8	1

Table 4. srd measure is computed using similarity, rank, distance

srd	1	2	3	4	5	6	7
1	8.7	6.9	6	4.1	-1.2	-4	-2.1
2	6.9	8.7	5.2	5.7	0.9	-1.7	-0.7
3	6.4	5.2	8.7	4.9	-1.9	-2.8	-2.4
4	4.1	6.3	4.9	8.7	1.6	-2.5	-2.4
5	-2.3	0.9	-2.7	1.9	8.7	6.1	4.8
6	-4.7	-1.1	-2.8	-1.1	6.1	8.7	5
7	-2.1	-0.2	-2	-0.6	4.8	5.8	8.7

Table. 4 shows srd measure for pervious example. Eq 6 finds the position of most similar-looking boid to boid i. for example $pos_5^2 = pos_6$, it means that the second most similar-looking boid to b_5 is b_6. It can be proven that $pos_i^1 = pos_i$ for $i = 1, 2 \ldots n$ using srd equation.

Using srd we can define N_i^k (Eq 8) as the set of the position first, second, until k^{th} "similar looking" mates to b_i.

$$N_i^k = \left\{ pos_i^j | pos_i^j = pos_{\hat{c}} \quad if \ srd_{i\hat{c}} = j^{th} \max_{c=1\ldots n} (srd_{ic}), \ for \ j = 1 \ldots k \right\} \quad (8)$$

For example:
$N_5^3 = \{pos_5^1, pos_5^2, pos_5^3\} = \{pos_5, pos_6, pos_7\}$
$N_4^5 = \{pos_4^1, pos_4^2, pos_4^3, pos_4^4, pos_4^5\} = \{pos_4, pos_2, pos_3, pos_1, pos_5\}$

When N_i^k was determined the center of similar-looking particles can be computed by Eq 9. Then, the velocity of b_i is modified by steering towards the center of similar-looking mates (Eq 10).

$$cen_i^{(t)} = \frac{\sum_{pos_i^j \in N_i^k} (pos_i^j)}{k} \quad (9)$$

$$velo_i^{(t+1)} = w_I \times velo_i^{(t)} + w_c \times s_i^k \times \left(cen_i^{(t)} - pos_i^{(t)} \right) \quad (10)$$

$velo_i^{(t)}$ is the velocity of boid b_i in the time t. w_I indicate the influence of pervious velocity in new one. s_i^k is the similarity between p_i and the last similar-looking boid in N_i^k and w_c indicates what fraction of position vector from b_i to cen_i effect the next velocity. Finally, the boids move according to their velocity(Eq 11).

$$pos_i^{(t+1)} = pos_i^{(t)} + velo_i^{(t+1)} \tag{11}$$

In order to keep the boids within the framework of the swarm space their velocity and position need to be clamped by means of Eq 12 and Eq 13.

$$\forall c \in \{1,2,3\}, \, velo_{i,c}^{(t+1)} = \max\left(\min\left(velo_{i,c}^{(t+1)}, V_{max}, -V_{max}\right)\right) \tag{12}$$

$$\forall c \in \{1,2,3\}, pos_{i,c}^{(t+1)} = \begin{cases} pos_{i,c}^{(t+1)} + 2 \times \left(-pos_{i,c}^{(t+1)}\right) & pos_{i,c}^{(t+1)} < 0 \\ pos_{i,c}^{(t+1)} + 2 \times \left(pos_{i,c}^{(t+1)} - X_{max}\right) & pos_{i,c}^{(t+1)} > X_{max} \\ pos_{i,c}^{(t+1)} & otherwise \end{cases} \tag{13}$$

Using these equations we are ready to present the algorithm:

```
Program SSC
Initialization
Place particle p_i in random position pos_i
    within swarming space
Set particle velocity velo_i = (v_1i, v_2i, v_3i) components
    to something between -v_max and v_max
Set w_c, w_i, k, t_max
Running the swarm
For t=1 to t_max do
    Find N_i^k by Eq 8 , for i=1 to n
    Compute cen_i by Eq 9 , for i=1 to n
    Revise velo_i by Eq 10, for i=1 to n
    Clamp velocity by Eq 12
    Move particle p_i by Eq 11 , for i=1 to n
    Clamp position by Eq 13
    t=t+1.
End For
Retrieve clusters from swarm space
End program
```

3 Experiment and Result

In this section we briefly explained the datasets, the six cluster quality metrics and finally presented the comparisons and results.

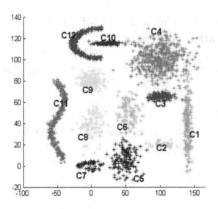

Fig. 5. R1 synthesized dataset

Table 5. Datasets characteristics summary

Datasets	ClusN	Dim
Iris	3	4
Wine	3	13
Ecoli	5	7
WBC	2	9
Dim5	9	5
A2	35	2
S2	15	2
R1	12	2

3.1 Data Sets

Four real-life and four synthesized datasets were used as the benchmark. The most popular and widely-used real-life datasets that many novel algorithms are evaluated by, are Iris, Wisconsin, Win and Ecoli. Beside those, we chose two larger datasets, s2 and a2, with 5000 and 5250 members that have pretty large number of clusters and members. The size of datasets is challenging for some algorithm and the excessive number of classes is appropriate test to see how well our proposed algorithm recognizes the correct number of classes. General characteristics of datasets were summarized in Table. 5. In addition, we synthesized R1 dataset for testing different aspect of SSC algorithm (Fig. 5). You observe that this dataset consists of heterogeneous type of classes in term of density, shape and also geometrical relationship of class to each other. For example class C4 is much expanded and sparse than the others, class C3 is relatively dense, class C2 has a few members and easily can be ignored by some algorithms, two classes C5 and C6 are slightly merged and may be considered as one, class C11 has a wavy shape and also stretched along three other classes. It means that the distance between a point in head and a point in the tail of the class is longer than distance between a point in a class C9 and a point in C7 that is two classes away. It is hard to recognize the unity of class C11 for many algorithms. Class C12 is lunette-shaped and class C10 went within it. In fact, it is a test for the ability of the algorithm to identify the two head of the lunette. It would be a big challenge because the distance between two points each within a head of lunette is almost twice longer than their distance to the nearest point in class C10.

3.2 Cluster Quality Metrics

We used six cluster quality metrics for evaluating an comparing the quality of clusters: Normalized variation of information (NVI) [13], Normalized information distance (NID) [14], V-measure [15], Mirkin metric [16], Adjusted Rand Index [17] and F-measure [18]. The lower values of NVI, NID and Mirkin metrics and

Table 6. Comparisons and results

	SSC	DSC	LF	KM	KHM	SOM	SSC	DSC	LF	KM	KHM	SOM
	R1						Ecoli					
NID	0.036	0.298	0.177	0.182	0.197	0.179	0.411	0.5	0.448	0.52	0.521	0.535
NVI	0.069	0.324	0.286	0.272	0.293	0.26	0.561	0.616	0.591	0.659	0.658	0.671
Mirkin	0.007	0.077	0.052	0.061	0.063	0.059	0.149	0.19	0.179	0.228	0.229	0.237
v_measure	0.964	0.807	0.833	0.842	0.827	0.851	0.609	0.553	0.579	0.509	0.51	0.495
ARI	0.967	0.519	0.745	0.684	0.673	0.686	0.644	0.571	0.553	0.391	0.381	0.361
F_measure	0.984	0.62	0.803	0.789	0.791	0.797	0.816	0.782	0.726	0.65	0.651	0.639
Num clust	12	25.2	12.7				4.6	3.4	5			
	Iris						Wine					
NID	0.245	0.345	0.225	0.275	0.255	0.258	0.579	0.633	0.574	0.571	0.584	0.573
NVI	0.361	0.446	0.361	0.415	0.4	0.402	0.725	0.75	0.727	0.727	0.737	0.728
Mirkin	0.121	0.167	0.102	0.138	0.119	0.123	0.287	0.313	0.282	0.28	0.288	0.281
v_measure	0.779	0.711	0.779	0.736	0.75	0.748	0.432	0.399	0.429	0.429	0.417	0.428
ARI	0.724	0.606	0.768	0.695	0.729	0.722	0.369	0.302	0.371	0.371	0.354	0.369
F_measure	0.875	0.78	0.912	0.871	0.894	0.891	0.697	0.626	0.681	0.69	0.672	0.689
Num clust	3.2	4	3				3	3.5	2.9			
	S2						a2					
NID	0.045	0.265	0.168	0.107	0.072	0.073	0.013	0.214	0.173	0.065	0.06	0.051
NVI	0.086	0.319	0.251	0.168	0.124	0.121	0.023	0.328	0.258	0.104	0.102	0.086
Mirkin	0.006	0.044	0.039	0.025	0.014	0.015	0.001	0.03	0.024	0.01	0.008	0.007
v_measure	0.955	0.81	0.856	0.907	0.934	0.936	0.988	0.803	0.852	0.945	0.946	0.955
ARI	0.952	0.577	0.721	0.821	0.889	0.885	0.98	0.517	0.628	0.836	0.854	0.874
F_measure	0.977	0.623	0.81	0.875	0.931	0.923	0.989	0.626	0.738	0.882	0.906	0.916
Num clust	15.1	32.3	15.4				34.8	37.6	34.4			
	Dim4						WBC					
NID	0	0.315	0.209	0.122	0	0.07	0.271	0.797	0.327	0.253	0.278	0.253
NVI	0	0.317	0.21	0.164	0	0.07	0.416	0.88	0.475	0.398	0.425	0.398
Mirkin	0	0.068	0.082	0.053	0	0.025	0.082	0.453	0.106	0.075	0.084	0.075
v_measure	1	0.811	0.88	0.91	1	0.964	0.737	0.213	0.687	0.751	0.73	0.751
ARI	1	0.533	0.702	0.768	1	0.886	0.834	0.088	0.785	0.849	0.83	0.849
F_measure	1	0.549	0.784	0.819	1	0.926	0.957	0.649	0.945	0.961	0.956	0.961
Num clust	9	26.2	7.1				2	2.2	2			

higher value of V-measure, Adjusted Rand Index and F-measure metrics show better cluster quality. For detail information on how they are computed and their weakness and advantages, please look at the references.

3.3   Results

For proofing the efficiency and performance of the proposed algorithm, we used five other algorithms: K-mean, K-harmonic, SOM, Data Swarm Clustering (DSC) and Lumer and Faiete. All the algorithms were implemented by MATLAB and ran 20 times. The average of quality metrics are presented in Table. 6. Remember that lower values of the first three metric and higher values of second next three metric represent higher clustering quality.

4   Discussion

At the following, we interpreted the result of SSC algorithm on each dataset case by case:

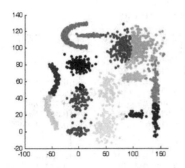

Fig. 6. KM failed to recognize the pattern

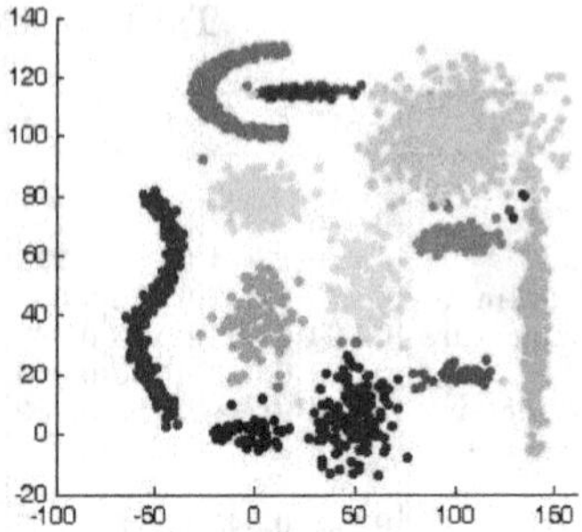

Fig. 7. SSC succeeded to recognize the pattern

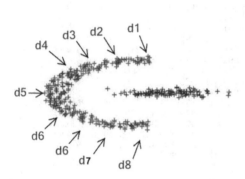

Fig. 8. d1 is related to d8 through many points between like d2 to d7

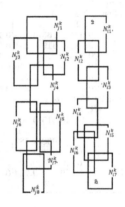

Fig. 9. Similar-looking neighbor sets connect like rings of chain

R1: KM, KHM and SOM are the kind of algorithms that well-suited for the datasets with Gaussian cluster structure. R1 does not have Gaussian structure. The strategy of these algorithms is to locate the center of clusters in a good spot and then divide the dataset space into regions that are most approximate to a cluster center. All the modifications of KM algorithm like KHM center around finding a better strategy for figuring out a better position for the cluster center. Fig. 6 shows the result of KM on R1. You see that KM failed to find the cluster structure.

Fig. 7 shows the result of running SSC on R1 dataset. You see that the result is a total success. Why did SSC succeed on finding the patters of R1 dataset and outperformed the other algorithms? The answer lies in the mechanism and strategy that SSC algorithm works on. Supposed that b_i and b_j are near and similar boids and N_i^k and N_j^k are their neighbor sets correspondingly. It can be shown that N_i^k and N_j^k have many boids in common. Suppose that b_t is again another boid near and similar to b_j then N_t^k and N_i^k will have many boids in common and so on. Therefore, The similar boids are connected to each other by their common neighbor in the four-wall of swarming space like rings of chine (Fig. 8).

Iris: The cluster quality metrics show that SSC algorithm did not work well on Iris dataset. The result is interpretable by investigating the dataset cluster structure an internal quality metric. There are two types of quality metric: external quality metric and internal quality metrics. External quality metric evaluates the cluster quality in respect to an external piece of information that is called label. It determines which data point belong to which one cluster. In contrast to internal quality metrics that assesses the cluster quality in respect to the clustering structure of dataset itself, especially by measuring intra-cluster similarity and inter-cluster dissimilarity and combining them into a formula. All quality metrics we used here are external. External quality metrics have one drawback: in case that class labeling does not confirm with the real clustering structure, even though the algorithm find the best of the cluster structure it could, the external quality metrics shows a poor clustering result. It is what happened for the Iris dataset. Fig. 10 presents first versus second dimensions of the four dimensional Iris dataset. A quick glimpse at Fig. 10 reveals that the diamond and circle classes are completely merged from this perspective and it does not have any recognizable cluster structure.

Table 7. Iris dataset silhouette value

	classes	Clusters
Squ	0.945	0.9502
Cir	0.5929	0.5975
Dia	0.4316	0.6761

Table 8. KM and SSC quality metric difference

Metric	diff
NID	-0.018
NVI	-0.018
Mirkin	-0.007
v_measure	0.014
ARI	0.015
F_measure	0.004

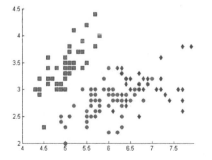

Fig. 10. Iris dataset-first versus second dimension

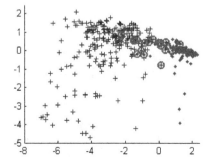

Fig. 11. WBC first and second principle component

In other words, point d1 is related to the point d8 through many other points between, like d2 to d7. In fact, the two head of lunette are related to each other by the other points that form the rest of the cluster. They are not connected to each other by the direct distance. Thus, they can bypass c10 class points. It is the key point that SSC can recognize the integrity of the class c12 even though c10 has gone in middle of lunette shape cluster. Therefore, every point in a class has many common neighbors with the other points of same class and no common neighbor elements with points belong to the other class.Fig. 9 shows the two disjoint sets of chain of neighbors.

In order to state the claim from a concrete ground, we measure how well the classification confirms with natural clustering structure using an internal cluster quality metric called silhouette value (Eq 14 Eq 15)

$$S(o_i) = \frac{\min_{k_j \in K} (b(o_i, k_j)) - a(o_i)}{\max(a(o_i), \min_{k_j \in K}(b(o_i, k_j)))} \qquad (14)$$

$$S_{k_j} = \frac{\sum_{o_i \in k_j} S(o_i)}{|k_j|} \qquad (15)$$

where $b(o_i, k_j)$ is the average distance from the o_i object to objects in another cluster k_j and and $a(o_i)$ is the average distance from the o_i point to the other points in its cluster. S_{k_j} is the average silhouette value of cluster k_j. In fact, silhouette value shows how much a point is well-suited within the context of its cluster. In another words, it measures how much the point is in the right cluster according to intra-similarity and inter-dissimilarity of clusters. The average silhouette value of SSC clustering result is greater than the average silhouette value of the classification represented by label (Table. 7). It shows that what SSC find as the clusters describe the cluster structure better than what is offered by classification.

WBC: The SSC algorithm exhibited a relative poor performance on WBC dataset and the best result is exhibited by KM. At the first place, the difference between the KM and SSC is not much, in fact less than 0.02. Table. 8 gives more details on the difference between each metric for SSC and KM algorithm. For the purpose of providing an account of what happened for WBC, a visual image of the dataset has been papered. Please note that WBC data has 9 dimensions. In order to visualize it on a 2-dimensional plane, we applied Principle Component Analysis (PCA) and plot first and second principal component (Fig. 11).

The points displayed by plus and diamond sign are the first and second classes. In addition, the points that have a red circle around, are the ones which were wrongly considered belonging to the other class by SSC algorithm. In other words, the pluses surrounded by the circle were assigned to diamond class and the diamonds surrounded by the circle were assigned to plus class by SSC algorithm.

We can conclude two issue form the image. Firstly, the two classes do not have any distinguishable boundary even though the diamond cluster is much denser than the plus cluster. In the worst case scenario, if you give an evenly distributed point to KM algorithm and want it to find two clusters, it would certainly do but if you give to SSC algorithm it would not. Secondly, the points that are mistakenly assigned belong to one of these two cases: 1) they are the points that are at the boundary of two classes or 2) they are the points that are nearer to the other class. In other words, they are the pluses at the diamond side or the diamonds at the plus side.

a2, s2, dim4: a2, s2 and dim4 are well-structure data set and SSC algorithm could successfully discover the pattern both in term of number of clusters and cluster identity and outperforms other algorithm. Fig. 12 shows the result of SSC on a2.

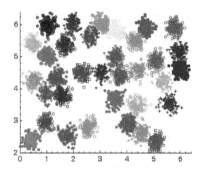

Fig. 12. SSC can recognize all 35 clusters of a2 dataset

5   Conclusion

In this paper we devised a new swarm-base clustering algorithm that has the interesting properties of swarm systems. The control is decentralized and therefor the algorithm has a high potential for distributed and parallel computing. It is completely fault-tolerant and would not break down by wrong movement of individual boids. If an individual move in the incorrect direction the other individuals move in way that alleviate that wrong movement.

The algorithm is not sensitive to initialization at all. It produces the same result in different runs. Of course, the different sub-swarms (clusters) may form in different region of swarm space but it is irrelevant to the final resulting clusters. That algorithm has many parameters which enable it to adjust to different condition but its performance is stable to a rather wide range of their changes. Although each individual moves in direction that is decided *locally* according to the other individuals' movement in the neighborhood but the cascaded effects in the whole swarm leads to a very desirable*global* behavior so that all similar boids, in middle, sides and eight corners can easily find each other and move base on one driving force and separated very well from the rest of the population. In fact, individual interaction leads to desired emergent collective behavior. Some swarm-base clustering algorithms convert the problem into an optimization problem and try to optimize the cost function by warm-base optimization approach. Our algorithm looks at the problem just as it is and solves it in its natural problems space. It makes it completely efficient as well as simple quit like the nature solves the problem. A high remarkable property of the algorithm is the power of recognizing true number of even overlapped clusters. It makes it superior to K-mean, K-harmonic and SOM that need the number in advance.

References

1. Razavi Zadegan, S., Mirzaie, M., Sadoughi, F.: Ranked k-medoids: A fast and accurate rank-based partitioning algorithm for clustering large datasets. Knowledge-Based Systems (2012)
2. Tari, L., Baral, C., Kim, S.: Fuzzy c-means clustering with prior biological knowledge. J. Biomed. Inform. 42(1), 74–81 (2009)
3. Datta, S., Datta, S.: Comparisons and validation of statistical clustering techniques for microarray gene expression data. Bioinformatics 19(4), 459–466 (2003)
4. Vesanto, J., Alhoniemi, E.: Clustering of the self-organizing map. ITNN 11(3), 586–600 (2000)
5. Wu, Z., Leahy, R.: An optimal graph theoretic approach to data clustering: Theory and its application to image segmentation. IEEE Transactions on Pattern Analysis and Machine Intelligence 15(11), 1101–1113 (1993)
6. Senthilnath, J., Omkar, S., Mani, V.: Clustering using firefly algorithm: Performance study. Swarm and Evolutionary Computation 1(3), 164–171 (2011)
7. Kalyani, S., Swarup, K.: Particle swarm optimization based k-means clustering approach for security assessment in power systems. Expert Syst. Appl. 38(9), 10839–10846 (2011)
8. Chen, L., Xu, X.H., Chen, Y.X.: An adaptive ant colony clustering algorithm. In: Proceedings of 2004 International Conference on Machine Learning and Cybernetics, vol. 3, pp. 1387–1392. IEEE (2004)
9. Veenhuis, C., Köppen, M.: Data swarm clustering. SIDM 34, 221–241 (2006)
10. Kazemian, M., Ramezani, Y., Lucas, C., Moshiri, B.: Swarm clustering based on flowers pollination by artificial bees. In: Abraham, A., Grosan, C., Ramos, V. (eds.) Swarm Intelligence in Data Mining. SCI, vol. 34, pp. 191–202. Springer, Heidelberg (2006)
11. Lumer, E., Faieta, B.: Diversity and adaptation in populations of clustering ants. In: Proceedings of the 3rd International Conference on Simulation of Adaptive Behavior: From Animals to Animats 3 (1994)
12. Reynolds, C.W.: Flocks, herds and schools: A distributed behavioral model. ACM SIGGRAPH Computer Graphics 21, 25–34
13. Reichart, R., Rappoport, A.: The nvi clustering evaluation measure. In: Proceedings of the Thirteenth Conference on Computational Natural Language Learning, pp. 165–173. Association for Computational Linguistics
14. Vitányi, P.M., Balbach, F.J., Cilibrasi, R.L., Li, M.: Normalized information distance, pp. 45–82. Springer (2009)

15. Rosenberg, A., Hirschberg, J.: V-measure: A conditional entropy-based external cluster evaluation measure. In: EMNLP-CoNLL, vol. 7, pp. 410–420
16. Mirkin, B.: Mathematical classication and clustering. Kluwer Academic Press (1996)
17. Rand, W.M.: Objective criteria for the evaluation of clustering methods. Journal of the American Statistical Association 66(336), 846–850 (1971)
18. Hripcsak, G., Rothschild, A.S.: Agreement, the f-measure, and reliability in information retrieval. Journal of the American Medical Informatics Association 12(3), 296–298 (2005)

Link and Annotation Prediction Using Topology and Feature Structure in Large Scale Social Networks

Burak Isikli[1,2], Fatih Erdogan Sevilgen[1], and Mustafa Kirac[2]

[1] Department of Computer Engineering Gebze Insititute of Technlogy,
Gebze, Kocaeli, Turkey
sevilgen@gyte.edu.tr
[2] Turkcell, Inc. Istanbul, Turkey
{burak.isikli,mustafa.kirac}@turkcell.com.tr

Abstract. Repeated patterns observed in graph and network structures can be utilized for predictive purposes in various domains including cheminformatics, bioinformatics, political sciences, and sociology. In large scale network structures like social networks, graph theoretical link and annotation prediction algorithms are usually not applicable due to graph isomorphism problem, unless some form of approximation is applied. We propose a non-graph theoretical alternative to link and annotation prediction in large networks by flattening network structures into feature vectors. We extract repeated sub-network pattern vectors for the nodes of a network, and utilize traditional machine learning algorithms for estimating missing or unknown annotations and links in the network. Our main contribution is a novel method for extracting features from large scale networks, and evaluation of the benefit each extraction method provides. We applied our methodology for suggesting new Twitter friends. In our experiments, we observed 11-27% improvement in prediction accuracy when compared to the simple methodology of suggesting friends of friends.

Keywords: social networks, data mining and knowledge discovery, big data, business intelligence, link prediction, graph processing, graph mining.

1 Introduction

Importance of network structures has been recently increasing, due to advancements in data collection, storage, and processing technologies. Social networks, mobile call networks, biological networks, and World Wide Web are some examples of network structures that many end users of online services and researchers deal with. Network structures can be utilized for revealing information that is difficult to obtain from nodes directly. For instance, discovery of disease related proteins within a protein interaction network [5], targeting potential customers with ads by using social networks [4], and looking for similar graph patterns

S. Kozielski et al. (Eds.): BDAS 2014, CCIS 424, pp. 238–249, 2014.
© Springer International Publishing Switzerland 2014

in chemical molecular structures are some tasks that can only be accomplished using a network oriented analytic methodology.

A node-annotated network consists of i) a list of nodes, ii) links representing some relationship between nodes, and iii) features associated with nodes. For example, consider the Facebook network: Facebook is a node-annotated social network where its nodes are people, friendship activities amongst people are links, and liked/favorited items are node features. Since the association of Facebook users with content items is voluntarily accomplished by the users themselves, not all users reveal all pieces of information about themselves. The problem we primarily attack in this paper is the problem of completing missing or incomplete node annotations. We present a novel methodology for predicting unknown annotation of network nodes. We propose a solution to network link prediction problem [12] by reducing it into another form of node annotation prediction problem as well.

Recent research shows that homophily [3] and contagion [3] in real world networks suggest shared features between connected network nodes. In addition, correlation between existing node features is shown to be employed for predicting missing or unknown features of network nodes [15]. Such properties of real world networks are utilized by learning correlations between existing node annotations into a data mining model, and applying such model to network nodes with missing annotations. D. Liben et al [9] showed that features from network topology of a co-authorship network can be used for supervised learning effectively. This is the first and common approach for solving the link prediction problem. Taskar et al. [13] fulfilled the relational Markov network algorithm. They aim to predict missing links in a network of web pages and a social network. Although it is not directly related to link prediction problem, users are linked by their common interests. Natural language processing (NLP) and text mining methods [7] are used to predict interests by finding tweet similarity. Because of multitude of common interests, users can be associated with each other. There are several other researches can be related to our work such as Goldberg and Roth [2]. They employed the neighborhood as a reliable property of small networks (e.g. protein-protein interactions) for confidence. Clauset et al. [1] developed a prediction algorithm on social and biological networks.

Our approach for predicting missing node annotations in a network has the following steps. First, the network is preprocessed (i.e., extract-transform-load) to collect for each node all available annotation patterns in the network neighborhood of that node, and create an integer feature vector representing the counts of annotation patterns observed in the whole network. This approach is similar to text mining [7], where text documents are converted into term frequency vectors. Next, we create a table from the count vectors of network nodes, and feed into machine learning algorithms in order to obtain models that explain complex relationships between node annotations. Finally, we utilize resulting models learned from the network for estimating whether a particular network node is highly likely to be associated with a target feature annotation.

The process summarized above requires processing large amount of data. This process depends on the ability of forming and counting annotation patterns in a large network data. In real world cases, network data is so huge that it does not fit in the memory of a single computer with commodity hardware. One option is building an expensive supercomputer, and applying a trivial pattern counting algorithm. Another option is making use of a smaller size, sample data. In the latter case, sampling network structures partitions it into disconnected components, resulting in information loss. We propose a hybrid approach: We compute network patterns in a disk-based distributed processing environment, and convert network data into a node-based feature vector sets. Then, we run machine learning algorithms on a single computation framework using data sampled from the node-based feature sets.

Main contributions of this paper are 1) a systematic way of converting network structures into flat vector sets, 2) a scalable methodology for computing feature vectors in a parallel DBMS environment, and 3) a novel solution to link prediction problem by formulating it as a form of node annotation prediction problem.

2   Methods and Algorithms

Definition 1 (Network): A directed network $G = (N, L, F, A)$ is a structure that consists of a set of nodes N, set of links L that connects two nodes ($l \in L$ then $l : n_1 \to n_2$ and $n_1, n_2 \in N$), set of features F that have information about node (e.g. person of interests, hobbies), and set of feature annotations A that associates nodes with features ($f : n \to a, a \in A, n \in N$, and $f \in F$).

Example 1: Fig. 1 illustrates an example network structure. Nodes that represent social network profiles are numbered from 1 to 6 and each node is assigned some features, such as Sailing, Swimming and Baseball. Each annotation represents existence of feature such as if a person like sailing, sailing's annotation is 1. Nodes are connected with links that represent friendship relationship between nodes.

In this paper, we study Twitter Social Network presented in [6]. Twitter graph is created in the form of a two-column table of user pairs. An excerpt from a Twitter follower table is shown in Tab. 1. Each column in the Twitter table denotes a Twitter user node, and each row in the Twitter table represents a directed link. Hence a Twitter table is a list of links in the Twitter network. Twitter network and further details can be obtained via Twitter Application Programming Interface (API) [14]. The Twitter dataset we employ consists of approximately 40 million nodes and 1.5 billion links.

Problem 1 (Network Link Prediction): Real world networks continuously evolve, as they gain or lose nodes, links, and annotation. Hence, a network $G = (N, L, F, A)$ is a snapshot of a real world network at time t_1, and the network

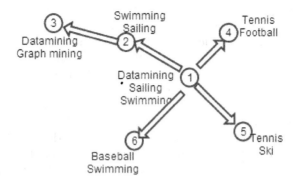

Fig. 1. Representing a graph. Each circle denotes a node, each arrow between them denotes a link and each node has a label which is representing person's interests.

Table 1. Twitter graph table. It consists of user ids that denote whom following. For example user 48954673, 49233593, 27433315, 60843485 are followers of user 43661838. User 27433315 is a follower of user 43661851.

To Id	From Id
43661838	48954673
43661838	49233593
43661838	27433315
43661838	60843485
43661851	27433315

evolves to $G' = (N', L', F', A')$ at t_2 where $t_2 > t_1$. **Network link prediction** problem is to obtain a new $l = n_1' \rightarrow n_2$ $(n_1, n_2 \in N, n_1, n_2 \in N')$ such that $l \notin L, l \in L$.

Problem 2 (Network Annotation Prediction): Real world network data does not perfectly represent the underlying network. Node annotations are not complete, and existing annotations may not be correct due to data quality issues. Let $G = (N, L, F, A)$ is a real world network, and $G' = (N', L', F', A')$ is the known representation of G in the data obtained. Network annotation prediction problem is to obtain the annotations in $A - A'$ with some probability and confidence. The problem can also be configured as **a network annotation correction**, where the existing annotations in $A - A'$ can be assigned of being noise or error with some probability and confidence.

Definition 2 (Network Neighborhood): A node $n \in N$ in graph $G = (N, L, F, A)$ has **incoming direct neighbor of** m such that $m \in N$, and $\bigvee l_1 \in L$, $l_1 = m \rightarrow n$, and has an **outgoing direct neighbor** o such that $o \in N$, and $\bigvee l_2 \in L$, $l_2 = n \rightarrow o$. A **directed path** of length z from $n \in N$ to $n_z \in N$ in G consists of links $l_1, l_2, \ldots, l_z$ such that $l_1 = n \rightarrow n_1$, $l_2 = n_1 \rightarrow n_2$,

..., and $l_z = n_{z-1} \rightarrow n_z$. All direct neighbors of a node **n** are connected to n through a directed path of length 1. **Network neighborhood** of a node n consists of nodes that have at least one directed path connected to **n**. When we limit the maximum length of directed paths between all nodes in the network neighborhood of node **n** and the node **n** by a fixed number **z**, we say that the **radius** of network neighborhood of **n** is **z**.

In this paper, we focus on predicting properties of Twitter user profiles. There are many ways of obtaining Twitter user interest. In [11], the text in the news feed of users, or the tweets of the users are processed through NLP and text mining to obtain the concepts the users interested in. In another work, physical locations and point of interests of the users are retrieved from their check in and shared locations. As locations reflect personal interests, it is also possible to generate location-based Twitter user interests [8]. In this work, we consider the celebrities that the Twitter users follow, as their interests. In Twitter, there is no distinction between celebrity and a regular user, as Facebook does (i.e., people and pages). Hence, we make use of a simple approach to make such distinction. When the follower count of a user is above some certain threshold, we classify the user as a celebrity. Activities of following celebrities are considered as interests of other regular users. In short, our aim in this paper is to predict interests (i.e., annotations) of regular users or equivalently, estimating whether regular users would follow (i.e., show interest) a celebrity user (e.g., a famous actress, a rock band, or an auto maker brand).

Definition 3 (Transformation of Network Link Prediction problem to Network Annotation Prediction problem): Real world networks follow a scale-free property [6] where some nodes have very high link counts whereas many nodes have quite few link counts. Usually, the nodes with very high link counts have a special real world meaning (i.e., celebrity people, or brands in the Twitter network). Hence, set of regular nodes (i.e., common users) are obtained by removing all nodes with very high link counts (i.e., above some certain threshold). Not to loose link information of removed nodes, the information is stored as node annotations; suppose that u is a celebrity node with dense link information, and v and w are common nodes with much less link counts. Further suppose that the network G has the links $v \rightarrow u$ and $w \rightarrow v$. The links $v \rightarrow u$ and $w \rightarrow v \rightarrow u$ are converted to two annotations of v and w as, v is assigned {outgoing_U}, and w is assigned {incoming_U} as new annotations. As we converted link information to annotation information, we can now formulate link prediction problem as an annotation prediction problem. In the beginning, a Twitter network dataset is a non-annotated directed graph. After we remove all the celebrity nodes from the network, and convert links of celebrity nodes into annotations of regular nodes, it becomes an annotated directed graph.

Definition 4 (Network Pattern):   Given a large directed annotated $\mathbf{G} = (\mathbf{N}, \mathbf{L}, \mathbf{F}, \mathbf{A})$, a **network pattern** is a sub-graph $\mathbf{p} = (\mathbf{N'}, \mathbf{L'}, \mathbf{F'}, \mathbf{A'})$ such that $\mathbf{N'} \subseteq \mathbf{N}$, $\mathbf{L'} \subseteq \mathbf{L}$, $\mathbf{F'} \subseteq \mathbf{F}$, $\mathbf{A'} \subseteq \mathbf{A}$, and **p** is a connected component where there exists at least one directed path between every $n_1, n_2 \in N'$. In the case when p is a much smaller sub-graph of G, there can be multiple isomorphic mappings between p and G. In that case, we say p is a **repeated network pattern** in G.

Definition 5 (Network Pattern Flattening): A network pattern $p = (N', L', F', A')$ in network $G = (N, L, F, A)$ can be flattened into a vector as follows; define annotation subset of a node n as $S(n) = \{s_i | s_i \rightarrow f, s_i \in A, f \in F,$ and $n \in N\}$. Then, we compute a list P of all annotation subsets of a node n and all other nodes in the neighborhood of n by radius z. It is possible that p can have repeated annotation subsets. We can group p by distinct annotation subsets and record the number of repetitions as counts. A vector that consists of counts of such network patterns is called a **flattened vector of a network pattern** (See Tab. 2).

Example 2: In Example 1, node 4 has annotations Tennis (T) and Football (F) hence annotation subsets $\{T, F, TF\}$. Node 4 has an incoming link from node 1 and node 1 has annotations data mining (M), Sailing (S), and swimming (W), hence annotation sets $\{M, S, W, MS, MW, SW, MSW\}$. Network neighborhood of Node 4 on radius 1 can be flattened as $\{T:1, F:1, TF:1, M:1, S:1, W:1, MS:1, MW:1, SW:1, MSW:1\}$ where T:1 denotes that pattern T appears only once. Length-1 patterns in this list are $\{T:1, F:1, M:1, S:1, W1\}$. Adding all other annotations (i.e., Graph Mining (G), Ski (SK), Baseball (B) from the network, final flattened vector becomes $\{T:1, F:1, M:1, S:1, W:1, G:0, SK:0, B:0\}$. It is also possible to separate a flattened vector to indicate whether a pattern came from an incoming neighbor or self and so on. A separated flattened vector for this example becomes { self:$\{T:1, F:1, G:0, SK:0, B:0\}$, incoming_radius_1: $\{M:1, S:1, W:1, G:0, S:0, B:0\}$ }.

Pattern discovery and counting is the step where we flatten network neighborhoods of nodes in a network into feature vectors. First we create a table including two columns which names are from_id and to_id, representing the Twitter following graph (TWITTER_GRAPH) using the dataset in [6]. We define a celebrity node table using a predefined threshold on count of links (CELEBRITY_NODES). The threshold is decided by looking up to the count data. Then top-k celebrity nodes are selected (k piece) from the data created in the previous step using ordering of the count data.

These selected top-k celebrity nodes are used as feature using TWITTER_GRAPH network (N, L, F, A) and then feature annotations are extracted from network nodes and links, and created a table named FEA-TURE_TABLE. The difference between celebrity node and feature is that celebrity node can be famous persons (singer, artist...etc.), or ordinary person with high link count but features can't be a famous person which means we decide which one is used so that they are subset of celebrity nodes excluding the famous persons.

We can compute the number of annotation subsets as support. For example, in order to compute length-1 patterns in the outgoing neighbors (denoted by o1)

Table 2. This table is explaining the each different kind of pattern type regarding the neighborhood in demand

Type	Description
t1	Patterns that links has 1 length (radius is fixed, for instance set to 1)
i1	Patterns that incoming links has 1 length (radius is fixed, for instance set to 1)
i2	Patterns that incoming links has 2 length (radius is fixed, for instance set to 1)
o1	Patterns that outgoing links has 1 length (radius is fixed, for instance set to 1)
o2	Patterns that outgoing links has 2 length (radius is fixed, for instance set to 1)
c1	Patterns that links has length 1(radius is fixed, for instance set to 1, i.e. t1-i1-o1)
c2	Patterns that links has length 2 (radius is fixed, for instance set to 1, i.e. t1-i2-o2)
c3	Patterns that links has length 3 (radius is fixed, for instance set to 1, i.e. t1-i3-o3)
n0	Patterns that links has radius 0 (length is predefined, for instance set to 1)
n1	Patterns that links has radius 1 (length is predefined, for instance set to 1)
n2	Patterns that links has radius 2 (length is predefined, for instance set to 1)
n3	Patterns that links has radius 3 (length is predefined, for instance set to 1)

of all nodes, node itself, and total link of each node which is matched from the feature table is used excluding the celebrity nodes from the TWITTER_GRAPH set.

More complex patterns can be obtained, say annotation subsets of length-2 in the outgoing neighborhood of distance 1. The hint is combining two different features to use as a single feature using the same table (FEATURE_TABLE) twice and same rules we explained above, in the query such as $feature_1$ || '-' || $feature_2$. Thus these new-forming features are permutation of features.

SQL provides a viable and scalable option to flatten network structures into feature vectors. After a bunch of transformation SQLs, we convert Twitter graph into a set of records, where each record represent a network node, flattened feature vector of that node, and known annotations of that network node. This tabular representation allows us to use various machine learning algorithms to make predictions.

There are several powerful classification machine learning algorithms that can be used in this work. Although their performances are comparable, one of them usually works better than others. In this work, we compared four supervised machine learning (classification) algorithms. Support Vector Machine (SVM) [10], Naive Bayes (NB), Generalized Linear Model (GLM), Decision Tree (DT). They are compared using accuracy and lift. All the algorithms are available in Oracle Data Miner (ODM) Library. Default ODM parameter values are used since they perform quite good. For all the algorithms, we used k-fold cross validation.

3   Experimental Results

All the experiments are performed on Oracle Exadata V2 using 64 bit Red Hat Linux 2.4 and the Oracle 11g database. Preprocessing is programmed with Structured Query Language (SQL) and Procedural Language/Structured Query Language (PLSQL).

As part of our experiment, the nodes having more than 90000 links are considered as celebrity. The nodes with link count from 90000 to 100000 are selected as features. There are 34 such nodes. Using these features, tag patterns are stored in a tag pattern table including type, node, feature, and count of pattern that has links (support). However, it's hard to calculate the table due to large number (1785) of different patterns. So, only patterns having more than 10000 links are selected. The patterns are enumerated and the lookup table that contains the information about patterns for each type with their support is created. From this, another table is created summing of the support for each pattern. Because our study is aimed to predict the existence of a link, Boolean support (if support is greater than 0 then 1, otherwise 0) is used instead of actual support when it's as target. Thus, count vectors of network nodes that we use to feed the machine learning algorithm, are formed. The generated data is fed to classifiers (SVM, NB, GLM, DT). To choose the machine learning algorithm, 10-fold cross validation is employed. After the machine learning algorithm that is chosen, 70% of data is used for training and the rest of data is used in tests.

3.1   Choosing a Machine Learning Algorithm

In this experiment, we built a dataset from network nodes by converting annotation existence (or counts) as an integer vector. For instance, suppose that {A, B, C, D} as the 4 annotations available in whole dataset, and node v is assigned {A, C}, we build a feature vector for v as {A:1, B:0, C:1, D:0}. Then, for each annotation, we build a learning set by choosing an annotation as target, and omitting the target from the input set. For instance, for testing prediction ability of whether v has annotation A, we convert its vector to {B:0, C:1, D:0} by omitting A.

Table 3. Comparison table for machine learning algorithms using accuracy and lift

Algorithm	Average Accuracy %	True Positive Accuracy %	True Negative Accuracy %	Lift Cumulative
Support Vector Machine (SVM)	66.68	99.89	7.66	4.06
Naive Bayes (NB)	87.49	98.81	51.44	4.92
Generalized Linear Model (GLM)	89.18	99.07	45.43	4.88
Decision Tree (DT)	89.91	99.26	38.39	4.43

Next, we build a dataset for every annotation in the network, and test prediction accuracy via 10-fold cross validation. Then, we repeated this procedure for 4 machine learning algorithms, and compared prediction accuracy metrics. We observed that Generalized Linear Models and Decision Trees have higher average accuracy in comparison with other two algorithms (i.e., Support Vector Machines, and Naive Bayes). In addition, we observed a better lift curve for GLM, hence we utilized GLM algorithm for the rest of the experiments.

3.2   Prediction Performance Comparison by Annotation Position

In this experiment, we observe how much prediction capability is supplemented by the known annotations of a node (say vector t1), annotations of the neighbors connected via incoming links (say vector i1), and annotations of the neighbors connected via outgoing links (say vector o1). Finally we built 4 datasets to compare prediction performance: t1, i1, o1, and $c1=t1|><|i1|><|o1$.

We observed (see Fig. 2) that, incoming links provide more prediction information than outgoing links. We computed that augmenting t1, i1, and o1 (i.e., dataset c1) provided only 11% improvement over using i1 alone, while augmentation supplemented 27% improvement over o1. In addition, providing neighborhood information boosted prediction performance by 362%.

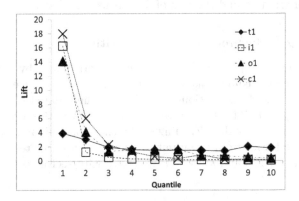

Fig. 2. Impact of annotation position (node itself, or incoming or outgoing neighbors) on prediction capability

3.3   Prediction Improvement by Increasing Pattern Complexity

In this experiment, we build feature vectors from annotations in the direct neighborhood of nodes (i.e., node itself and its direct incoming and outgoing links). We divide direct neighborhood annotation information into 3 distinct vectors. 1) Annotations of the node itself, namely t, 2) annotations of the node's incoming links, namely i, and 3) annotations of the nodes are outgoing links, namely o. For each vector, we build feature sets by computing frequency of annotation patterns. For example, if a node has three annotations, namely, A, B, and C, we build the following feature vectors: $\{A, B, C\}$ as length-1 patterns, $\{AB, BC, AC\}$ as length-2 patterns, and $\{ABC\}$ as a length-3 pattern. We name feature vectors of node and its direct links by their corresponding pattern lengths. For instance, t1, i1, and o1 correspond to length-1 pattern counts observed in a node v, observed among the nodes pointing to v, and observed among the nodes pointed by v, respectively. Similarly, t2, i2, and o2 represent length-2 patterns, and t3, i3, and o3 represent length-3 patterns.

Different learning sets are considered to compute the impact of pattern lengths in prediction capability:

$c1 = t_1|><|i_1|><|o_1$, $c2 = c1|><|t_2|><|i_2|><|o_2$, and $c3 = c2|><|t_3|><|i_3|><|o_3$, where $|><|$ is the joining operator.

In other words, all feature vectors is generated corresponding to patterns of the same length together.

Note that, we omit all patterns that contain the target annotation from the t1, t2, and t3 portions of the learning sets. For example, if we are attempting to predict annotation A, and t2 vector of a node has patterns $\{AB, BC, AC\}$, we omit $\{AB, AC\}$ from t2, as such patterns already indicate the existence of A.

Predictive capability of c1, c2, and c3 datasets are presented in Fig. 3. An improvement of 8% is observed when length-2 patterns' feature vectors are added, and improvement is only 3.5% when length-3 patterns feature vectors added (according to the lift values at the first quantile). We conclude that more complex patterns in the feature vectors definitely improve prediction capability. However, such improvement gets less significant while the complexity is increasing.

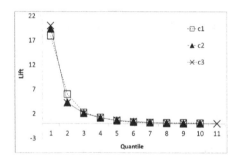

Fig. 3. Impact of pattern complexity on prediction capability

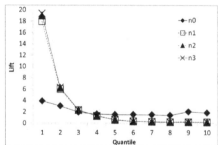

Fig. 4. Impact of network neighborhood radius on prediction capability

3.4   Prediction Information Coming from Indirect Neighbors

In this experiment, we study whether adding data from network neighborhoods with larger radiuses supplement prediction capability. In other words, we first build a feature vector using only the annotations from a node v (say, neighborhood n_0). Then we test with additional annotations coming from direct neighbors of v (say, neighborhood n_1 that is both incoming and outgoing neighbors). Next, we increase radiuses to n_2, n_3, and n_4 as well. In order to augment network neighborhood annotations into a single vector, we inner join feature vectors of each radius. For instance n_3 means, self-annotations of node v (i.e., n_0), joined with annotation vector of direct neighbors of v (i.e., r_1), joined with indirect neighbors in distance 2 and 3 of v, namely r_2 and r_3: $n_i = n_0 \mathbin{|><|} r_1 \mathbin{|><|} \ldots \mathbin{|><|} r_i$.

Note that we employ annotation patterns of length 1 for this experiment. In this experiment, we observed that, annotations of a node and annotations of direct neighbors provide the most of the prediction power. Radiuses 2 and higher do not bring additional prediction capability worth the computational effort (see Fig. 4). Lift improvement at the first quantile for n_3 over n_2 is 1.7%, and n_2 over n_1 is 5.8%. For the second quantile, improvement for n_3 over n_2 is 1%, and n_2 over n_1 is 1.04%.

4 Conclusions

In this paper, we proposed a methodology for converting network links into network annotations, and predicting missing links and annotations. Our approach is scalable in the sense that network data is preprocessed into a dense and simple feature vector before consumed by complicated machine learning algorithms. We observed in our experimental evaluation that augmenting network-based features to a simple correlation-based prediction algorithm provides 11-27% improvement in prediction accuracy. We plan to extend our work by evaluating use of more complex patterns, and evaluating scalability tradeoffs of those.

Acknowledgments. We would like to thank Zeki Erdem for his comments and support during the development of this work.

References

1. Clauset, A., Moore, C., Newman, M.E.J.: Hierarchical structure and the prediction of missing links in networks. Nature 453(7191), 98–101 (2008)
2. Goldberg, D., Roth, F.: Assessing experimentally derived interactions in a small world. Proc. Natl. Acad. Sci. U.S.A. (2003)
3. Golub, B., Jackson, M.O.: How homophily affects the speed of learning and best-response dynamics. Quarterly Journal of Economics (2012)
4. Gupta, P., Goel, A., Lin, J., Sharma, A., Wang, D., Zadeh, R.: WTF: The Who to Follow Service at Twitter. In: Proceedings of the 22nd International Conference on World Wide Web, WWW 2013, pp. 505–514 (2013)
5. Kirac, M., Ozsoyoglu, G., Yang, J.: Annotating proteins by mining protein interaction networks. Bioinformatics 22(14) (2008)
6. Kwak, H., Lee, C., Park, H., Moon, S.: What is Twitter, a Social Network or a News Media? In: Proceedings of the 19th International Conference on World Wide Web, WWW 2010, pp. 591–600 (2010)
7. Lee, D.: Document ranking and the vector-space model. IEEE Computer Society 14(2) (1997)

8. Lee, R., Sumiya, K.: Measuring geographical regularities of crowd behaviors for twitter-based geo-social event detection. In: LBSN (2010)

9. Liben-Nowell, D., Kleinberg, J.M.: The link prediction problem for social networks. In: LinkKDD (2004)

10. Milenova, B., Yarmus, J., Campos, M.: SVM in Oracle Database 10g: Removing the Barriers to Widespread Adoption of Support Vector Machines. In: Very Large Databases, VLDB (2005)

11. Pennacchiotti, M., Popescu, A.M.: A machine learning approach to twitter user classification. In: AAAI Conference on Weblogs and Social Media (2011)

12. Sen, P., Namata, G., Bilgic, M., Getoor, L., Gallagher, B., Eliassi-Rad, T.: Collective classification in network data. AI Magazine 29(3) (2008)

13. Taskar, B., Wong, M.F., Abbeel, P., Koller, D.: Link prediction in relational data. In: Proceeding of Neural Information Processing Systems (2003)

14. Twitter Inc.: Twitter rest api, https://dev.twitter.com/docs/api

15. Zhou, T., Lu, L., Zhang, Y.C.: Predicting missing links via local information. The European Physical Journal B 71(4) (2009)

Time Series Forecasting with Volume Weighted Support Vector Machines

Kamil Żbikowski

Institute of Computer Science
Faculty of Electronics and Information Technology
Warsaw University of Technology
ul. Nowowiejska 15/19, 00-665 Warsaw, Poland
kamil.zbikowski@ii.pw.edu.pl

Abstract. Accurate prediction of financial time series or their direction of changes may result in highly profitable returns. There are many approaches to build such models. One of them is to apply machine learning algorithms i.e. Neural Networks or Support Vector Machines. In this paper we would like to propose a modified version of Support Vector Machine classifier, Volume Weighted Support Vector Machine which has the ability to predict short term trends on the stock market. Modification is based on the assumption that incorporating transaction volume into penalty function may lead to better future trends forecasting. Experimental results obtained on the data set composed of daily quotations from 420 stocks from S&P500 Index showed that proposed method gives statistically better results than basic algorithm.

Keywords: Support Vector Machines, stock trading, time series, investment strategies backtesting.

1 Introduction

For decades, investors and scientists have been looking for a method to successfully predict trends on the stock market. Due to its nonstationary nature this is a very challenging task. To accomplish it the state of the art techniques should be applied. Despite of this, the results have often high level of uncertainty.

There are plenty of techniques to analyse financial times series data. Plain statistical approach to this problem was presented in [5]. Recently much more attention got machine learning algorithms which have been successfully applied in other fields such as healthcare or transportation. An example of application of fuzzy decision trees to forecast trading signals from financial data is described in [4]. Neural networks and their modifications are also widely used to extract trading recommendations from time series data. One of many examples is described in [2], where the circular back-propagation neural network was employed to predict values of future prices. Comparison of Neural Networks and Support Vector Machines for futures contracts prediction was described in [9]. It showed that SVM gives promising results and can be further investigated.

S. Kozielski et al. (Eds.): BDAS 2014, CCIS 424, pp. 250–258, 2014.
© Springer International Publishing Switzerland 2014

Machine learning algorithms do no require any assumption about distribution of underlying data. This is their great advantage over other solutions taken i.e. from econometrics such as autoregressive model(AR) or autoregressive moving average model (ARMA).

As an input vector for prediction models different indicators are usually selected. Most of them are taken from technical analysis which is based on the study of past prices and volume of securities. Transaction volume analysis is especially important when it comes to confirmation of price movements [6]. Because of this fact the idea of incorporate it into machine learning algorithm, such as SVM, seems to be justified and worth examination.

2   Support Vector Machines for Classification

In [10] Vapnik and Chervonenkis described the structural risk minimization principle. This was a novel approach which emphasises the importance of minimizing error on test data. It was showed that this error increases when complexity of the statistical model grows. It adapts well to training data whereas accuracy on unseen sample is getting lower. The boundary of an error R^f on the test set for a given model is defined with the probability $1 - \eta$ as follows:

$$R^f \leq R_{emp} + \phi(\frac{h}{m}, \frac{\log(\eta)}{m}), \tag{1}$$

where ϕ is the measure of model capacity defined as:

$$\phi(\frac{h}{m}, \frac{\log(\eta)}{m}) = \sqrt{\frac{h(log(\frac{2m}{h} + 1)) - log(\frac{\eta}{4})}{m}}, \tag{2}$$

where h is the Vapnik–Chervonenkis dimension computed for linear classifiers in $\mathbb{R}^n$ as $h = n + 1$, m is the number of examples in the training set and R_{emp} is an empirical error on the training set.

Practical implementation of this principle is SVM classifier that was originally proposed in 1995 by Vapnik and Cortes [3]. The basic idea of SVM is to minimize margin of decision hyperplane:

$$f(\mathbf{x}) = \mathbf{w}\varphi(\mathbf{x}) + b, \tag{3}$$

where $\mathbf{x}$ is an input vector and φ is non-linear function mapping $\mathbf{x}$ to high dimensional feature space. This operation would be time consuming but the kernel transformation can be applied [12]. It allows to compute the inner product in the high dimensional feature space without ever transforming input vector to that space:

$$K(x_1, x_2) = \varphi(x_1)\varphi(x_2). \tag{4}$$

Then margin shown in Fig. 1 is defined as follows:

$$M = \frac{1}{||\mathbf{w}||}. \tag{5}$$

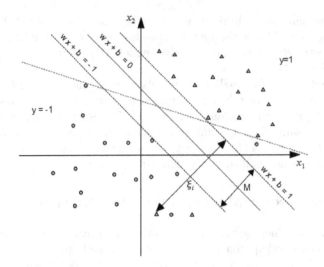

Fig. 1. Classification with basic SVM classifier

SVM can be used in classification as well as in regression analysis. In the classification task the optimization problem has the following form:

$$\min_{\mathbf{w} \in \mathcal{H}, b \in \mathcal{R}} \frac{1}{2} ||\mathbf{w}||^2 + C \sum_{i}^{m} \xi_{\mathbf{i}} \tag{6}$$

subject to:

$$y_i(\mathbf{w}\mathbf{x_i} - b) \geq 1 - \xi_i, \tag{7}$$

$$\xi_i \geq 0, \tag{8}$$

where ξ_i are slack variables that defines the degree of misclassification, C is the
penalty parameter, x_i is the ith input vector and y_i is the corresponding target
value. Expression 6 allows to incorporate into the model the fact that problem
is not always separable. Problem 6 can be reformulated using Lagrange function
wit multipliers α_i and μ_i:

$$L(\mathbf{w}, b, \alpha) = \frac{1}{2} ||\mathbf{w}||^2 + C \sum_{i=1}^{m} \xi_i - \sum_{i=1}^{m} \alpha_i(y_i(\mathbf{w}\mathbf{x_i} - b) - 1 + \xi_i) - \sum_{i=1}^{m} \mu_i \xi_i. \tag{9}$$

Then partial derivatives of L are computed to obtain optimality constraints:

$$\frac{\partial L(\mathbf{w}, b, \alpha, \mu)}{\partial \mathbf{w}} = 0 \Rightarrow \mathbf{w} = \sum_{i}^{m} \alpha_i y_i \mathbf{x_i}, \tag{10}$$

$$\frac{\partial L(\mathbf{w}, b, \alpha)}{\partial b} = 0 \Rightarrow \sum_{i=1}^{m} \alpha_i y_i = 0, \tag{11}$$

$$\frac{\partial L(\mathbf{w}, b, \alpha, \mu)}{\partial \xi_i} = 0 \Rightarrow C = \mu_i + \alpha_i. \tag{12}$$

Based on constraints 10, 11 and 12 equation 9 can be stated as the dual problem, which has the following form:

$$Q(\alpha) = \sum_{i=1}^{m} \alpha_i - \frac{1}{2} \sum_{i=1}^{m} \sum_{j=1}^{m} \alpha_i \alpha_j y_i y_j \varphi(\mathbf{x_i}) \varphi(\mathbf{x_j}). \tag{13}$$

The dual problem $Q(\alpha)$ has to be maximized according to following conditions:

$$\sum_{i=1}^{m} \alpha_i y_i = 0, \tag{14}$$

$$\alpha_i \geq 0. \tag{15}$$

3   Volume Weighted Support Vector Machines

SVM classifier in its basic formulation gives all examples equal weights. As was shown in [8] assigning different weights to particular examples may result in better prediction capability for Support Vector Regression. In Volume Weighted Support Vector Machines classifier (VW-SVM) loss function is transformed as follows:

$$\min_{\mathbf{w} \in \mathcal{H}, b \in \mathcal{R}} \frac{1}{2} ||\mathbf{w}||^2 + \sum_{i=1}^{m} C_i \xi_i, \tag{16}$$

where penalty term is different for each of input vectors. Generally it is of the form:

$$C_i = v_i C, \tag{17}$$

where v_i is the weight for each example. For each training phase the input matrix $\mathbb{X}$ and the corresponding vector of weights $\mathbb{V}$ are defined as follow:

$$\mathbb{X} = \begin{bmatrix} x_{t-m,1} & x_{t-m,2} & \cdots & x_{t-m,n} \\ x_{t-m+1,1} & x_{t-m+1,2} & \cdots & x_{t-m+1,n} \\ \vdots & \vdots & \ddots & \vdots \\ x_{t,1} & x_{t,2} & \cdots & x_{t,n} \end{bmatrix} \quad \mathbb{V} = \begin{bmatrix} v_{t-m} \\ v_{t-m+1} \\ \vdots \\ v_t \end{bmatrix}, \tag{18}$$

where $x_{t,j}$ denotes jth feature for the moment t and m is the training set size. In the experiment presented in this paper each weight v_t for the moment t takes the following form:

$$v_t = \frac{\sum_{k=0}^{d} V_{t-k}}{\sum_{i=0}^{m} W_{t-i}}, \tag{19}$$

where V_{t-k} is the real transactional volume for particular example from training set for the moment t with delay of k periods. The length d is defined as length of the longest technical indicator which occurs in the jth input vector. The number of plain quotations which are used to compute features values for the training set is equal $m + d$ i.e. to compute value of x_{t-m} it is required to use quotations from the range between $t - m - d$ and $t - m$. Volume W_i is defined as a sum of volumes from all d periods taken into consideration for the moment t.

After introducing the Lagrange multipliers and computing derivatives the dual problem of 16 takes the following form:

$$Q(\alpha) = \sum_{i=1}^{m} \alpha_i - \frac{1}{2} \sum_{i=1}^{m} \sum_{j=1}^{m} \alpha_i \alpha_j y_i y_j \varphi(\mathbf{x_i})\varphi(\mathbf{x_j}). \qquad (20)$$

As the kernel function for the testing purpose the radial basis function was used:

$$K(x_1, x_2) = \exp(\frac{\|x_1 - x_2\|^2}{\gamma}), \qquad (21)$$

where γ was the number of features in the input vector.

As was showed in 20 modification 16 does not require any changes in the optimization procedure. It can be still solved by the sequential minimal optimization algorithm [7]. However, several changes were made in LIBSVM package in order to adapt this SVM library to use examples weighting.

4 Experimental Results

4.1 Data Set

In order to determine whether application of VW-SVM provides significantly better results 420 time series were chosen. All come from S&P500 index and have sufficient history to provide daily quotation for period from 2003-01-01 to 2013-10-21.

Common approach in machine learning is to divide dataset into three parts: training, test and validation. As was described in [1] financial data series due to their nonstationarity require different methods. For purpose of this experiment the walk forward procedure was conducted. It is based on applying moving window to available time series data (see Fig. 2). First, model is optimized on the window of length oWL and then tested on next tWL samples. Next optimization window is shifted by tWL and next optimization occurs. The procedure is repeated until the end of data is reached. According to the construction of the testing procedure the whole data set can be logically divided in two subsets. First one consists of oWL examples, marked on the 2 as DFL, is a data feed necessary from the optimization perspective but no actual predictions occurs during this period. The second one, marked as TW, is the subset when predictions take place.

Each input vector consists of three indicators: 10 days Relative Strength Index (RSI^{10}), 10 days rate of return (R^{10}) and 10 days Williams %R oscillator

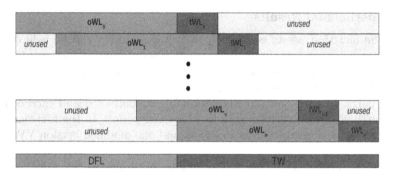

Fig. 2. Walk forward analysis

$(\%R^{10})$. RSI^n measures velocity and magnitude of price movement. It is defined as follows:

$$RSI^n = 100 - \frac{100}{1 + RS^n}, \tag{22}$$

$$RS^n = \frac{\sum_{t=0}^{n} U_t}{\sum_{t=0}^{n} D_t}, \tag{23}$$

where U_t and D_t are respectively upward and downward trend indicators with the following definition:

$$U_t = \begin{cases} P_t - P_{t-1} & \text{if } P_t \geq P_{t-1} \\ 0 & \text{if } P_t < P_{t-1} \end{cases}, \quad D_t = \begin{cases} P_{t-1} - P_t & \text{if } P_t < P_{t-1} \\ 0 & \text{if } P_t \geq P_{t-1} \end{cases}, \tag{24}$$

where P_t is end of day price for particular stock at the day t. N-day simple rate of return is computed as follows:

$$R^n = \frac{P_t - P_{t-n}}{P_{t-n}}. \tag{25}$$

$\%R^n$ oscillator which expresses aberration of the price series from its maximal value for a given period of time is defined as:

$$\%R^n = 100 \frac{\max\{P_t, P_{t-1}, \ldots, P_{t-n}\} - P_t}{\max\{P_t, P_{t-1}, \ldots, P_{t-n}\} - \min\{P_t, P_{t-1}, \ldots, P_{t-n}\}}. \tag{26}$$

There were two classes for classifiers to predict, both of them describing trend: *Upward* and *Downward*. They depend on the rate of return for next 5 days. If R^5 is greater or equal than 0 than target class should be *Upward*. In case it is less than 0 the input vector should be assigned to *Downward* class.

The input matrix from 18 takes the following form in this particular case:

$$\mathbb{X} = \begin{bmatrix} RSI_{t-oWL}^{10} & R_{t-oWL}^{10} & \%R_{t-oWL}^{10} \\ RSI_{t-oWL+1}^{10} & R_{t-oWL+1}^{10} & \%R_{t-oWL+1}^{10} \\ \vdots & \vdots & \vdots \\ RSI_{t}^{10} & R_{t}^{10} & \%R_{t}^{10} \end{bmatrix}. \tag{27}$$

4.2   Experimental Results

Due to the fact that assets considered in this experiment can be held only as long positions it is reasonable to analyse only *Upward* trend detections. One can imagine a trading algorithm which opens position after model predicts *Upward* trend in the next 5 days and closes it after this period. Experiments were divided into two parts. In each one 420 simulations were conducted and each of them goes through 2577 time series points for the given period. In the first one basic SVM algorithm was used whereas in the second one modified version VW-SVM was tested.

Results are presented in Tab. 1. To compare both models the following ratio was used:

$$\psi = \frac{\sum_{i=0}^{d} Tp_i}{\sum_{i=0}^{d} Fp_i}, \tag{28}$$

where Tp_i and Fp_i are respectively numbers of true and false positives; d equals 420 which is the number of experiments which were held for each classifier.

It can be seen that results for VW-SVM are better than for basic SVM. Both means: true positives $\bar{T}p$ and false negatives $\bar{F}p$ have change in the desired direction. In particular, number of true positives increased by 27 on average. On the other hand, the amount of false positives decreased by 28 on average. In Fig. 3 histograms of ψ ratio are compared. It can be seen that mean value is shifted to the right for the VW-SVM classifier.

Table 1. SVM and VW-SVM results in Upward trend prediction. $\bar{T}p$ - mean *true positives* rate which denotes how many of predicted *Upward* trends was actually from this class; $\bar{F}p$ - mean *false positives* rate which denotes how many of predicted *Upward* trends was not correctly classified;ψ - mean of true positives to mean of false positives ratio among all simulations; σ_ψ - standard deviation of ψ.

•	$\bar{T}p$	$\bar{F}p$	ψ	σ_ψ
SVM	860	591	1.4749	0.3019
VW-SWVM	887	563	1.6147	0.3985
Δ	27	-28	0.1398	0.0966

Presented numbers raise a question of statistical significance of those results. In order to investigate this problem Welch's t test [11] was used. It's null hypothesis is that means of both populations are equal. It does not assume that two distributions have the same variance or sample size which is appropriate for our case. The t statistics is defined as following formula:

$$t = \frac{\bar{X}_1 - \bar{X}_2}{\sqrt{\frac{\sigma_1}{N_1} + \frac{\sigma_2}{N_2}}} \tag{29}$$

Tp/Tn ratio histograms

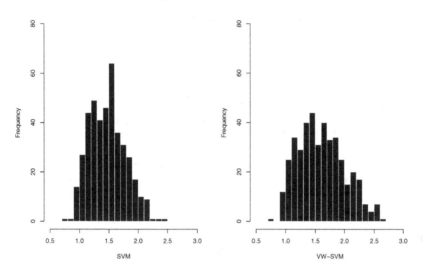

Fig. 3. Histogram of ψ ratio for both test cases: SVM and VW-SVM

and has Student t distribution with ν degrees of freedom that can be computed as follows:

$$\nu \approx \frac{\left(\frac{\sigma_1^2}{N_1} + \frac{\sigma_2^2}{N_2}\right)}{\frac{\sigma_1^4}{N_1^2(N_1-1)} + \frac{\sigma_2^4}{N_2^2(N_2-1)}}, \tag{30}$$

where $\bar{X}_1$, $\bar{X}_2$ are means of two populations; σ_1, σ_2 are their standard deviations; N_1 and N_2 denotes sizes of particular samples.

Value of computed Welch's statistics is 5.73 which allows us to reject the null hypothesis about equality of means of those two samples in favor of the alternative hypothesis at the 99% confidence level.

5   Conclusion

We have proposed a modification of standard SVM classifier which uses volume weights for training examples. We showed that it can be still optimized with the Sequential Minimal Optimization algorithm due to the fact that dual optimization problem does not change after introducing new penalty function. As the input for the presented model several technical indicators were used. As the target value two classes were chosen which both indicated short term trends: *Upward* and *Downward*.

VW-SVM effectively forecast short term trends on the stock market. Its results are better than basic algorithm. There are two issues worth examining in the further work. Firstly, it is interesting whether predictions about short term rate of return can be converted to profitable investment strategy. It is quite promising because of the high accuracy of trends forecasting presented in this paper. Secondly, it should be examined whether there exists other weight functions that would give better results than presented volume based weights.

References

1. Ładyżyński, P., Żbikowski, K., Grzegorzewski, P.: Stock Trading with Random Forests, Trend Detection Tests and Force Index Volume Indicators. In: Rutkowski, L., Korytkowski, M., Scherer, R., Tadeusiewicz, R., Zadeh, L.A., Zurada, J.M. (eds.) ICAISC 2013, Part II. LNCS, vol. 7895, pp. 441–452. Springer, Heidelberg (2013)
2. Chen, S., Dai, Q.: Discounted least squares-improved circular back-propogation neural networks with applications in time series prediction. Neural Computing and Applications 14(3), 250–255 (2005),
http://link.springer.com/10.1007/s00521-004-0461-9
3. Cortes, C., Vapnik, V.: Support-vector networks. Machine Learning 20(3), 273–297 (1995)
4. Lai, R.K., Fan, C.Y., Huang, W.H., Chang, P.C.: Evolving and clustering fuzzy decision tree for financial time series data forecasting. Expert Systems with Applications 36(2), 3761–3773 (2009),
http://linkinghub.elsevier.com/retrieve/pii/S0957417408001474
5. Momot, A., Momot, M.: Perspektywy zastosowań metod statystycznych w konstrukcji strategii działania na rynkach kapitałowych – wykorzystanie systemów hierarchicznych oraz regularyzacji. Studia Informatica 34(2A), 263–274 (2013)
6. Murphy, J.J.: Technical Analysis of the Financial Markets. New York Institute of Finance (1999)
7. Platt, J.C.: Fast Training of Support Vector Machines Using Sequential Minimal Optimization. In: Advances in Kernel Methods, pp. 185–208. MIT Press (1999)
8. Tay, F.E.H., Cao, L.J.: Modified support vector machines in financial time series forecasting. Neurocomputing 48, 847–861 (2002)
9. Tay, F.E.H., Cao, L.: Application of support vector machines in financial time series forecasting. Omega 29, 309–317 (2001)
10. Vapnik, V., Chervonenkis, A.J.: On The Uniform Convergence of Relative Frequencies of Events to their Probabilities. Theory of Probability and its Applications 16(2), 264–280 (1971)
11. Welch, B.L.: The generalization of "student's" problem when several different population variances are involved. Biometrika 34, 28–35 (1947)
12. Yu, H., Kim, S.: SVM Tutorial – Classification, Regression and Ranking. In: Rozenberg, G., Bäck, T., Kok, J.N. (eds.) Handbook of Natural Computing, pp. 479–506. Springer, Heidelberg (2012),
http://link.springer.com/10.1007/978-3-540-92910-9

Application of Ordered Fuzzy Numbers in a New OFNAnt Algorithm Based on Ant Colony Optimization

Jacek M. Czerniak[1], Łukasz Apiecionek[2], and Hubert Zarzycki[3]

[1] Kazimierz Wielki University in Bydgoszcz, Institute of Technology
ul. Chodkiewicza 30, 85-064 Bydgoszcz, Poland
jczerniak@ukw.edu.pl
[2] Foundation for Development of Mechatronics,
ul. Jeynowa 19, 85-343 Bydgoszcz, Poland
lapiecionek@mechatronika.org.pl
[3] Wroclaw School of Applied Informatics "Horyzont",
ul. Wejherowska 28, 54-239 Wroclaw, Poland
hzarzycki@horyzont.eu

Abstract. This paper describes the results of experiments concerning the optimization method called OFNAnt. As the benchmarks, the author used the set of files from TSPlib repository which includes well known samples of the travelling salesman problem. The innovation of the proposed method consists in implementation of Ordered Fuzzy Numbers to the decision-making process of individual ant agents. This also made it possible to correlate the colony development optimization with the trend. Previous implementations of the fuzzy logic to such meta heuristics like ant systems came down to fuzzy control over the decision-making process of an ant or fuzzy control of the pheromone release mechanism. Thanks to the proposed method, it was possible to expand the family of solutions with the solutions represented by ants moving outside the main circulation. The improvement was possible thanks to better stress that, according to the OFN arithmetic, was put on their participation in the process as compared to the conventional approach, as the direction of their movement has been opposed to the trend followed by majority of colonies. Final conclusions of the experiment indicate to superiority of methods based on Ant Colony Optimization, and in particular the superiority of OFNAnt method over heuristic methods.

Keywords: fuzzy logic, fuzzy numbers, ordered fuzzy numbers, ant colony optimization, fuzzy ant, OFNAnt, OFN, ACO, TSP.

1 Introduction

Observation of living organisms is an interesting research field not only for biologists. A new current within Artificial Intelligence called Swarm Intelligence acquired significance in the 1990-ties [17]. His studies were inspired by observation of animals and insects living in colonies. We have finally got successful

S. Kozielski et al. (Eds.): BDAS 2014, CCIS 424, pp. 259–270, 2014.
© Springer International Publishing Switzerland 2014

experiments and methods based on ant or termite colony observation [6]. Observations of birds in V-formation inspired many researchers to create and to develop the concept of Particle Swarm Optimization [19].

1.1   Swarm Intelligence

Those studies in the field of AI were also inspired by information obtained from marine biologists on collective intelligence of a shoal of fish or plankton. Other sources of inspiration stemmed from the development of industry, in particular the automotive industry in that case. Particle swarm optimization was created thanks to studies on, among others, sand-blasting of a car body or other corroded metal parts. Hence, generally, this branch of AI has been called swarm intelligence [17]. Conversion of those intelligence mechanisms prevailing amongst simple individuals into the field of computer systems resulted in creation of the current called sometimes Computational Swarm Intelligence. It exists parallel to the branch of science called Multi-agent systems and those two fields often overlap one another. Although they are often not directly based on associations with colonies of living organisms but they are often similar as regards the rules of their operation. They enable creation of interesting implementations in the domain of Parallel computing. Development of Swarm intelligence [8] was preceded by the development of multiple-valued logic, in particular the fuzzy logic.

1.2   Fuzzy Logic

The author of Fuzzy logic is an American professor of the Columbia University in New York City and of Berkeley University in California Lotfi A. Zadeh, who published the paper entitled Fuzzy sets in the journal Information and Control in 1965 [20]. He defined the term of a fuzzy set there, thanks to which imprecise data could be described using values from the interval (0,1). The number assigned to them represents their degree of membership in this set. It is worth mentioning that in his theory L. Zadeh used the article on 3-valued logic published 45 years before by a Pole Jan ukasiewicz [16]. That is why many scientists in the world regard this Pole as the father of fuzzy logic. Next decades saw rapid development of fuzzy logic. As next milestones of the history of that discipline one should necessarily mention L-R representation of fuzzy numbers proposed by D.Dubois and H. Prade [7,4,10], which enjoys great successes today. Coming back to the original analogy, an observer can see a trend, i.e. general increase during rising tide or decrease during low tide, regardless of momentary fluctuations of the water surface level. This resembles a number of macro and micro-economic mechanisms where trends and time series can be observed. The most obvious example of that seems to be the bull and bear market on stock exchanges, which indicates to the general trend, while shares of individual companies may temporarily fall or rise.

1.3   Ordered Fuzzy Numbers and Ant Colony Optimization

The aim is to capture the environmental context of changes in the economy or another limited part of reality. Changes in an object described using fuzzy logic seem to be thoroughly studied in many papers. But it is not necessarily the case as regards linking those changes with a trend. Perhaps this might be the opportunity to apply generalization of fuzzy logic which are, in the opinion of authors of that concept, W. Kosiński [3,13] and his team [12,14], Ordered Fuzzy Numbers. There are already interesting studies available published by well-known scientists [18] which present successful implementation of fuzzy logic to swarm intelligence methods, including methods inspired by ant and termite colonies. However, according to the best knowledge of the authors of this paper, nobody has published studies on implementation of Ordered Fuzzy Logic into Ant Colony Optimization so far. This fact was one of the reasons for execution of the research described in this paper The main emphasis in this paper is on application of a new, hybrid method of ant colony optimization (ACO) with implemented decision logic of an ant calculated in OFN domain in order to solve the optimum route selection problem. To make a comparison, authors selected several well-known ant methods and several heuristic methods dedicated for solving the same problem, the methodology of which does not use either swarm intelligence or, in particular, ACO.

2 Application of Ant Colony Algorithms in Searching for the Optimal Route

Ant Colony Optimization (ACO) is one of currently the best known ant colony algorithms. It was first defined by Marco Dorigo, Di Caro and Gambardell in 1999 as a method used to solve discrete optimization problems. ACO was presented as the algorithm, which can find good route using a graph. It is the algorithm inspired by foraging theory [5] both for ant colonies and for discrete optimization problems. This algorithm is designed for solving two kinds of static and dynamic optimization problems. In the general case, the ant colony optimization is performed according the following diagram (Fig. 1).

If screenshots are necessary, please make sure that you are happy with the print quality before you send the files.

Studies on ant colony algorithms were commenced based on observation of ant colony environment. The scientists noticed interesting fact that ants communicate mainly using chemical substances produced by them. As it has already been mentioned, the key matter in this algorithm is indirect foraging communication represented by pheromone trace. The advantage of the evaporation of that pheromone is that it can prevent convergence for local optimum solutions. Assuming that there is no evaporation issue, each time each path selected by first artificial agents would be treated in the same way and would be equally attractive, which would make it inapplicable to optimization problems. Thus, when one ant finds good path from the colony to the food source, this path becomes more preferable for other ants. The idea behind the ACO [9] algorithm is to follow that behavior using artificial agents moving within the frame of a graph in order to solve a given problem. The ACO algorithm has been used for solving TSP problem. This algorithm has an advantage over genetic algorithms

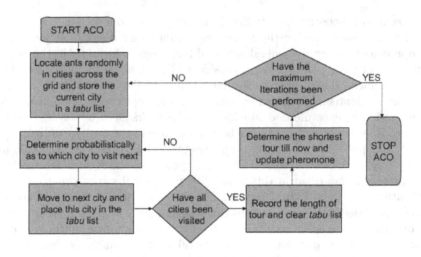

Fig. 1. ACO block diagram

or simulated annealing algorithm. Its important feature is that for a dynamically changing graph, ACO algorithm can work continuously and it can adapt to the changes in real time. Thanks to such properties, it has been applied for the method of solving the problem of network routing and urban transportation systems. Route selection an ant shall travel the distance from point i to point j with the probability of:

$$p_{ij} = \frac{(\tau_{i,j}^{\alpha})(\eta_{i,j}^{\beta})}{\sum(\tau_{i,j}^{\alpha})(\eta_{i,j}^{\beta})} \tag{1}$$

where:
$\tau_{i,j}$ - quantity of pheromone on the route i, j,
$\eta_{i,j}$ - defines attraction of the route i, j,
α - parameter used for effect control $\tau_{i,j}$,
β - parameter used for effect control $\eta_{i,j}$.

Pheromone update this issue is represented by the following formula

$$\tau_{i,j} = p\tau_{i,j} + \Delta\tau_{i,j} \tag{2}$$

where:
$\tau_{i,j}$ - quantity of pheromone on the route i, j,
$\Delta\tau_{i,j}$ - represents the quantity of left pheromone,
P pheromone evaporation scale.

Below, we presented more detailed pseudocode of one of the numerous ant colony algorithms, called ACS (Ant Colony System), i.e. the ant colony optimization.

The pseudocode of ACS

```
Initialize
Repeat {
 Place each ant in a randomly chosen city;
 For each ant
   Repeat {
     Choose NextCity (each ant);
     Update pheromone levels using a local rule;
   } Until (No more cities to visit);
   Return to the initial cities;
   Compute the length of the Tour found by each ant;
 End For;
 Update pheromone level using a global rule;
}
Print Best Path;
```

The following tables present the most important ant colony optimization algorithms dedicated for TSP in chronological order of their publishing. In the methodological sense, all of the algorithm listed below and described in the following section are direct successors of Ant System. This is due to an obvious reason, i.e. the Ant System method, which has become the foundations for the entire new branch of knowledge (ACO) was the first worldwide success of then young scientist Marco Dorigo. Now professor M. Dorigo [6] is a world class expert in the field of Swarm Intelligence. The set of methods presented below is in chronological order.

Table 1. *ACO* algorithms which have already been applied to the *TSP*

ACO method	Authors
Ant System (AS)	Dorigo 1992; Dorigo, Manizzo, Colorni 1996;
Elitist AS (EAS)	Dorigo 1992; Dorigo, Manizzo, Colorni 1996;
Ant-Q (AQ)	Gambardella, Dorigo 1995-96;
Ant Colony System (ACS)	Dorigo, Gambardella 1997;
Max-Min AS (MMAS)	Sttzle 1999; Sttzle, Hoos 2000;
Rank-base AS (ASrank)	Bullnheimer, Hartl, Strauss 1997-99;

3   Ordered Fuzzy Numbers

First attempts to redefine new operations on fuzzy numbers were undertaken at the beginning of the 1990-ties by Witold Kosiński and his PhD student P. Sysz [13]. Further studies of W. Kosiński published in cooperation with P. Prokopowicz and D. Ślęzak [11,3,12,14] led to introduction of the ordered fuzzy numbers model OFN.

Definition 1. *An ordered fuzzy number A was identified with an ordered pair of continuous real functions defined on the interval* $[0,1]$, *i.e.,* $A = (f, g)$ *with* $f, g \in [0,1] \to R$ *as continuous functions. We call f and g the up and downparts of the fuzzy number A, respectively. In order to comply with the classical denotation of fuzzy sets (numbers), the independent variable of both functions f and g is denoted by y, and their values by x* [13].

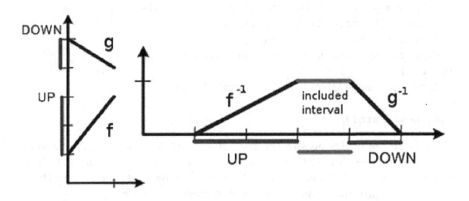

Fig. 2. OFN presented in a way referring to fuzzy numbers

Continuity of those two parts shows that their images are limited by specific intervals. They are named respectively: UP and DOWN [12,14,4]. The limits (real numbers) of those intervals were marked using the following symbols:
$UP = (l_A, l_A^-)$ and $DOWN = (l_A^+, p_A)$
If both functions that are parts of the fuzzy number are strictly monotonic, then there are their inverse functions x_{up}^{-1} and x_{down}^{-1} defined in respective intervals UP and $DOWN$ and the following assignment is valid:

$$l_A : x_{up}(0)l_A^- := x_{up}(1)l_A^+ := x_{down}(1)p_A := x_{down}(0) \tag{3}$$

If a constant function equal to 1 is added within the interval $[l_A^-, l_A^+]$ we get UP and $DOWN$ with one interval (Fig. 3), which can be treated as a carrier. Then the membership function $\mu_A(x)$ of the fuzzy set defined on the R set is defined by the following formulas:

$$\mu_A(x) = 0 \quad for \quad x \notin [l_A, p_A]$$
$$\mu_A(x) = x_{up}^{-1}(x) \quad for \quad x \in UP \tag{4}$$
$$\mu_A(x) = x_{down}^{-1}(x) \quad for \quad x \in DOWN$$

The fuzzy set defined in that way gets an additional property which is called **order**. Whereas the following interval is the carrier:

$$UP \cup [l_A^-, l_A^+] \cup DOWN \tag{5}$$

The limit values for up and down parts are:

$$\mu_A(l_A) = 0, \mu_A(l_A^-) = 1, \mu_A(l_A^+) = 1, \mu_A(p_A) = 0 \tag{6}$$

Generally, it can be assumed that ordered fuzzy numbers are of trapezoid form. Each of them can be defined using four real numbers:

$$A = (l_A \quad l_A^- \quad l_A^+ \quad p_A) \tag{7}$$

The figures below show sample ordered fuzzy numbers including their characteristic points.

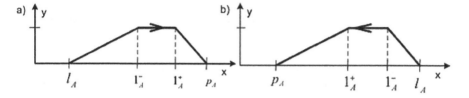

Fig. 3. Fuzzy number that is ordered a) positively b) negatively

Functions fA,gA correspond to parts upA , downA R2 respectively, so that:

$$up_A = (f_A(y), y) : y \in [0, 1] \tag{8}$$

$$down_A = (g_A(y), y) : y \in [0, 1] \tag{9}$$

The orientation corresponds to the order of graphs f_A and g_A.

4   OFNAnt - A New Ant Colony Algorithm

Implementation of OFN to the ant colony system consists mainly in determination of the trend and in establishing relationship to the order of OFN number. This order is used in OFNAnt in two ways. On the one hand, it is related to pheromone evaporation on the route and its mathematical description. On the other hand, it concerns the decision-making process of a single ant. The pheromone quantity on the route is updated in accordance with OFN arithmetic. If the pheromone trace (quantity) on the route increases, then this trend is marked as a positive order trend, whereas if this quantity decreases - it is marked as a negative order trend. Each pass of k-th ant, which is associated with placing a pheromone trace results in the update of the pheromone trace on the route by the amount left by the ant resulting in positive order on the route with increasing trend and with negative order for decreasing trend of the route. The above relationship is pursuant to the following formula:

$$\tau_{ij}[l_A, l_A^-, l_A^+, p_A] \leftarrow \tau_{ij}[l_A, l_A^-, l_A^+, p_A] + \sum_{k=1}^{m} \Delta\tau_{ij}^k[l_K, l_K^-, l_K^+, p_K] \qquad (10)$$

Every ant constructs a complete route, and the ants make a decision at each stage of the route construction. This creates a multi-stage process of fuzzy control. When talking about the route construction, we usually refer to the situation when an ant located in the town i wants to go to the town j and makes a decision based on the following information:

1. A parameters defining effect of the pheromone trace $\tau_{i,j}^\alpha$

2. A parameters defining effect of heuristic information $\eta_{i,j}^\alpha$, used to estimate attraction of the route,

3. N_i^k parameter, which represents the list of k available neighbors of an ant. The "available" neighbors mean the towns that have not been visited yet.

The decision-making process taken by an ant at each node of the route is associated with calculation of fuzzy probability in OFN sense. The probability is calculated pursuant to the following redefined formula of the route selection probability.

$$P_{ij}^k[l_A, l_A^-, l_A^+, p_A] = \frac{\left|\tau_{ij}[l_A, l_A^-, l_A^+, p_A]\right|^\alpha \left|\eta_{ij}[l_A, l_A^-, l_A^+, p_A]\right|^\beta}{\sum_{l \in N_i^k} \left|\tau_{ij}[l_A, l_A^-, l_A^+, p_A]\right|^\alpha \left|\eta_{ij}[l_A, l_A^-, l_A^+, p_A]\right|^\beta} \qquad (11)$$

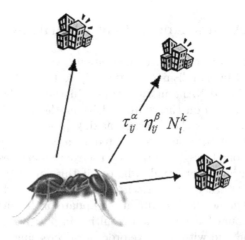

Fig. 4. An ant located in the town i selects next town j

5 Experiment

5.1 Experiment Execution Method

In this paragraph, the author shall compare effectiveness of heuristic methods, meta-heuristic methods [8] and the new hybrid method OFNAnt. All those methods shall be tested using 10 benchmarks for their performance in solving NP-hard problem, such as TSP. Thus, it shall be a comparison of well-known algorithms with a completely new approach represented by OFN arithmetics implemented to control over an ant colony in order to solve optimizing problems. They will be tested according to the following principles:

1. As regards Ant Colony Algorithms, a program with implemented method shall be run three times at $t = 10$ for each problem, while for implementation of heuristic algorithms a program shall also be run three times, but without additional parameters.
2. Effectiveness of a given algorithm shall be assessed as follows:
   (a) by specification of the obtained result (route length),
   (b) as a percentage i.e.: optimum achieved in x%, as presented in the table including the set of benchmarks,
3. A graph showing the effectiveness of individual algorithms shall be presented for each of the 10 problems.
4. Each such a graph shall be provided with a short summary where the obtained results will be discussed.
5. An overall graph showing the effectiveness of all algorithms shall be presented at the end. The value "optimum achieved in x" shall be totaled up for each algorithm, and thus the overall score per 1000 available points shall be calculated. Such way of data presentation shall allow easy assessment of the hierarchy of all the algorithms on the basis of 10 benchmarks used for tests.

5.2 Software Used for Experiment

Author's own implementation of ant colony methods developed in JAVA language was used in the experiments and the results obtained by the implementation were verified on the basis of ACOTSP [17]. The author's OFNAnt method was added to the implementation. The CONCORDE application was developed to solve Symmetric TSP type problems and other problems of network optimization [15,1]. The application is supported by the Office of Naval Research, National Science Foundation and by the School of Industrial and System Engineering at Georgia Tech. This program uses the Cutting Planes algorithm. The interface of the program shows the optimum solution searching process which is displayed at the end of each main iteration. The edges are coloured according to currently calculated LP value (Linear Programing Relaxation). At the moment when a new, better solution is found, the colour of edges is changed to red. The program includes several algorithms designed to create edges which are used by the program to search for the optimum solution. The program includes also several heuristic [2] algorithms used to solve TSP problem [18]. Those algorithms include:

1. Greedy algorithm (GR),
2. Boruvka algorithm (BOR),
3. Quck Boruvka algorithm (QBOR),
4. Nearest neighbour algorithm (NN),
5. Lin-Keringhana algorithm (LK).

5.3   Experimental Data

The following table shows 10 benchmarks thoroughly selected from TSPlib library of TSP problems, including the expected optimum value for each of them. They were applied in a way described in the previous paragraph as a set of benchmarks for testing well known algorithms and a new OFNAnt method.

Table 2. List of analyzed problems including their optimum values

No	Problem	Optimum	Description	Author
1	Eil51	426	Problem for 51 towns	Christofides / Eilon
2	D198	15780	Represents the Dribling Problem (198 holes)	Reinelt
3	Gil262	2378	Problem for 262 towns	Gillet/Johnson
4	Lin318	42029	Problem for 318 towns	Lin / Kernighan
5	Pcb442	50778	Represents the Dribling Problem (442 holes)	Groetschel/Juenger/Reinelt
6	Rat783	8806	Problem of 783 points in the power network	Pulleyblank
7	Pcb1173	56892	Represents the Dribling Problem (1173 holes)	Juenger/Reinelt
8	D1291	50801	Represents the Dribling Problem (1291 holes)	Reinelt
9	Nrw1379	56638	The problem for 1379 towns in Westphalia	Bachem/Wottawa
10	Pr2392	378032	Problem for 2392 towns	Padberg/Rinaldi

6   Results of Experiment

A number of tests were performed according to above specified rules, using ten selected problems. Results of individual tests were presented below in the table with optimum percentage for each of them.

7   Summary and Conclusions

Having performed a number of experiments according to the rules specified in the above chapters, one can be certain about the superiority of ant colony algorithms over classical algorithms. There was only one case out of ten studied samples, where the Lin-Kernighan (LK) algorithm achieved better result than all other known methods, including ant colony methods. This could have resulted from the nature of the problem, i.e. nrw1379. In that very case the results obtained by LK algorithm were slightly worse than the results of OFNAnt algorithm only.

Table 3. Summary of classical algorithms effectiveness based on the sample of 10 problem

Data sets		ACOTSP				Concorde TSP		
*.tsp file	optimum	AS	ASRK	OFNAnt	ACS	GR	QBOR	LK
eil51	426	426	426	426	426	521	480	426
d198	15780	15781	15780	15780	15780	18399	18140	15828
gil262	2378	2380	2378	2378	2378	2846	2818	2380
lin318	42029	42091	42029	42029	42029	49744	54090	42272
pcb442	50778	50964	50883	50778	50778	61891	58695	51071
rat783	8806	8833	8812	8808	8806	10294	10402	8831
pcb1173	56892	57612	56950	57040	56897	65829	66493	57063
d1291	50801	51020	50824	50870	50820	59293	57228	52729
nrw1379	56638	57281	56859	56917	56770	66371	66110	56756
pr2392	378032	386541	382089	381077	379602	444853	448641	383277

For the remaining files, LK algorithm outpaced, at best, only older ant colony methods, i.e. AS and EAS The remaining algorithms from the group of heuristic methods performed definitely much worse than the leading algorithms. They fulfilled the optimum solution within the range from 79% to 89%, which is far from the results of the leading algorithms. The noticeable feature of the studied group of algorithms is their tendency for worse results with the increase of the problem magnitude. Clear example of that tendency is the seventh tested problem, i.e. pcb1173. This statement is confirmed by the problem d1291 and further large data sets. The diagram presented above, which summarizes all performed tests, shows the hierarchy of all algorithms and their respective scores. The maximum available score is 1000 points. The scores closest to the maximum were achieved by representatives of ant colony algorithms, including OFNAnt with the score of 999,31 points. It is worth noting, that first four places on the list of optimum solution searching efficiency are taken by ant colony algorithms. Subsequent places on the list are taken by representatives of heuristic methods with their definite leader - LK algorithm, which is widely regarded as one of the best methods for solving the traveling salesman problem. Ant colony algorithms represent a new generation of optimizing algorithms using meta-heuristic approach to NP-hard problems, i.e. the approach which gave excellent results. Ant colony algorithms find many more applications other than TSP. Those applications include many real life fields. Based on the results of experiments with the new method using trend and fuzzy logic, one can expect obtaining interesting solutions also for other problems than those where ant colonies have already been successfully applied. The new method, OFNAnt, which is a hybrid combination of ACO and OFN, and which introduces fuzzy decision of an ant, is the first known attempt to implement the arithmetic of ordered fuzzy numbers to ant colony optimization. Performed experiments confirmed efficiency of that method in solving TSP problems. Currently, there are works ongoing on application of the modification of that method for solving problems of other classes.

Acknowledgments. The authors would like to express their thanks to employees of AIRlab - Artificial Intelligence and Robotics Laboratory at Casimir the Great University in Bydgoszcz for their commitment and help during research and tests performed within this study.

References

1. Adamus, E., Klesk, P., Kolodziejczyk, J., Korzen, M., Piegat, A., Plucinski, M.: Significance of condition attributes in child well-being analysis. Studies and Materials in Applied Computer Science 3(4), 11–16 (2011)
2. Angryk, R.A., Czerniak, J.: Heuristic algorithm for interpretation of multi-valued attributes in similarity-based fuzzy relational databases. International Journal of Approximate Reasoning 51(8), 895–911 (2010)
3. Chwastyk, A., Kosiński, W.: Fuzzy calculus with applications. Mathematica Applicanda 41(1), 47–96 (2013)
4. Czerniak, J.M., Dobrosielski, W.T., Angryk, R.A.: Comparison of two kinds of fuzzy arithmetic, lr and ofn, applied to fuzzy observation of the cofferdam water level. Computer Science 14(3), 443–457 (2013)
5. Dorigo, M., Gambardella, M.: Ant colony system: A cooperative learning approach to the traveling salesman problem. IEEE Transactions on Evolutionary Computation (53-66) (1997)
6. Dorigo, M., Stutzle, T.: Ant Colony Optimization. The MIT Press, Cambridge (2004)
7. Dubois, D., Prade, H.: Fuzzy elements in a fuzzy set. Fuzzy Set. Soft Computing 12(165-175) (2008)
8. Engelbrecht, A.: Fundamentals of Computational Swarm Intelligence. Wiley (2005)
9. Helsgann, K., Ngassa, J.L., Kierkegaard, J.: ACO and TSP. Roskilde University (2007)
10. Klir, G.: Fuzzy arithmetic with requisite constraints. Fuzzy Sets and Systems - Special Issue: Fuzzy Arithmetic Archive 91(2), 165–175 (1997)
11. Kosiński, W., Prokopowicz, P., Rosa, A.: Defuzzification functionals of ordered fuzzy numbers. IEEE Trans. Fuzzy Systems 21(6) (2013)
12. Kosinski, W., Prokopowicz, P., Slezak, D.: Ordered fuzzy number. Bulletin of the Polish Academy of Sciences, Ser. Sci. Math. 53(3), 327–338 (2003)
13. Kosiński, W., Słysz, P.: Fuzzy numbers and their quotient space with algebraic operations. Bull. Polish Acad. Sci. Ser. Tech. Sci. 41, 285–295 (1993)
14. Kosiński, W., Chwastyk, A.: Ordered fuzzy numbers in financial stock and accounting problems. In: Proceedings of the 2013 Joint IFSA World Congress and NAFIPS Annual Meeting, IFSA/NAFIPS 2013, pp. 546–551 (2013)
15. Kováč, D., Vince, T., Molnár, J., Kováčová, I.: Modern internet based production technology. In: Joo Er, M. (ed.) New Trends in Technologies: Devices, Computer, Communication and Industrial Systems, pp. 145–164. SCIYO (2010)
16. Łukasiewicz, J.: On three-valued logic. Ruch Filozoficzny 5(170-171) (1920) (in Polish)
17. Merkle, D.: Swarm Intelligence: Introduction and Application. Springer Verlag Gmbh (2008)
18. Rozin, V., Margaliot, M.: The fuzzy ant. IEEE Computational Intelligence Magazine 2, 18–28 (2007)
19. Vince, T., Hricko, J.: Lego mindstorms robot controlled by android smartphone. In: XV International PhD Workshop OWD 2013, October 19–22, pp. 62–65 (2013)
20. Zadeh, L.A.: Fuzzy sets. Information and Control 8(3), 338–353 (1965)

Fuzzy Interface for Historical Monuments Databases

Krzysztof Czajkowski and Piotr Olczyk

Institute of Telecomputing, Faculty of Physics, Mathematics and Computer Science,
Cracow University of Technology, Cracow, Poland

Abstract. Issues concerning the conservation of historic monuments can be be supported by information systems. Databases offer significant opportunities for structuring knowledge, but they need to provide the users with effective tools. Efficiency of these tools depends largely on functionality and flexibility of the interfaces. This paper presents a proposal for a new interface access to databases of historical buildings using fuzzy sets.

Keywords: historical monument, database, fuzzy set, interface.

1 Introduction

Nowadays, various dangers affect national heritage in a broad sense, and this is the reason why special attention is given to its protection. The threats include air pollution, natural disasters, as well as exploitation. Many countries have extremely rich and highly diverse historical legacy, of which one of the more obvious components are the historic monuments. The problem of monuments protection is very complex due to the diversity of objects classified as monuments, and it requires the use of different approaches, techniques and solutions depending on the needs.

In Poland, there is over 65,000 historical monuments (divided into 14 major categories) [15], and their number is still growing. Due to the complexity of the matter, and the number of monuments, this paper focuses on the historical tenements (one of the types of historic residential buildings, which number exceeds 17,000).

The role of information systems in the monuments conservation may be very important. The complexity of aspects of the restoration and maintenance requires a structured approach. That is why considerable attention is given to applications that use the databases. Numerous studies are conducted on the use of such solutions [2]. However, these works are usually related to a single issue, such as specialized and detailed description of the materials that are made of elements of historic buildings [6] or encompass the wide range of topics that describe the multimedia and the GIS data [7]. Many problems still remain unsolved and research in many aspects seems not to be carried out.

A bibliography review and consultations with experts in conservation studies showed no complete solutions in this field. Simultaneously, there is great need

S. Kozielski et al. (Eds.): BDAS 2014, CCIS 424, pp. 271–279, 2014.

for the availability of such tools. Existing applications are fragmentary, covering only some aspects of the problem and do not represent sufficient support for the wider protection of national heritage. This work presents optimal solutions, which allow efficient retrieval of information from large data sets.

2   Searching Similar Objects

The usefulness of the system that collects information largely depends on options in the searching process. The method of extraction of information determines whether the application meets the requirements [11]. In the protection of monuments, it is particularly important aspect. Majority of this systems users do not have any knowledge in programming. Often they are not even the engineers, and a large part of them can be artists.

The historical subject matter is always fraught with a lack of complete precision. As an example the issue of objects dating may be consider. A complete and documented history of an object or element is rarely available. Even if the complete data is given, it may still be the subject of discussion [11]. Also other, seemingly more tangible aspects, are not entirely clear. For example, there is a problem with the dimensions of individual components. Although it is easy to measure, the use of precise results will be useless if there is a need to search the corresponding object. In the past, many (if not all) elements were performed by different masters using their own developed methods and techniques, making it impossible to accurately compare the results of their work. In the case of dimensions of objects and their comparison, paradoxically, the greater the accuracy, the less effective search. In modern building techniques, it is significantly easier to use the precision data. Nowadays, objects are usually constructed with typical materials of defined characteristics, their components are standardized, cataloged and only minor deviations from standards are admitted. Quite different is the case of historic structures. The older objects are considered, the less precise criteria are assumed.

In summary, it is impossible to accurately compare data that are either of debatable precision (such as creation date, modification, repair, maintenance), or unique (dimensions of structural elements). Searching the database have to be characterized by inexactness in the formulation of questions, because that is the nature of the data.

The necessity of development of that kind of technology is undeniable [4,3]. Many examples of its usage can be enumerated. For instance, if there is a need to renovate a particular architectural element which, in the case of a given monument, has been destroyed, it can be very useful (especially in the absence of full documentation) to rely on other objects of similar structure, construction time, dimensions and purpose. Another example would be a situation in which conservators join their work to prevent further degradation of the monument and they would like to have the possibility to choose the best method of preservation. Relying on methods that have worked well with other objects (because of the chosen technology, materials, etc.), is always extremely valuable. Again, the key is to find similar objects (similar in different points of view) and verify the methods used in their maintenance. Different example is a situation where the conservator would like to explore the risks concerning specific elements, because of their design or the materials that they were made of. Analysis of other similar objects can give important clues in this regard.

The key in all these cases is the word "similar".

3   Fuzzy Interface

Since the fundamental problem is the lack of ability to perform precise searches in the area of historical monuments databases, it is necessary to provide users with the possibility of formulating vague queries. In presented work, the theory of fuzzy sets [12,1] has been used. It is a starting point to model the interface that would find the usage in applications which support historic restorers, architects and artists.

There are also many papers that discuss design and implementation issues of solutions to formulate queries in natural language and processing them in relational databases [8,9]. Fuzzy sets theory has been used in the systems for monuments preservation, such as fuzzy number ranking in project selection [13] or reliability in archaeological virtual reconstruction [10]. It has not been used in the area of interfaces for calculation of similarity between monuments.

Our approach focuses on one category of historic buildings - a historical tenement. It is also possible to generalize the method, after its verification, to other types of buildings. Focusing on one category of objects enables simplification of the process of implementation, without losing the practical aspect.

As a part of a historic building five main types of structural elements can be distinguished [5]:

- The foundation
- The facade
- The roof
- Stairs
- Doors

This solution is a part of a larger project concerning the comprehensive system of historic buildings, and therefore each of those items in the database has a few or more instances, for example, two facades: the front and rear. The solution of the fuzzy search for similar objects is independent from the construction of given building, so the complexity of the building structure is not important.

For each structural element the following attributes are considered:

- The construction period
- The state of preservation
- Dimensions
- The material (that was used)
- The design (type or its properties)

The dictionaries have been prepared for each type of structural element. Various materials (including types of plaster, wood, roofing materials), types and characteristics of the structure (including resistance to weather conditions, load) had been taken into consideration.

The system has a set of various defined types of features. Each of them has assigned default values that can be edited by an expert. It is also possible to create new features to extend the functionality of the system. The features are based on fuzzy sets. The degree of belonging to the fuzzy set is the subjective assessment. Therefore, the possibility of free edition is fully justified. It should be taken into consideration that one of the experts may regard a similar period of construction within + / - 20 years, but the other one may consider a value of + / - 50 years. A similar situation is noticed in case of the strength of the structure, materials and all other features. The interface designed for the definition of features for the roof element is shown in Fig. 1.

The complexity of the interface is a compromise between the possibility of a detailed specification of the similarity by using multiple keywords, and the simplicity of the usage. We intentionally proposed a simpler interface, allowing the users to extend it for the creation of more precise criteria.

In case when the user searches for similar objects, and selects the characters by which the comparison should be carried out, then the degree of similarity should be determined (similar or very similar) - Fig. 2. Moreover, the user may indicate the importance (weight) of data characteristics.

The application that analyses the correlations between selected details of buildings is based on fuzzy sets. This solution allows searching the database for a pre-defined similarity. In presented application, the implementation uses a symmetric triangular membership function [12].

An example of calculation relative to the attribute "construction period" has been presented below.

1. A historical tenement built in 1600 has been selected as an input object (for which similar objects are searched). According to the opinion of an expert, construction period of tenements that are similar in terms of year of construction equals + / - 40 years.
2. One of the objects in database is the tenement built in 1590.
3. It is possible to calculate the wanted similarity of the tenement by using triangular membership function [12] and taking into consideration the values defined by the expert:

$$
\mu\left(x : a, b, c\right) = \begin{cases} 0, & x \leq a \\ \frac{x-a}{b-a}, & for\ a < x \leq b \\ -\frac{x-c}{c-b}, & for\ b < x \leq c \\ 0, & x > c \end{cases} \tag{1}
$$

where: a, b, c - parameters that define the similarity border and $(a < x \leq b < x \leq c)$

Roof

Rule	Level of similarity		
Construction period	Very similar(+/-)	20	years
Construction period	Similar (+/-)	40	years
Preservation	Very similar (+/-)	7	%
Preservation	Similar (+/-)	12	%
Dimensions	Very similar (+/-)	0,7	m
Dimensions	Similar (+/-)	2	m

Material

Similarity [0-1]	Sandringham interlocking clay tile	Shingle camber clay tile	Metal shingle slates panel	Roofting felt	Asbestos roof
Sandringham interlocking clay tile	1.0	0.9	0,5	0,2	0.4
Shingle camber clay tile	-	1.0	0,5	0,2	0.5
Metal shingle slates panel	-	-	1.0	0.3	0,5
Roofting felt	-	-	-	1.0	0,1
Asbestos roof	-	-	-	-	1.0

Construction

Rule	Trussed	Trussed rafter	Couple	Collar
High strength to weight of snow	0,8	0,5	0,4	0,6
Good thermal protection	0,7	0,8	0,4	0,5
High resistance to wind	0,8	0,7	0,5	0,7

Fig. 1. Interface of features defining for the roof

Construction period	Very similar		Weight	0,6
Preservation	Very similar		Weight	0,8
Dimensions	Very similar		Weight	0,5
Material	Similar		Weight	1,0
Construction	Similar		Weight	1,0
			Limit value	0.5

Search

Fig. 2. Defining criteria for given monument

4. Supposing that we have the expert opinion ($1560 < x \leq 1600 < x \leq 1640$), the membership function takes the following form:

$$\mu(x : a, b, c) = \begin{cases} 0, & x \leq 1560 \\ \frac{x}{40} - 39, & for \ 1560 < x \leq 1600 \\ -\frac{x}{40} + 41, & for \ 1600 < x \leq 1640 \\ 0, & x > 1640 \end{cases} \tag{2}$$

5. Memberships of the tenement of 1590 equals 0.75.

The degree of belonging to a given fuzzy set determines the similarity to the reference element.

The problem of similarity of elements, such as the type of construction of foundation or roof, has been solved in a different way. Fuzzy sets that describe certain characteristics or properties of a given element have been defined in this application. The task of an expert is to determine the degree of belonging of a considered element to a selected set of structures. For instance, if a set is defined by the rule "the high durability on the weight of snow", a specialist can evaluate that the trussed roof belongs to the set at 0.8 degree. However, trussed rafter roof considered as less robust, belongs to the set at only 0.6 degree.

The paper [14] gives a variety of methods for calculating the similarity between elements belonging to fuzzy sets. In most cases, the concept of geometric distance between these elements is used. The similarity between element A and B may,

Name	Address	Monument class	Building date	Present usage	Photography		Similarity
Długosz House	Kanonicza Street No. 25	I	05-03 -1575	The Pontifical University of John Paul II in Cracow		Show report	0.724
Under The Angels	Florianska Street No. 1	II	01-30 -1486	Residential building		Show report	0.722
Dobrodziejskich House	Florianska Street No. 5	II	01-18 -1478	Jewelry factory		Show report	0.632
Under The God's eye	Grodzka Street No. 6	III	03-17 -1735	Residential building		Show report	0.614
Margrabska Tenement	Slawkowska Street No. 2	I	07-13 -1574	Residential building		Show report	0.547

Fig. 3. The list of tenements with degree of similarity

for example, be determined by the equation (3) for a fuzzy set defined by the class of triangular membership function.

$$S_{A,B} = 1 - \frac{1}{n} \sum_{i=1}^{n} |a_i - b_i| \qquad (3)$$

where: A, B - fuzzy sets; a, b - elements belonging to fuzzy sets.

It is possible to compare the data structure under certain traits by using the expert-defined features visible in Fig. 1. If trussed roof and trussed rafter roof is considered by doing an analysis of their individual characteristics and then average the results is made, final similarity between structures is obtained.

In the query in Fig. 3 criteria were defined: construction period 1600 (weight 0,8), preservation 78% (weight 0,8), dimension very similar (referential: 12mx9m, weight 0,5). material similar (referential: roofing felt, weight 1,0). The final result of the application is a list of tenements arranged in descending order in terms of similarity of the selected elements. Apart from general information about building products, the list contains references to reports that precisely discuss the historic details.

4   Conclusion

The paper discusses the issues of information retrieval about historical monuments in the databases. In our solution, the theory of fuzzy sets was used to design the interface to such databases. The authors focused on developing the most universal way to formulate imprecise queries. An important assumption was the simplicity of querying and the possibility of relying on the characteristics of various kinds. Another important goal was to enable users such as preservationists, architects to redefine the fuzzy numbers as well as extend interface attributes. The authors focused on the development of the most useful interface to formulate imprecise queries for simple integration with existing database systems.

Approaches presented in [8], [9] are proposals of universal solutions which can be implemented in standard database environments. Universal implementation of fuzzy queries can be very useful and gives possibility to construct a lot of combination of queries to any data. However, such a solution requires implementation in given database server. Each database server vendor provides its own procedural language and each database can be hermetic environment with limited access to make such implementation. General solution could be a great improvement in given integrated system, for example in data warehouse [9].

As an alternative to such solutions, this paper presents dedicated approach. It is possible to identify some advantages of proposed solution. First, it is not necessary to add any extension to existing databases, which have own security policies, because the implementation is performed in the application layer. In case of the need to search for data in multiple databases, universal solution have to be implemented in each database, often from different providers. Finally, in approach presented in this paper, the change of database servers is very easy (for example: from open-source to commercial or vice versa), also there is no need to move the implementation of fuzzy queries.

Based on a review of literature, no available solution in this area can be found (even the consistent concept or prototype). In the aforementioned papers, only theoretical analysis of usefulness of fuzzy sets is presented. A few examples describe a ranking procedure among various conservational and enhancement projects that may be defined for an archaeological site [13] or in describing virtual models of monuments [10].

We have developed the application that verifies the approach presented in the paper. Tests performed with the use of the application confirmed the utility of this approach.

References

1. Bosc, P., Kacprzyk, J.: Fuzziness in Database Management Systems. Physica Verlag, Heidelberg (1995)
2. Doehne, E., Price, C.A.: Stone Conservation: An Overview of Current Research. The Getty Conservation Institute, Los Angeles (2010)
3. Drabowski, M., Czajkowski, K.: Internet database application for heritage preservation. In: Proceedings of the IADIS International Conference on Computer Science and Information Systems (2005)
4. Drabowski, M., Czajkowski, K.: The internet information system for heritage preservation. In: Kozielski, S., Malysiak, B., Kasprowski, P., Mrozek, D. (eds.) Bazy Danych - Model, Technologie, Narzedzia - Analiza Danych i Wybrane Zastosowania, pp. 349–356. Transport and Communication Publ., Warszawa (2005)
5. Feilden, B.: Conservation of Historic Buildings. Technical studies in the arts, archeology and architecture. Architectural Press (2003),
   http://books.google.pl/books?id=tVyPXDJfqgAC
6. Hyslop, E., et al.: Building stone databases in the UK: a practical resource for conservation. Engineering Geology 115, 143–148 (2010)
7. Klamma, R., et al.: A hypermedia afghan sites and monuments database. In: Stefanakis, E. (ed.) Geographic Hypermedia: Concepts and Systems. Lecture Notes in Geoinformation and Cartography, pp. 59–73. Springer, Berlin (2006)
8. Kowalczyk-Niewiadomy, A., Pelikant, A.: Fuzzy queries processing by means of meta-natural language using fuzzy algorithms. Studia Informatica 31, 479–488 (2010)
9. Małysiak-Mrozek, B., Mrozek, D., Kozielski, S.: Processing of crisp and fuzzy measures in the fuzzy data warehouse for global natural resources. In: García-Pedrajas, N., Herrera, F., Fyfe, C., Benítez, J.M., Ali, M. (eds.) IEA/AIE 2010, Part III. LNCS, vol. 6098, pp. 616–625. Springer, Heidelberg (2010),
   http://dl.acm.org/citation.cfm?id=1945955.1946026
10. Niccolucci, F., Hermon, S.: A fuzzy logic approach to reliability in archaeological virtual reconstruction. In: Proc. of the 32nd International Conference on Computer Applications and Quantitative Methods in Archaeology (2004)
11. Pawlicki, B.M., Drabowski, M., Czajkowski, K.: Monitoring stanu zachowania obiektow zabytkowych przy wykorzystaniu wspolczesnych systemow informatycznych. Wydawnictwo Politechniki Krakowskiej, Krakow (2004)

12. Piegat, A.: Fuzzy Modeling and Control. Springer, Heidelberg (2001)
13. Sanna, U., Atzeni, C., Spanu, N.: A fuzzy number ranking in project selection for cultural heritage sites. Journal of Cultural Heritage 9, 311–316 (2008)
14. Sridevi, B., Nadarajan, R.: Fuzzy similarity measure for generalized fuzzy numbers. International Journal of Open Problems Compt. Math. 2, 241–253 (2009)
15. Szmygin, B.: System ochrony zabytkow w Polsce - analiza, diagnoza, propozycje. Polski Komitet Narodowy ICOMOS, Warszawa (2011)

Optimization of Mechanical Structures Using Artificial Immune Algorithm

Arkadiusz Poteralski

Institute of Computational and Mechanical Engineering,
Faculty of Mechanical Engineering,
Silesian University of Technology, ul. Konarskiego 18a, 44-100 Gliwice, Poland
arkadiusz.poteralski@polsl.pl

Abstract. The paper is devoted to application of the artificial immune systems (artificial immune algorithm) to selected shape and topology optimization problems of structures.

Keywords: artificial immune system (AIS), optimization, finite element method (FEM), computational intelligence, shape and topology optimization, artificial immune algorithm.

1 Introduction

Shape and topology structural optimization is a very active research area. Several competing approaches for topology optimization exist. Intelligent optimal design techniques based on evolutionary algorithms (EA) have found applications to structural optimization problems [3,2,5]. The evolutionary methods are based on the theory of evolution. The main feature of those methods is to simulate biological processes based on heredity principles (genetics) and the natural selection (the theory of evolution) to create optimal individuals (solutions) presented by single chromosomes. More recently, other bio-inspired approaches, alternative to EA, as the Particle Swarm Optimizers (PSO) [9,16] or **the Artificial Immune Systems (AIS)** have gained popularity. The papers is devoted to applications of **the artificial immune algorithm** to selected shape and topology optimization problems of structures analyzed by finite element method FEM [20]. A short descriptions of biological aspect of natural immune systems is described in the context of optimization procedures. The clonal selection algorithm which represents one of the main features of the artificial immune system is described. Standard and modified versions of artificial immune systems and its applications in different optimization problems of mechanical structures were widely presented by the authors [1,12,13,17,14]. In the present paper the applications of this algorithm to topology optimization problems of structures is demonstrated.

2 Artificial Immune Systems

The artificial immune systems (AIS) are developed on the basis of a mechanism discovered in biological immune systems [15]. An immune system is a complex

S. Kozielski et al. (Eds.): BDAS 2014, CCIS 424, pp. 280–289, 2014.

system which contains distributed groups of specialized cells and organs. The main purpose of the immune system is to recognize and destroy pathogens - funguses, viruses, bacteria and improper functioning cells. The lymphocytes cells play a very important role in the immune system. The lymphocytes are divided into several groups of cells. There are two main groups B and T cells, both contains some subgroups (like B-T dependent or B-T independent). The B cells contain antibodies, which could neutralize pathogens and are also used to recognize pathogens. There is a big diversity between antibodies of the B cells, allowing recognition and neutralization of many different pathogens. The B cells are produced in the bone marrow in long bones. A B cell undergoes a mutation process to achieve big diversity of antibodies. The T cells mature in thymus, only T cells recognizing non self cells are released to the lymphatic and the blood systems. There are also other cells like macrophages with presenting properties, the pathogens are processed by a cell and presented by using MHC (Major Histocompatibility Complex) proteins. The recognition of a pathogen is performed in a few steps. First, the B cells or macrophages present the pathogen to a T cell using MHC, the T cell decides if the presented antigen is a pathogen. The T cell gives a chemical signal to B cells to release antibodies. A part of stimulated B cells goes to a lymph node and proliferate (clone). A part of the B cells changes into memory cells, the rest of them secrete antibodies into blood. The secondary response of the immunology system in the presence of known pathogens is faster because of memory cells. The memory cells created during primary response, proliferate and the antibodies are secreted to blood. The antibodies bind to pathogens and neutralize them. Other cells like macrophages destroy pathogens. The number of lymphocytes in the organism changes, while the presence of pathogens increases, but after attacks a part of the lymphocytes is removed from the organism. The artificial immune systems [12,14] take only a few elements from the biological immune systems. The most frequently used are the mutation of the B cells, proliferation, memory cells, and recognition by using the B and T cells. The artificial immune systems have been used to optimization problems in classification and also computer viruses recognition. The cloning algorithm presented by von Zuben and de Castro [6,7,8] uses some mechanisms similar to biological immune systems to global optimization problems. The unknown global optimum is the searched pathogen. The memory cells contain design variables and proliferate during the optimization process. The B cells created from memory cells undergo mutation. The B cells evaluate and better ones exchange memory cells. In Wierzchoń S. T. [19] version of Clonalg the crowding mechanism is used - the diverse between memory cells is forced. A new memory cell is randomly created and substitutes the old one, if two memory cells have similar design variables. The crowding mechanism allows finding not only the global optimum but also other local ones. The presented approach is based on the Wierzchoń S. T. algorithm [19], but the mutation operator is changed. The Gaussian mutation is used instead of the nonuniform mutation in the presented approach [12]. The Fig. 1 presents the flowchart of an artificial immune system.

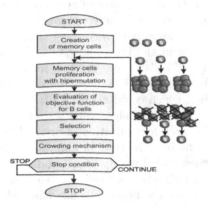

Fig. 1. The algorithm of an artificial immune system

The memory cells are created randomly. They proliferate and mutate creating B cells. The number of clones created by each memory cell is determined by the memory cells objective function value. The objective functions for B cells are evaluated. The selection process exchanges some memory cells for better B cells. The selection is performed on the basis of the geometrical distance between each memory cell and B cells (measured by using design variables). The crowding mechanism removes similar memory cells. The similarity is also determined as the geometrical distance between memory cells. The process is iteratively repeated until the stop condition is fulfilled. The stop condition can be expressed as the maximum number of iterations. The unknown global optimum is represented by the searched pathogen. The memory cells contain design variables and proliferate during the optimization process.

3 Topology Optimization

The distribution of the mass density $\rho(X), (X) \in \Omega_t$ in the structure is described by a hyper surface $W_\rho(X), \ (X) \in H^3$ [3,2,4,5]. The hyper surface $W_\rho(X)$ is stretched under $H^3 \subset E^3$ and the domain Ω_t is included in H^3 i.e. $(\Omega_t \subseteq H^3)$. The shape of the hyper surface $W_\rho(X)$ is controlled by parameters $d_j, j = 1, 2, \ldots, G$, which create a B-cell receptor:

$$B - cell = \langle d_1, d_2, ..., d_j, ..., d_G \rangle , \quad d_j^{\min} \leq d_j \leq d_j^{\max} \qquad (1)$$

where: $d_j^{\min}$, $d_j^{\max}$ are minimum and maximum values of parameters of B-cell. B-cell parameters are the values of the function $W_\rho(X)$ in the control points $(X)_j$ of the hyper surface, i.e. $d_j = W_\rho\left[(X)_j\right]$, j=0,1,...,G. The finite element method is applied in analysis of the structure [20]. The domain Ω of the structure is discretized using the finite elements, $\Omega = \bigcup_{e=1}^{E} \Omega_e$. The assignation of the mass density to each finite element Ω_e, $e = 1, 2, ..., E$ is adequately performed by the mappings $\rho_e = W_\rho[(X)_e]$, $(X)_e \in \Omega_e$, e=1,2,...,E. It means that each finite element can have the different mass density. When the value of the mass density for the e-th finite element is included in the interval $0 \leq \rho_e < \rho_{min}$ the finite element is eliminated and the void is created and when in the interval $\rho_{min} \leq \rho_e < \rho_{max}$, the finite element remains. The illustration of this idea of immune optimization for a 2-D structure is presented in paper [4,1].

In the next step the Youngs modulus for the e-th finite element is evaluated using the following equation

$$E_e = E_{max}\left(\frac{\rho_e}{\rho_{max}}\right)^r \tag{2}$$

where: E_{max}, ρ_{max} - Youngs modulus and mass density for the same material, respectively, and r parameter which can change from 1 to 9 (depending on the type of material).

The bio-inspired process proceeds in the environment in which the structure fitness is described by the minimization of the mass of the structure

$$J = \int_\Omega \rho d\Omega \tag{3}$$

with constraints imposed on equivalent stresses σ_{eq} and displacements u of the structure

$$\sigma_{eq}(x, y, z) \leq \sigma^{ad}, \ (x, y, z) \in \Omega \tag{4}$$

$$|u(x, y, z)| \leq u^{ad}, \ (x, y, z) \in \Omega \tag{5}$$

Parameterization is the key stage in the structural optimization. The great number of design variables causes that the optimization process is not effective. A connection between design variables (parameters of B cell receptor) and number of finite element leads to poor results. The better results can be obtained when the surface (or hyper surface $W_\rho(X)$) of mass density distribution is interpolated by suitable number of values given in control points $(X)_j$. This number, on the one hand, should provide the good interpolation, and on the other hand the number of design variables should be small. Two different types of the interpolation procedures were applied. First the multinomial interpolation and the second one interpolation bases on the neighbourhood of the elements described for 3-D structure were introduced [4].

4   Immune Optimization Examples

Optimization of 3D structures for various set of parameters of immune algorithm is presented in this chapter. Topology optimization of 3D structure (Fig. 2) by the minimization of the mass of the structure and with imposed stress or displacement constraints [4]. The structures are considered in the framework of the theory of elasticity. The input data to the optimization task are included in Tab. 1. The parameters of the immune algorithm are presented in the Tab. 2. In the first step the influence of the number of clones are checked (Tab. 3 - variants

Table 1. The input data to optimization task of a 3-D structure

a x b x c	Maximal displacement [mm]	Maximal stress [MPa]	Q [kN]	range of ρ [g/cm3] existence or elimination of the finite element
10 x 10 x 10	0.05	35	36.6	$0 \leq \rho_e < 3.14$ elimination
				$0 \leq \rho_e < 3.14$ elimination

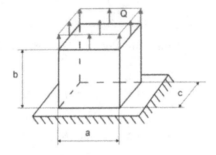

Fig. 2. 3-D structure: geometry and scheme of loading

Table 2. Parameters of artificial immune system

Crowding factor	Gaussian mutation
0.5	50%

Table 3. Variants of artificial immune parameters

Variant (example)	Number of memory cells	Number of the clones
1	5	5
2	5	10
3	5	15
4	5	20
5	3	10
6	5	10
7	10	10
8	15	10

Table 4. Results after optimization process

Fitness function [g]	Number of iterations	Number of fitness function evaluations
	Variant 1	
2308	47	940
	Variant 2	
1303	48	1680
	Variant 3	
1334	33	1650
	Variant 4	
1378	42	2730
	Variant 5	
1806	49	1129
	Variant 6	
1303	48	1680
	Variant 7	
1288	49	3185
	Variant 8	
1392	48	4560

Table 5. Results after optimization process change of number of clones

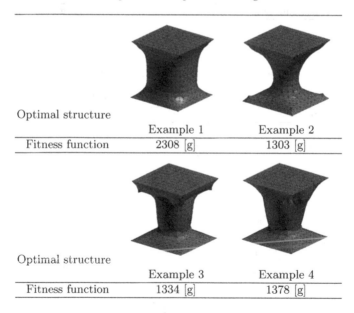

Optimal structure		
	Example 1	Example 2
Fitness function	2308 [g]	1303 [g]

Optimal structure		
	Example 3	Example 4
Fitness function	1334 [g]	1378 [g]

14). In the next step the influence of the memory cells are checked (Tab. 3 - variants 58). Numerical results for 8 variants of immune parameters are presented in the Tab. 4, 5 and 6. In last step the influence of mechanical constraints are checked (Tab. 7). Numerical results for 4 variants of mechanical constraints are presented in the Tab. 8 and Tab. 9.

Table 6. Results after optimization process change of memory cells

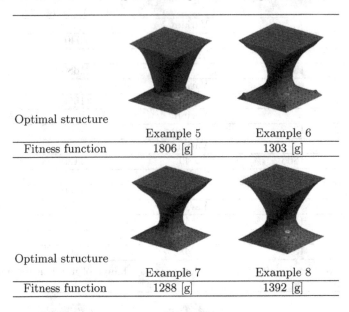

Optimal structure		
	Example 5	Example 6
Fitness function	1806 [g]	1303 [g]

Optimal structure		
	Example 7	Example 8
Fitness function	1288 [g]	1392 [g]

Table 7. Variants of mechanical constraints

Variant (example)	Equivalent stresses σ [MPa]	Displacements u [mm]
1 (9)	35	0.01
2 (10)	35	0.03
3 (11)	35	0.05
4 (12)	55	0.03

Table 8. Results after optimization process

Fitness function [g]	Number of iterations	Number of fitness function evaluations
	Variant 1	
2068	46	1610
	Variant 2	
1521	48	1680
	Variant 3	
1303	48	1680
	Variant 4	
1485	49	1715

Table 9. Results after optimization process change of mechanical constraints

Optimal structure		
	Example 9	Example 10
Fitness function	2068 [g]	1521 [g]

Optimal structure		
	Example 11	Example 12
Fitness function	1303 [g]	1485 [g]

5   Conclusions

In the paper, a description of algorithm of artificial immune approach is presented and applied to optimization of structures. This bio-inspired approach can be simply implemented because they need only the values of objective functions. An important feature of this approach is a strong probability of finding the global optimal solutions. The described approach is free from limitations of classic gradient optimization methods. This approach belong to methods based on population of solutions and they have some interesting features which can be considered as alternative to evolutionary algorithms. Described approach has applied to simultaneous shape, topology and material optimization of 3D structures. The influence of the immune parameters and mechanical constraints on the final solutions are presented. Efficiency of presented approaches and comparison with evolutionary algorithms is presented in the paper [18,13]. There are possibilities of further efficiency improvement of the proposed method, e.g. by the application of adjoint variable method in the sensitivity analysis. Also, the application of another hybridized global optimization algorithms, like hybrid artificial immune system, would be interesting. Efficiency of the proposed method can be also improved by distributing computations in multi-agent systems [10]. A set of solutions can be increased through the use of fuzzy methods [11].

References

1. Burczyński, T., Bereta, M., Poteralski, A., Szczepanik, M.: Immune computing: Intelligent methodology and its applications in bioengineering and computational mechanics. In: Comput. Meth. Mech. Advanced Structured Materials, vol. 1, Springer, Heidelberg (2010)
2. Burczyński, T., Długosz, A., Kuś, W., Orantek, P., Poteralski, A., Szczepanik, M.: Intelligent computing in evolutionary optimal shaping of solids. In: Proceedings of the 3rd International Conference on Computing, Communications and Control Technologies, vol. 3, pp. 294–298 (2005)
3. Burczyński, T., Kuś, W., Długosz, A., Poteralski, A., Szczepanik, M.: Sequential and distributed evolutionary computations in structural optimization. In: Rutkowski, L., Siekmann, J.H., Tadeusiewicz, R., Zadeh, L.A. (eds.) ICAISC 2004. LNCS (LNAI), vol. 3070, pp. 1069–1074. Springer, Heidelberg (2004)
4. Burczyński, T., Poteralski, A., Szczepanik, M.: Genetic generation of 2-d and 3-d structures. In: Second M.I.T. Conference on Computational Fluid and Solid Mechanics. Massachusetts, Institute of Technology, Cambridge (2003)
5. Burczyński, T., Poteralski, A., Szczepanik, M.: Topological evolutionary computing in the optimal design of 2d and 3d structures. Eng. Optimiz. 39(7), 811–830 (2007)
6. de Castro, L.N., Timmis, J.: Artificial immune systems as a novel soft computing paradigm. Soft Computing 7(8), 526–544 (2003)
7. de Castro, L.N., Von Zuben, F.J.: Immune and neural network models: theoretical and empirical comparisons. International Journal of Computational Intelligence and Applications (IJCIA) 1(3), 239–257 (2001)
8. de Castro, L.N., Von Zuben, F.J.: Learning and optimization using the clonal selection principle. IEEE Transactions on Evolutionary Computation, Special Issue on Artificial Immune Systems 6, 239–251 (2002)
9. Kennedy, J., Eberhart, R.: Swarm Intelligence. Morgan Kaufmann (2001)
10. Momot, A., Małysiak-Mrozek, B., Kozielski, S., Mrozek, D., Hera, Ł., Górczyńska-Kosiorz, S., Momot, M.: Improving performance of protein structure similarity searching by distributing computations in hierarchical multi-agent system. In: Pan, J.-S., Chen, S.-M., Nguyen, N.T. (eds.) ICCCI 2010, Part I. LNCS, vol. 6421, pp. 320–329. Springer, Heidelberg (2010)
11. Mrozek, D., Małysiak-Mrozek, B.: An improved method for protein similarity searching by alignment of fuzzy energy signatures. International Journal of Computational Intelligence Systems 4(1), 75–88 (2011)
12. Poteralski, A., Szczepanik, M., Dziatkiewicz, G., et al.: Immune identification of piezoelectric material constants using bem. Inverse Problems in Science and Engineering 19(1), 103–116 (2011)
13. Poteralski, A., Szczepanik, M., Dziatkiewicz, G., Kuś, W., Burczyński, T.: Comparison between PSO and AIS on the basis of identification of material constants in piezoelectrics. In: Rutkowski, L., Korytkowski, M., Scherer, R., Tadeusiewicz, R., Zadeh, L.A., Zurada, J.M. (eds.) ICAISC 2013, Part II. LNCS, vol. 7895, pp. 569–581. Springer, Heidelberg (2013)
14. Poteralski, A., Szczepanik, M., Ptaszny, J., Kuś, W., Burczyński, T.: Hybrid artificial immune system in identification of room acoustic properties. Inverse Problems in Science and Engineering (2013)
15. Ptak, M., Ptak, W.: Basics of immunology. Jagiellonian University Press, Cracow (2000) (in Polish)

16. Reynolds, C.W.: Flocks, herds, and schools, a distributed behavioral model. Computer Graphics 21, 25–34 (1987)
17. Szczepanik, M., Poteralski, A., Długosz, A., Kuś, W., Burczyński, T.: Bio-inspired optimization of thermomechanical structures. In: Rutkowski, L., Korytkowski, M., Scherer, R., Tadeusiewicz, R., Zadeh, L.A., Zurada, J.M. (eds.) ICAISC 2013, Part II. LNCS, vol. 7895, pp. 79–90. Springer, Heidelberg (2013)
18. Szczepanik, M., Poteralski, A., Ptaszny, J., Burczyński, T.: Hybrid particle swarm optimizer and its application in identification of room acoustic properties. In: Rutkowski, L., Korytkowski, M., Scherer, R., Tadeusiewicz, R., Zadeh, L.A., Zurada, J.M. (eds.) SIDE 2012 and EC 2012. LNCS, vol. 7269, pp. 386–394. Springer, Heidelberg (2012)
19. Wierzchoń, S.: Artificial Immune Systems, Theory and Applications. EXIT, Warsaw (2001) (in Polish)
20. Zienkiewicz, O., Taylor, R.: The finite element method, vol. I. McGraw-Hill (1989), vol. II (1991)

Global Decisions Taking Process, Including the Stage of Negotiation, on the Basis of Dispersed Medical Data

Małgorzata Przybyła-Kasperek

University of Silesia, Institute of Computer Science,
Będzińska 39, 41-200 Sosnowiec, Poland
malgorzata.przybyla-kasperek@us.edu.pl
http://www.us.edu.pl

Abstract. The article discusses the issues related to the decision-making system using dispersed knowledge. In the proposed system, the classification process of the test object can be described as follows. In the first step, we investigate how particular classifiers classify a test object. We describe this using probability vectors over decision classes. We cluster classifiers with respect to similarities of the probability vectors. In the paper a new approach has been proposed in which the clustering process consists of two stages and three types of relations between classifiers: friendship, conflict and neutrality are defined. In the first step initial groups are created. Such a group contains classifiers that are in friendship relation. In the second stage, classifiers which are in neutrality relation are attached to the existing groups. In experiments the situation is considered in which medical data from one domain are collected in many medical centers. We want to use all of the collected data at the same time in order to make a global decisions.

Keywords: decision-making system, global decision, relation of friendship, relation of conflict, negotiations, conflict analysis.

1 Introduction

In the modern world we are inundated with the huge data sets. Also, large volumes of medical data are collected, often they are stored in a dispersed form. Support the decision-making process in such situations is a great challenge and a difficult task. Traditional methods of data analysis are not sufficient in this situation. Therefore it is necessary to develop new methods of dealing with dispersed knowledge, which is available in the form of many different knowledge bases. Issues concerning the use of dispersed knowledge was considered by the author in earlier papers [8,9,7,13]. In these papers decision-making system using dispersed knowledge and some methods to deal with inconsistencies of knowledge and analysis of the conflict have been proposed. In this paper, we develop the concepts discussed earlier. Major innovation, which is proposed, consists in the formation the structure of a dispersed decision support system. The new approach uses the negotiation stage in the process of creating the structure of

S. Kozielski et al. (Eds.): BDAS 2014, CCIS 424, pp. 290–299, 2014.

the system. The main assumptions of the proposed approach are as follows. We assume that, in the system, knowledge is available in the form of many local knowledge bases. Between local knowledge bases three types of relations: friendship, conflict and neutrality are defined. Then a two-stage process of connecting local knowledge bases in coalitions - groups of bases that are agreed on the classification of the test object, is implemented. In the first step of this process the local knowledge bases, that remaining in a friendship relation, are combined into groups. The second step consists of re-examine the relations between the created initial coalitions and knowledge bases that are not connected into any coalition. The negotiation process is implemented, in which to the initial coalitions the local knowledge bases remain in a neutrality relation are included.

The concept of distributed decision making is widely discussed in the paper [10]. The concept of taking a global decision on the basis of local decisions is also used in issues concerning the multiple model approach. Examples of the application of this approach can be found in the literature [1,12]. Also in many other papers [3,11], the problem of using distributed knowledge is considered. This paper describes a different approach to the global decision-making process. We assume that the set of local knowledge bases that contain information from one domain is pre-specified. The only condition which must be satisfied by the local knowledge bases is to have common decision attributes.

2   Notations and Definitions

We assume that the set of local knowledge bases that contain dispersed medical data from one domain is pre-specified. The only condition which must be satisfied by the local knowledge bases is to have common decision attributes. We assume that each local knowledge base is managed by one agent, which is called a resource agent.

Definition 1. *We call ag in $Ag = \{ag_1, \ldots, ag_n\}$ a resource agent if it has access to resources represented by a decision table $D_{ag} := (U_{ag}, A_{ag}, d_{ag})$, where U_{ag} is the universe; A_{ag} is a set of conditional attributes, V_{ag}^a is a set of values of the attribute a; d_{ag} is a decision attribute, V_{ag}^d is called the value set of d_{ag}.*

Each resource agent $ag \in Ag$ can independently determine the value of the decision for a test object for which the values on the set of attributes A_{ag} are defined. The agents which agree on the classification for a test object into the decision classes will be combined in the group. The approach proposed in this paper uses the dynamic structure of the system. In this approach three types of relations: friendship, conflict and neutrality are considered. Clustering process can be divided into two stages. The first step is to create groups of agents remaining in the relation of friendship. The second step is to use issues related to negotiations and join agents which are neutral to the existing coalition.

2.1   Relations of Friendship, Conflict and Neutrality

In this part of the paper definitions of relations between agents are given. Then, the clustering process is discussed and the definition of dispersed decision-making system will be presented. The definitions of the relations, as well as a method for determining the intensity of the conflict are modeled on the basis of the concepts given in the papers of Pawlak [6,5].

Let there be given a test object $\bar{x}$ for which we want to generate a global decision. Let for the object $\bar{x}$ the values of conditional attributes belonging to the set $\bigcup_{i=1}^{n} A_{ag_i}$ be defined. In order to determine groups of agents, from each decision table of a resource agent $D_{ag_i}, i \in \{1, \ldots, n\}$ and from each decision class $X_v^{ag_i}, v \in V^{d_{ag_i}}$, the smallest set containing at least m_1 objects is chosen, for which the values of conditional attributes bear the greatest similarity to the test object. The value of the parameter m_1 is selected experimentally. The subset of relevant objects is the union of the sets of objects selected from all decision classes. In order to determine the subset of relevant objects, the measure of similarity is used. In the proposed system any similarity measures could be applied. Since the data sets, which are examined in experiments, have qualitative, quantitative and binary attributes, the Gower similarity measure [13] is used. The next stage in the process of generating groups of agents is to determine the vectors of values specifying the classification of the test object made by the agents. So, for each resource agent, the vector that indicates the level of certainty with which the decisions are taken by the agent for the test object is generated. Each coordinate of the vector is determined on the basis of relevant objects that were previously selected from the decision table of the resource agent. Thus, for each resource agent $i \in \{1, \ldots, n\}$, a c-dimensional vector $[\bar{\mu}_{i,1}(\bar{x}), \ldots, \bar{\mu}_{i,c}(\bar{x})]$ is generated, where the value $\bar{\mu}_{i,j}(\bar{x})$ means the certainty with which the decision $v_j \in V^d, j \in \{1, \ldots, c\}, c = card\{V^d\}$ is made about the object $\bar{x}$ by the resource agent ag_i. The value $\bar{\mu}_{i,j}(\bar{x})$ is defined as follows:

$$\bar{\mu}_{i,j}(\bar{x}) = \frac{\sum_{y \in U_{ag_i}^{rel} \cap X_{v_j}^{ag_i}} s(\bar{x}, y)}{card\{U_{ag_i}^{rel} \cap X_{v_j}^{ag_i}\}}, i \in \{1, \ldots, n\}, j \in \{1, \ldots, c\},$$

where $c = card\{V^d\}$, $U_{ag_i}^{rel}$ is the subset of relevant objects selected from the decision table D_{ag_i} of a resource agent ag_i and $X_{v_j}^{ag_i}$ is the decision class of the decision table of resource agent ag_i; $s(x, y)$ is the measure of similarity between objects x and y. On the basis of the vector of values defined above a vector of rank assigned to the values of the decision attribute is specified. The vector of rank is defined as follows: rank 1 is assigned to the values of the decision attribute which are taken with the maximum level of certainty. Rank 2 is assigned to the values of the decision attribute that have the maximum level of certainty in the set of decisions that have not received the rank 1, etc. Proceeding in this way for each resource agent $ag_i, i \in \{1, \ldots, n\}$, the vector of rank $[r_{i,1}(\bar{x}), \ldots, r_{i,c}(\bar{x})]$ will be defined. The definitions of friendship relation, conflict relation and neutrality relation are given next. Relations between agents are defined by their views on the classification of the test object x to the decision class. We define the function $\phi_{v_j}^x$ for the test object x and each value of the decision attribute

$$v_j \in V^d; \phi_{v_j}^x : Ag \times Ag \rightarrow \{0, 1\}; \phi_{v_j}^x(ag_i, ag_k) = \begin{cases} 0 & \text{if } r_{i,j}(x) = r_{k,j}(x) \\ 1 & \text{if } r_{i,j}(x) \neq r_{k,j}(x) \end{cases} \text{ where}$$

$ag_i, ag_k \in Ag.$

Definition 2. *Agents* $ag_i, ag_k \in Ag$ *are in a friendship relation due to the object* x *and decision class* $v_j \in V^d$, *which is written* $R^+_{v_j}(ag_i, ag_k)$, *if and only if* $\phi^x_{v_j}(ag_i, ag_k) = 0$. *Agents* $ag_i, ag_k \in Ag$ *are in a conflict relation due to the object* x *and decision class* $v_j \in V^d$, *which is written* $R^-_{v_j}(ag_i, ag_k)$, *if and only if* $\phi^x_{v_j}(ag_i, ag_k) = 1$.

We also define the intensity of conflict between agents using a function of the distance between agents. We define the distance between agents ρ^x for the test object x: $\rho^x : Ag \times Ag \rightarrow [0, 1]$;

$$\rho^x(ag_i, ag_k) = \frac{\sum_{v_j \in V^d} \phi^x_{v_j}(ag_i, ag_k)}{card\{V^d\}}, \quad \text{where } ag_i, ag_k \in Ag.$$

Definition 3. *Let* p *be a real number, which belongs to the interval* $[0, 0.5)$. *We say that agents* $ag_i, ag_k \in Ag$ *are in a friendship relation due to the object* x, *which is written* $R^+(ag_i, ag_k)$, *if and only if* $\rho^x(ag_i, ag_k) < 0.5 - p$. *Agents* $ag_i, ag_k \in Ag$ *are in a conflict relation due to the object* x, *which is written* $R^-(ag_i, ag_k)$, *if and only if* $\rho^x(ag_i, ag_k) > 0.5 + p$. *Agents* $ag_i, ag_k \in Ag$ *are in a neutrality relation due to the object* x, *which is written* $R^0(ag_i, ag_k)$, *if and only if* $0.5 - p \leq \rho^x(ag_i, ag_k) \leq 0.5 + p$.

2.2   The Process of Clusters Creating

The first step in the process of clusters creating is to define the initial group of agents remaining in the friendship relation.

Definition 4. *Let* Ag *be the set of resource agents. The initial cluster due to the classification of object* x *is the maximum, due to the inclusion relation, subset of resource agents* $X \subseteq Ag$ *such that* $\forall_{ag_i, ag_k \in X} \ R^+(ag_i, ag_k)$. *Thus, the initial cluster is the maximum, due to inclusion relation, set of resource agents that remain in the friendship relation due to the object* x.

After the first stage of clusters creating we obtain a set of initial clusters and a set of agents which are not included in any cluster. In the group of agents which were not joined to any clusters there are agents which remained undecided. So those which are in neutrality relation with agents belonging to some initial clusters. For each agent, which is not attached to any clusters, relations between this agent and the generated initial clusters and other agents without coalition are analyzed. In the second stage, agents without coalition are connected to each initial cluster, with which his relations will be good enough. Also new clusters, consisting of agents without coalition, which are in a sufficiently good relations, are creating. Now we will proceed to the formal description of the second stage of cluster creating process. Let $C_1, \ldots, C_k$ be a set of initial clusters and $Ag \setminus \bigcup_{i=1}^{k} C_i$ be a set of agents which are not attached to any clusters. As it is known the goal of negotiation process is to reach a compromise by accepting some concessions by the parties involved in a conflict situation. Also, in the considered situation some concessions were accepted. In the second stage of clustering process a generalized distance function between agents is defined. This definition assumes that during the negotiation, agents put the greatest emphasis on compatibility of the ranks assigned to the decisions with the highest ranks. That is the values of the decisions that are most significant for the agent. Compatibility of the ranks assigned to the less meaningful decision is omitted during the second stage of clustering process.

We define the function ϕ_G^x for the test object x; $\phi_G^x : Ag \times Ag \to [0, \infty)$;

$$\phi_G^x(ag_i, ag_j) = \frac{\sum_{v_l \in Sign_{i,j}} |r_{i,l}(x) - r_{j,l}(x)|}{card\{Sign_{i,j}\}}$$

where $ag_i, ag_j \in Ag$ and $Sign_{i,j} \subseteq V^d$ is the set of significant decision values for the pair of agents ag_i, ag_j. In the set $Sign_{i,j}$ there are the values of the decision, which the agent ag_i or agent ag_j gave the highest rank. During the second stage of the clusters creating process - the negotiation process, the intensity of the conflict between the two groups of agents is determined by using the generalized distance. We define the generalized distance between agents ρ_G^x for the test object x; $\rho_G^x : 2^{Ag} \times 2^{Ag} \to [0, \infty)$

$$\rho_G^x(X, Y) = \begin{cases} 0 & \text{if } card\{X \cup Y\} \leq 1 \\ \frac{\sum_{ag, ag' \in X \cup Y} \phi_G^x(ag, ag')}{card\{X \cup Y\} \cdot (card\{X \cup Y\} - 1)} & \text{else} \end{cases} \quad \text{where } X, Y \subseteq Ag.$$

As it can be easily seen the value of the generalized distance function for two sets of agents X and Y is equal to the average value of the function ϕ_G^x for each pair of agents ag, ag' belonging to the set $X \cup Y$. This value can be interpreted as the average difference of the ranks assigned to significant decisions within the combined group of agents consisting of the sets X and Y. Then the agent ag is included to all initial clusters, for which the generalized distance does not exceed a certain threshold, which is set by the system's user. Also agents without coalition, for which the value of the generalized distance function does not exceed the threshold, are combined into a new cluster. Proper selection of the threshold value is very important. If this value is too low then the few agents will be added to the initial cluster, in extreme cases, no change will be made. If the threshold is too high it may happen that the agents which significantly differ in their views on the classification of the test object will be connected to one cluster. The threshold value will be chosen experimentally. After completion of the second stage of the process of clustering we get the final form of clusters. The proposed decision-making system has a hierarchical structure. The resource agents that are connected into clusters are located at the lowest level of the hierarchy. For each cluster that contains at least two resource agents, a superordinate agent is defined, which is called a synthesis agent, as_j, where j is the number of cluster. The synthesis agent, as_j, has access to knowledge that is the result of the process of inference carried out by the resource agents that belong to its subordinate group. As is a finite set of synthesis agents. Now, we can provide a formal definition of a dispersed decision-making system.

Definition 5. *By a dispersed decision-making system (multi-agent system) with dynamically generated clusters we mean* $WSD_{Ag}^{dyn} = \langle Ag, \{D_{ag} : ag \in Ag\}, \{As_x : x \text{ is a classified object}\}, \{\delta_x : x \text{ is a classified object}\}\rangle$ *where Ag is a finite set of resource agents;* $\{D_{ag} : ag \in Ag\}$ *is a set of decision tables of resource agents;* As_x *is a finite set of synthesis agents defined for clusters dynamically generated for the test object* x, $\delta_x : As_x \to 2^{Ag}$ *is a injective function that each synthesis agent assigns a cluster generated due to classification of the object* x.

2.3 Elimination of Inconsistencies in the Knowledge and Conflict Analysis

On the basis of the knowledge of agents from one cluster, local decisions are taken. An important problem that occurs when taking a global decision is to eliminate inconsistencies in the knowledge stored in different knowledge bases. This problem stems from the fact that the system has the general assumptions and we do not require that the sets of conditional attributes of decision tables are disjoint. We understand inconsistency of knowledge to be situations in which, on the basis of two different knowledge bases that have common attributes and for the same values for common attributes using logical implications, conflicting decisions are made. In previous papers some methods of elimination inconsistencies in the knowledge have been proposed [8,13]. In this paper, one of these methods - the approximated method of the aggregation of decision tables, will be used. Conflict analysis is implemented after completion of the process of inconsistencies elimination in knowledge, because then the synthesis agents have access to the knowledge on the basis of which they can independently establish the value of a local decision to just one cluster. Two methods to resolve the conflict analysis will be used in this paper: the method of weighted voting and the method of a density based algorithm. These methods allow the analysis of conflicts and enable to generate a set of global decisions. In the case of the density-based method the generated set will contain not only the value of the decisions that have the greatest support of knowledge stored in local knowledge bases, but also those for which the support is relatively high. These methods were discussed in detail in the paper [13].

3 Experiments

The aim of the experiments is to examine the quality of the classification made on the basis of dispersed medical data by the decision-making system with dynamically generated clusters and the stage of negotiation. An additional objective is to compare the effectiveness of this system with the results obtained in the paper [9] (where a different approach to the dynamically generate clusters was proposed) and the papers of other authors [2,4] who have used the medical data in non-dispersible form. For the experiments the following data, which are in the UCI repository (archive.ics.uci.edu/ml/), were used: Lymphography data set, Primary Tumor data set. Both sets of data was obtained from the University

Medical Centre, Institute of Oncology, Ljubljana, Yugoslavia (M. Zwitter and M. Soklic provided this data). In order to determine the efficiency of inference of the proposed decision-making system with respect to the analyzed data, each data set was divided into two disjoint subsets: a training set and a test set. A numerical summary of the data sets is as follows: Lymphography: # The training set - 104; # The test set - 44; # Conditional - 18; # Decision - 4; Primary Tumor: # The training set - 237; # The test set - 102; # Conditional - 17; # Decision - 22. We will consider a situation in which medical data from one domain are collected in different medical centers. We want to use all of the collected data at the same time in order to make a global decisions. This approach not only allows the use of all available knowledge, but also should improve the efficiency of inference. In order to consider the discussed situation it is necessary to provide the knowledge stored in the form of a set of decision tables. Therefore, the training set was divided into a set of decision tables. Divisions with a different number of decision tables were considered. For each of the data sets used, the decision-making system with five different versions (with 3, 5, 7, 9 and 11 resource agents) were considered. For these systems, we use the following designations: WSD_{Ag1}^{dyn} - 3 resource agents; WSD_{Ag2}^{dyn} - 5 resource agents; WSD_{Ag3}^{dyn} - 7 resource agents; WSD_{Ag4}^{dyn} - 9 resource agents; WSD_{Ag5}^{dyn} - 11 resource agents. Note that the division of the data set was not made in order to improve the quality of the decisions taken by the decision-making system, but in order to store the knowledge in a distributed form. We consider the situation, that is very common in life, in which data are collected in different medical centers as separate knowledge bases. The measures of determining the quality of the classification are: *estimator of classification error e* in which an object is considered to be properly classified if the decision class used for the object belonged to the set of global decisions generated by the system; *estimator of classification ambiguity error e_{ONE}* in which object is considered to be properly classified if only one, correct value of the decision was generated to this object; *the average size of the global decisions sets $\overline{d}_{WSD_{Ag}^{dyn}}$* generated for a test set. In the description of the results of experiments for clarity some designations for algorithms have been adopted: $A(m_2)$ - the approximated method of the aggregation of decision tables; W - the method of weighted voting; $G(\varepsilon, MinPts)$ - the method of a density-based algorithm. During experiments influence of the parameter p, which occurs in the definition 3 of friendship, conflict and neutrality relations, on the effectiveness of inference of a dispersed decision-making system was analyzed. Five different values of the parameter p were examined. The analyzed values are $p = 0.05$, $p = 0.1$, $p = 0.2$, $p = 0.3$, $p = 0.4$. In tables presented below the best results, obtained for values of the parameter p, are given.

The results of the experiments with the proposed approach and the Lymphography data set are presented in the first part of Tab. 1. In the table the following information is given: the name of multi-agent decision-making system (System); the algorithm's symbol (Algorithm); the three measures discussed earlier $e, e_{ONE}, \overline{d}_{WSD_{Ag}^{dyn}}$; the time t needed to analyse a test set expressed in minutes. For comparison, in the second part of Tab. 1, the results of the experiments that

Table 1. Experiments results with the Lymphography data set

The proposed approach with the stage of negotiation

System	Algorithm	e	e_{ONE}	$d_{WSD_{Ag}^{dyn}}$	t
WSD_{Ag1}	$A(1)G(0.0268; 2)$	**0.068**	0.568	1.523	0.01
$m_1 = 2, p = 0.3$	$A(1)G(0.0004; 2)$	0.159	0.182	1.023	0.01
WSD_{Ag2}	$A(3)G(0.042; 2)$	**0.091**	0.682	1.591	0.01
$m_1 = 2, p = 0.3$	$A(3)G(0.0128; 2)$	0.136	0.318	1.182	0.01
WSD_{Ag3}	$A(1)G(0.0515; 2)$	**0.114**	0.523	1.409	0.01
$m_1 = 1, p = 0.05$	$A(1)G(0.0005; 2)$	0.159	0.273	1.114	0.01
WSD_{Ag4}	$A(1)G(0.0625; 2)$	**0.114**	0.591	1.477	0.01
$m_1 = 1, p = 0.05$	$A(1)G(0.052; 2)$	0.136	0.5	1.364	0.01
WSD_{Ag5}	$A(2)G(0.058; 2)$	**0.159**	0.568	1.409	0.07
$m_1 = 5, p = 0.3$	$A(2)G(0.0292; 2)$	0.182	0.545	1.364	0.07

Dynamically generated disjoint clusters - results presented in the paper [9]

System	Algorithm	e	e_{ONE}	$d_{WSD_{Ag}^{dyn}}$	t
WSD_{Ag1}	$A(1)G(0.0624; 2)$	0.091	0.591	1.545	0.01
$m_1 = 2$	$A(1)G(0.0092; 2)$	0.182	0.295	1.159	0.01
WSD_{Ag2}	$A(1)G(0.0775; 2)$	0.136	0.636	1.500	0.01
$m_1 = 2$	$A(1)G(0.029; 2)$	0.159	0.364	1.205	0.01
WSD_{Ag3}	$A(1)G(0.0858; 2)$	0.136	0.591	1.455	0.01
$m_1 = 2$	$A(1)G(0.0006; 2)$	0.159	0.273	1.114	0.01
$WSD_{Ag4}, m_1 = 2$	$A(1)G(0.0702; 2)$	0.136	0.455	1.318	0.01
WSD_{Ag5}	$A(1)G(0.084; 2)$	0.159	0.614	1.477	0.07
$m_1 = 1$	$A(1)G(0.0672; 2)$	0.182	0.545	1.364	0.07

are presented in the paper [9] are given. These results were obtained using different method to define the structure of dispersed decision-making system, which is different from the approach presented in this paper. In this approach agents, which are in friendship relation due to the classification of the test object, are connected in clusters. However, the major difference between these approaches lies in the definition of the relations between the agents. In the approach proposed in the paper [9,7] only two types of relations were considered: the relation of friendship and the relation of conflict. Moreover, in the process of generating clusters the negotiations process did not occur. The results of the experiments with the Primary Tumor data set are presented in Tab. 2. In the first part of the table the results for the proposed approach are described. For comparison, in the second part of Tab. 2, the results of the experiments that are presented in the paper [9] are given. Based on the results of the experiments given in Tab. 1 and Tab. 2 the following conclusions can be drawn. For the Lymphography data set the proposed approach generates better results than the approach proposed in the paper [9]. For the Primary Tumor data set the proposed approach generates comparable results with the approach proposed in the paper [9].

The papers [2,4] also shows the experiments with the Lymphography and the Primary Tumor data set. Data in the non-dispersible form were examined.

Table 2. Experiments results with the Primary Tumor data set

The proposed approach with the stage of negotiation

System	Algorithm	e	e_{ONE}	$\bar{d}_{WSD_{Ag}^{dyn}}$	t
WSD_{Ag1}, $m_1 = 5$, $p = 0.05$	$A(1)G(0.00546; 2)$	**0.373**	0.814	3.020	0.01
WSD_{Ag2}, $m_1 = 3$, $p = 0.1$	$A(2)G(0.00001; 2)$	**0.343**	0.814	3.029	0.02
WSD_{Ag3}, $m_1 = 2$, $p = 0.05$	$A(1)G(0.00001; 2)$	**0.373**	0.902	3.745	0.02
WSD_{Ag4}, $m_1 = 4$, $p = 0.1$	$A(2)G(0.00001; 2)$	0.353	0.882	3.686	0.05
WSD_{Ag5}, $m_1 = 2$ $p = 0.2$	$A(3)G(0.00001; 2)$	**0.314**	0.892	4.245	0.33

Dynamically generated disjoint clusters - results presented in the paper [9]

System	Algorithm	e	e_{ONE}	$\bar{d}_{WSD_{Ag}^{dyn}}$	t
WSD_{Ag1}, $m_1 = 5$	$A(2)G(0.00549; 2)$	**0.373**	0.814	3.020	0.01
WSD_{Ag2}, $m_1 = 17$	$A(3)G(0.0003; 2)$	0.353	0.814	2.990	0.02
WSD_{Ag3}, $m_1 = 5$	$A(5)G(0.00573; 2)$	0.373	0.912	3.755	0.02
WSD_{Ag4}, $m_1 = 4$	$A(3)G(0.0063; 2)$	**0.343**	0.902	3.667	0.05
WSD_{Ag5}, $m_1 = 6$	$A(1)G(0.0003; 2)$	0.333	0.941	4.294	0.33

Tab. 3 presents the results given in these papers. Presented, in this paper results cannot be compared uniquely with the results shown in Tab. 3, because the decision-making system described in the paper generates a set of decisions, while Tab. 3 shows the results of the algorithms that generate one decision. It should be noted that for the Lymphography data set the average size of the global decisions sets is small, since it is close to the value 1. In the case of the Primary Tumor data set the average size of the global decisions sets is between 3 and 4, note that there are 22 decision classes. This means that this result may be considered as a quite good result. However, the quality of classification has significantly improved in comparison with the results shown in Tab. 3. Moreover, very important advantage of the proposed decision-making system is the possibility of using dispersed knowledge, which are collected in different medical centers.

Table 3. Results of experiments from other papers

Lymphography				Primary Tumor			
Algorithm	Error rate	Algorithm	Error rate	Algorithm	Error rate	Algorithm	Error rate
Bayes	0.17	AQ15	0.18	Bayes	0.61	AQ15	0.59
AQR	0.24	Human Experts	0.15	AQR	0.65	Human Experts	0.58
CN2	0.22	Random Choice	0.75	CN2	0.63	Random Choice	0.95

4 Conclusion

In this paper, a new approach to the organization of the structure of a dispersed decision-making system that operates on the basis of dispersed knowledge is proposed. In this approach a two-stage process of dynamically generating clusters are used. Between agents three types of relations: friendship, conflict and neutrality are defined. In the experiments, which are presented in the article, dispersed medical data have been used: Lymphography data set, Primary Tumor data set. The usage of dispersed medical data is very important, because in many medical centers, information from one domain, are collected. Thus, these data are in the dispersed form. Based on the presented results of experiments it can be concluded that the proposed decision-making system achieves good results for dispersed medical data.

References

1. Bazan, J.G., Peters, J.F., Skowron, A., Nguyen, H.S., Szczuka, M.S.: Rough set approach to pattern extraction from classifiers. Electr. Notes Theor. Comput. Sci. 82(4), 20–29 (2003)
2. Clark, P., Niblett, T.: Induction in noisy domains. In: EWSL, pp. 11–30 (1987)
3. Delimata, P., Suraj, Z.: Feature selection algorithm for multiple classifier systems: A hybrid approach. Fundam. Inform. 85(1-4), 97–110 (2008)
4. Michalski, R.S., Mozetic, I., Hong, J., Lavrac, N.: The multi-purpose incremental learning system AQ15 and its testing application to three medical domains. In: AAAI, pp. 1041–1047 (1986)
5. Pawlak, Z.: On conflicts. International Journal of Man-Machine Studies 21(2), 127–134 (1984)
6. Pawlak, Z.: An inquiry into anatomy of conflicts. Inf. Sci. 109(1-4), 65–78 (1998)
7. Przybyla-Kasperek, M.: Wieloagentowy system decyzyjny z dynamicznie generowanymi rozlacznymi klastrami. ZN Pol. Sl. Studia Informatica 34(2A(111)), 275–294 (2013)
8. Przybyla-Kasperek, M., Wakulicz-Deja, A.: Application of reduction of the set of conditional attributes in the process of global decision-making. Fundam. Inform. 122(4), 327–355 (2013)
9. Przybyła-Kasperek, M., Wakulicz-Deja, A.: Global decisions taking on the basis of dispersed medical data. In: Ciucci, D., Inuiguchi, M., Yao, Y., Ślęzak, D., Wang, G. (eds.) RSFDGrC 2013. LNCS, vol. 8170, pp. 355–365. Springer, Heidelberg (2013)
10. Schneeweiss, C.: Distributed decision making. Springer, Berlin (2003)
11. Skowron, A., Wang, H., Wojna, A., Bazan, J.: Multimodal classification: Case studies. In: Peters, J.F., Skowron, A. (eds.) Transactions on Rough Sets V. LNCS, vol. 4100, pp. 224–239. Springer, Heidelberg (2006)
12. Slezak, D., Wroblewski, J., Szczuka, M.S.: Neural network architecture for synthesis of the probabilistic rule based classifiers. Electr. Notes Theor. Comput. Sci. 82(4), 251–262 (2003)
13. Wakulicz-Deja, A., Przybyla-Kasperek, M.: Application of the method of editing and condensing in the process of global decision-making. Fundam. Inform. 106(1), 93–117 (2011)

UMAP - A Universal Multi-Agent Platform for .NET Developers

Dariusz Mrozek*, Bożena Małysiak-Mrozek, and Igor Waligóra

Institute of Informatics, Silesian University of Technology
Akademicka 16, 44-100 Gliwice, Poland
{dariusz.mrozek,bozena.malysiak}@polsl.pl,
waligora@gmail.com
http://zti.polsl.pl/dmrozek/umap.htm

Abstract. Multi-agent systems allow to build custom software solutions that work on the basis of groups of cooperating agents pursuing a common goal. In the paper, we present a universal multi-agent platform (UMAP) for .NET developers. UMAP provides ready to use interface for the implementation of software agents, leaving the programmers only the necessity to implement business logic of agents, according to the expected deliverables. UMAP is compliant with standards of the Foundation for Intelligent Physical Agents (FIPA) and unlike the popular systems, such as JADE or SPADE, it has been developed using the Microsoft .NET Framework. In the paper we show architecture of the UMAP, communication and message transport system, agent cataloging and management services, and runtime environment. The platform is available for free from: `http://zti.polsl.pl/dmrozek/umap.htm`

Keywords: multi-agent system, software agent, multi-agent platform, .NET Framework.

1 Introduction

The development of informatics, from the time of its birth, allows the creation of more and more complex systems. This complexity requires a different approach to software engineering. Both, the concept of designing programs and paradigms of their development have changed. Imperative (procedural) programming is no longer sufficient, because it does not allow to create complex programs in clear and legible way. The essence of the problem was, inter alia, the difficulty in describing, implementing and testing the project as a whole. Therefore, a natural step has been made toward a modular and object-oriented programming that enabled breaking the problem into smaller parts. The rapid growth of computer networks, both global and local, gave users the ability to disperse the systems into multiple logical subsystems. Such an approach facilitates the design and implementation of the system. However, it should be mentioned that this approach also generates some problems, such as communication or its reliability.

* This work was supported by the European Union through the European Social Fund (grant agreement number: UDA-POKL.04.01.01-00-106/09).

S. Kozielski et al. (Eds.): BDAS 2014, CCIS 424, pp. 300–311, 2014.

A glance at the problem of complex systems as a group of independent, but communicating with each other, entities leads to the formation of another programming paradigm, agent programming. In this approach, we construct an information system composed of many independent and decision making agents communicating with each other in order to pursue joint ventures. Agent is characterized by two properties, the ability to take action and communicate with other agents. The set of multiple agents working together to achieve a common goal is called a multi-agent system (MAS) [6, 25].

Multi-agent systems play a special role in many decision support systems or expert systems, where the data necessary to make decisions are scattered among databases that differ in the domain of stored information. Many solutions in applied industrial informatics are also based on agents, because such systems are fault tolerant and able to operate on incomplete data, and in special cases, make the right decisions, even if they receive incorrect data [21]. Spectrum of applications of multi-agent systems is very broad. Multi-agent systems are widely used to solve many real-world problems, like urban traffic signal control [2], air traffic management [24], evaluation of electricity market rules [16], knowledge discovery from databases [19], testing database performance by replaying the specified workload [15], frequency assignment in cellular radio networks [1], optimization of the supply chain [27], hierarchical control with self-organizing database [5], control and monitoring of a complex process control system [23], protein structure similarity searching [12–14].

The greatest potential for multi-agent systems results from the fragmentation of knowledge and functions between its various components. This makes it possible to solve complex problems using domain-specific resources, or systems that solve different classes of problems. However, the challenge does not remain in the integration, but in mutual communication between different cooperating systems. This can be achieved by creating standards for methods and formats of communication between these systems and agents [17, 18, 25].

In the paper, we present a new, universal multi-agent platform (UMAP) for .NET developers. UMAP offers ready to use solutions for the implementation of software agents, leaving the programmers only the necessity to implement business logic of agents, according to the expected deliverables. UMAP platform provides a unified agent interface, which constitutes a base for any agent solution. UMAP agent has a fixed interface that enables communication and management of agents within the platform. We also show theoretical considerations related to the functioning of the UMAP platform, its architecture, programming interface and runtime environment.

2 Related Works

Multi-agent systems are used to solve various problems, ranging from industrial process control systems to decision support systems or systems for the analysis of currency markets. Depending on user's needs and requirements, the systems are created from scratch or based on ready-made platforms, which provide necessary components responsible for the functioning of the system, leaving only the need to implement the logic of the agents. This chapter presents examples of competitive multi-agent platforms.

JADE (Java Agent Development Framework) [3] is the most developed project of multi-agent platform, implemented in accordance with the FIPA specifications. The system was developed at Telecom Italia Lab (TILAB). It is a system with open access to the source code. It is distributed under the LGPL license (Lesser General Public License Version 2). The Project Management Committee consists of five members: Telecom Italia, Motorola, Whitestein Technologies AG, Profactor GmbH and France Telecom R&D. JADE is implemented in Java programming language and it is possible to use any version of the Java platform: J2SE, J2EE and J2ME. JADE platform agents can communicate with each other according to the ACL standard (Agent Communication Language).

SPADE (Smart Python multi-Agent Development Environment) [22] is a multi-agent platform, which is based on the XMPP/Jabber technology. It is implemented in Python. SPADE platform provides users with features that facilitate the construction of multi-agent systems. Examples of such features include defined communication channels and extensible communication protocol based on XML, similarly to FIPA-ACL. It should be noted that SPADE is the first multi-agent platform that is based on the XMPP technology. XMPP (Extensible Messaging and Presence Protocol) is a technology that allows real-time communication, which provides a wide range of applications, such as instant messaging, shared conversations, and voice/video services [26].

FIPA-OS (FIPA Open Source) [7] is described as a set of tools for efficient implementation of multi-agent systems compliant with FIPA specifications. FIPA-OS is a system with an open access to the source code. This system was first released in August 1999. The FIPA-OS is implemented using Java language. There are two versions of the system available: Standard FIPA-OS and MicroFIPA-OS. The first version includes two distributions, compatible with Java 2 and Java 1.1. MicroFIPA-OS extends the Standard version compatible with Java 1.1. This version was created by researchers at the University of Helsinki. It is designed to work with mobile devices, such as Pocket PC or Compaq iPaq.

3   UMAP Platform

UMAP platform (*Universal Multi-Agent Platform*) has been developed in the Institute of Informatics at the Silesian University of Technology in Gliwice, Poland. This chapter presents the main features of the platform, its architecture, basic concepts, component modules, description of communication between agents and runtime environment.

3.1   Architecture of the UMAP Platform

General architecture of the UMAP multi-agent platform is shown in Fig. 1. The architecture consists of several UMAP agents working inside containers. Containers are logical units that group agents. There can be many agents working within one container. There is no limit to the number of containers running on the same machine, but we must remember that communication between agents within a single container is more efficient than between agents operating in two separate containers. There is also a special process called an Agent Management System (AMS) associated with each container. One container is associated with only one AMS. The AMS provides various services, such as Directory Facilitator (DF) for cataloging agents, and Message Transport System (MTS) for communication between agents. But above all, the Agent Management System allows to run and supervise the work of agents.

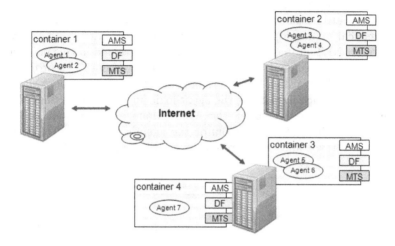

Fig. 1. General architecture of the UMAP multi-agent platform

Agents operate inside the platform that is able to supervise their work, including also a completion of a particular agent's action. Agents can communicate directly with each other within a single container. Communication with agents acting on a remote computer (or physically the same, but inside another container) is possible through the Message Transport System (MTS). Each container also provides a catalog of available agents through the Directory Facilitator (DF), which supplies an information on known agents and their description. Catalog of agents is also available for remote agents through the Message Transport System. There can be many containers hosted on a single computer. Communication between agents operating inside different containers is carried out by network sockets; the messages are sent in the XML format.

3.2   Agent Management System

In the UMAP platform, we decided to create the Agent Management System (AMS), which provides a runtime environment for software agents operating inside containers. Message transport services (MTS) and directory services (DF)

act as components that are executed and operate directly under the control of the AMS. The AMS has been designed according to the FIPA standard [8].

The AMS is also responsible for loading **Agent** classes delivered in DLL libraries implemented by a developer of a particular multi-agent solution for the UMAP platform. In order to ensure the dynamic loading of new **Agent** classes at run time, we used the reflection mechanism of the Microsoft .NET Framework. By using the reflection mechanism, we can dynamically create instances of these classes, and since we know the structure of the agent class by inheriting from an abstract class **Agent**, we can refer to the appropriate methods and properties.

In order to start the agent loaded from an external DLL library we need to use the reflection mechanism again. Starting the agent in this way differs somewhat from the standard instantiation of objects, whose definitions are available at the time of application development. This follows from the strict control of types in the Microsoft .NET Framework. Object, i.e. agent, created in this manner must be properly initialized and registered within the container. According to the adopted approach regarding the functioning of UMAP agents, the AMS module creates threads, which implement the agent logic and process messages (Fig. 2). Once an agent has been created, the Agent Management System preserves a reference to it, which allows to monitor the agent's status, and in particular, terminate its threads and release resources used by the agent.

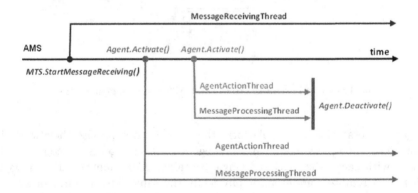

Fig. 2. Creation of threads in a container of the UMAP platform

3.3   Directory Facilitator

In order to enable efficient operation of the UMAP multi-agent platform we had to provide a mechanism for identification of local and remote agents. In the UMAP this functionality is offered by the Directory Facilitator (DF), which single instance is always created during the creation of a container. The DF provides then information for agents running locally. Providing the information on known agents, working within the local container or in a remote containers, is actually the primary function of the Directory Facilitator. To make this possible, each agent, during its initialization, must register with the local DF. The information about remote agents is delivered when two containers, e.g. local and remote, are interrelated.

In order to enable effective management and communication between agents, it is necessary to clearly identify agents not only within a single container, or even the platform, but a global network within which they are able to connect. This means that in order to connect any two existing agent systems the agents' names must be unique. The best solution turned out to be a hierarchical construction of an agent identifier that contains the name of the platform, the name of the container within which the agent is running, and the name of the agent, which should be unique within its container. The construction of an agent identifier (AgentID) is as follows:

`<agent_name>@<container_name>/<platform_name>`

for example:`alice@container1/UMAP`

For regular agents working inside a container the DF is visible as any other agent. This means that DF has its own agent identifier with a fixed syntax `df@<container_name>/UMAP`. Therefore, the DF as an agent has the ability to receive and send messages, which is used to receive and broadcast information about the agents working on the entire platform.

In order to keep the agent directory up to date, the DF of each container broadcasts to associated containers the information about all agents running locally. Messages received by the Message Transport System that are addressed to the agent with the name **df** are transferred to a local directory of agents, which, depending on their content, records or deletes information about a remote agent, responds to a request for a list of active agents in a local container, or registers the remote container.

3.4   Communication between Agents

Message transport services are one of the main functions of frameworks for creating multi-agent systems. Messaging, in addition to the runtime environment, is the main functionality of each multi-agent platform. The communication between agents working under control of the UMAP platform is built based on guidelines of the corresponding FIPA standards [9, 10].

In Fig. 3 we show the location of the Message Transport System (MTS) within the UMAP container, which can group multiple working agents, and the way of communicate with agents and other components of the system. Each container of the UMAP platform has one instance of the Message Transport System, which is directly available for agents in this container. The MTS module also allows the communication with remote containers through the Agent Communication Channel (ACC). This is accomplished by a TCP/IP service listening on the port defined for the container.

From the technical point of view, functions performed by the Message Transport System of the UMAP platform can be divided into three categories:

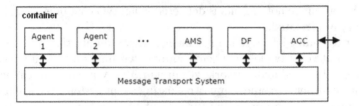

Fig. 3. Message Transport System (MTS)

- providing methods for sending messages to agents inside and outside the container,
- listening for incoming messages from agents from local or remote containers,
- serialization and deserialization of messages to/from XML format.

Both the structure of the message and the process of sending messages must be unified in order to ensure compatibility between agents operating on heterogeneous platforms. During the development of the UMAP platform, we implemented the structure of message described in the standard [10] and we created a mechanism for encoding messages.

The Message Transport System (MTS) of the UMAP platform provides messaging via shared memory (for local agents) or through the TCP/IP protocol using the XML technology (for remote agents). Such a solution enables efficient communication between agents operating within a single container, and provides, compliant with the FIPA standards, methods for communication between agents operating on remote machines, even in heterogeneous environments.

The method of sending messages through the MTS is different depending on its destination, namely whether the message must be sent beyond the local container. The course of sending messages is shown in Fig. 4. For each of the specified recipients the MTS decides whether the message should be encoded into XML format and then sent over the network to a remote container, or added to the message queue of the local agent, if the recipient acts in the local container.

4 Discussion

Providing a ready-made platform, such as UMAP, that allow controlling and cataloging agents, and exchanging messages in a structured manner, is very important for the development of systems that base on interacting agents. In this area the UMAP platform provides the same key functionalities comparing to platforms, such as JADE and SPADE. However, for the functioning of agents it is also very important to provide the runtime environment in which agents can act and complete their jobs.

When we were designing the UMAP platform we had to define how agents will be executed and how they will be running inside the system. Choosing the right course of action for the agent is very important, because it determines

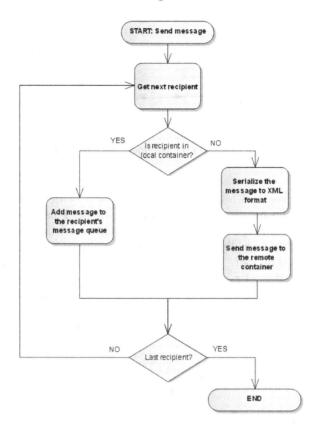

Fig. 4. Algorithm of sending messages between agents of the UMAP platform

the agent's ability to communicate with the Agent Management System, the ability to communicate inside the container, and above all, the ability to control the agent's work by the container, with which it is logically related. Analysis of existing solutions shows us at least two approaches to the problem of running agents.

Under the control of the JADE platform agent classes are loaded from the specified files, and their execution occurs in the child process of the container, to which they belong. This approach provides some control over the correctness of the loaded agent, enables a very efficient communication (via shared memory), and also gives complete control over the active agent. The idea of the agents working on the JADE platform is shown in Fig. 5.

For the sake of simplicity the system components (DF, MTS, AMS) are marked as a single AMS. Control over the process or thread, in which the agent runs is very important, because FIPA documentation requires the container to be able to control, and in special cases, interrupt the agent regardless of his actions. Therefore, it is essential to posses such a control over the agent, which will allow implementation-independent and immediate interruption of the agent.

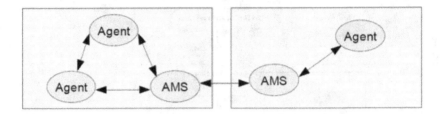

Fig. 5. Agents working inside containers under control of the JADE platform

A different approach is adopted by the developers of the SPADE platform, where the platform is not responsible for starting agents. This approach is shown in Fig. 6. Agents are programs that run independently of the platform or container, to which they logically belong. The only requirement is a way of communication with the container. This approach gives more flexibility when it comes to the implementation of agents and their locations on the physical machines. On the other hand, it limits the ability to control agents, complicates the implementation of efficient communication between agents operating on the same computer, which is an important issue in terms of system performance.

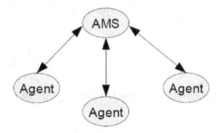

Fig. 6. Independent agents working on the SPADE platform

Taking into account the possibilities and limitations of the .NET programming framework, we decided to adopt the strategy known from the JADE platform, that is, running agents under the control of containers. The problem that had to be solved in this case was the mechanism for the creation of classes of objects, whose definition or even declaration is not known at the compilation time. It was also necessary to specify the way of communication between created objects and the container, within which they operate.

In order to let agents operate in a similar way as it is in the case of the JADE platform, we have designed and implemented the Agent Management System (AMS), which is able to load and run classes from the libraries loaded at run time. Consequently, we have identified how to create a library that is compatible with the UMAP platform and designed the abstract base class for an agent (`Agent`), which defines a common interface for communication between the agent and the platform.

The agent interface designed in such a way allows any implementation of an agent - as a program running in the background, as a windows application, or a program wrapping an existing application. The designer of the agent logic does not need to focus on how to implement the communication between agents and on the monitoring of agent's work, so he can focus just on the logic of the system that he develops. Other similarities and differences between compared platforms are presented in Tab. 1.

Table 1. Comparison of selected features of UMAP and two other platforms

Feature	JADE	SPADE	UMAP
FIPA standards compliance	Yes	Yes	Yes
Platform implementation language	Java	Python	C#
Operating systems	Platform independent	Platform independent	MS Windows, other operating systems through Mono platform
Free access to source code	Yes	Yes	Yes
Programming language for agents	Java (mainly)	Python, Java, C++, others	All languages compiled to CIL (e.g. C#, VB.NET,C++/CLI)
Direct dependence of agents from platform	Yes	No	Yes
P2P communication between agents	Yes	Yes	Yes
Agent communication language	ACL	XMPP	ACL
Media transfer protocols	XML	XMPP, HTTP, SIMBA	XML
Efficient communication through shared memory	Yes	No	Yes
GUI for agents/ platform control	Stand-alone	HTTP Web-based	Stand-alone

5 Concluding Remarks

Multi-agent systems are used in various fields, and their development is associated with standardization of certain behaviors of agents in order to allow the cooperation of programs working in different environments. While creating the UMAP platform, we aimed to provide the framework of the system, which simplifies the implementation of agents and offers ready-made environment in which agents operate and communicate with each other. We adopted a very important assumption that UMAP should not be limited to any specific area of the multi-agent system application. As a result, based on a ready UMAP platform it is possible to create solutions in many fields, e.g. knowledge gathering, decision support, or personal assistance.

For users of the UMAP platform and programmers developing solutions with the use of the platform, an equally important feature of the system is a clear and understandable programming interface, through which they can create their own agents. The structure of the UMAP platform reflects the architecture proposed by the FIPA standard, including agent management, message transport system and agent cataloging service. Agent Management System was developed as a simple application that controls the runtime environment of agents. By reducing the number of steps required to implement the methods of an agent, the UMAP platform provides a layer of abstraction, so that it is not necessary to focus on the details of implementation of the multi-agent system itself. This makes the development of new applications much easier.

Future works will cover more sophisticated methods for controlling the activity of UMAP agents, including e.g., artificial immune systems, like in [11]. This bio-inspired technique is widely used to solve many optimization problems [4, 20]. We believe that it will be a correct step toward the coordination of distributed systems containing a large number of agents - systems that could be built with the use of the UMAP platform.

References

1. Abril, J., Comellas, F., Cortés, A., Ozón, J., Vaquer, M.: A multiagent system for frequency assignment in cellular radio networks. IEEE Transactions on Vehicular Technology 49(5), 1558–1565 (2000)
2. Balaji, P., Srinivasan, D.: Multi-agent system in urban traffic signal control. The IEEE Computational Intelligence Magazine 5(4), 43–51 (2010)
3. Bellifemine, F., Caire, G., Poggi, A., Rimassa, G.: JADE, A White Paper, http://jade.tilab.com/papers/2003/WhitePaperJADEEXP.pdf (accessed February 02, 2014)
4. Burczyński, T., Poteralski, A., Szczepanik, M.: Genetic generation of 2-d and 3-d structures. In: Proc. of the Second M.I.T. Conference on Computational Fluid and Solid Mechanics, pp. 1–10, Massachusetts Institute of Technology, Cambridge (2003)
5. Choiński, D., Nocoń, W., Metzger, M.: Multi-agent system for hierarchical control with self-organising database. In: Nguyen, N.T., Grzech, A., Howlett, R.J., Jain, L.C. (eds.) KES-AMSTA 2007. LNCS (LNAI), vol. 4496, pp. 655–664. Springer, Heidelberg (2007)
6. Ferber, J.: Multi-Agent System: An Introduction to Distributed Artificial Intelligence. Addison Wesley Longman, Harlow (1999)
7. FIPA-OS: Developers Guide, http://fipa-os.sourceforge.net/docs/Developers_Guide.pdf (accessed on February 02, 2014)
8. FIPA Standard: FIPA Agent Management Specification, http://www.fipa.org/specs/fipa00023/ (accessed on February 02, 2014)
9. FIPA Standard: FIPA Communicative Act Library Specification, http://www.fipa.org/specs/fipa00037/ (accessed on February 02, 2014)

10. FIPA Standard: FIPA Message Structure Specification, http://www.fipa.org/specs/fipa00061/ (accessed on February 02, 2014)
11. Lau, H., Wong, V.: An immunity-based distributed multi-agent control framework. IEEE Transactions on System, Man and Cybernetics - Part A 36(1), 91–108 (2006)
12. Małysiak-Mrozek, B., Momot, A., Mrozek, D., Hera, Ł., Kozielski, S., Momot, M.: Scalable system for protein structure similarity searching. In: Jędrzejowicz, P., Nguyen, N.T., Hoang, K. (eds.) ICCCI 2011, Part II. LNCS, vol. 6923, pp. 271–280. Springer, Heidelberg (2011)
13. Momot, A., Małysiak-Mrozek, B., Kozielski, S., Mrozek, D., Hera, Ł., Górczyńska-Kosiorz, S., Momot, M.: Improving performance of protein structure similarity searching by distributing computations in hierarchical multi-agent system. In: Pan, J.-S., Chen, S.-M., Nguyen, N.T. (eds.) ICCCI 2010, Part I. LNCS, vol. 6421, pp. 320–329. Springer, Heidelberg (2010)
14. Mrozek, D., Małysiak, B., Augustyn, W.: Agent-supported protein structure similarity searching. In: Ghose, A., Governatori, G., Sadananda, R. (eds.) PRIMA 2007. LNCS, vol. 5044, pp. 49–61. Springer, Heidelberg (2009)
15. Mrozek, D., Małysiak-Mrozek, B., Mikołajczyk, J., Kozielski, S.: Database under pressure - testing performance of database systems using universal multi-agent platform. In: Gruca, A., Czachórski, T., Kozielski, S. (eds.) Man-Machine Interactions 3. Advances in Intelligent Systems and Computing, vol. 242, pp. 637–648. Springer, Heidelberg (2014)
16. Nan-Peng, Y., Chen-Ching, L., Price, J.: Evaluation of market rules using a multi-agent system method. IEEE Transactions on Power Systems 25(1), 470–479 (2010)
17. Nwana, H.: Software agents: An ovierview. Knowledge Engineering Review 11(3), 1–40 (1996)
18. Padgham, L., Winikoff, M.: Developing Intelligent Agent Systems: A Practical Guide. Halsted Press, New York (2004)
19. Popa, H., Pop, D., Negru, V., Zaharie, D.: Agentdiscover: A multi-agent system for knowledge discovery from databases. In: Proc. of the International Symposium on Symbolic and Numeric Algorithms for Scientific Computing, SYNASC, pp. 275–282 (2007)
20. Poteralski, A., Szczepanik, M., Ptaszny, J., Kuś, W., Burczyński, T.: Hybrid artificial immune system in identification of room acoustic properties. Inverse Problems in Science and Engineering 21(6), 957–967 (2013)
21. Shoham, Y., Leyton-Brown, K.: Multiagent Systems: Algorithmic, Game-Theoretic, and Logical Foundations. Cambridge University Press (2009)
22. SPADE2: Project home page, http://code.google.com/p/spade2/ (accessed on February 02, 2014)
23. Tan, V., Yoo, D., Shin, J., Yi, M.: A multiagent system for hierarchical control and monitoring. J. UCS 15(13), 2485–2505 (2009)
24. Tumer, K., Agogino, A.: Improving air traffic management with a learning multi-agent system. IEEE Intelligent Systems 24(1), 18–21 (2009)
25. Wooldridge, M.: An Introduction to Multiagent Systems, 2nd edn. John Wiley & Sons (2009)
26. XMPP: technology overview, http://xmpp.org/about-xmpp/technology-overview/ (accessed on February 02, 2014)
27. Zarandi, M., Avazbeigi, M., Turksen, I.: An intelligent fuzzy multi-agent system for reduction of bullwhip effect in supply chains. In: Proc. of the IEEE Annual Meeting of the North American Fuzzy Information Processing Society, NAFIPS, pp. 1–6 (2009)

Multivariate Estimation of Resource Utilization Bounds of any Variable Schedule in a Computing System

Susmit Bagchi

Department of Informatics, Gyeongsang National University
Jinju, South Korea 660 701
susmitbagchi@yahoo.co.uk

Abstract. Estimations of resource availability and utilizations in a computer system are necessary to achieve fair allocation of resources and to measure performance of a system. The software system designers often look for computing a limiting bound of CPU-utilizations given a schedule in a concurrent multitasking system, where tasks may have different CPU-affinity and IO-affinity. In complex software systems, the variations of CPU-utilizations due to variable scheduling quanta of the concurrent tasks are difficult to estimate through global time-averaging. This paper proposes a computing model of multivariate functional estimation of limiting bound of CPU-utilizations in a concurrent multitasking system comprised of heterogeneous tasks. An analytical model is formulated to compute dynamics of variable scheduling quanta and CPU-utilizations. The relation between continuous single-variable estimation and sampled multi-variable estimation is established. The integral remainder terms along with values of converging polynomials denoting estimation errors are computed.

Keywords: Multivariate function, performance estimation, concurrency, task scheduling.

1 Introduction

The estimation of resource consumptions by various executing processes in a computing system is an important step in order to ensure fair distribution of resources to executing processes and to enhance overall resource utilization. In large scale software systems such as, web servers, clusters and computing grid, the resource management techniques require knowledge of resource estimation and resource demands of applications [16]. One of the main computing resources of a system is CPU and estimation of CPU-utilizations by various executing processes is necessary to analyze execution statistics and performance of the computing system [14]. Often, the software systems are extended by updating the functionalities and by modifying architectural designs. In such cases, the analysis of variations of CPU-utilizations is required at the time of software

S. Kozielski et al. (Eds.): BDAS 2014, CCIS 424, pp. 312–322, 2014.

re-designing and deploying new software platforms [1]. In general, direct measurements of CPU-utilizations are made by software instrumentation. However, the direct and continuous instrumentation for estimating CPU-utilization incurs overheads [20]. In addition, direct estimation of CPU-utilization based on time-variable at discrete intervals may not provide an accurate estimate. It is often required by the software designers to compute the limiting bound of resource utilization in a system given any schedule of concurrent multitasks. The direct instrumentation methods as well as passive single-variable functional estimations of CPU-utilizations are not completely adequate to compute the limiting bounds of time-varying CPU-utilizations in a system. The complexity of estimation increases if the scheduling quanta of tasks become variable. Hence, a multivariate model is necessary to analyze and compute limiting bounds of CPU-utilizations in heterogeneous multitasking systems.

1.1 Motivation

The overall computation in a system involves execution of tasks on CPU and frequent IO calls made to the devices. However, the CPU-utilization values differ widely for CPU-bound tasks, IO-bound tasks and the CPU/IO tasks depending on different natures of executions. In general, the continuous estimation of CPU-utilizations through global averaging method without considering IO calls, context switches and associated IO intervals introduces approximations to the estimated values. It is often required to estimate the limiting bound of overall CPU-utilizations by the tasks in a schedule irrespective of unpredictable IO interruptions. An efficient method would be to periodically sample CPU-utilizations at different time-intervals within the scheduled quanta of the executing tasks. The limiting bound of CPU-utilizations by the tasks can be found employing multivariate functional estimation method involving multiple independent variables. The limiting bound of aggregated CPU-utilizations by different tasks can be found by integrating the CPU-utilizations during IO intervals as a separate entity. Hence, by using multivariate functional estimation of CPU-utilizations through sampling during execution of a task and by using estimation of CPU-utilizations by tasks during IO intervals, it is possible to compute the limiting bound of overall CPU-utilization in a system. The static analysis of the CPU-utilization can be made following multivariate functional estimation model if a deterministic execution schedule is known. This paper proposes the computing model for estimating resource utilizations by tasks following the sampled multivariate functional dynamics involving two independent variables. The multivariate functional estimation determines the limiting upper bound of the overall resource utilizations in a system executing concurrent heterogeneous processes. Rest of the paper is organized as follows. Section 2 describes the mathematical model of multivariate estimation and analyzes limiting bound. Section 3 illustrates evaluation of the model and section 4 describes related work. Section 5 concludes the paper.

2   Estimation Model

Let, the instantaneous values of CPU-utilizations by a task in a system is given by the set $U = \{u : 0 \leq u \leq 1\}$. The tasks in the system under consideration have a constant maximum scheduling quantum equals to $\tau > 0$. The execution time instances t are measured on real-axis R within the interval τ of a task and is represented as a set $X = \{x : x = t/\tau, 0 \leq x \leq 1\}$. An arbitrary continuous function $g(t)$ measures the absolute value of CPU-utilization at time t in the system. Let, the function ξ is defined on set of positive integers as, $\xi : I \to U$ such that ξ is continuous in $(0, +\infty)$ and $\forall i \in I, x_i \in X$ for the corresponding $\xi(i) = (u_i \in U)$. A multivariable function $z = f(x, u)$ is defined in domain $D = (X \times U)$ for any point $P(x, u) \in D$ such that, $f : (x, u) \to D_z$ where $D_z \subset R$. If for a point $P(x_a, u_a) \in D$ it is true that $f(P) \in D_z$ then, the neighborhood of P is given by, $N_P = \{(x, u) : [(u - u_a)^2 + (x - x_a)^2]^{0.5} < \delta, \delta > 0\}$. If there exists a real L and a small value $\epsilon > 0$ such that, $\forall (x, u) \in N_P$ in D, $|f(x, u) - L| < \epsilon$ then, L is the limiting value at $f(P)$ in N_P. Let, $Q(x_b, u_b) \in D$ such that, $x_b - x_a > 0$. Furthermore, $\{f(P), f(Q)\} \in D_z$ and $N_{PQ} \in D$ where, $N_{PQ} \subset \{N_P \cap N_Q\}$ in the system under consideration. If $f(x, u)$ is continuous function throughout N_P and N_Q then, $N_{PQ} \neq \{\emptyset\}$ if $0 < (x_b - x_a) < \delta$.

2.1   Analysis for CPU-bound Tasks

The analysis of CPU-utilization values requires definition of z. Let, the definition of z is formulated as,

$$z = \left\{ \begin{array}{l} (u/x) : x > 0 \\ 0 : x = 0, x < 0 \end{array} \right\} \tag{1}$$

The average value of CPU-utilization by a strictly CPU-bound task in a singular time quantum can be computed as,

$$\Psi = \frac{1}{\tau} \int_0^\tau g(t) dt \tag{2}$$

If there exists $n > 0$ such that, $x_b - x_a = (\delta/n)$ then, $t_b - t_a = (\tau\delta/n)$. In case of CPU-bound tasks, the function z will be continuous in D_z. If z has a globally unique limiting value then, $\forall f(x_a, u_a) \in D_z$ the limiting bound is given by,

$$\lim_{(x,u) \to (x_a, u_a)} z = L \tag{3}$$

For example, a single non-preemptive CPU-bound task will have a globally unique limiting value of CPU-utilization. However, a preemptive task may have several local maxima and a global maximum value of CPU-utilizations. In the case of CPU-bound tasks, if any two points $P(x_a, u_a) \in D$ and $Q(x_b, u_b) \in D$ exist, where $f(x_a, u_a), f(x_b, u_b) \in D_z$ such that Eq. 3 holds for an unique L, then $(u_b - u_a) = (\delta u_a / n x_a)$. This verifies the fact that, in sufficiently small time duration the CPU-utilization cannot change abruptly in a system. If one considers two consecutive intervals where, $b - a = 1$ and $a = i - 1$, then $\xi(i) - \xi(i-1) =$

$(\delta/n)f(x_{i-1}, u_{i-1})$. Considering the estimation of CPU-utilization through M-points samples, it can be derived as,

$$\sum_{i=1}^{M} \xi(i) - \xi(i-1) = \frac{\delta}{n} \sum_{j=1}^{M} f(x_{j-1}, u_{j-1}) \tag{4}$$

However, the sum of $\xi(i)$ in $[0, M]$ in Eq. 4 can be computed as a continuous function having finitely many integral points resulting in the following equation,

$$\int_{0}^{M} \xi(i)di = \frac{\delta}{n} \sum_{j=1}^{M} f(x_{j-1}, u_{j-1}) \tag{5}$$

Again, the accurate estimation of CPU-utilization in a scheduling quantum requires the following condition to be satisfied in a system considering the scale conversion,

$$\frac{1}{M} \int_{0}^{M} \xi(i)di = 0.01\Psi \tag{6}$$

By combining Eqs. 2, 5 and 6, the aggregated CPU-utilization by a CPU-bound task in singular scheduling quantum can be computed as,

$$\int_{0}^{\tau} g(t)dt = \left(\frac{100\tau\delta}{Mn}\right) \sum_{j=1}^{M} f(x_{j-1}, u_{j-1}) \tag{7}$$

However, according to Euler-Maclaurin integral theorem, Eq. 7 can be reduced in the discrete domain as,

$$\sum_{t=0}^{\tau} g(t) = \left(\frac{100\tau\delta}{Mn}\right) \sum_{j=1}^{M} f(x_{j-1}, u_{j-1}) + Rg \tag{8}$$

In Eq. 8, Rg represents the remainder term comprised of error as well as rapidly converging Bernoulli polynomial. Hence, Eq. 8 defines the limiting value of CPU-utilization of a system, which can be computed by observing discrete values of the multivariate function $f(x, u)$ in D. In other words, using $f(x, u)$ the passive observations can be made during CPU-execution of a task providing an estimate of the limiting upper bound of CPU-utilization.

2.2 Analysis for CPU/IO Tasks

In any system, not all tasks are strictly CPU-bound throughout the execution. In reality, tasks often perform data IO during executions resulting in frequently switching between CPU and IO devices. These tasks can be termed as CPU/IO tasks. If a task is not strictly CPU-bound, then the function $f(x, u)$ is not continuous for the corresponding task during the interval τ. Hence, the discrete form of estimation of CPU-utilization by a CPU/IO task in a quantum can be represented as follows, considering $E(\tau)$ be the average CPU-utilization during IO execution, where symbols s, q, and r represent integer counts,

$$\int_0^\tau g(t)\mathrm{d}t = \left(\frac{100\tau\delta}{Mn}\right)\left[\sum_{j=1}^s f(x_{j-1},u_{j-1}) + \sum_{j=s+q}^r f(x_{j-1},u_{j-1}) + \ldots\right] + E(\tau)$$

(9)

In case of CPU/IO tasks, there can be different local maxima of $f(x,u)$. Let, $f(x,u)$ is continuous in two disjoint intervals represented by, (x_a, x_b) and (x_c, x_d). Furthermore, let the function $f(x,u)$ has two limiting values at $x_{ab}(x_a < x_{ab} < x_b)$ and at $x_{cd}(x_c < x_{cd} < x_d)$ in the two intervals represented by, L_{ab} and L_{cd}, respectively. If there exist two constants $K_1 > 0$ and $K_2 > 0$, then in the two disjoint intervals following equation holds,

$$K_1\left[\lim_{(x,u)\to(x_{ab},u_{ab})} z\right] = K_2\left[\lim_{(x,u)\to(x_{cd},u_{cd})} z\right]$$

(10)

If the traces of execution of a task are such that, $K_1/K_2 \approx 1$, then $f(x_{ab}, u_{ab}) - f(x_{cd}, u_{cd}) = L_{ab} - L_{cd}$. Let, a task has invoked h-instances of IO in a τ and, the aggregated value of CPU-utilization is represented by $C(0,1)$ in the corresponding τ in set X, then the average value of CPU-utilization during IO in a single time quantum τ can be computed by,

$$E(\tau) = \frac{1}{h}C(0,1)$$

(11)

Thus, the value of the average CPU-utilization ($\Psi|_{\mathrm{CPU/IO}}$) by a CPU/IO task in a quantum can be computed by Euler-Maclaurin integral equation as follows, where R_τ represents integral remainder terms,

$$\sum_{t=0}^\tau g(t) - R_\tau = \left(\frac{100\tau\delta}{Mn}\right)\left[\sum_{j=1}^s f(x_{j-1},u_{j-1}) + \sum_{j=s+q}^r f(x_{j-1},u_{j-1}) + \ldots\right] + E(\tau)$$

(12)

If one considers a singular task model or if all the tasks are executing IO in a multitasking model, then $\Psi|_{\mathrm{CPU/IO}} < \Psi$.

3   Evaluation

The evaluation of the computing model is carried out on the data collected from the execution traces of concurrent CPU/IO tasks in a multitasking system. The schedule is composite in nature involving multiple concurrent CPU/IO tasks. The snapshot of schematic representation of the execution traces of a deterministic schedule for three concurrent tasks is illustrated in Fig. 1, where instantaneous CPU-utilization values are indicated.

The instantaneous values of CPU-utilizations by multiple concurrent tasks are measured online at discrete intervals during runtime and a global average

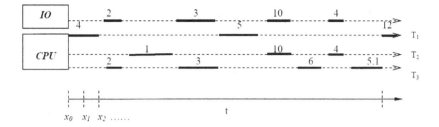

Fig. 1. Execution traces of 3 concurrent CPU/IO tasks

Table 1. Various execution and estimation parameters

Quanta(τ)	Sampling(M)	IO calls(h)
0.3	02	01
0.5	03	02
0.7	04	03
01	06	04

is computed for comparison. In another method, the CPU-utilizations are measured continuously at runtime for the same deterministic schedule of the concurrent tasks considering pair-wise linear averaging of CPU-utilizations in short
periodic intervals. In case of multivariate sampling method, the CPU-utilization
values of a task in the quantum are measured considering IO intervals of the
corresponding task as separate functional entities. In case of other two conventional methods, the CPU-utilizations are measured at discrete time intervals
throughout the composite schedule. The execution traces are collected for increasing execution quanta of the tasks, where the scheduling quantum is varied
from extremely small value to a relatively large value. The sampling and multivariate functional estimations are carried out at discrete time intervals in each
quantum of a task during execution on CPU. The quanta values, sample intervals and number of IO calls made by tasks in a schedule are illustrated in
Tab. 1. The sampling instances are kept at uniform low frequency in order to
avoid increasing space complexity for increasing sampling frequency. The variations of computed limiting bound of CPU-utilizations by multiple CPU/IO
tasks and the corresponding global averaging as well as pair-wise linear estimation methods are illustrated in Fig. 2. According to Fig. 2, the estimated
CPU-utilizations by different methods converge at lower scheduling quantum of
the tasks. However, monotonically increasing scheduling quanta of the tasks in
a system introduces increasing approximations into estimated values. The estimation of CPU-utilizations by global averaging performs poorly than others
at mid-range quanta. The low-frequency samples along with multivariate functional estimation provide the limiting bound of CPU-utilizations in a system
at different scheduling quanta. It is evident from Fig. 2 that, for deterministic
schedule of CPU/IO multitasks, the overall CPU-utilization exhibits saturation

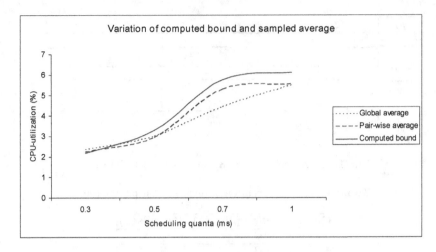

Fig. 2. Estimations for different methods at uniform sampling

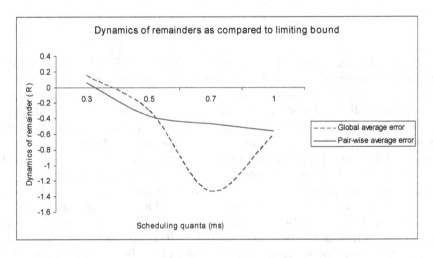

Fig. 3. Variations of remainders at uniform sampling

effect at higher scheduling quanta, which is consistent to the scheduling theory of multitasking. The variations of errors in estimations as compared to limiting bounds of CPU-utilizations for the given schedule are illustrated in Fig. 3, where y-axis represents R_r. In this case, the sampling rate is kept uniform throughout the estimation period. According to Fig. 3, the estimation of CPU-utilizations by global averaging method performs poorly as compared to pair-wise linear estimation method. The percentage (%) variations of absolute estimation errors as compared to computed limiting values are presented in Fig. 4. In order to understand the error dynamics at higher sampling frequency having higher space complexity, the sampling frequency is monotonically increased with increasing

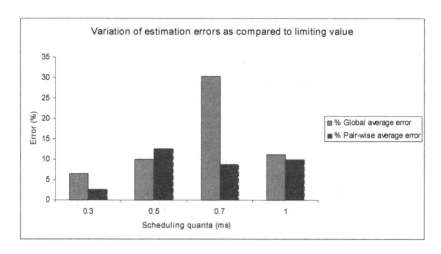

Fig. 4. The % variations of absolute errors

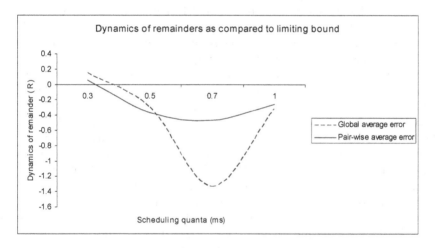

Fig. 5. Variations of estimation errors at higher sampling frequency

scheduling quanta of tasks. The sampling instances in every estimation inter-vals are nearly doubled as compared to the uniform sampling rate employed in earlier experiments. The samples are measured using multivariate functional es-timation and integral remainders are computed. The resulting error dynamics are presented in Fig. 5, where y-axis represents R_τ. The results indicate that, rel-ative errors of global averaging method remain insensitive to changing sampling frequency. However, the relative errors of continuous pair-wise linear averaging method as compared to multivariate functional estimation method tend to re-duce if the sampling frequency is monotonically increased at higher scheduling quanta of the tasks. Thus, a moderately low-frequency sampling and multivariate functional estimation of the CPU-utilization method is efficient in determining the limiting bound of resource utilization in a computing system while executing multiple concurrent tasks. The moderately low sampling frequency reduces the space complexity of the estimator. The multivariate functional method can be employed to determine scheduling quantum for near-optimal resource utilization by the tasks for a particular schedule while conducting static analysis.

4   Related Work

One of the metrics for measuring performance of software systems is the resource utilization. The resource allocation and management in large software systems need precise knowledge about demands from applications and performance [16]. Researchers have proposed a model of dynamic estimation of CPU-demands of web serving software systems [16]. The dynamic estimation is modeled following linear regression involving external parameters such as, request rates generated by applications [16,22,17]. Regression models are used to measure CPU-demands in multitier systems, where as splines are used to model CPU-service demands [10,22]. It can be argued that regression models may be inefficient to accurately estimate CPU-utilizations in highly complex and time-varying systems. However, the resource capacity estimation is carried out following trace-analysis, workload characterization and by employing queuing model in end-to-end measurement setup [12,15]. On the other hand, the CPU-performance counters are highly effective to estimate power consumption in a system [7,4]. The CPU-performance based power estimation model uses trickle-down model [6]. The estimation methodology covers IO calls, CPU-utilizations and disk bandwidth. The CPU-demands and utilizations may vary within a software system. For example, an analytical method to estimate CPU-utilization and its variation over time are measured in DBMS [14,18]. In this case, the CPU-utilizations are estimated by measuring query time for each query submitted to the DBMS. However, the performance of the model is based on the predicted value, which is prone to deviations. Often static analysis is performed to estimate resource utilization in a computing system. The CPU-utilization and power-utilization estimations are realized by conducting program analysis of mobile applications based on Android platform [13,3]. In general, two different techniques are employed to estimate CPU and power consumptions such as, cycle-accurate simulators and system-call based estimation [8,11]. However, the simulator-based estimations are not applicable in time-varying computing systems having large combinatorial execution space. On the contrary, the rule-based systems exhibit deterministic execution pathways with limited combinatorial execution space. Researchers have proposed various models to estimate CPU-utilizations in deterministic rule-based software systems [1,2,9]. In these models, the capacity modeling, tuning and capacity estimation methods are employed. However, these models may not guarantee an accurate estimation in dynamic time-varying computing systems. The application-level instrumentation techniques are proposed to estimate CPU-utilization in large-scale software systems such as, Web applications, Internet Computing and Java middleware servers [20,5,19]. The Kalman filter model is applied to estimate CPU-utilizations in case of multitier Web applications. Following a different approach, researchers have proposed to employ kernel-level instrumentation techniques to estimate CPU-utilizations by capturing event-log data [21]. However, the dynamic instrumentation techniques have relatively large computing overhead.

5 Conclusion

The single-variable based estimation of CPU-utilizations in a concurrent multitasking software system may not provide accurate estimates over time. The continuous estimation through instrumentation would enhance overhead. The multivariate functional estimation employing independent variables in discrete form would provide an improved estimation and limiting bound of CPU-utilizations can be computed. The dynamics of CPU-utilizations with respect to variable scheduling quanta of concurrent heterogeneous tasks can be computed by multivariable functional estimation. The error terms including converged polynomial values in global time-averaging and continuous pair-wise averaging can be estimated with respect to limiting bound for a given schedule of heterogeneous concurrent tasks in a software system.

References

1. Avritzer, A., Ros, J., Weyuker, E.: Estimating the CPU utilization of a rule-based system. In: Proceedings of the International Conference (WOSP). ACM (2004)
2. Avritzer, A., et al.: The automatic generation of load test suites and the assessment of the resulting software. IEEE Trans. on Software Engineering 21(9), 705–716 (1995)
3. Balasubramanian, N., Balasubramanian, A., Venkataramani, A.: Energy consumption in mobile phones: A measurement study and implications for network applications. In: The 9th ACM SIGCOMM Internet Measurement Conference (IMC). ACM (2009)
4. Bellosa, F.: The benefits of event-driven energy accounting in power-sensitive systems. In: Proc. of the 9th ACM SIGOPS European Workshop (Beyond the PC: New Challenges for the Operating System) (2000)
5. Binder, W., Hulaas, J.: A portable CPU-management framework for Java. IEEE Internet Computing 8(5), 74–83 (2004)
6. Bircher, W.L., John, L.K.: Complete system power estimation: A trickle-down approach based on performance events. In: Proceedings of the IEEE International Symposium on Performance Analysis of Systems and Software. IEEE (2007)
7. Bircher, W.L., John, L.K.: Complete system power estimation using processor performance events. IEEE Transactions on Computers 61(4), 563–577 (2012)
8. Brooks, D., et al.: Wattch: a framework for architectural-level power analysis and optimizations. ACM Computer Architecture News 28, 83–94 (2000)
9. Cheng, A.M., Chen, J.R.: Response time analysis of OPS5 production systems. IEEE Transactions on Knowledge and Data Engineering 12(3), 391–409 (2000)
10. Courtois, M., Woodside, M.: Using regression splines for software performance analysis. In: Proceedings of the 2nd Intl. Workshop on Software and Performance (WOSP). ACM (2000)

11. Dong, M., Zhong, L.: Sesame: self-constructive system energy modeling for battery-powered mobile systems. In: Proceedings of the 9th International Conference on Mobile Systems, Applications and Services (MobiSys). ACM (2011)

12. Gmach, D., Rolia, J., Cherkasova, L., Kemper, A.: Capacity management and demand prediction for next generation data centers. In: Proceedings of the International IEEE Conference on Web Services (ICWS). IEEE (2007)

13. Hao, S., Li, D., Halfond, W., Govindan, R.: Estimating android applications CPU energy usage via bytecode profiling. In: Proceedings of IEEE Intl. Conference (GREENS). IEEE (2012)

14. Hwang, H.Y., Yu, Y.T.: An analytical method for estimating and interpreting query time. In: Proceedings of the 13th International Conference on Very Large Data Bases (VLDB). Morgan Kaufmann (1987)

15. Liu, Z., et al.: Parameter inference of queuing models for IT systems using end-to-end measurements. Performance Evaluation 63(1), 408–409 (2006)

16. Pacifici, G., Segmuller, W., Spreitzer, M., Tantawi, A.: CPU demand for web serving: Measurement analysis and dynamic estimation. Performance Evaluation 65, 531–553 (2008)

17. Rolia, J., Vetland, V.: Correlating resource demand information with ARM data for application services. In: 1st International Workshop on Software and Performance (WOSP). ACM (1998)

18. Stonebraker, M., et al.: Performance enhancements to a relational database system. ACM Transactions on Database Systems 8(2), 167–185 (1983)

19. Wang, W., et al.: A statistical approach for estimating CPU consumption in shared Java middleware server. In: Proceedings of the IEEE 35th Annual Conference on Computer Software and Applications (COMPSAC). IEEE (2011)

20. Wang, W., et al.: Application-level CPU consumption estimation: Towards performance isolation of multi-tenancy web applications. In: Proceedings of the IEEE 5th International Conference on Cloud Computing. IEEE (2012)

21. Yaghmour, K., Dagenais, M.: Measuring and characterizing system behavior using kernel-level event logging. In: The USENIX Annual Technical Conference, USENIX (2000)

22. Zhang, Q., Cherkasova, L., Smirni, E.: A regression-based analytic model for dynamic resource provisioning of multitier applications. In: Proceedings of the 4th IEEE International Conference on Autonomic Computing (ICAC). IEEE (2007)

An Effective Way of Storing and Accessing Very Large Transition Matrices Using Multi-core CPU and GPU Architectures

Bożena Wieczorek[1], Marcin Połomski[1], Piotr Pecka[2], and Sebastian Deorowicz[3]

[1] Institute of Electrical Engineering and Informatics,
Silesian University of Technology, Gliwice, Poland
{bozena.wieczorek,marcin.polomski}@polsl.pl
[2] Institute of Theoretical and Applied Informatics, Polish Academy of Sciences,
Gliwice, Poland
piotr@iitis.pl
[3] Institute of Informatics, Silesian University of Technology, Gliwice, Poland
sebiastian.deorowicz@polsl.pl

Abstract. The topic of probabilistic model checking algorithms of Markov processes makes the problem of storing and processing of very large transition matrices crucial. While the symbolic approach utilises very effective storage and access methods, like Multi–terminal Binary Decision Diagrams, the traditional explicit method of representing and solving the model using sparse matrices has many advantages, especially when we concired the flexibility of computing of different properties. We show and examine a compact representation of a very large transition matrix in the sparse form. The technique employs an effective method of compression, storage, and accessing of the matrix and is suitable for use in mutli-core CPU and GPU environments. Such large transition matrices consume a lot of space and cannot be considered as typical types of data for storage in database systems. Nevertheless, their storage and efficient access to them could be beneficial for solving of Markov models, so constructing specialised compression methods is obviously the way to go.

1 Introduction

The topic of automatic checking of models [3,7] has a very broad field of applications like testing of the correctness of algorithms implemented in a high–level language [20] or analysis of performance of a cloud computing system [8]. This paper presents our work on an effective handling of very large amounts of data to be used for the probabilistic model checking of Markov chains (PMC) [9] that can offer an insight into stochastic systems and is particularly connected with the probabilistic analysis of computer networks. A traditional approach in model checking is to use an explicit method [6]. A sparse matrix in that methodology holds rates of all transitions of the model. This leads to two problems: the time of building the matrix and the size of the memory for matrix storage. Both time

S. Kozielski et al. (Eds.): BDAS 2014, CCIS 424, pp. 323–334, 2014.

and space can be very extensive, despite employing mathematical ways of solving linear systems like the various direct or indirect projection methods [19], for example the Purcell's algorithm with addressed stability issues.

To overcome these limitations some PMC systems like Prism [9] also offer symbolic methods, based on, for example, Binary Decision Tress [4] or derivatives like Multi-Terminal Binary Decision Diagrams (MTBDD) [13]. These approaches allow building the model in a symbolic, indirect form as opposed to the discussed sparse matrix.

A symbolic model checker can have much smaller memory requirements and it can be faster if the model's representation, like previously mentioned MTBDD, is compact. However, depending on the model, MTBDD or other decision diagrams can also be very slow and large and they can introduce an overhead caused by the implicit storage of the model transitions. This led Prism developers to create a hybrid method that aims at having the speed comparable to an explicit sparse matrix method while still maintaining substantial memory savings.

In this paper we specialise in a pure sparse matrix approach, as an alternative to the often time–consuming symbolic method [9] and the perfect simulation [16,15]. The basic problems to deal with would be the construction time of the sparse matrix and the memory volume needed to store the matrix. If the checked model exhibits an extensive 'state explosion', there is virtually no way for using the traditional explicit method. Each transition of the model's state would need to be separately found and stored, making it easy to require days and hundreds of gigabytes of operational memory.

Our solution is a two–way one:

1. An effective compression of the sparse matrix containing the model's transitions using finite state automata holding billions of transitions with relatively complex values of rates, due to its extreme redundancy, could be compressed to below ten megabytes in several minutes [5,14].
2. An effective parallel decompression and access to rows of the matrix using multi-core CPU or many-core GPU architecture.

The paper is organised as follows. Section 2 discusses general processing at graphics processing units. Section 3 discusses the properties of a transition matrix, and presents the specialised compression algorithm designed for it. Section 4 briefly introduces basic numeric algorithms used to process a transition matrix, in order to solve a Markov model. Then, it presents two ways of accessing the matrix: multi-core CPU and GPU-based access, which make it convenient and effective to apply the algorithms to the matrix. In section 5 we study a set of versatile models with our method. The last section concludes the paper.

2   General Processing at GPUs

Graphics Processing Units have become a powerful choice for accelerating the scientific and engineering computations [1]. They contain a set of streaming multiprocessors (SMs) and each SM hosts a set of processor cores (SPs). GPUs support thousands of light-weight threads running in parallel and operating on many data elements simultaneously in order to hide off-chip memory latency and provide massive throughput for parallel computation.

Parallel programming interfaces for general purpose computations, such as Compute Unified Device Architecture (CUDA) allowed programmers to write GPU code using extensions to standard programming languages like C, C++ and Fortran. Programs in CUDA C consist of a host program running on CPU and CUDA kernel (or kernels) invoked by the host and executed on a fixed number of threads in parallel on GPU. Threads are organised in blocks and thread blocks are grouped in a grid. All threads in a single thread block run on the same multiprocessor and share on-chip memory. The number of thread blocks and the number of threads in one block are specified while launching the kernel. All threads in a kernel access the GPU global memory.

The GPU architecture adapts a Single Instruction Multiple Threads programming model in which threads are executed in groups of 32 called warps. All threads within a warp execute the same instruction at a time. Each thread block consists of an integral number of warps and the computational resources of a multiprocessor unit are shared among the active blocks on that unit. Creation of threads, warp scheduling and management is performed in hardware.

3   Efficient Storage of the Matrix

3.1   Transition Matrix

The transition matrices in Markovian processes have some properties that allow to propose a specialised compression algorithm for them. The matrices are very rare (often only a few or a few tens of elements in a row are non zero). Moreover, due to the properties of the processes they describe, the rows are highly similar.

Our main aims of the compression method for effective storage and access of sparse matrices in the Markovian processes are:

- fast serial access to rows,
- no random access queries support,
- in compression, the rows must be given one by one,
- ability to start decompressing the rows from many positions (for effective decompression in multi-threaded programs).

3.2   Differential Encoding of the Rows

The analysis of the Markovian sparse matrices allowed us to identify two kinds of redundancies. The first is the observation that many rows are the same, when the column indices are encoded differentially according to the row id. E.g., for cells of 1st row: $\langle col\_id, value \rangle$):

$$\langle 1, 0.7 \rangle, \langle 3, 0.2 \rangle, \langle 17, 0.1 \rangle$$

we obtain:

$$\langle 0, 0.7 \rangle, \langle 2, 0.2 \rangle, \langle 16, 0.1 \rangle.$$

Therefore, instead of explicit storage of rows we construct a dictionary of unique differentially encoded rows. The dictionary can be stored in a plain form or can be compressed, e.g., using finite state machine compression techniques [5]. For simplicity, let us assume uncompressed representation of the dictionary. The matrix is now a sequence of integers from 0 to the size of the dictionary subtracted by 1.

3.3 Recursive Compression of the Matrix

The second observation was that the sequence of indices has "fractal-like" similarities. I.e., there can be found identical regions at different "levels", from a few values to even millions of values. To effectively represent such sequence we proposed *MetaRLE* algorithm[1] [14]. Its basic idea is to encode the input integer sequence as a sequence of 2- or 3-tuples. A 2-tuple represents a literal: $\langle 0, value \rangle$, where 0 is a binary flag and value is an integer form the input sequence. A 3-tuple represents a repetition: $\langle 1, no\_of\_rep, rep\_len \rangle$, where 1 is a binary flag, rep_len tells how many following tuples should be repeated during decompression, and no_of_rep denote how many times the repetition occurred.

For example, for the sequence

$$5, 5, 6, 5, 5, 6, 6, 5$$

we obtain the following tuple sequence:

$$\langle 1, 2, 3 \rangle, \langle 1, 2, 1 \rangle, \langle 0, 5 \rangle, \langle 0, 6 \rangle, \langle 0, 6 \rangle, \langle 0, 5 \rangle.$$

To better understand the algorithm let us take a look at the decompression of the above tuple sequence. The first 3-tuple $\langle 1, 2, 3 \rangle$ says that next 3 tuples should be repeated 2 times, so the partially uncompressed sequence is:

$$\langle 1, 2, 1 \rangle, \langle 0, 5 \rangle, \langle 0, 6 \rangle, \langle 1, 2, 1 \rangle, \langle 0, 5 \rangle, \langle 0, 6 \rangle, \langle 0, 6 \rangle, \langle 0, 5 \rangle.$$

Now we see two fragments $\langle 1, 2, 1 \rangle, \langle 0, 5 \rangle$ denoting that the value 5 should be repeated 2 times, so the uncompressed sequence is:

$$\langle 0, 5 \rangle, \langle 0, 5 \rangle, \langle 0, 6 \rangle, \langle 0, 5 \rangle, \langle 0, 5 \rangle, \langle 0, 6 \rangle, \langle 0, 6 \rangle, \langle 0, 5 \rangle.$$

Removing the binary flags gives us the initial sequence.

In real implementation the tuples are stored in 32-bit fields each, so the value can be up to $2^{31} - 1$, the repetition length (stored in 16 bits) up to 2^{16} and the number of repetitions (stored in 15 bits) up to 2^{15}.

The compression algorithm processes the input sequence symbol by symbol and for each symbol appends it to the output sequence as a literal tuple. Then,

---

[1] RLE – Run-Length Encoding.

it tries to identify the repetition of from 1 to 2^{16} previous symbols. If such a repetition can be found (if many, the longest one is chosen), the algorithm replaces the repeated symbols by a 3-tuple describing the repetition and the repeating symbols. What is important, the compression algorithm makes a compression on-line, symbol by symbol, when they are added to the matrix.

3.4   Access to the Compressed Matrix

A serial access to the rows of the matrix is relatively easy. In the above section we showed the decompression for the complete sequence, but in practice, symbols are decompressed one by one and the only data that are necessary to maintain beside the compressed tuple sequence is a stack describing the state of the decompressor. The stack contains the information about the current position in the tuple sequence and the positions of current number of repetitions at successive levels of the recursive representation of repetitions. Thus, to get a single matrix row, the algorithm decompresses a single value, take the related differentially-encoded row, reverts the differentially encoding.

The situation becomes a bit more complicated for multi-threading access. In our current implementation (proposed in this paper) each thread can start decompressing the matrix serially but from some specified position. E.g., when the matrix has 1000 rows and there are 4 threads, the 1st of them starts from the 0th row, the 2nd from the 250th row, the 3rd from 500th row, and finally the 4th from 750th row. Each thread decompresses only 250 rows of the matrix. To allow such an access pattern, before a decompression some preprocessing is necessary. When the number of the decompressing threads is known we need to perform one-time decompression of values sequence only (not the rows) and store the state of the stack in so-called bookmarks at positions from which the successive threads should start the decompression. Then, each thread can start decompression from its related row thanks to the stored stack state.

While accessing the matrix on GPU, each warp has its own bookmark to decompress the assigned part of rows. Bookmarks for specified number of warps are prepared on host and copied to GPU device.

4   Solving Markov Models in Compressed Domain

4.1   Basics of the Numeric Processes to Solve a Markov Model

In computational mathematics the are two basic methods to solve sparse linear equations:

- direct methods,
- iterative methods.

The direct methods try to solve equations with a constant number of steps and in case of lack of rounding errors they deliver an exact solution. These methods are efficient only for small linear sparse systems (about 10^5 states) because of their large memory requirement and rounding errors that increase proportionally to the size of the matrix. To solve large linear systems (more than 10^6 states) the most efficient methods are iterative ones. They find the solution processing an initial vector until the convergence test is satisfied. Among numerical solvers there are simple methods like the Jacobi and Gauss-Seidel method which just multiplies the matrix by an initial vector to pass the convergence test [19]. Such simple methods often reach convergence after many iterations. In many cases this never happens. More sophisticated methods like projection methods, e.g., GMRES or CGS require additional memory space [17]. They are more efficient than the ones mentioned above and need fewer iterations to obtain correct results. For the GMRES method extra memory required is $n \times k$, where n is the dimension of a matrix and k is the dimension of Krylov space (usually ~ 15).

Our research focuses on the analysis of Markovian models for transient state [18]. The transient state solvers are a certain modification of the steady state solvers mentioned above. One of the most effective ways to obtain a transient state solution is the uniformization method. This method is simple to implement, requires relatively little memory and often achieves better results comparing to other methods (e.g., projection). It needs only two vectors and a matrix (in our case the matrix in compressed form consumes less space than the vectors [14]) that gives a marginal cost of memory usage. The most time-consuming operation in this algorithm is a sparse matrix-vector multiplication (SpMV). The decoding process in our experiments gives an additional cost.

Uniformization algorithm

1. Compute number of terms K
   - $K = 0;\ \xi = 1;\ \sigma = 1;\ \eta = (1 - \epsilon)//e^{-\Gamma t}$
   - **while** $\sigma < \eta$ **do**
       $K = K + 1;\ \xi = \xi \times (\Gamma t)/K;\ \sigma = \sigma + \xi$
2. Approximate $\pi(t)$
   - $\pi = \pi(0);\ y = \pi(0)$
   - **for** $k = 1$ **to** K
       $y = yP \times (\Gamma t)/k;\ \pi = \pi + y$
   - $\pi(t) = e^{-\Gamma t}\pi$

4.2 Solving at Multi-core CPU

As we mentioned above, the most important and time-consuming operation in iterative methods of solving large linear systems is a sparse matrix-vector multiplication $(\boldsymbol{A} \cdot \boldsymbol{x} = \boldsymbol{b})$. To reach convergence, hundreds or even thousands of multiplications are required. In order to investigate how effective the compressed matrices can be accessed and processed, multi-core CPU and many-core GPU implementations of iterative decompression and a matrix-vector multiplication are proposed.

Our multi-core CPU implementation uses OpenMP application interface [12]. It starts with preparing bookmarks for all threads. Having the state of decompressor stack, each thread can serially access its own segment of rows. Threads iteratively decompress a single value, take the related row, revert the differentially encoding and perform row-vector multiplication. After computing the resultant vector b all threads need to synchronise their work and then they can start a new SpMV operation.

4.3   Solving at Many-core GPU

In our GPU implementation a computational kernel performs a decompression and a single matrix-vector multiplication [2,10]. Each warp is assigned to the specified part of the compressed matrix so it has its own bookmark with the state of stack to perform decompression. The number of iterations each warp has to execute is equal to the number of rows in its part of the matrix. In every iteration one thread within a warp decompresses a single value and takes the related differentially-encoded row. Having a single matrix row, 32 threads in a warp revert the differentially encoding of 32 elements in a row and perform a single multiplication until all nonzero elements are processed[2]. After that, a warp-wide parallel reduction is performed to sum the results together, one thread within a warp fills in the resultant vector b with calculated value and the next row can be accessed. Kernels are invoked one by one performing a sequence of multiplications.

The dictionary of unique rows, bookmarks, the stack of decompressor, x and b vectors are stored in off-chip global memory of the GPU. Other variables essential for the process of decompression and multiplication are kept in on-chip shared memory as they are accessed only by threads within a block (Fig. 1, 2).

5   Experimental Results

Parallel CPU and GPU implementations of accessing compressed matrices were evaluated experimentally using 4-core Intel® Core™2 Quad Q6600, 2.40 GHz processor and GeForce GTX 770 equipped with 8 SMs, 192 cores each. We implemented the algorithms in C++ language with double precision floating point arithmetic. The CUDA kernels were compiled using the NVIDIA CUDA Compiler 5.0. CPU implementation was developed in the Visual Studio environment with OpenMP support at -O2 optimization level.

The sparse matrices used in our experiments describe CTMP (Continuous-Time Markov Processes) Markovian model of synchronous slotted-ring network with the optical packet switching for a different size of buffers. The details of this analytical model are described in [11]. The experimental details of the matrices are presented in Tab. 1. The column R refers to a number of rows, U is a number of unique rows stored in the dictionary and Nz is a maximum number of

---

[2] In our matrices no row has more than 32 nonzero values so each thread performs just one multiplication.

```
1   __global__ void kernel(t_csa_elem** rowPointers, unsigned int* rowSizes,
2   tag_elem* mrleBuffer, unsigned int mtxRowsCount, unsigned int uniqueRowsCount,
3   CuBookmark* dev_CuBookmarks, float* x, float* b)
4   {
5       __shared__ int row_id[NR_WARPS];
6       __shared__ int isRow[NR_WARPS];
7       __shared__ float values[NR_WARPS][MAX_ROW_LEN];
8       __shared__ int indices[NR_WARPS][MAX_ROW_LEN];
9       __shared__ float cache[NR_WARPS][32];
10      __shared__ t_csa_elem* row[NR_WARPS];
11      __shared__ unsigned int row_len[NR_WARPS];
12      __shared__ int row_no[NR_WARPS];
13      int tid = threadIdx.x;                    // thread index within a block
14      int bid = blockIdx.x;                     // block index
15      int gtid = blockDim.x * bid + tid;        // global thread index
16      int wid = gtid / 32;                      // warp index
17      int wib =  wid & (NR_WARPS - 1);          // warp index within a block
18      int ltid = gtid & (32 - 1);               // local thread id
19      int start_pos = dev_CuBookmarks[wid].start_pos;
20      int len = dev_CuBookmarks[wid].no_el_in_part;
21      int stop_pos = start_pos + len - 1;
22
23      isRow[wib] = 1;
24      row_no[wib] = start_pos;
25      while(isRow[wib])
26      {
27          // row decompression
28          if(ltid == 0)
29          {
30              isRow[wib] = get_row_id(wid, row_id[wib], dev_CuBookmarks, mrleBuffer);
31              if(isRow[wib])
32              {
33                  row[wib] = rowPointers[row_id[wib] - 1];
34                  row_len[wib] = rowSizes[row_id[wib] - 1];
35                  for(int i=0; i < row_len[wib]; ++i)
36                      values[wib][i] = row[wib][i].value;
37              }
38          }
39          __syncthreads();
40          // differential encoding and multiplication
41          cache[wib][ltid] = 0.0;
42          int idx = ltid;
43          float temp = 0.0;
44          while(idx <= row_len[wib])
45          {
46              indices[wib][idx] = row[wib][idx].col_id + row_no[wib];
47              temp += values[wib][idx] * x[indices[wib][idx]];
48              idx += 32;
49          }
50          cache[wib][ltid] = temp;
51          __syncthreads();
52          // reduction
53          int red = 16;
54          while(red != 0)
55          {
56              if(ltid < red)
57                  cache[wib][ltid] += cache[wib][ltid + red];
58              __syncthreads();
59              red /= 2;
60          }
61          __syncthreads();
62          // saving the result in b, incrementing row number
63          if(ltid == 0 && isRow[wib])
64          {
65              b[row_no[wib]] = cache[wib][0];
66              row_no[wib]++;
67          }
68      }
69  }
```

Fig. 1. The kernel code for a decompression and multiplication algorithm

```
1    __device__ bool get_row_id(int warp_id, int &row_id, CuBookmark*
2    dev_CuBookmarks, tag_elem* mrleBuffer)
3    {
4        cu_buffer_stack_t &b_buffer_stack = dev_CuBookmarks[warp_id].stack_state;
5        int &b_no_elements_in_part = dev_CuBookmarks[warp_id].no_el_in_part;
6        int &b_counter = dev_CuBookmarks[warp_id].counter;
7
8        if(b_counter >= b_no_elements_in_part)
9            return false;
10
11       if(mrleBuffer[b_buffer_stack.top().cur_pos].n_rep == 0)
12       {
13           row_id = mrleBuffer[b_buffer_stack.top().cur_pos].value;
14           b_buffer_stack.top().cur_pos++;
15           while(!b_buffer_stack.empty() && b_buffer_stack.top().cur_pos >=
16                             b_buffer_stack.top().stop_pos)
17           {
18               if(--b_buffer_stack.top().n_rep)
19               {
20                   b_buffer_stack.top().cur_pos = b_buffer_stack.top().start_pos;
21                   break;
22               }
23               else
24                   b_buffer_stack.pop();
25           }
26           b_counter++;
27           return true;
28       }
29       int cur_pos = b_buffer_stack.top().cur_pos;
30       tag_bs_elem tmp;
31       b_buffer_stack.top().cur_pos += mrleBuffer[cur_pos].value + 1;
32       tmp.n_rep = mrleBuffer[cur_pos].n_rep;
33       tmp.start_pos = cur_pos + 1;
34       tmp.stop_pos = tmp.start_pos + mrleBuffer[cur_pos].value;
35       tmp.cur_pos = tmp.start_pos;
36       b_buffer_stack.push(tmp);
37       return get_row_id(warp_id, row_id, dev_CuBookmarks, mrleBuffer);
38   }
```

Fig. 2. Helper function used in the kernel to decompress next row id

nonzero elements in a row. As the process of decompression and multiplication a sparse matrix by a vector is performed many times, the columns $T40$, $T100$ and $T300$ show the processing times of sequential algorithm in which the number of multiplications is equal to 40, 100, and 300 respectively. The amount of memory which is required for storing the matrices in widely used CSR format and proposed in this paper compressed format is given in Tab. 2.

The results of experiments for parallel implementations are shown in Tab. 3. In CPU implementation 4 ($CPU4$) and 8 ($CPU8$) threads are used to perform the computations. On GPU the best results are obtained when each kernel comprises of 8192 thread blocks and each block consists of 64 threads (2 warps). For other configurations the execution times are longer as you can see in Fig. 3, where results for ring_n_4 and 100 multiplications are shown.

It can be observed that the parallel implementations has some kind of limitations in reaching a good speedup. It might be connected with the way rows are decompressed. It is very likely, that decompressing different rows takes different time and some threads wait at the points of synchronization for the other ones. We can also consider an access to the dense vector x. On GPU threads of a warp while multiplying nonzero elements of a row need to access non-contiguous words (elements of the vector) in memory. This non-coalesced access results in separate memory requests, and effectively results in serialisation of the threads. On CPU memory-intensive computations with irregular accesses to vector x can also be the bottleneck of processing.

Table 1. Description of sparse matrices

Name	R	U	Nz	T40 [s]	T100 [s]	T300 [s]
ring_n_2	656 100	51 386	15	2.0	5.0	15.0
ring_n_3	3 686 400	124 334	15	13.2	32.6	97.8
ring_n_4	14 062 500	268 700	15	46.5	115.9	345.9
ringn5p2	16 796 160	520 948	15	92.8	229.1	686.2
ringn5p3	25 194 240	520 948	15	140.2	354.4	1 072.2
ringn5	41 990 400	520 948	15	69.3	170.2	510.3

Table 2. Comparison of storage requirements

Name	CSR format [MB]	Compressed format [MB]	Compression ratio
ring_n_2	35.8	3.8	9.5
ring_n_3	215.8	11.1	19.4
ring_n_4	856.8	29.4	29.1
ringn5p2	1113.8	69.5	16.0
ringn5p3	1617.7	69.5	23.3
ringn5	2625.4	69.5	37.8

Table 3. Speedups achieved for multi-core CPU and GPU implementations (N – number of decompressions and multiplications)

Matrix	$N = 40$			$N = 100$			$N = 300$		
	CPU4	CPU8	GPU	CPU4	CPU8	GPU	CPU4	CPU8	GPU
ring_n_2	1.16	1.89	2.07	1.14	1.79	2.18	1.28	1.97	2.23
ring_n_3	1.13	1.83	2.57	1.29	1.88	2.60	1.45	2.02	2.63
ring_n_4	1.49	1.72	2.49	1.40	2.01	2.52	1.53	1.95	2.53
ringn5p2	1.43	1.96	2.93	1.60	1.88	2.92	1.32	1.87	2.94
ringn5p3	1.25	1.82	2.64	1.42	1.92	2.69	1.47	1.84	2.71
ringn5	1.45	1.84	2.13	1.30	1.90	1.86	1.43	1.84	1.87

GPU implementation achieves only slightly better speedups comparing to multi-core CPU method. Except for inefficient memory access to vector x, the reason might be related to a small amount of work each thread has to do between requesting memory and the idleness of threads while multiplying since the maximum number of nonzero elements in each row is 15.

To compare the performance of our GPU implementation with other SpMV implementations the experiments with CUSPARSE library and the matrices in CSR format have been performed. For two largest matrices there was not enough device memory to perform the multiplication. For other matrices CUSPARSE library routine was 50% or 3 times faster (*ring_n_2*, *ringn5p2* and *ring_n_3*, *ring_n_4* respectively).

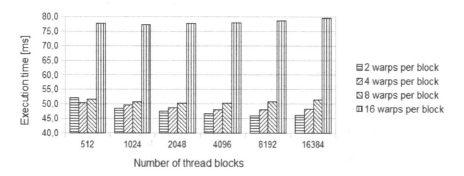

Fig. 3. Execution time for different number of thread blocks and 2, 4, 8, and 16 warps per block

6   Conclusions

The vast applications of modern database management systems require employing of more and more specialised types of records—simple numeric and text fields are not sufficient any more. We have shown, that a dedicated approach to the storage and access of transition matrices offers very substantial time and memory gains, if compared to a plain binary record.

In this work we have presented two parallel implementations of accessing and processing compressed matrices for multi-core CPU and many-core GPU architecture. The obtained speedups of parallel calculation of sparse matrix by vector (SpMV) product seem to be rather disappointing, as at 4-core (8 virtual cores due to hyper-threading) CPU the speedups are below 2 and for 1536-core GPU the speedups are less than 3. For both architectures the most probable explanation of such small speedups are highly random accesses to memory and few computations between consecutive accesses typical in SpMV product computation. Especially for GPUs, to obtain large speedups there should be significantly more computations than accesses to global memory. Moreover, the global memory accesses should be made in a coalesced mode to fully utilise the memory bandwidth. Both conditions are clearly unfulfilled in the considered problem. We, however, plan to perform further experiments to explore how the speedups scale for GPUs with larger memory bandwidths. We also plan to change the organization of threads in warps to increase the possibility of coalesced memory accesses.

Acknowledgements. The work was supported by the Polish National Science Center upon decision no. 4796/B/T02/2011/40.

References

1. Baskaran, M., Bordawekar, R.: Optimizing sparse matrix-vector multiplication on GPUs using compile-time and run-time strategies. Tech. Rep. RC24704 (W0812-047), IBM Research Report (2008)

2. Bell, N., Garland, M.: Efficient sparse matrix-vector multiplication on cuda. Tech. Rep. NVR-2008-004, NVIDIA Technical Report (2008)
3. Berard, B., Bidoit, M., Finkel, A.: Systems and Software Verification: Model-Checking Techniques and Tools. Springer, Berlin (2001)
4. Bryant, R.: Graph-based algorithms for boolean function manipulation. IEEE Transactions on Computers 35, 677–691 (1986)
5. Ciura, M., Deorowicz, S.: How to squeeze a lexicon. Software–Practice and Experience 31, 1077–1090 (2001)
6. Deavours, D., Sanders, W.: An efficient disk-based tool for solving very large markov models. In: Marie, R., Plateau, B., Calzarossa, M.C., Rubino, G.J. (eds.) TOOLS 1997. LNCS, vol. 1245, pp. 58–71. Springer, Heidelberg (1997)
7. Edmund, M., Clarke, J., Grumberg, O., Peled, D.: Model Checking. MIT Press (1999)
8. Kikuchi, S., Matsumoto, Y.: Performance modeling of concurrent live migration operations in cloud computing systems using PRISM probabilistic model checker. In: Proceedings of the 4th International Conference on Cloud Computing, pp. 49–56 (2011)
9. Kwiatkowska, M., Norman, G., Parker, D.: PRISM 4.0: Verification of probabilistic real-time systems. In: Gopalakrishnan, G., Qadeer, S. (eds.) CAV 2011. LNCS, vol. 6806, pp. 585–591. Springer, Heidelberg (2011)
10. Mukunoki, D., Takahashi, D.: Optimization of sparse matrix-vector multiplication for CRS format on NVIDIA kepler architecture GPUs. In: Murgante, B., Misra, S., Carlini, M., Torre, C.M., Nguyen, H.-Q., Taniar, D., Apduhan, B.O., Gervasi, O. (eds.) ICCSA 2013, Part V. LNCS, vol. 7975, pp. 211–223. Springer, Heidelberg (2013)
11. Nowak, M., Pecka, P.: Reducing the number of states for markovian model of optical slotted ring network. In: Balandin, S., Dunaytsev, R., Koucheryavy, Y. (eds.) ruSMART 2010. LNCS, vol. 6294, pp. 231–241. Springer, Heidelberg (2010)
12. OpenMP 2002: OpenMP architecture review board, OpenMP C and C++ application program interface (2002), http://www.openmp.org/mp-documents/cspec20.pdf
13. Parker, D.: Implementation of Symbolic Model Checking for Probabilistic Systems. Ph.D. thesis, University of Birmingham (2002)
14. Pecka, P., Deorowicz, S., Nowak, M.: Efficient Representation of Transition Matrix in the Markov Process Modeling of Computer Networks. In: Czachórski, T., Kozielski, S., Stańczyk, U. (eds.) Man-Machine Interactions 2. AISC, vol. 103, pp. 457–464. Springer, Heidelberg (2011)
15. Rabih, D., Gorgo, G., Pekergin, N., Vincent, J.: Steady state property verification of very large systems. International Journal of Critical Computer-Based Systems 2, 309–331 (2011)
16. El Rabih, D., Pekergin, N.: Statistical model checking using perfect simulation. In: Liu, Z., Ravn, A.P. (eds.) ATVA 2009. LNCS, vol. 5799, pp. 120–134. Springer, Heidelberg (2009)
17. Saad, Y.: Projection methods for solving large sparse eigenvalue problems. In: Adelsberger, H.H., Lazanský, J., Mařík, V. (eds.) Education in CIM 1995. LNCS, vol. 973, pp. 121–144. Springer, Heidelberg (1995)
18. Sidje, R.: Parallel Algorithms for Large Sparse Matrix Exponentials: application to numerical transient analysis of Markov processes. Ph.D. thesis, University of Rennes (1994)
19. Stewart, W.: Introduction to the Numerical Solution of Markov Chains. Princeton University Press (1994)
20. Visser, W., Havelund, K., Brat, G., Park, S., Lerda, F.: Model checking programs. Automated Software Engineering Journal 10, 203–232 (2001)

An Improved Algorithm for Fast and Accurate Classification of Sequences

Jolanta Kawulok* and Sebastian Deorowicz**

Institute of Informatics, Silesian University of Technology
Akademicka 16, 44-100 Gliwice, Poland
{jolanta.kawulok,sebastian.deorowicz}@polsl.pl

Abstract. Understanding of biocenosis derived from environmental samples can help understanding the relationships between organisms and the environmental conditions of their occurrence. Therefore, the classification of DNA fragments that are selected from different places is an important issue in many studies. In this paper we report how to improve (in terms of speed and qualification accuracy) the algorithm of fast and accurate classification of sequences (FACS).

Keywords: sequence classification, Bloom filter, metagenomics.

1 Introduction

In recent years, metagenomics, a relatively new field of genomics, has been developing rapidly. The aim of the metagenomic study is to characterize qualitative and quantitative composition of the environment, as well as to check relationships between species composition and environmental conditions. Within these studies, microbial genomes from the environment or patient samples are analyzed precisely. The biggest advantage of metagenomics is that it is not necessary to isolate and culture organisms in the laboratory to study them. Therefore, we can discover novel genes, derived from uncultured species, that are responsible for some useful functions [5,9]. The analysis of sets of genomes coming from various environments is desirable in a variety of areas such as medicine, engineering, agriculture, and ecology [6,11].

The metagenomics analysis consists of several steps [12], namely: 1) sampling from the environment; 2) extracting DNA; 3) cloning and construction of libraries; 4) DNA sequencing; 5) computational analysis. The DNA sequencing is the determination of the precise nucleotide sequence within a DNA molecule. After this process, we obtain a huge set of DNA fragments, called reads, derived from thousands of unknown species. The average size of these sequences

---

* This work was supported by the European Union from the European Social Fund (grant agreement number: UDA-POKL.04.01.01-00-106/09). This research was supported in part by PL-Grid Infrastructure.
** This work was supported by the Polish National Science Centre under the project DEC-2012/05/B/ST6/03148.

S. Kozielski et al. (Eds.): BDAS 2014, CCIS 424, pp. 335–344, 2014.

depends on the sequencing method used, and it can vary from ~ 50 to ~ 1000 nucleotides. In computational analysis, the important step is the taxonomical or functional classification of DNA sequences, which allows us to answer basic questions such as: "who is there?", "in what proportions?", and "what are they doing?". The basic strategy to answer the above questions is to compare the obtained DNA fragments to the reference sequences, e.g., located in the GenBank database [2]. Comparing thousands of DNA fragments to a huge database is very time consuming. Therefore, in order to effectively search the databases, some special techniques are used, including compression and indexing. Stranneheim et al. proposed an algorithm for fast and accurate classification of sequences (FACS) [10], which classifies reads to one of many reference sequences or regards it as "novel". Their algorithm uses Bloom filters [3] for compressed representation of the reference sequences of species. The experiments suggest that FACS is relatively fast and accurate.

In this paper we improve the last version (2.1) of FACS algorithm both in terms of classification speed and quality of the results. In addition, we analyze the influence of the choice of some internal parameters of FACS on the classification accuracy.

The paper is organized as follows. The Bloom filter and FACS algorithm with proposed improvements are described in detail in section 2. The comparative experiments are reported in section 3. The paper is concluded in section 4.

2   Algorithms

The original FACS program consists of two Perl scripts: 1) building a compressed reference database using the Bloom filter; 2) classifying metagenomic reads according to the references.

2.1   Bloom Filters

A Bloom filter is a simple and cost-effective memory structure that represents a given set of elements. It is used for rapid determination whether the argument belongs to this set of items. This probabilistic technique, introduced by Bloom in 1970 [3], exhibits no false negative test results, but it is susceptible to false positives. However, often the gain from small memory requirements is much higher than the loss caused by rare incorrect answers.

In essence, the Bloom filter is a bit vector of length m with a set of some number of hash functions representing an n-element set $S = x_1, x_2, ..., x_n$. Initially, all bits are set to 0. The elements are inserted into the vector using κ independent hash functions $(h_1, ..., h_\kappa)$. For each $x_i \in S$, every bit at index $h_j(x_i)$ (for all $1 \leq j \leq \kappa$) is changed to 1. Setting the bit can occur repeatedly, but only the first assignment changes its value from 0 to 1. Verification whether some element y belongs to the set S involves checking whether all bits at positions $h_j(y)$ for all valid j are 1. If not, we are sure that y is absent in S. Otherwise, it is reported that $y \in S$, but there is a small probability that this answer is wrong. An example of adding and querying procedures is presented in Fig. 1. The bits are changed from 0 to 1, after inserting two elements (x_1 and x_2) to the vector using two hash functions (represented by arrows). Elements y_1, y_2, and y_3 represent the three possible results of the query: y_1 is absent in S and the filter correctly reports that (example of a true negative, $h_2(y_1) = 0$); y_2 is present in S and the filter correctly reports that (example of a true positive, all bits at indexes $h_j(y_2)$ are 1); y_3 is absent in S and the filter is wrong reporting it is present (example of a false positive, all bits at indexes $h_j(y_3)$ are 1, but they were derived from different elements x_1 and x_2).

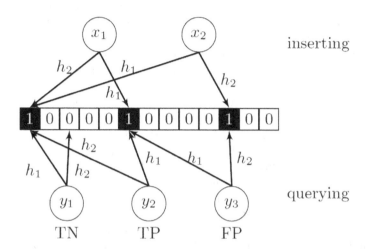

Fig. 1. Example of using the Bloom filter

From Fig. 1, it can be noticed that the probability of error decreases with increasing the number of 0s in the bit vector. Thus, the chance of a wrong answer decreases with increasing m/n, unfortunately at the expense of memory. Both too small and too large number of hash functions affects negatively the probability of error. A smaller number of κ increases the proportion of the value 0 in the bit array, on the other hand, greater number of κ increases chance to hit at bit 0, which is the correct negative response too. The probability of obtaining an incorrect positive response can be determined in a simple way by assuming that the hash functions are characterized by a perfect dissipation of information [4]. The probability that a specific bit remains 0 is $(1 - 1/m)^{\kappa n}$ ($\approx e^{-\kappa n/m}$), therefore the probability of false positive is $p_f \approx (1 - e^{-\kappa n/m})^\kappa$. For fixed n and m, this is minimized for $\kappa = \frac{m}{n} \ln 2$ and κ have to be integer. To insert n elements into the filter and allowing the probability of error p_f, the length of the bit vector should be $m = -\frac{n \ln p_f}{(\ln 2)^2}$.

The Bloom filters have many applications, including lexicon representation, networking [8], filtering of XML packets [1]. In bioinformatics, it was used (beside FACS), also for counting k-mer in DNA sequences [7]. The k-mers are every possible substrings of length k in a DNA sequence. Counting them is applicable for the *de novo* genome assemblers, error correction of sequence reads.

2.2 FACS

The FACS algorithm [10] classifies reads to one of several reference sequences. Its two scripts (for building the Bloom filter and for classification) are based on a Perl module Bloom-Faster from the Comprehensive Perl Archive Network.

In the preprocessing, Bloom filters must be constructed for all r reference sequences that are to be searched. This database is built by inserting each possible subsequence of length k (k-mer) to the Bloom filter. The appropriate k-mer length and the value of false positive probability for the Bloom filter (p_f) have to be selected before this stage. They define the length of a bit vector (m). It should be emphasized that the parameter p_f is an error for checking k-mer, rather than the error of whole sequence classification, as reads are classified according to all k-mers they contain.

During classification, all q reads are compared with the first reference sequence (R_1). On the output, we obtain a set containing c_1 reads assigned to this sequence, the remaining ($q - c_1$) are compared with the next reference sequence (R_2). The process is repeated, thus ($q - c_1 - \ldots - c_{i-1}$) reads are compared with the i-th reference sequence (R_i, $1 < i \leqslant r$). The classification ends when all reads are classified ($c_1 + \ldots + c_j = q$ and $1 \leqslant j \leqslant r$) or each reference sequence is checked. In the latter case, a set of ($q - c_1 - \ldots - c_r$) reads remains unclassified and those sequences are termed as "novel".

In order to assign the query read to a reference sequence, it has to accumulate more match scores than a chosen cut-off value (MC), which is based on sequence similarity. In a given read, the successive k-mers are obtained using the 1-base sliding window. All possible subsequent k-mers are checked for occurrence in the reference sequences, and a read obtains 1 point for each base from k-mer with a positive hit.

Before classification, Quick Pass filtering is performed to quickly remove non-significant reads. This filter uses a k-base offset sliding window approach on the query. The process with 1-base offset sliding window is started, when at least one k-mer is matched.

2.3 The Improvements of FACS

The original FACS written in Perl has a serious limit for memory. The authors in their article indicate that the maximal size of the Bloom filter is about 312 MB. Therefore, the sequences longer than $\sim 2.5 \cdot 10^9$ nucleotides (for $p_f = 0.0005$) have to be split into a number of subsequences that satisfy this limit, and Bloom filters have to be created separately for each of them. We rewrote FACS (2.1) scripts in C++ language using highly optimized implementation of Bloom filters by Arash Partow[1]. These changes made it possible to increase the capacity of the Bloom filters, accelerate the classification significantly, and improve its accuracy.

---

[1] Source with Common Public License was downloaded from http://www.partow.net

We also noticed that in the FACS algorithm the counting of points assigned to reads is incorrect (due to a small bug in implementation). In some cases, this causes that the match scores for a query are even higher than its length. This bug has been corrected in the reported work.

Moreover, the original FACS algorithm treats the unknown nucleotides ("N"s) in the same way as the normal bases (A, T, C, G). Hardly can this approach be justified, especially for reads containing a lot of "N"s (such reads are present in the FACS evaluation set of reads). We propose to check all sequences, including those that contain 'N', and only ignore those k-mers that contain even a single "N", in such a way that it does not affect the ranking. . Hence, FACS is run exclusively using such k-mers, whose all nucleotides are defined.

3   Experimental Results

In our experiments, we compared three programs: the original FACS 2.1 scripts written in Perl language (P_P), the FACS scripts rewritten in C++ language (P_{C1}), and the FACS scripts that use another implementation of the Bloom filter in C++ language (P_{C2}). Subsequently, we introduced the modifications proposed in section 2.3 to P_{C1} and P_{C2}. The experiments were conducted on a computer equipped with an Intel Xeon 2.67 GHz (96 GB RAM with 12 cores).

We assessed the quality of the sequence classification taking into account:

Time: Processing time of classification.
Classified: The overall percentage of sequences that were classified ($\frac{TP+FP}{all}$).
Sensitivity: The number of correctly classified sequences to the total number of sequences in the data set ($\frac{TP}{all}$).
Precision: The number of correctly classified sequences to the number of all classified sequences ($\frac{TP}{TP+FP}$).

The experimental study was carried out using a simulated metagenomic set, which was downloaded from the FACS web site. This data set contains 100 000 sequences of an average length of 269 bases. The sequences are from 25 various species: 17 bacterial genomes, 3 viral genomes, 3 archaea genomes, and 2 human chromosomes. Originally, many reads in this set comprise more than 50% of unknown nucleotides. For such data, comparison of the classification scores may be unreliable. Therefore, in addition to the whole data set, the results were also presented for a reduced set, in which the sequences containing more than 50% of unknown nucleotides were excluded, leaving 93 653 reads.

The classification was performed using different parameters of probability of false positive for Bloom filter (p_f), match cut-off for sequence similarity (MC) and length of k-mers. Influence of p_f on classification results (sensitivity and precision) is given in Fig. 2. We can notice that regardless of the MC value and k-mer length, sensitivity (Fig. 2 A and B) and precision (Fig. 2 C and D) grow with decreasing values of p_f to 0.0005, and then the classification score almost does not change, therefore $p_f = 0.0005$ is used for all further analyses. Parameters of MC and k do not have such a clear impact on the quality of the

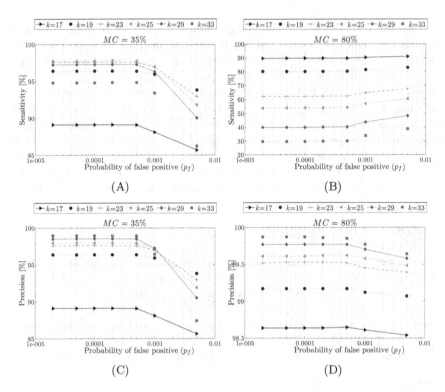

Fig. 2. Influence of p_f on sensitivity (A and B) and precision (C and D)

classification. With the increase in the length of the k-mer, the precision increases too, however fewer sequences are classified, thus the sensitivity decreases from a value of $k \approx 21$ (Fig. 3 A and B). The impact of growth of MC on the classification results also is dependent on the k-mer length. The precision always grows with the increase of MC (Fig. 3 D). However, for small values of k ($\lesssim 15$), the sensitivity increases with the MC, whilst for large k ($\gtrsim 24$) it immediately drops. For k between 15 and 24, the sensitivity at the beginning grows and then declines (Fig. 3 C). This behavior can be explained in the following way. Short k-mers occur in the reference sequences with a large probability. Thus, they are classified to one of the first reference sequences, just the increase of the match cut-off increases chances of assigning them to subsequent correct sequences. For long k-mers, the match is very accurate, as the k-mers are more unique for each reference sequence. However, in this case, the read is poorly resistant to changes located close to each other (mismatch, insertion, deletion). Read gets a few points for similarity to reference, so with the increase of MC, the number of classified sequences (as well as precision) decreases.

We tried to reproduce the results of Stranneheim's article, however we received different results, despite using their scripts and the same parameters. We compared the original FACS program P_P in Perl with two programs P_{C1} and P_{C2} in C++ without any changes to the algorithm (Tab 1). The biggest differences

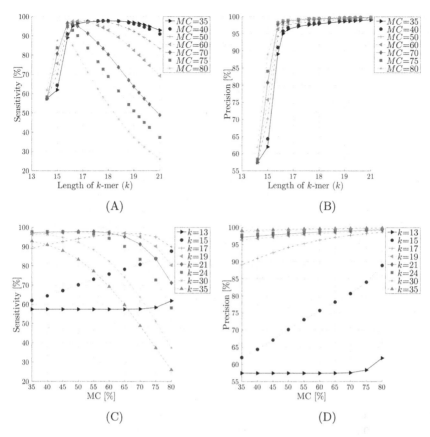

Fig. 3. Influence of MC and length of k-mers on sensitivity (A and C) and precision (B and D)

are in the classification time. Rewriting the FACS program to C++ causes the acceleration of 1.5-3 times (for P_{C1}), whereas when we used another implementation of the Bloom filter, P_{C2} worked about 3-5 times faster than the original FACS. Differences in the quality of the classification depend on the actual values of the parameters. For short k-mers, there are no significant differences in the results. For longer k-mers, our algorithms have less misclassifications, thus the precision increases. Number of correctly classified reads (TP) for P_P and P_{C1} is close to each other (so the sensitivity is the same), but a greater number of false positives in P_P causes the decrease in the number of all classified reads for P_{C1}. P_{C2} has more TP than the original FACS and, in this case, the sensitivity and the overall fraction of sequences that were classified increases. In summary, the new implementation of the Bloom filter (P_{C2}) returns the best results due to the largest number of correctly classified fragments and the smallest number of errors. Therefore, we focused on P_{C2} when comparing the improved version of the FACS.

Table 1. Classification results obtained for: P_P – the original FACS scripts written in Perl; P_{C1} – the FACS scripts rewritten in C++; P_{C2} – the FACS scripts using a different implementation of the Bloom filter in C++

| | Program P_P | | | Program P_{C1} | | | Program P_{C2} | | |
k	Time [mm:ss]	Class [%]	Sens [%]	Prec [%]	Time [mm:ss]	Class [%]	Sens [%]	Prec [%]	Time [mm:ss]	Class [%]	Sens [%]	Prec [%]
						MC = 65%						
13	02:56	100.00	55.14	55.14	01:36	100.00	55.14	55.14	00:51	100.00	55.14	55.14
17	03:36	100.00	90.78	90.78	01:34	99.94	90.73	90.78	01:04	99.99	90.78	90.78
21	02:58	99.86	92.99	93.13	01:23	99.73	92.93	93.18	00:59	99.78	92.97	93.18
25	02:53	99.11	92.19	93.01	01:14	98.32	92.14	93.71	00:56	99.55	92.89	93.31
35	02:46	92.34	79.33	85.91	01:06	84.51	79.31	93.84	00:43	98.42	91.80	93.27
						MC = 70%						
13	03:04	100.00	55.14	55.14	01:39	100.00	55.14	55.14	00:55	100.00	55.14	55.14
17	03:35	99.99	91.34	91.35	01:37	99.93	91.29	91.36	01:04	99.98	91.34	91.36
21	03:07	99.67	92.96	93.27	01:24	99.53	92.91	93.34	00:48	99.57	92.95	93.35
25	02:53	98.33	91.41	92.96	01:18	97.40	91.36	93.81	00:43	99.20	92.72	93.46
35	03:02	88.69	75.30	84.90	01:03	80.34	75.29	93.72	00:44	98.10	91.65	93.42
						MC = 75%						
13	03:00	100.00	55.16	55.16	01:34	100.00	55.14	55.14	00:57	100.00	55.14	55.14
17	03:43	99.96	91.78	91.81	01:38	99.91	91.72	91.80	00:54	99.96	91.77	91.81
21	03:04	99.30	92.73	93.39	01:26	99.17	92.69	93.46	01:02	99.20	92.72	93.47
25	03:03	96.93	90.18	93.03	01:17	96.04	90.14	93.86	00:52	98.66	92.35	93.61
35	03:11	84.05	70.64	84.04	01:13	75.51	70.64	93.55	00:42	97.58	91.29	93.55

Tab. 2 shows the effects of introducing the proposed modifications to P_{C2}. Specifically, the table presents the scores obtained without (O/N and I/N) and with (O/Y and I/Y) rejecting the k-mers that include "N"s (see section 2.3) as well as using original (O/N and O/Y) and corrected (I/N and I/Y) counting of the match points. Also, two metagenomic sets were used: the set that contains all of the sequences available at the FACS web site (A), and the filtered set (F) which does not include the reads that comprise more than 50% of unknown nucleotides. Results obtained for $I/N(F)$ and $I/Y(F)$ are identical, hence $I/N(F)$ scores are not presented in the table. Best two classification results for each version are shown in boldface. Corrected counting of the match points improves the classification score, however lower values of MC should be applied after the correction. Overall, the improved match points counting reduced the number of incorrectly classified reads, therefore the precision was increased. For the reduced set (F), rejecting the k-mers, which contain a single "N", does not influence the

Table 2. Classification results obtained for different versions of P_{C2}. Symbols explanation: "O" – original counting match points; "I" – improved counting match points; "N" – no filtration of k-mers with "N"; "Y" – filtration of k-mers with "N"; "(A)" – all reads in the metagenomic set; "(F)" – reads with more than 50% of unknown nucleotides filtered out.

	Sensitivity [%]							Precision [%]						
k	O/N (A)	O/Y (A)	I/N (A)	I/Y (A)	O/N (F)	O/Y (F)	I/Y (F)	O/N (A)	O/Y (A)	I/N (A)	I/Y (A)	O/N (F)	O/Y (F)	I/Y (F)
						MC = 50%								
13	55.14	53.80	55.14	53.80	57.44	57.44	57.44	55.14	57.44	55.14	57.44	57.44	57.44	57.44
17	88.09	86.75	89.52	88.18	92.62	92.57	94.15	88.09	92.63	89.52	94.15	92.62	92.62	94.15
21	92.57	91.23	**93.02**	**91.68**	97.41	97.35	**97.89**	92.60	97.44	**93.25**	**98.15**	97.43	97.43	**98.14**
25	92.65	91.30	92.91	91.57	97.49	97.43	97.77	92.74	97.59	93.31	98.23	97.59	97.59	98.22
35	91.57	90.22	91.82	90.49	96.34	96.28	96.61	92.76	97.67	93.15	98.10	97.67	97.67	98.10
						MC = 55%								
13	55.14	53.80	55.14	53.80	57.44	57.44	57.44	55.14	57.44	55.14	57.44	57.44	57.44	57.44
17	89.20	87.86	90.59	89.25	93.81	93.75	95.29	89.20	93.81	90.62	95.33	93.81	93.80	95.32
21	92.74	91.40	**92.91**	**91.57**	97.59	97.54	**97.77**	92.80	97.66	**93.48**	**98.41**	97.66	97.65	**98.41**
25	92.77	91.43	92.65	91.31	97.62	97.57	97.50	92.92	97.80	93.52	98.48	97.79	97.79	98.47
35	91.69	90.35	91.74	90.40	96.47	96.41	96.52	92.93	97.86	93.35	98.33	97.85	97.85	98.33
						MC = 65%								
13	55.14	53.80	55.15	53.81	57.44	57.44	57.45	55.14	57.44	55.15	57.45	57.44	57.44	57.45
17	90.78	89.43	92.03	90.69	95.49	95.44	96.83	90.78	95.50	92.30	97.14	95.50	95.50	97.14
21	**92.97**	**91.63**	90.50	89.17	**97.84**	**97.78**	95.21	**93.18**	**98.07**	93.76	98.87	**98.07**	**98.06**	98.87
25	92.89	91.54	89.57	88.23	97.74	97.69	94.21	93.31	98.22	93.77	98.91	98.22	98.22	98.91
35	91.80	90.45	90.10	88.75	96.58	96.52	94.76	93.27	98.24	93.65	98.76	98.24	98.24	98.76
						MC = 70%								
13	55.14	53.80	55.26	53.92	57.44	57.44	57.57	55.14	57.44	55.26	57.57	57.44	57.44	57.57
17	91.34	90.00	91.89	90.55	96.09	96.04	96.68	91.36	96.11	92.82	97.72	96.11	96.11	97.71
21	**92.95**	**91.61**	86.70	85.38	**97.81**	**97.76**	91.17	**93.35**	**98.26**	93.74	99.05	**98.26**	**98.26**	99.05
25	92.72	91.37	85.42	84.00	97.56	97.51	89.78	93.46	98.40	93.74	99.09	98.40	98.40	99.08
35	91.65	90.30	87.19	85.85	96.42	96.37	91.66	93.42	98.41	93.66	98.94	98.41	98.41	98.94

classification results significantly. The differences appear for the whole set – the precision is increased by about 5%, however the sensitivity decreases by about 1%. Comparing the results for the whole (A) and reduced (F) set, it can be seen that for the latter, the classification quality is much better. Sequences consisting of many unknown nucleotides often do not contain sufficient information to be classified, therefore such data should not be used as a benchmark set for testing the classifiers. Such sequences can be classified as a separate group, with the assumption that there is a large uncertainty of the obtained results.

4 Conclusions

In this paper we discussed our improved algorithm for fast and accurate classification of sequences. Rewriting FACS scripts in C++ language and using another implementation of Bloom filters accelerated the process of classification 3-5 times. Moreover, the improved match points counting and filtration of those k-mers, which contain unknown nucleotides, reduces the number of incorrectly classified fragments. In addition, we showed that the k-mer length has a large influence on the final score. Increasing this value has a positive effect on the precision, but the number of correctly classified sequences decreases.

It is difficult to exploit FACS to compare fragments with a large set of sequences (e.g., coming from GenBank), so in the further works, we plan to improve the algorithm, so that large database could be used. In addition, our future work includes investigating whether the k-mer length can be adapted based on the length of the reference sequence, which would improve the flexibility of the algorithm.

References

1. Antonellis, P., Kontopoulos, S., Makris, C., Plegas, Y., Tsirakis, N.: Semantic xml filtering on peer-to-peer networks using distributed bloom filters. In: WEBIST 2013 - Proceedings of the 9th International Conference on Web Information Systems and Technologies, pp. 133–136 (2013)
2. Benson, D.A., Cavanaugh, M., Clark, K., Karsch-Mizrachi, I., Lipman, D.J., Ostell, J., Sayers, E.W.: GenBank. Nucleic Acids Research 41(D1), D36–D42 (2013)
3. Bloom, B.H.: Space/time trade-offs in hash coding with allowable errors. Commun. ACM 13(7), 422–426 (1970)
4. Broder, A., Mitzenmacher, M.: Network Applications of Bloom Filters: A Survey. Internet Mathematics 1(4), 485–509 (2004)
5. Handelsman, J.: Metagenomics: application of genomics to uncultured microorganisms. Microbiology and Molecular Biology Reviews 68(4), 669–685 (2004)
6. Kennedy, J., O'Leary, N., Kiran, G., Morrissey, J., O'Gara, F., Selvin, J., Dobson, A.: Functional metagenomic strategies for the discovery of novel enzymes and biosurfactants with biotechnological applications from marine ecosystems. Journal of Applied Microbiology 111(4), 787–799 (2011)
7. Melsted, P., Pritchard, J.K.: Efficient counting of k-mers in DNA sequences using a bloom filter. BMC Bioinformatics 12 (2011)
8. Mishra, A.K., Turuk, A.K.: Efficient mechanism to exchange group membership identities among nodes in wireless sensor networks. IET Wireless Sensor Systems 3(4), 289–297 (2013)
9. Simon, C., Daniel, R.: Metagenomic Analyses: Past and Future Trends. Applied and Environmental Microbiology 77(4), 1153–1161 (2011)
10. Stranneheim, H., Käller, M., Allander, T., Andersson, B., Arvestad, L., Lundeberg, J.: Classification of DNA sequences using Bloom filters. Bioinformatics 26(13), 1595–1600 (2010)
11. The NIH HMP Working Group, Peterson, J., Garges, S., et al.: The NIH Human Microbiome Project. Genome Research 19(12), 2317–2323 (2009)
12. Wooley, J.C., Godzik, A., Friedberg, I.: A primer on metagenomics. PLoS Computational Biology 6(2) (2010)

Methods of Gene Ontology Term Similarity Analysis in Graph Database Environment

Łukasz Stypka[1,2] and Michał Kozielski[1]

[1] Institute of Informatics, Silesian University of Technology
Akademicka 16, 44-100 Gliwice, Poland
{lukasz.stypka,michal.kozielski}@polsl.pl
http://adaa.polsl.pl
[2] Future Processing, Gliwice, Poland
lstypka@future-processing.com

Abstract. The article presents and analyses three graph processing issues that can be identified in three methods of GO term similarity evaluation. The solutions of these problems are implemented in Neo4j graph database environment. Each of the issues can be solved directly by a single Cypher query or can be divided into several queries which results have to be merged. The comparison of the introduced solutions is presented in terms of time and memory effectivness. The results show how to implement the effective solutions of this class of issues.

Keywords: graph database, Neo4j, Gene Ontology, GO term similarity.

1 Introduction

Gene Ontology (GO) [2] is a widely used knowledge base that is continuously developed and corrected. GO enables annotation of gene products to ontology terms representing biological process, molecular function or biological component. Gene Ontology terms are connected by means of relations of different types, such as e.g., *is a, part of* or *regulates*.

Gene Ontology is an important source of knowledge utilized in several research projects and analysis [2]. It is modeled as a directed acyclic graph where ontology terms are graph nodes and the edges are defined by the relations between terms. One of the important issues that can be approached in Gene Ontology analysis is evaluation of GO terms similarity. Several methods refering to this problem were introduced [8] and many of these methods require different Gene Ontology graph processing.

Recently, several new types of database management systems have been introduced. One of the interesting and well received by the market is a concept of graph database. A very popular representative of this type of systems is Neo4j [7]. It offers, among others, graph data model, graph oriented query language *Cypher* and Java based API. Recently, a version 2.0 of this system has been released, what has solved several previously existing issues (e.g., efficient memory

S. Kozielski et al. (Eds.): BDAS 2014, CCIS 424, pp. 345–354, 2014.
© Springer International Publishing Switzerland 2014

use). Neo4j has been already reported as an interesting environment that can support Gene Ontology graph analysis [5].

The goal of this article is to present and analyse three graph processing issues that can be identified in three methods of GO term similarity evaluation. The solutions of these problems are implemented in Neo4j graph database environment. Each of the issues can be solved directly by a single Cypher query or can be divided to several queries which results have to be merged. The comparison of the introduced solutions is presented in the article and also their time and memory effectivness are evaluated. The results show how to implement the effective solutions of this class of issues.

The structure of this work is as follows. Section 2 presents the similarity measures which require different graph processing approaches. The solutions that were introduced and implemented, and the results of the effectiveness analysis are presented in section 3. Conclusions of the work are presented in section 4.

2   Similarity Measures

Three GO term similarity measures are analysed in this study. They have different characteristics and they require different approaches to GO graph processing.

The first two approaches are classified as semantic similarity measures and utilize the concept of *Information Content* $\tau(a)$ of an ontology term a given by the following formula:

$$\tau(a) = -log(P(a)),\qquad(1)$$

where $P(a)$ is a ratio of a number of annotations to a term a, to a number of analysed genes.

The basic semantic similarity measure was proposed by Resnik [9] and it takes under consideration only the *Information Content* of the common ancestor $\tau_{ca}(a_i, a_j)$ of the compared terms a_i and a_j:

$$s_A^{(R)}(a_i, a_j) = \tau_{ca}(a_i, a_j).\qquad(2)$$

Fig. 1 A presents the common ancestors (terms 1, 2, 5 and 6) of terms 4 and 8.

Other popular approaches requireing identification of the common ancestor term were introduced by Jiang and Conrath [4] and by Lin [6]. Each of these methods was introduced with a purpose other then Gene Ontology term analysis.

The approach based on semantic similarity and taking into account the specificity of GO was presented in [3]. GraSM introduced by Couto et al. [3] extends the above similarity measures by taking into consideration different paths leading to a common ancestor what can also result in different terms regarded as common ancestor. GraSM takes into consideration all such common disjunctive ancestors and instead of the information content of a single common ancestor used in semantic similarity measures it calculates the average of the information content of a disjunctive common ancestors [3]. In that way all the "classical" semantic similarity measures [9,4,6] can be extended to GraSM versions. Fig. 1 B presents the disjunctive common ancestors (terms 1 and 6) of terms 4 and 8.

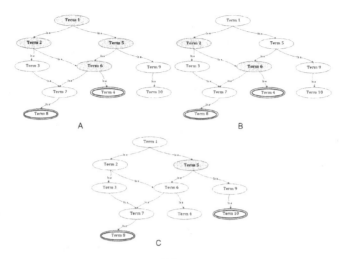

Fig. 1. Illustration of common ancestors (A), disjunctive common ancestors (B), lowest common ancestor (C) of a given pair of terms

The last approach that is considered here defines the distance between two terms a_i and a_j on the basis of a length $l(a_i, a_j)$ of the shortest path between them. Calculating shortest paths in Gene Ontology it has to be taken into consideration that the ontology graph is a directed one. Therefore, the length of a path between two ontology terms that are not connected by a parent-child relation can be set as infinity or it can be calculated as a sum of path lengths leading to the lowest common ancestor. The latter approach was chosen in the work presented. When the shortest paths are calculated then the similarity of the two GO terms can be defined as exponential dependency on a path length $l(a_i, a_j)$ [1]:

$$s_A^{(P)}(a_i, a_j) = e^{-\lambda l(a_i, a_j)}, \tag{3}$$

where λ is a parameter setting the strength of the path length influence on the similarity value. Fig. 1 C presents the lowest common ancestor (term 5) of terms 8 and 10.

3   Experiments and Results

The main goal of the experiments was to compare average query execution time and memory usage level for three types of problems: finding common ancestors, finding the lowest common ancestor and finding disjunctive common ancestors. Each of the issues listed above can be associated with one of the similarity measures presented in section 2 - Resnik, shortest path and GraSM similarity measure respectively.

Twenty terms located deeply in the graph of the gene ontology were selected for the experiments. These were the terms with the significant number of ancestors what enables a proper evaluation of the compared methods.

Table 1. No. of ancestors of the twenty selected terms

Term Id	No. of ancestors	Term Id	No. of ancestors	Term Id	No. of ancestors
GO:0039542	126	GO:0039551	121	GO:0039560	99
GO:0039559	121	GO:0039544	117	GO:0039550	104
GO:0039558	113	GO:0039575	113	GO:0039545	103
GO:0039546	112	GO:0039583	111	GO:0039549	102
GO:0039543	109	GO:0039555	109	GO:0039548	99
GO:0039541	109	GO:0039540	108	GO:0039557	99
GO:0039554	108	GO:0039561	107		

Database Neo4j in version 2.0.6, embedded, was used during the test. In accordance with the recommendations of the creators of the database the following database and the Java virtual machine settings have been used:

Listing 1.1. Database system settings

```
−server −XX:+UseConcMarkSweepGC −Xms200m −Xmx512m −XX:MaxPermSize=512m −XX:+
    UseGCOverheadLimit
```

3.1 Finding Common Ancestors

Each semantic similarity of GO terms (e.g., Resnik similarity) requires identification of a list of common ancestors. The solution of this problem in Neo4j database environment can be approached in two ways. The first one is a comprehensive query of Cypher language referred further as *single*. The example of this query is presented below on listing 1.2.

Listing 1.2. *Single* query identifying a list of common ancestors

```
START firstTerm=node({0}), secondTerm=node({1}) MATCH firstTerm −[*]−>ancestor
    <−[*]− secondTerm RETURN distinct ancestor
```

The second method, presented below on listing 1.3, is based on separate ancestors finding for each term, and then calculating the intersection of the sets. Elements of the result set form a set of common ancestors. This approach is refered further as *divide*.

Listing 1.3. *Divide* query identifying a list of common ancestors

```
Set<Term> findCommonAncestors(Long firstTermId, Long secondTermId) {
  return intersection(ancestors(firstTermId), ancestors(secondTermId));
}

@Query("START term=node({0}) MATCH a−[*]−>ancestor RETURN distinct ancestor")
Set<Term> ancestors(Long term);
```

Both approaches can be compared according to their execution time and memory usage.

Tab. 2 shows the execution time expressed in milliseconds for both queries presented on lisitings 1.2 and 1.3. Tab. 2 and the rest of the tables presenting

Table 2. Execution time [ms] of *single* (listing 1.2) and *divide* (listing 1.3) queries

Single / Divide	GO:0039542	GO:0039559	GO:0039551	GO:0039544	GO:0039558	GO:0039575	GO:0039546	GO:0039583	GO:0039555	GO:0039543
GO:0039542		8187	8877	8425	7642	8714	7700	8550	8151	7990
GO:0039559	155		7797	7708	7032	8038	7189	8164	7660	7658
GO:0039551	147	136		7704	7037	8009	7185	8155	7652	7662
GO:0039544	177	149	141		6879	7859	7076	7975	7527	7536
GO:0039558	173	126	127	131		7245	6544	7516	7001	7114
GO:0039575	186	136	142	138	122		7413	8402	7888	7896
GO:0039546	164	137	137	137	115	127		7477	7066	7041
GO:0039583	163	138	138	139	117	131	128		8042	8034
GO:0039555	178	148	136	150	129	139	144	137		7475
GO:0039543	174	142	135	150	127	140	133	135	149	

results contain the comparison of only 10 out of 20 terms analysed, what is a consequence of a lack of space. However, this number of measurements presents clearly the existing dependencies.

In the case of a comprehensive query (*single*, 1.2) the average execution time was 7686.5 ms, while in case of the latter approach (*divide*, 1.3) only 142.2 ms. The execution time of the *divide* query is thus on average more than 54 times shorter than it is in the case of a comprehensive query.

The Fig. 2 shows the level of memory usage during the performed experiments. It is worth noting that in the case of the first query (1.2) the value of the maximum memory usage is 150 MB, while in the case of the second one (1.3) it equals 125 MB, what is almost 17% less.

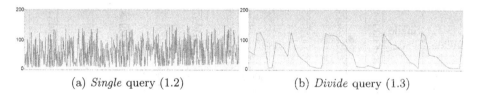

(a) *Single* query (1.2)   (b) *Divide* query (1.3)

Fig. 2. Level of memory usage [MB] for method of finding common ancestors

3.2   Finding the Lowest Common Ancestor

The second issue analysed in this work is the problem of a search for the lowest common ancestor in GO graph. This problem can be encountered in the algorithms based on paths and it is based on searching for such ancestor, for which the length of the path between it and the two terms is the shortest. Again, two approaches are presented below. The first of them is based on a *single* query presented on listing 1.4.

Listing 1.4. *Single* query identifying the lowest common ancestor

```
START firstTerm=node({0}), secondTerm=node({1}) MATCH path=firstTerm −[*]−>
    commonAncestor <−[*]−secondTerm RETURN commonAncestor    order by length(path)
    limit 1
```

In this query, the mechanism adjusting the pattern in graph with the simultaneous memorizing of its structure was used. Next, common ancestors are sorted by the length of the path, and the resulting set is narrowed down to only one term with the shortest path.

The second method (listing 1.5) relies on finding separate list of ancestors of two terms, the calculation of the intersection of the sets, and then calculating path lengths between terms and common ancestor.

Listing 1.5. *Divide* query identifying the lowest common ancestor

```
Term findTheLowestAncestor (Long firstTermId, Long secondTermId) {
    Set<Term> commonAncestors = intersection(ancestors(firstTermId), ancestors(
        secondTermId));
    Integer theBestPathLength = Integer.MAX_VALUE;
    Term theLowestAncestor = null;
    for (Term commonAncestor : commonAncestors) {

        int totalPathLength = shortestPath(firstTermId, commonAncestor.getId()) +
            shortestPath(secondTermId, commonAncestor.getId())

        if ( theBestPathLength < totalPathLength ) {
            theBestPathLength = totalPathLength; theNearestAncestor = commonAncestor;
        }
    }
    return theLowestAncestor;
    }
}

@Query("START firstTerm=node({0}), secondTerm=node({1}) MATCH p = shortestPath(
    firstTerm −[*]−>secondTerm) RETURN p")
Path shortestPath (Long firstTerm , Long secondTerm);
```

The results of the execution time comparison of the queries 1.4 and 1.5 are presented in tab. 3.

Table 3. Execution time [ms] of *single* (listing 1.4) and *divide* (listing 1.5) queries

Single / Divide	GO:0039542	GO:0039559	GO:0039551	GO:0039544	GO:0039558	GO:0039575	GO:0039546	GO:0039583	GO:0039555	GO:0039543
GO:0039542		9734	10228	10002	9020	10391	9235	10268	9739	9716
GO:0039559	192		9164	9206	8301	9529	8526	9662	9151	9129
GO:0039551	178	201		9203	8296	9535	8533	9655	9151	9141
GO:0039544	196	184	167		8278	9497	8518	9592	9125	9129
GO:0039558	234	176	162	157		8636	7745	8742	8287	8278
GO:0039575	250	179	166	168	145		8872	10022	9481	9491
GO:0039546	235	173	155	159	137	151		8947	8504	8525
GO:0039583	226	178	169	159	137	159	142		9636	9632
GO:0039555	240	178	169	168	149	163	158	160		9056
GO:0039543	236	177	163	171	144	165	155	159	166	

The results obtained show unambiguously that the average time of *divide* method is more than 52 times shorter. The average execution time of a *single* approach (1.4) was 9178.2 ms, whereas in case of *divide* approach (1.5) it was only 174.6 ms.

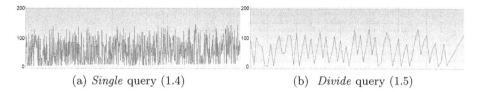

(a) *Single* query (1.4)   (b) *Divide* query (1.5)

Fig. 3. Level of memory usage [MB] for method of finding the lowest common ancestor

Fig. 3 shows that the maximum value of the memory usage for the *single* approach is equal 150 MB. Whereas for *divide* approach only 130 MB, what is about 13% less comparing to the *single* approach to this issue.

3.3   Finding Disjunctive Common Ancestors

The last issue considered in this paper is the problem of searching for disjunctive common ancestors. This issue is encountered in the GraSM algorithm. Analogously to the two previous problems, also in this case, the two distinct ways to achieve the goal were verified. The first of them, *single* is shown below:

Listing 1.6. *Single* query identifying the disjunctive common ancestors

```
START  firstTerm=node({0}),  secondTerm=node({1})  MATCH firstTerm −[firstTermPaths
   *]−>ancestor<−[secondTermPaths *]−secondTerm  WITH  firstTermPaths ,
   secondTermPaths ,  ancestor  WHERE  NOT  ANY( path  IN  firstTermPath  WHERE  path  IN
   secondTermPath)  RETURN  distinct  ancestor
```

This query finds common ancestors, memorizes a path connecting the first term with the ancestor and the second term with the ancestor. In the next step, the ancestors for which there is identical path between the first term and the common ancestor, and the second term and the common ancestor, are filtered out.

The *divide* method (listing 1.7) relies on finding common ancestors, next finding all paths between the term and the ancestor, and finally determining whether any path is included both for the first and for the second term.

The results of the execution time comparison of the queries 1.6 and 1.7 are presented in tab. 4.

The average time of finding the common disjunctive ancestors is equal to 24726.1 ms in case of a *single* query (1.6) and 744.9 ms in case of *divide* query (1.7). It means that an average *divide* query was executed in over 33 times shorter time then a *single* query.

Listing 1.7. *Divide* query identifying the disjunctive common ancestors

```
Set<Term> findDisjunctiveAncestors(Long firstTermId, Long secondTermId) {
    Set<Term> disjunctiveCommonAncestors = {};
    Set<Term> commonAncestors = intersection(ancestors(firstTermId), ancestors(
        secondTermId));
    for(Term commonAncestor : commonAncestors) {
        Set<Path> pathsBetweenFirstTermAndAncestor = findAllPaths(commonAncestor.
            getId(), firstTermId);
        Set<Path> pathsBetweenSecondTermAndAncestor = findAllPaths(commonAncestor.
            getId(), secondTermId);
        if (!doTheyHaveACommonPath(pathsBetweenFirstTermAndAncestor,
            pathsBetweenSecondTermAndAncestor) {
            disjunctiveCommonAncestors.add(commonAncestor);
        }
    }
    return disjunctiveCommonAncestors;
}

@Query("START firstTerm=node({0}), secondTerm=node({1}) MATCH p = (firstTerm −[*]−>
    secondTerm) RETURN p")
Set<Path> findAllPaths(Long firstTerm , Long secondTerm);
```

Table 4. Execution time [ms] of *single* (listing 1.6) and *divide* (listing 1.7) queries

Single / Divide	GO:0039542	GO:0039559	GO:0039551	GO:0039544	GO:0039558	GO:0039575	GO:0039546	GO:0039583	GO:0039555	GO:0039543
GO:0039542		27386	27341	18163	25310	30355	23362	30408	25973	18162
GO:0039559	787		23883	25892	18265	26221	23117	26042	25978	25976
GO:0039551	743	965		25865	21748	26241	23117	25936	25742	25751
GO:0039544	920	684	674		24109	28892	21797	28634	24381	16449
GO:0039558	793	848	828	628		23810	21024	23669	23357	23382
GO:0039575	896	737	718	680	661		25345	27492	28204	28209
GO:0039546	906	649	640	867	576	622		25306	21120	21069
GO:0039583	794	853	852	676	686	746	618		27982	27945
GO:0039555	949	690	669	854	603	650	775	642		24264
GO:0039543	963	678	661	854	604	657	770	644	813	

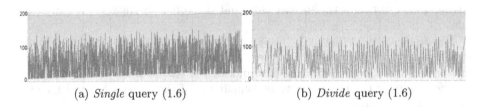

(a) *Single* query (1.6)   (b) *Divide* query (1.6)

Fig. 4. Level of memory usage [MB] for method of finding disjunctive common ancestors

The results presented on Fig. 4 show that the level of memory usage for *divide* query slightly exceeds 125 MB, whereas for *single* query it slightly exceeds 150 MB, what is about 13% difference.

3.4 Average Execution Time and Its Standard Deviation

All of the previously presented experiments were repeated for 5% of randomly selected terms for the three Gene Ontology subgraphs: Biological Process (1267 terms), Mollecular Function (481 terms) oraz Cellular Component (164 terms). For each pair of terms belonging to the same subgraph of GO the average execution time and standard deviation were measured.

Table 5. Average time [ms] (Avg) and standard deviation [ms] (St. dev.) of execution time for 5% populations of three GO subgraphs : Biological Process (BP), Cellular Component (CC), Molecular Function (MF)

		Common ancestors		Lowest Common Ancestor		Disjunctive common ancestors	
		divide	single	divide	single	divide	single
BP	Avg	9.38	53.66	9.84	58.07	11.99	149.95
	St. dev.	21.13	388.72	21.37	433.77	26.52	1317.15
CC	Avg	6.58	40.86	7.04	48.29	11.05	123.41
	St. dev.	10.01	142.66	10.26	174.60	17.54	512.70
MF	Avg	11.52	42.42	12.07	48.45	14.83	130.42
	St. dev.	52.26	741.01	52.59	934.97	68.57	2435.99

The obtained results are consistent with the results from the previous experiments. The results presented in Tab. 5 clearly indicate, that the *divide* method has a much smaller execution time and standard deviation. It is worth noting that the standard deviation for the *single* method is many times larger, so this approach can be considered as less stable and more sensitive to the location of the terms in the GO graph.

4 Conclusions

The article presented two different approaches, applied in graph database Neo4j, to solve the issues related to Gene Ontology term similarity analysis. The issues that were taken into consideration cover finding common ancestors, the nearest common ancestor and disjunctive common ancestors.

The first approach was based on a single query of Cypher language, which is integral part of Neo4j database. The second approach divided the query into subqueries and additional merging processing.

The experiments performed proved that *divide* queries perform significantly better. The execution time analysis showed that they can be from 33 to 54 times faster then *single* queries. It was also shown that *divide* queries are more stable (in terms of execution time) and less dependent on the terms' depth in GO graph. Additionally, the memory usage was verified in the experiments. Increase of the memory usage is often observed during the time optimization in a classic optimization approach. However, this trend was not observed in the experiments performed. Moreover, the figures presented show that the maximal memory usage is always reduced when *divide* queries are executed.

This work presents how to implement efectively the three different classes of problems that can be encountered during Gene Ontology analysis. The conclusions can be generalized, as we can point a certain type of query that is a common feature of the analysed issues. This query that has the following form in Cypher language "->common_node<-" is faster realised by Neo4j when it is divided into separate queries. This conclusion does not disqualify graph database systems and Neo4j being their representative in this work as an interesting and profitable environment that can be applied to GO-based analysis, as it was pointed in other works [5].

Acknowledgments. The work was supported by National Science Centre (decision DEC-2011/01/D/ST6/07007), by Ministry of Science and Higher Education as a Statutory Research Project (decision 8686/E-367/S/2013) and by POIG.02.03.01-24-099/13 grant: "GCONiI - Upper-Silesian Center for Scientific Computation".

References

1. Al Mubaid, H., Nagar, A.: Comparison of four similarity measures based on go annotations for gene clustering. In: IEEE Symposium on Computers and Communications, ISCC 2008, pp. 531–536. IEEE (2008)
2. Ashburner, M., et al.: Gene Ontology: tool for the unification of biology. Nat. Genet. 25(1), 25–29 (2000)
3. Couto, F.M., Silva, M.J., Coutinho, P.M.: Measuring semantic similarity between gene ontology terms. Data & Knowledge Engineering 61(1), 137–152 (2007)
4. Jiang, J., Conrath, D.: Semantic similarity based on corpus statistics and lexical ontology. In: Proc. on International Conference on Research in Computational Linguistics, pp. 19–33 (1997)
5. Kozielski, M., Stypka, Ł.: Gene ontology based gene analysis in graph database environment. Studia Informatica 34(2A), 111 (2013)
6. Lin, D.: An information-theoretic definition of similarity. In: ICML, vol. 98, pp. 296–304 (1998)
7. Neo4j: Graph database: http://www.neo4j.org
8. Pesquita, C., Faria, D., Falcao, A.O., Lord, P., Couto, F.M.: Semantic similarity in biomedical ontologies. PLoS Computational Biology 5(7), e1000443 (2009)
9. Resnik, P.: Semantic similarity in a taxonomy: An information-based measure and its application to problems of ambiguity in natural language. Journal of Artificial Intelligence Research 11, 95–130 (1999)

Mining of Eye Movement Data
to Discover People Intentions

Pawel Kasprowski

Institute of Informatics,
Silesian University of Technology, Gliwice, Poland
pawel.kasprowski@polsl.pl

Abstract. The process of face recognition is a subject of the research
in this paper. 1430 recordings of participants eye movements while they
were observing faces were analyzed statistically and various data mining
techniques were used to extract information from eye movements signal.
One of the findings is that the process of face recognition is different for
different subjects and therefore formulating general rules for face recognition process may be difficult. The hypothesis was that it is possible
to analyze eye movements signal to predict if the subject observing the
face recognizes it. A model that automatically differentiates observations
of recognized and unrecognized faces was built and the results are encouraging. One of the contributions of the paper is a conclusion that the
optimal set of attributes of eye movement signal for such classification is
individually specific and different for different people.

Keywords: eye movement, face recognition, data mining.

1 Introduction

Face recognition is one of the first abilities of a newborn child and is the basic
ability of every human being. Therefore, there is no surprise that the process
of face recognition is an intensively studied subject. However to the authors
knowledge there is no published information regarding an existence of a successful algorithm that automatically recognizes whether a person observing the
face knows this face. The paper describes an attempt to use advanced data mining techniques to recognize subjects familiarity of the observed face, basing on
eye movements of that subject, recorded during the observation. The other contribution of the paper is the idea of distinguishing the personal differences in a
faces recognition process for different subjects.

2 Related Research

There are a lot of studies concerning the way humans recognize faces. It may
be divided into two main categories: analyzing neural aspects of face recognition
what parts of our brain are responsible for it [11] and analyzing what elements

S. Kozielski et al. (Eds.): BDAS 2014, CCIS 424, pp. 355–363, 2014.
© Springer International Publishing Switzerland 2014

of a face are taken into account in a recognition process [1]. Eye movements information proved to be very useful for the latter. The face observation research may be divided into several fields briefly described below.

Models. People are able to recognize faces from the early beginnings of their life. However, there are still doubts about the way the "algorithm" for face recognition works. There are two different theories: holistic and analytical [18]. According to the holistic theory, people recognize faces acquiring an image of the face that is matching the pattern seen with the pattern remembered previously. A face is perceived as a whole a pattern is a combination of all specific features [16]. The analytical theory concentrates on peoples ability to decompose a face into different features. It suggests that the recognition process uses information about special faces properties as shape and color of eyes, a nose and a mouth. According to this theory people separately compare patterns of specific parts of a face with patterns recorded in their brain. The evidence to this theory is that eye tracking data recorded during face recognition concentrate on specific parts of the face (like eyes and nose) [1,8]. There were many interesting experiments conducted, like showing only a part of a face [17] or automatically removing a face after specific number of fixations [7].

Regions of Interest. Eye movement studies show that the first fixation is usually placed in the middle of the face but near the upper part. Second fixation typically goes to the left side (in most cases near the right eye on the face) [5]. People concentrate on the upper part of the face because it contains more personally distinctive features.

Familiarity. There are also some studies searching for the differences in how people observe familiar and unfamiliar faces. These studies are particularly interesting for our work. According to [15] the length of the first fixation is different when observing familiar and unfamiliar faces. Van Belle at al [18] analyses these differences with the conclusion that the differences are the most significant towards the end of the observation and that the last fixation should be taken into account. Barton et al [1] concludes that known faces are analyzed in more simplified way only to confirm familiarity. On the contrary, unknown faces are always scanned for all interesting features. Differences among people. The studies concerning differences in how people ob-serve faces are not widely adopted. Rozhokova et al [14] suggested as one of the conclusions that fixation position when recognizing faces demonstrates significant inter-individual variability. Blais at all [3] proved that scan patterns are different for different races comparing Western Caucasian and East Asian observers. Rigas et al [13] were able to identify people basing on the way they observed faces on a computer screen.

Eye Movement Data Mining. The latter paper is one of the examples of using data mining techniques to extract information from an eye movement signal. Other interesting examples include mostly people identification [9,10] but also people intentions [2]. To the best of authors knowledge there are no published attempts to automatically recognize observers familiarity of the face being observed.

3 Experiment and Dataset

The main hypothesis of this research was that it is possible to predict if people recognize presented faces basing on their eye movements characteristic. To check that hypothesis a dataset of eye movements samples recorded during faces observations was built. A head mounted Jazz-Novo eye tracker that records eye positions with 1000Hz frequency was used. 34 participants took part in the experiment. One session consisted of initial nine points of regard calibration and subsequent faces presentations. Between presentations the system was always recalibrated with a simplified three points calibration. The participants task was to look at a face on the screen and asses, by pressing one of two possible buttons, if they recognize the face or not. After pressing the button the face disappeared. This simple task will be named the observation in the subsequent text. Every face appearing on the screen was cropped, so that eyes were in the same place for every picture. No further processing was applied and faces were just photographs of different people.

Every person took part in at least one session a sequence of face observations. There were overall 56 sessions provided. 22 participants took part in two sessions with one week interval between them. The first session consisted of 24 observations (24 different faces) and the second session of 27 observations (27 faces different than in the first session). The total number of separate observations was 1430. The number of observations for which participants decision was positive (i.e. the face is recognized) was 418 and the number of observations with not recognized decision (referred later as negative observations) was 1012.

It is worth mentioning that familiarity of the face is something that is not a binary property [18]. We can divide familiar faces into: famous faces, personally familiar faces (i.e. known from real life), familiarized faces (e.g. familiarized during the previous sessions), artificial faces (painted, drawn or produced by a computer). There are also a number of factors that are not measurable or very difficult to measure. Unknown face may be similar to some known face, known person may be photographed in unusual circumstances (like strange hair cut or strange face expression). That is why we should rather say about a level of familiarity. This level is personally dependent and may differ for different people.

It is also very important to notice that in the experiment it wasnt checked if the subject really knows the observed face. It was only checked if the subject recognizes the face. It was assumed that eye movements characteristic is different when subjects decide that they know the face. However, it could likely happen that the subject made a mistake. For instance only 85% of subjects recognized the face of Arnold Schwarzenegger. The faces of people organizing the experiment were recognized by only 63% of the subjects. So, in the subsequent analyses, the eye movements recordings were used to predict the tested subjects decision about the face familiarity and not the ground truth about the familiarity.

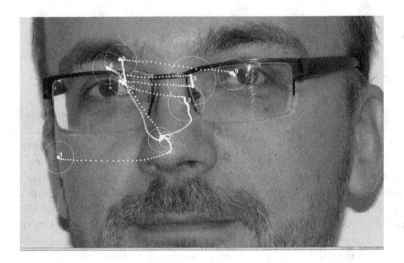

Fig. 1. Example of typical scan-path recorded during face observation

4 Data Preparation

After gathering the data, there were some initial studies performed, based on the previously published findings to see if there were features of the registered observations that might be used. At first the observation time was examined, hoping that unfamiliar faces were observed longer. It occurred that there was no significant difference between recognized and not recognized faces with 2.31 sec. and 2.48 sec (p=0.1). The shortest observation time was recorded for the most known face of Barack Obama. However, the second shortest observation time was recorded for a *face 11* not known by anybody. The observation time also depends on practically immeasurable similarity of the face to the most known pictures of the same person. It was the case of the actress Sandra Bullock, for which a picture was presented with an unusual face expression what resulted in longer observation time. The most important finding, regarded significant differences in observation time between participants, with average values between 1.26 and 5.93 sec. The ANOVA test for different subjects and observation lengths showed (with F(33,1396)=20.24, p<1*10-10) that hypothesis that the length of observation was independent of the participants cannot be proved. Interestingly, it occurred that there existed a strong negative correlation between the observation time and the number of a presented face in the sequence during the session (-0.62). It means that subsequent observations during the same session were shorter the observer was able to make decisions faster. Probably the observers got used to their task and became more focused on the observation. It showed that the length of observation is personally specific and depends on many other elements like the length of a session, similarity of a face to other known faces and so on.

There were several additional features compared for positive and negative observations. For instance the difference in fixation number was checked with

7.63 fixations for negative (unrecognized face) and 7.25 fixations for positive (recognized face) observations. For most of the features used, the differences were not significant with p>0.05. But it occurred that - similarly to observation length - for nearly all the features it was impossible to prove independency of the participant. This important finding was used in further experiments.

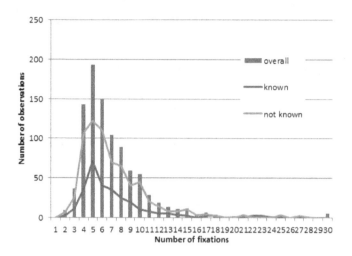

Fig. 2. Histogram of number of fixations per observation

4.1   Features Extraction

Because it occurred that it is difficult to find one feature of eye movement signal that indicates if a subject will choose known or unknown option, the fusion of different features to classify observations was checked. The hypothesis was that combination of a set of weak classifiers may produce satisfactory results.

Because the focus was on differences between positive and negative observations, the most interesting part of the eye movement signal was after the moment when a person had already made her/his decision. That is why the eye movement recorded at the end of the observation was mostly used assuming that it would be more meaningful [18]. Surprisingly, it occurs that the decision is made relatively fast. According to [7] people are able to decide about familiarity of the face just after two fixations. All subsequent fixations are done only for confirming their primary in most cases final assessment. According to [15] even the length of the first fixation may be different for known and not known faces.

It was decided to use different feature sets to see how it was possible to predict the subjects decision. There were both fixation related and signal related features used. There were seven groups of features built with different parameters.

- Observation length
- Histogram of velocities during the last X ms (with X= 500, 1000 and 1500ms) 8 values

- Histogram of movement directions during the last X ms (with X= 500, 1000 and 1500ms) 8 values
- Number of fixations during the last X ms (with X = 500, 1000, 1500, 2000 and 2500 ms)
- Length of the first N fixations (with N = 2, 3, 4, 5)
- Length of the last N fixations (with N = 1, 2, 3, 4, 5)
- X and Y position of the last N fixations (with N = 1 and 2)

Next, there were feature sets created for every combination of groups with different values of parameters. It resulted in 212 feature sets.

4.2 Building a Classification Model

The main intention was to use the data described in the previous sections to build a model that automatically classifies observations as positive (recognized face) or negative. There are plenty of possible classification algorithms that could be used. There were several the most popular algorithms used and its performance was checked against available data. There were different algorithms implemented in WEKA library used [6]:

- Nave Bayes
- Random Forest
- J48
- K Nearest Neighbors (with k = 1, 3 and 7)
- SVM (with kernels poly(1), poly(2) and RBF)

There was every feature set combination with every classification algorithm used. It resulted in creation of 1908 different combinations. For every combination it was possible to build a classification model using some training data (face observations with known/unknown labels) and check its performance for some testing data. Because the number of negative observations was higher than positive observations, the positive observations were weighted.

The process of building a model for a given dataset consisted of two nested cross-validation steps. At first, the dataset was divided into 10 folds. Then, for every possible combination of feature sets and classifiers, a model using every possible set of 9 folds as examples was built and evaluated against samples in the remaining fold. The result of each such step was a collection of models with information about its performance. Area under ROC curve was used to evaluate the model performance [4]. The next step was choosing 10 combinations that gave the best results to create new models using all samples from the training set. The models were then used for classification of samples from the testing set. For every testing sample the results of models were summarized and normalized to build the systems answer in the range <0,1>. Because previous findings showed that features values were significantly different for different subjects it was additionally decided to repeat this procedure separately for every subject.

5   Results

The same procedure was executed for the whole dataset and independently for observations of 22 subjects that took part in both sessions. The results of the classifications are presented in Tab. 1. Although accuracy seems to be the most obvious result of classification, it should be interpreted with care because it doesnt say much about results in two class experiments when samples distribution between classes is not even [12]. That is why there were two other metrics used that gave more reliable estimation about the real power of the model. AUC stands for Area Under ROC Curve and EER stands for Equal Error Rate.

Table 1. Results for different subjects

Sid	AUC	EER	Accuracy
All	0.56	0.46	0.70
s9	0.92	0.13	0.88
s4	0.91	0.20	0.90
s33	0.89	0.22	0.84
s10	0.89	0.22	0.88
s30	0.87	0.25	0.88
s13	0.86	0.14	0.94
s17	0.86	0.22	0.78
s14	0.85	0.17	0.84
s12	0.85	0.26	0.82
s27	0.84	0.26	0.84
s31	0.83	0.25	0.84
s21	0.82	0.29	0.88
s7	0.78	0.23	0.82
s32	0.76	0.28	0.82
s2	0.75	0.34	0.75
s20	0.74	0.35	0.73
s5	0.73	0.34	0.73
s15	0.73	0.24	0.76
s3	0.71	0.35	0.73
s18	0.71	0.36	0.80
s28	0.70	0.40	0.73
s29	0.65	0.29	0.84

As it is visible in Tab. 1, the results for the whole dataset are not encouraging. AUC equal to 0.56 is only a fraction better than a random guess. So, in this field the experiment failed. But when we analyze the results achieved independently for different subjects, it occurs that the method works. The average AUC is 0,8 what seems to be a quite good value as for the first attempt. Similarly, the average EER is 26% what is quite encouraging.

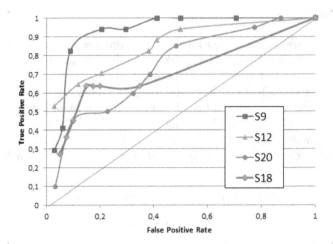

Fig. 3. Examples of ROC curves for four different subjects

6   Conclusion

The possibility of predicting the subjects decision about the familiarity of the observed face was analyzed in the paper. The prediction was made basing on eye movements recordings. It occurred that the task was not trivial and it may be difficult to propose one universal classification model that works with every observer. Traditional features as observation length, last fixation length, fixation position and so on were not enough to reliably estimate decision for each person being examined. However, it occurred that there were significant differences between people in the way they observe faces. This observation is in agreement with [13]. Having this in mind, it was possible to optimize classification models to work independently for each subjects observations. The encouraging results were achieved for the examined observers.

There are multiple possible extensions of this work. Of course the results should be checked on larger population (because of possible model over-fitting). As mostly famous faces were studied in this work, the personally familiar faces could be more interesting for future work. There also could be additional feature sets proposed, which may work better for less successful examples, like features derived for instance from heatmaps or scan-paths.

References

1. Barton, J.J., Radcliffe, N., Cherkasova, M.V., Edelman, J., Intriligator, J.M.: Information processing during face recognition: The effects of familiarity, inversion, and morphing on scanning fixations. Perception 35, 1089–1105 (2006)
2. Bednarik, R., Vrzakova, H., Hradis, M.: What do you want to do next: a novel approach for intent prediction in gaze-based interaction. In: Proceedings of the Symposium on Eye Tracking Research and Applications, pp. 83–90. ACM (2012)

3. Blais, C., Jack, R.E., Scheepers, C., Fiset, D., Caldara, R.: Culture shapes how we look at faces. PLoS One 3(8), e3022 (2008)
4. Fawcett, T.: An introduction to roc analysis. Pattern Recognition Letters 27(8), 861–874 (2006)
5. Guo, K., Smith, C., Powell, K., Nicholls, K.: Consistent left gaze bias in processing different facial cues. Psychological Research 76(3), 263–269 (2012)
6. Hall, M., Frank, E., Holmes, G., Pfahringer, B., Reutemann, P., Witten, I.H.: The weka data mining software: an update. ACM SIGKDD Explorations Newsletter 11(1), 10–18 (2009)
7. Hsiao, J.H.W., Cottrell, G.: Two fixations suffice in face recognition. Psychological Science 19(10), 998–1006 (2008)
8. Itier, R.J., Alain, C., Sedore, K., McIntosh, A.R.: Early face processing specificity: It's in the eyes! Journal of Cognitive Neuroscience 19(11), 1815–1826 (2007)
9. Kasprowski, P., Ober, J.: Enhancing eye-movement-based biometric identification method by using voting classifiers. In: Defense and Security, pp. 314–323. International Society for Optics and Photonics (2005)
10. Komogortsev, O.V., Jayarathna, S., Aragon, C.R., Mahmoud, M.: Biometric identification via an oculomotor plant mathematical model. In: Proceedings of the 2010 Symposium on Eye-Tracking Research & Applications, pp. 57–60. ACM (2010)
11. Pitcher, D., Dilks, D.D., Saxe, R.R., Triantafyllou, C., Kanwisher, N.: Differential selectivity for dynamic versus static information in face-selective cortical regions. Neuroimage 56(4), 2356–2363 (2011)
12. Provost, F.J., Fawcett, T., Kohavi, R.: The case against accuracy estimation for comparing induction algorithms. In: ICML, vol. 98, pp. 445–453 (1998)
13. Rigas, I., Economou, G., Fotopoulos, S.: Biometric identification based on the eye movements and graph matching techniques. Pattern Recognition Letters 33(6), 786–792 (2012)
14. Rozhkova, G., Ognivov, V.: Face recognition and eye movements: landing on the nose is not always necessary. Perception 38, 77 (2009)
15. Ryan, J.D., Hannula, D.E., Cohen, N.J.: The obligatory effects of memory on eye movements. Memory 15(5), 508–525 (2007)
16. Tanaka, J.W., Gordon, I.: Features, configuration and holistic face processing. In: The Oxford Handbook of Face Perception pp. 177–194 (2011)
17. Van Belle, G., De Graef, P., Verfaillie, K., Busigny, T., Rossion, B.: Whole not hole: Expert face recognition requires holistic perception. Neuropsychologia 48(9), 2620–2629 (2010)
18. Van Belle, G., Ramon, M., Lefèvre, P., Rossion, B.: Fixation patterns during recognition of personally familiar and unfamiliar faces. Frontiers in Psychology 1, 20 (2010)

Fast and Accurate Hand Shape Classification

Jakub Nalepa[1,2] and Michal Kawulok[1,*]

[1] Silesian University of Technology, Gliwice, Poland
{jakub.nalepa,michal.kawulok}@polsl.pl
[2] Future Processing, Gliwice, Poland
jnalepa@future-processing.com

Abstract. The problem of hand shape classification is challenging since a hand is characterized by a large number of degrees of freedom. Numerous shape descriptors have been proposed and applied over the years to estimate and classify hand poses in reasonable time. In this paper we discuss our parallel, real-time framework for fast hand shape classification. We show how the number of gallery images influences the classification accuracy and execution time of the algorithm. We present the speedup and efficiency analyses that prove the efficacy of the parallel implementation. Different methods can be used at each step of the proposed parallel framework. Here, we combine the shape contexts with the appearance-based techniques to enhance the robustness of the algorithm and to increase the classification score. An extensive experimental study proves the superiority of the proposed approach over existing state-of-the-art methods.

Keywords: hand shape classification, gesture recognition, parallel algorithm.

1 Introduction

Hand gestures constitute an important source of non-verbal communication, either complementary to the speech, or the primary one for people with disabilities. The problem of hand gesture recognition has been given a considerable research attention due to a wide range of its practical applications, including human-computer interfaces [4, 23], virtual reality [21], telemedicine [25], videoconferencing [15], and more [7, 17]. The proposed approaches can be divided into hardware- and vision- based methods. The former utilize sensors, markers and other equipment to deliver an accurate gesture recognition, but they lack naturalness and are of a high cost. Vision-based methods are contact-free, but require designing advanced image analysis algorithms for robust classification. Thus, an additional effort is needed for applying these techniques in real-time applications.

Numerous algorithms for hand shape classification have emerged over the years. In the contour-based approaches, the shape boundary of a detected hand

---

* This work has been supported by the Polish Ministry of Science and Higher Education under research grant no. IP2011 023071 from the Science Budget 2012–2013.

S. Kozielski et al. (Eds.): BDAS 2014, CCIS 424, pp. 364–373, 2014.

is considered to represent its geometric features. These methods include, among others, a very time-consuming approach based on the shape contexts analysis and estimating similarity between shapes [1], recently optimized by reducing the search space by using the mean distances and standard deviations of shape contexts [14], and Hausdorff distance-based methods [9]. The main drawback of these contour-based approaches lies in their limited use in case of missing contour information. In the appearance-based methods, not only is the contour utilized for shape features extraction, but also the shape's internal region is analyzed. For example, an input color or greyscale image is processed in the orientation histograms approach [6] or an entire hand mask can be fed as an input to various template matching and moment-based methods [22]. An interesting and thorough survey on vision-based hand pose estimation methods was published by Erol et al. [5].

In this paper we discuss a fast parallel algorithm for hand shape classification. We show how the parallelization affects the classification time, and makes it possible to apply for searching large hand gesture sets in reasonable time. We present the speedup and efficiency of the parallel algorithm. Moreover, we show how to increase the classification score by combining the shape contexts with the appearance-based methods.

The paper is organized as follows. The hand shape classification algorithm is described in detail in section 2. The experimental study is reported in section 3. The paper is concluded in section 4.

2   Parallel Hand Shape Classification Algorithm

In this section we describe our parallel algorithm (PA) for hand shape classification [18]. First, the input image I_i is subject to skin segmentation, only if necessary (Alg. 1, lines 2–5). This step is undertaken if the shape features are to be extracted from the skin map of I_i. There exist a number of skin detection and segmentation techniques applicable in real-time applications [10–12,19]. A thorough survey on current state-of-the-art skin detection approaches has been published recently [13]. Then, the image, either the skin mask or the original one, is normalized (line 6). The normalization procedure is presented in Fig. 1. An input image (A) or the skin map (B) is rotated (C) based on the position of wrist points, so as the hand is oriented upwards. Pixels below the wrist line are discarded, the image is cropped and downscaled to the width w_M (D). Once the image is normalized, hand shape features are calculated in parallel (Alg. 1, lines 7–9), and the input image I_i is compared with the gallery images, also in parallel (lines 10–12). Finally, the classification result is returned (line 13). Noteworthy, the classification procedure can be executed for a number of input images in parallel. Thus, larger databases of input hand images can be analyzed significantly faster than using a sequential approach.

In the shape features calculation and shape classification stages we utilized the following state-of-the-art methods: shape contexts analysis (SC) [1], template matching (TM), Hausdorff distance analysis (HD) [9], comparison of the

---

Algorithm 1. Parallel hand shape classification

1: **parfor** $I_i \leftarrow I_1$ **to** I_N **do**
2:     $SkinMapIsAnalyzed \leftarrow$ CheckIfSkinMapIsAnalyzed(I_i);
3:     **if** $SkinMapIsAnalyzed$ **then**
4:         Detect skin and create skin map;
5:     **end if**
6:     Normalize image;
7:     **parfor** $H_j \leftarrow H_1$ **to** H_M **do**
8:         Calculate j-*th* hand shape feature;
9:     **end parfor**
10:    **parfor** $G_j \leftarrow G_1$ **to** G_g **do**
11:        Compare I_i with j-*th* gallery image;
12:    **end parfor**
13:    **return** final hand shape classification;
14: **end parfor**

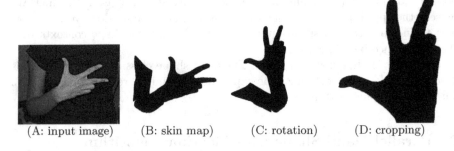

(A: input image)   (B: skin map)   (C: rotation)   (D: cropping)

Fig. 1. Example of the hand skin map normalization

orientation histograms (HoG) [6], Hu moments analysis (HM) [8], and two approaches combining the SC with the appearance-based methods: SC combined with the distance transform (SCDT) and SC enhanced by the orientation histograms analysis (SCH). In the TM algorithm we used the following summation for comparing the overlapped patches of images I_1 and I_2, of size $w_1 \times h_1$ and $w_2 \times h_2$, respectively:

$$R(i,j) = \sum_{a,b} (I_1(a,b) - I_2(i+a, j+b))^2, \tag{1}$$

where $0 \leq a \leq w_2 - 1$ and $0 \leq b \leq h_2 - 1$.

Let κ and λ be two contours compared in the SC method. For each point p_i^κ and p_i^λ, $i \in \{1, \ldots, m\}$, where m is the number of contour points, belonging to κ and λ respectively, the coarse log-polar histogram h_i, i.e., the shape context, is calculated. It depicts the distribution of the remaining $(m-1)$ points for each p_i. Let C_{ij} denote the cost of matching the points p_i^κ and p_j^λ, given as a chi-square distance between the corresponding shape contexts. Then, the total matching cost of two contours C is given as $C = \sum_i C(p_i^\kappa, p_{\pi(i)}^\lambda)$, where π is a permutation of the contour points. Clearly, the minimization of C is an instance of the bipartite matching problem. It can be solved in $O(n^3)$ time, where n is the number of sampled contour points, using the Hungarian method [14]. To speed up the SC, we sample and analyze a subset of M_{SC}, $M_{SC} \ll m$, contour points of a shape. Additionally, the distance transform (DT) of the hand mask from the contour is performed. Given the DT, its histogram H_i is calculated for the image I_i. Then, the distance between the histograms H_1 and H_2 of two images I_1 and I_2 is found using the chi-square metric:

$$d(H_1, H_2) = \sum_B \frac{(H_1(B) - H_2(B))^2}{H_1(B) + H_2(B)}. \tag{2}$$

The final cost of shapes matching of the SCDT is given as: $C' = \alpha C + \beta d(H_1, H_2)$, where α and β are the weights. Values of C and $d(H_1, H_2)$ are normalized, therefore $0.0 \leq C \leq 1.0$ and $0.0 \leq d(H_1, H_2) \leq 1.0$. Similarly, the shape contexts are combined with the orientation histograms approach [6] using the same values of weights α and β.

3   Experimental Results

The PA was implemented in C++ language using the OpenMP interface. The experiments were conducted on a computer equipped with an Intel Xeon 3.2 GHz (16 GB RAM with 6 physical cores and 12 threads) processor having the following cache hierarchy: 6×32 kB of L1 instruction and data cache, 6×256 kB L2 cache and 12 MB of L3 cache. The settings used in both stages of the PA are as follows: $\alpha = 0.17$, $\beta = 1.0$, $w_M = 100$, $M_{SC} = 20$.

3.1   Database of Hand Gestures

The experimental study was carried out using our database of 499 color hand images of 15 gestures presented by 12 individuals[1]. Each gesture was presented n times, $27 \leq n \leq 39$. The images are associated with ground-truth binary masks indicating skin regions along with the ground-truth hand feature points. In this study, we omitted the skin segmentation and wrist localization stages, and used the ground-truth data for fair assessment of investigated techniques applied at other algorithm steps. Examples of ground-truth binary masks are presented in Fig. 2. Here, each gesture (1, 2, 4, H, K, N) was presented by five individuals (I–V). It is easy to note that the difference between masks representing the same gesture (i.e. inner-class difference) may be significant, e.g. due to the hand rotation – see Fig. 2(N).

3.2   Classification Accuracy Analysis

The data set was split into a gallery (G) and a probe set (P) [20]. The gallery contains exactly g, $g \geq 1$, sample images per each gesture in the data set.

---

[1] For more details see http://sun.aei.polsl.pl/~jnalepa/BDAS2014

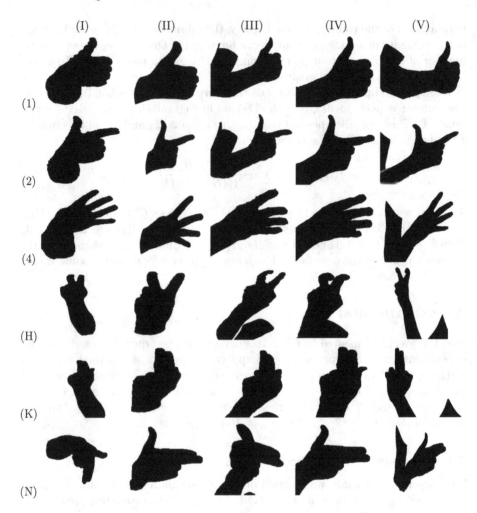

Fig. 2. Examples of ground-truth binary masks of various gestures (1, 2, 4, H, K, N) presented by five individuals (I–V)

Then, the similarities of the images from P to those in G were found using the techniques outlined in section 2. Classification effectiveness is assessed using its *rank* (R), $1 \leq R \leq |G|$. The rank is the position of the correct label on a list of gallery images sorted in the descending order by the similarity. If an image is classified correctly, then its rank is 1. The classification effectiveness for a given set is a percentage of correctly classified images. The analysis of the classification efficacy is performed based on cumulative response curves (CRCs).

The best CRCs for a single image per gesture in G are given in Fig. 3. Here, we performed 27 classification experiments with no overlaps between the galleries. Then, the results were averaged and the best set of gallery images was determined. It is easy to see that the shape context methods, SCDT, SCH and

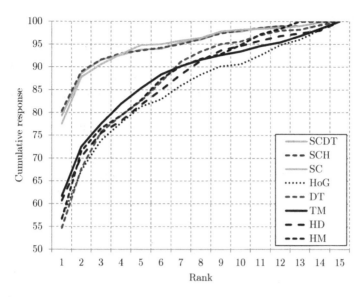

Fig. 3. Best CRCs for a single image per gesture in G

Table 1. Average CR along with the standard deviation σ (best CR shown in boldface) for various number of gallery images g: (A) $g = 1$, (B) $g = 3$, (C) $g = 5$

Method ↓		Rank 1 CR ± σ	Rank 2 CR ± σ	Rank 3 CR ± σ	Rank 4 CR ± σ
	SCDT	**69.30** ± 5.84	**78.04** ± 2.20	**82.92** ± 1.80	**85.95** ± 1.28
	SCH	69.02 ± 6.33	77.82 ± 2.19	82.38 ± 1.82	85.40 ± 1.17
	SC	69.04 ± 5.75	77.97 ± 2.27	82.63 ± 1.88	85.82 ± 1.05
(A)	HoG	44.14 ± 7.64	57.56 ± 3.05	65.25 ± 1.49	70.37 ± 1.52
	DT	41.95 ± 6.71	56.86 ± 2.76	66.34 ± 1.86	72.95 ± 1.47
	TM	51.38 ± 7.16	63.68 ± 2.40	69.48 ± 1.82	74.05 ± 1.32
	HD	49.10 ± 7.44	59.81 ± 2.01	66.39 ± 1.62	71.21 ± 1.73
	HM	46.69 ± 7.07	60.86 ± 3.48	68.46 ± 2.52	74.57 ± 1.62
	SCDT	76.66 ± 2.84	**82.17** ± 1.43	85.71 ± 0.90	**88.47** ± 1.19
(B)	SCH	**77.10** ± 1.56	82.11 ± 0.87	**86.23** ± 0.61	88.07 ± 0.70
	SC	75.74 ± 2.45	81.00 ± 1.47	84.23 ± 0.88	86.75 ± 0.73
	SCDT	**81.18** ± 1.25	85.97 ± 0.87	**88.35** ± 0.74	**90.33** ± 0.64
(C)	SCH	80.59 ± 2.30	**86.06** ± 1.38	88.28 ± 0.84	89.80 ± 0.65
	SC	80.17 ± 1.69	85.18 ± 1.57	87.63 ± 0.86	89.21 ± 0.55

SC, outperformed other techniques by at least ca. 15%, considering the correct classification (see rank 1). Additionally, the algorithms enhanced by the appearance-based approaches, SCDT and SCH outperformed standard SC by 2% and 3%.

Tab. 1 shows the average CR values for 4 initial ranks along with their standard deviations σ. In the average case for a single image per gesture in G (Tab. 1(A)), it is the SCDT method which turned out to be the best among the investigated techniques, resulting in the highest CR values for each rank. Clearly, the choice of the image to G has a strong impact on the later classification score, and selecting more distinctive images significantly affects the final results (see σ in Tab 1(A)). Although the standard deviation of the rank 1 is the smallest for the shape context based algorithms (SCDT, SCH, SC), it is still noticeable and proves the methods to be quite sensitive to the choice of the gallery images.

Fig. 4 presents the CRCs for three ($g = 3$) most discriminative images, i.e. these that gave the best score for $g = 1$ for each method, per gesture in G. Providing multiple gallery entries improved the correct results in the initial ranks by at least 6% (SCDT, SCH and DT), up to 12% for the HD (see Fig. 3 and Fig. 4). On the one hand, the appearance-based methods (HoG and DT) performed poorly for both $g = 1$ and $g = 3$. On the other hand, combining them with the contour-based shape contexts technique resulted in the best responses. Therefore, these methods are complementary. Noteworthy, combining the SC with other contour-based methods did not improve the classification score. Tab. 1(B) and Tab. 1(C) presents the average (out of 20 experiments) CR and its corresponding σ for $g = 3$ and $g = 5$ for SC-based methods. The enhanced approaches outperformed the SC significantly. Moreover, adding new images to the gallery (i.e. increasing g) made the algorithms more independent from the choice of gallery images (the σ values dropped).

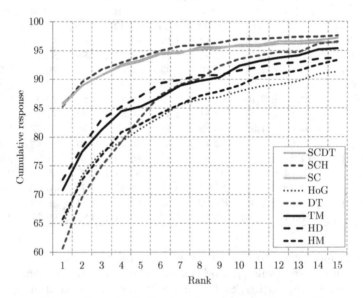

Fig. 4. Best CRCs for three images per gesture in G

Table 2. Average analysis time τ, speedup $\mathcal{S}$ and efficiency E of the PA for various numbers of threads T and gallery images g: (A) $g = 1$, (B) $g = 3$, (C) $g = 5$

Method ↓		$T=1$	$T=2$			$T=4$			$T=8$		
		τ	τ	$\mathcal{S}$	E	τ	$\mathcal{S}$	E	τ	$\mathcal{S}$	E
	SCDT	32	16	1.96	0.98	9	3.45	0.86	6	5.22	0.65
	SCH	39	19	2.01	1.01	11	3.56	0.89	7	5.29	0.66
(A)	SC	32	16	1.96	0.98	9	3.47	0.87	6	5.24	0.66
	HoG	6	3	1.76	0.88	2	3.08	0.77	1	4.68	0.59
	DT	4	2	1.70	0.85	1	3.02	0.76	1	4.61	0.58
	TM	22	12	1.80	0.9	7	3.28	0.82	4	5.11	0.64
	HD	83	41	2.05	1.03	21	3.89	0.97	14	6.00	0.75
	HM	6	3	1.88	0.94	2	3.39	0.85	1	4.84	0.61
	SCDT	38	19	1.99	1.00	11	3.43	0.86	7	5.46	0.68
	SCH	45	22	2.01	1.01	12	3.61	0.90	9	5.17	0.65
(B)	SC	37	19	1.97	0.99	11	3.41	0.85	7	5.36	0.67
	HoG	6	3	1.71	0.86	2	2.94	0.74	1	4.58	0.57
	DT	4	2	1.69	0.85	1	3.33	0.83	1	5.55	0.69
	TM	54	31	1.76	0.88	17	3.22	0.81	11	5.15	0.64
	HD	218	115	1.90	0.95	64	3.42	0.86	39	5.59	0.70
	HM	6	3	1.82	0.91	2	3.32	0.83	1	4.25	0.53
	SCDT	52	25	2.09	1.05	13	3.93	0.98	9	6.04	0.76
(C)	SCH	58	27	2.15	1.08	15	3.86	0.97	10	5.84	0.73
	SC	52	26	1.96	0.98	13	3.84	0.96	9	5.89	0.74

3.3 Speedup and Efficiency Analysis

In order to assess the performance of the PA, we measured the analysis time $\tau(1)$ of the sequential algorithm and of the PA, $\tau(T)$, for various numbers of threads T, and calculated the speedup $\mathcal{S} = \tau(1)/\tau(T)$, along with the efficiency $E = \mathcal{S}/T$ [16]. The analysis time τ consists of the feature extraction time τ_F and classification time τ_C, thus $\tau = \tau_F + \tau_C$.

We investigated the execution time of the sequential algorithm and the PA for both $g = 1$ and $g = 3$ for each technique. Also, we measured the analysis time for $g = 5$ for the best approaches, i.e. giving the best CR for a smaller number of images per gesture in G (SCDT, SCH and SC). In the latter case, we run the PA 20 times, with 5 random images representing a given gesture, using each mentioned approach. The average analysis time τ, along with the speedup $\mathcal{S}$ and efficiency E of the PA for various number of parallel threads T, are shown in Tab. 2. The HD is the most time-consuming classification approach. Although we significantly reduced the number of contour points considered in the SC, SCDT and SCH techniques, their sequential analysis time is still relatively high. The HM, HoG and DT turned out to be very fast for both $g = 1$ and $g = 3$. Providing new images to G increased the sequential analysis time of the TM and HD algorithms significantly.

The experiments performed for various number of threads T showed that the sequential algorithm can be speeded up almost linearly in case of more computationally intensive approaches. It is worth to mention that we experienced the superlinear speedup [3], i.e. $S > T$ and $E > 1.0$, while executing the PA with the HD, SCDT and SCH on two parallel threads ($T = 2$). Our preliminary studies indicated the local core caches as the source of superlinearity, however this issue requires further investigation. Applying the PA allows for increasing the G with a very fast analysis time and analyzing larger hand gesture databases. Thus, a more accurate classification can be performed in real-time (at more than 100 frames per second rate) using the available processor resources, e.g. the execution time of the SCDT ($g = 5$, $T = 8$) is more than 3.5 times lower than for a single thread and $g = 1$ with a very significant increase in the classification score.

4   Conclusions and Future Work

In this paper we discussed our parallel algorithm for fast hand shape classification. Introducing the parallelism allowed for decreasing the execution time of the sequential algorithm significantly. Moreover, we showed that the classification score can be boosted without increasing the execution time if the available processor resources are utilized. We experienced the superlinear speedup, which indicates high efficacy of the parallelization. Furthermore, we presented how the selection of gallery images influences the classification score.

Our ongoing research includes increasing the classification accuracy of the proposed parallel algorithm. We consider using radial Chebyshev moments here as they occurred to be very effective for image retrieval purposes [2]. Also, we plan to investigate the fusing schemes of contour-based and appearance-based techniques to enhance the final classification. Additionally, we aim at applying the proposed approach for searching the space of the parameters that control a 3D hand model [24].

References

1. Belongie, S., Malik, J., Puzicha, J.: Shape matching and object recognition using shape contexts. IEEE TPAMI 24(4), 509–522 (2002)
2. Celebi, M.E., Aslandogan, Y.: A comparative study of three moment-based shape descriptors. In: Proc. IEEE ITCC, vol. 1, pp. 788–793 (2005)
3. Chapman, B., Jost, G., van der Pas, R.: Using OpenMP: Portable Shared Memory Parallel Programming. The MIT Press (2007)
4. Czupryna, M., Kawulok, M.: Real-time vision pointer interface. In: 2012 Proceedings of ELMAR, pp. 49–52 (2012)
5. Erol, A., Bebis, G., Nicolescu, M., Boyle, R.D., Twombly, X.: Vision-based hand pose estimation: A review. Comp. Vis. and Im. Underst. 108(1-2), 52–73 (2007)
6. Freeman, W.T., Roth, M.: Orientation histograms for hand gesture recognition. Tech. rep., MERL (1994)
7. Grzejszczak, T., Nalepa, J., Kawulok, M.: Real-time wrist localization in hand silhouettes. In: Burduk, R., Jackowski, K., Kurzynski, M., Wozniak, M., Zolnierek, A. (eds.) CORES 2013. AISC, vol. 226, pp. 439–449. Springer, Heidelberg (2013)

8. Hu, M.K.: Visual pattern recognition by moment invariants. IRE Trans. on Inf. Theory 8(2), 179–187 (1962)
9. Huttenlocher, D., Klanderman, G., Rucklidge, W.: Comparing images using the hausdorff distance. IEEE TPAMI 15(9), 850–863 (1993)
10. Kawulok, M.: Fast propagation-based skin regions segmentation in color images. In: Proc. IEEE FG, pp. 1–7 (2013)
11. Kawulok, M., Kawulok, J., Nalepa, J.: Spatial-based skin detection using discriminative skin-presence features. Pattern Recognition Letters 41, 3–13 (2014), http://dx.doi.org/10.1016/j.patrec.2013.08.028
12. Kawulok, M., Kawulok, J., Nalepa, J., Papiez, M.: Skin detection using spatial analysis with adaptive seed. In: Proc. IEEE ICIP, pp. 3720–3724 (2013)
13. Kawulok, M., Nalepa, J., Kawulok, J.: Skin detection and segmentation in color images. In: Celebi, M.E., Smolka, B. (eds.) Advances in Low-Level Color Image Processing, Lecture Notes in Computational Vision and Biomechanics, vol. 11, pp. 329–366. Springer Netherlands (2014), http://dx.doi.org/10.1007/978-94-007-7584-8_11
14. Lin, C.C., Chang, C.T.: A fast shape context matching using indexing. In: Proc. IEEE ICGEC, pp. 17–20 (2011)
15. MacLean, J., Pantofaru, C., Wood, L., Herpers, R., Derpanis, K., Topalovic, D., Tsotsos, J.: Fast hand gesture recognition for real-time teleconferencing applications. In: Proc. IEEE ICCV Workshop on Recognition, Analysis, and Tracking of Faces and Gestures in Real-Time Systems, pp. 133–140 (2001)
16. Nalepa, J., Czech, Z.J.: A parallel heuristic algorithm to solve the vehicle routing problem with time windows. Studia Informatica 33(1), 91–106 (2012)
17. Nalepa, J., Grzejszczak, T., Kawulok, M.: Wrist localization in color images for hand gesture recognition. In: Gruca, A., Czachórski, T., Kozielski, S. (eds.) Man-Machine Interactions 3. AISC, vol. 242, pp. 79–86. Springer, Heidelberg (2014)
18. Nalepa, J., Kawulok, M.: Parallel hand shape classification. In: Proc. IEEE ISM, pp. 401–402 (2013)
19. Papiez, M., Kawulok, M.: Adaptive skin detection in colour images using error signal space. Studia Informatica 34(2A), 365–377 (2013)
20. Phillips, P., Wechsler, H., Huang, J., Rauss, P.: The FERET database and evaluation procedure for face recognition algorithms. Im. and Vis. Comp. J. 16(5), 295–306 (1998)
21. Shen, Y., Ong, S.K., Nee, A.Y.C.: Vision-based hand interaction in augmented reality environment. Int. J. Hum. Comput. Interaction 27(6), 523–544 (2011)
22. Thippur, A., Ek, C.H., Kjellstrom, H.: Inferring hand pose: A comparative study of visual shape features. In: Proc. IEEE FG, pp. 1–8 (2013)
23. Ul Haq, E., Pirzada, S.J.H., Baig, M.W., Shin, H.: New hand gesture recognition method for mouse operations. In: 2011 IEEE 54th International Midwest Symposium on Circuits and Systems (MWSCAS), pp. 1–4 (2011)
24. Šarić, M.: Libhand: A library for hand articulation, version 0.9 (2011), http://www.libhand.org/
25. Wachs, J., Stern, H., Edan, Y., Gillam, M., Feied, C., Smith, M., Handler, J.: A real-time hand gesture interface for medical visualization applications. In: Tiwari, A., Roy, R., Knowles, J., Avineri, E., Dahal, K. (eds.) App. of Soft Comp. AISC, vol. 36, pp. 153–162. Springer, Heidelberg (2006), http://dx.doi.org/10.1007/978-3-540-36266-1_15

Content-Based Image Indexing by Data Clustering and Inverse Document Frequency

Rafał Grycuk, Marcin Gabryel, Marcin Korytkowski, and Rafał Scherer

Institute of Computational Intelligence, Częstochowa University of Technology
Al. Armii Krajowej 36, 42-200 Częstochowa, Poland
{rafal.grycuk,marcin.gabryel,marcin.korytkowski,rafal.scherer}@iisi.pcz.pl
http://iisi.pcz.pl

Abstract. In this paper we present an algorithm for creating and searching large image databases. Effective browsing and searching such collections of images based on their content is one of the most important
challenges of computer science. In the presented algorithm, the process
of inserting data to the database consists of several stages. In the first
step interest points are generated from images by e.g. SIFT, SURF or
PCA SIFT algorithms. The resulting huge number of key points is then
reduced by data clustering, in our case by a novel, parameterless version
of the mean shift algorithm. The reduction is achieved by subsequent
operation on generated cluster centers. This algorithm has been adapted
specifically for the presented method. Cluster centers are treated as terms
and images as documents in the term frequency-inverse document frequency (TF-IDF) algorithm. TF-IDF algorithm allows to create an indexed image database and to fast retrieve desired images. The proposed
approach is validated by numerical experiments on images with different
content.

Keywords: CBIR, content based image retrieval, image database, keypoints, clustering, inverse document.

1 Introduction

Content image retrieval is one of the greatest challenges of modern computer
science. Along with the development of the Internet and the ability of capturing images by devices such as digital cameras and scanners, image databases
are growing very rapidly. Effective browsing and retrieving images by users is
required in many various fields of life e.g. medicine, architecture, forensic, publishing, fashion, archives and many others. In order to meet those expectations,
many general systems of retrieving have been already presented [1,10,12,21].

In the process of image recognition users search through databases which
consist of thousands, even millions of images. The aim can be the retrieval of a
similar image or images containing certain objects. Retrieving mechanisms use
image recognition methods. This is a sophisticated process which requires the
use of algorithms from many different areas such as computational intelligence

S. Kozielski et al. (Eds.): BDAS 2014, CCIS 424, pp. 374–383, 2014.
© Springer International Publishing Switzerland 2014

[4,9,17], mathematics [8] and image processing [7]. Those algorithms do not use raster graphics, on the contrary they find certain distinguishing features. In the literature it is possible to find many diverse methods of extracting those features. They extract for example primary features such as: colour, texture, shape and their position on the image. The features can be subjected to additional algorithms which are able to reduce their number. It is possible thanks to background elements exclusion, to eliminate the features absent in other images with the same object etc. These methods facilitate data storing in the database and accelerate future retrieving process.

The presented algorithm was based on the well known and commonly used in the literature algorithms: mean shift [3], TD-IDF [19] and SIFT [13]. The major task of the SIFT algorithm is extraction from an image appropriate local features around key points, independent from size and position of the objects. Those key points are saved as vectors with constant length. Most frequently it is a vector of 128 numbers. Despite this data reduction, the number of key points obtained from all images which are supposed to be saved in the database is still huge. Thus, we resort to another possibility to decrease number of the points, i.e. clustering. There are various commonly used clustering algorithms (k-means, SOM, k-nn). Nevertheless, they need initial number of groups to be determined. In the approach presented here, we use our novel version of the mean shift algorithm for which we do not need this type of knowledge. Certain number of key points will be assigned to each group. These points vectors values will be replaced with the number of the group they belong to. Thanks to this operation, considerable number of data saved in databases will be reduced. Only vectors of the centres of groups and numbers assigned to key points will be saved. It is definitely more advantageous than storing thousands or millions of key points in databases.

Database index as well as quick retrieval are created thanks to the TF-IDF algorithm. The algorithm is widely used as a method of assessment how relevant a given document is. The evaluation takes place for example in Internet search engines. In our case, terms will be treated as centers of groups and documents as images. In this way we obtain a mechanism for quick and efficient similar images retrieval.

In the literature we can find many similar solutions to our approach [11,15,16,20]. Each of them uses some combination of algorithms for image feature extraction, clustering and indexing. Our approach is distinguished by a new, modified method of clustering, which allows to be adjusted the kernel parameters of the mean shift algorithm to the data that are available in the database.

The paper consists of several sections. In further sections we will present available algorithms, modifications in the mean shift algorithm as well as the proposed algorithm. The last section presents the numerical simulations on an original software written in .NET.

2 Clustering Algorithm

The mean shift clustering algorithm [2,5,6] is a method that does not require any parameters such as cluster number or shape. The number of parameters of the algorithm is limited to the radius h, that determine the range of the clusters [3]. Mean shift determines the points in d-dimensional space as a probability density function, where the denser regions correspond to local maxima. For each data point in the feature space, one performs a gradient ascent procedure on the local estimated density until convergence. Points assigned to one cluster (stationary point) are considered to be a part of the cluster. Given n points $x_i \in R^d$, multivariate kernel density function $K(x)$ is expressed using the following equation

$$\widehat{f}(x) = \frac{1}{nh^d} \sum_{i=1}^{n} K\left(\frac{x - x_i}{h}\right), \tag{1}$$

where h defines the radius of the kernel function. Kernel function is defined as

$$K(x) = c_k k(\| x \|^2), \tag{2}$$

where c_k represents a normalization constant. With a density gradient estimator we can make the following calculations:

$$\nabla \widehat{f}(x) = \frac{2c_{k,d}}{nh^{d+2}} \underbrace{\left[\sum_{i=1}^{n} g\left(\| \frac{x - x_i}{h} \|^2\right)\right]}_{term1} \underbrace{\left[\frac{\sum_{i=1}^{n} x_i g\left(\| \frac{x-x_i}{h} \|^2\right)}{\sum_{i=1}^{n} g\left(\| \frac{x-x_i}{h} \|^2\right)} - x\right]}_{term2}, \tag{3}$$

where $g(x) = -k'(x)$ denotes the derivative a function of the selected kernel. The first $term1$ allows to determine the density, while the second $term2$ defines a mean shift vector $m(x)$. Where $m(x)$ is vector and t is algorithm step. Points toward the direction of maximum density and proportional to the density gradient can be determined at the point x, obtained with the kernel function K. The algorithm can be represented in the following steps:

1. Determine the mean shift vector, expressed by the formula $m(x_t^i)$,
2. Translate density estimation window: $x_{t+1}^i = x_t^i + m(x_t^i)$,
3. Repeat first and second step until: $\nabla f(x_i) = 0$.

3   TF-IDF Algorithm

TF-IDF algorithm (term frequency, inverse document frequency) [19] is an algorithm dedicated to fast document search. In the paper it is used to image key point indexing. TD-IDF algorithm determines the frequency of specific words in a given document and taking account of its occurrence in all documents. This value is calculated from the following formula

$$(tf - idf)_{i,j} = tf_{i,j} \times idf_i . \tag{4}$$

The left side of the equation contains the two components tf and idf. The first term is the frequency of the words, expressed by formula

$$tf_{i,j} = \frac{n_{i,j}}{\sum_k n_{k,j}}, \tag{5}$$

where $n_{i,j}$ is the number of occurrences of the term t_i in the document d_j. The denominator contains the sum of occurrences all the words in the selected document d_j. The second component from (4) is idf_i and it is expressed by the following formula:

$$idf_i = \log \frac{|D|}{|\{d : t_i \in d\}|}, \tag{6}$$

where $|D|$ is the total number of documents and $\{d : t_i \in d\}$ the number of documents that contain at least one instance of document [18,19].

4   SIFT Algorithm

SIFT (Scale-invariant feature transform) is an algorithm used to detect and describe local features of an image. It was presented for the first time in [13] and is now patented by the University of British Columbia. For each key point, which describes the local image feature, we generate feature vector, that can be used for further processing. SIFT contains four main steps [14]:

1. Extraction of potential key points by scanning the entire image,
2. Selection of stable key points (resistant to change of scale and rotation),
3. Finding key point orientation resistant to the image transformation,
4. Generating vectors describing key point.

SIFT key point consists of two vectors. First one contains: point position (x,y), scale (Detected scale), response (Response of the detected feature, strength), orientation (Orientation measured anti-clockwise from +ve x-axis), laplacian (Sign of laplacian for fast matching purposes). The second one contains descriptor of 128 length. In order to generate key points SIFT require one input parameter minHessian. In our experiments we set this variable in 400. This value was obtained empirically.

5   Proposed Method for Fast Image Retrieval

Proposed algorithm in Fig.1 consists of several stages. The first step is to create feature matrix ($n \times 128$), where 128 is the length of the vector generated by the SIFT algorithm and N is the total number of key points from all the images. Then we execute the mean shift clustering algorithm. This algorithm has been modified to our needs by adding a method to automatically match the cluster window. As a result of the algorithm, we obtain a set of features, assigned to specific cluster. The next step is to create a dictionary structure. The dictionary contains: key - cluster number obtained from the mean shift algorithm, value - sorted array of image numbers.

Next, for each dictionary value we calculate $tf - idf$. Values directly show the frequency of each feature in the cluster. Search stage compares the key points of input image with indexed key points stored in the database. Input data of the algorithm is a key points matrix of the new image. Next, each key point is assigned to a cluster number. Subsequently we calculate $tf - idf$ value. The last step is to compare the $tf - idf$ values with all clusters found on the query image. Another important step in the proposed algorithm is the ability to extend the created index by new images. This method is similar to the previous steps. First, we generate key points from a new image. Next, the algorithm generates new clusters (if needed) for the specified descriptors. Newly acquired clusters are added to the dictionary. The algorithm can be described in the form of steps. The first six steps create an database index that allows to perform content based search.

INPUT: Binary images
OUTPUT: Index for content based image search
foreach $image \in images$ **do**
    Generate Descriptors using SURF,SIFT or PCA SIFT;
    Load each descriptor into $featureArray$;
end
Estimate and set algorithm window (h parameter, see section 5.1)
Generate clusters using mean shift;
foreach $cluster \in clusters$ **do**
    Add Word($wordID, documentID$) to dictionary, where $wordID$ is cluster and $documentID$ is image ID;
end
foreach $cluster \in clusters$ **do**
    Calculate $tfidf$;
end

Algorithm 1. The algorithm steps

In result we obtain database index ready for searching (see Algorithm 2).

INPUT: Query image (model image)
OUTPUT: Retrieved, similar images
Generate descriptors for Query image;
foreach $descriptor \in QueryImageDescriptor$ **do**
    if $descriptor \in clusterCenters$ **then**
        Add cluster number into variable $similarGroups$;
    end
end
Remove distinctive values of $similarGroups$; **foreach** $cluster \in similarGroups$ **do**
    Get dictionary[$cluster$] value passing cluster number; Add these value into $similarImageIDs$;
end
Remove distinctive values of $similarImagesIDs$; Show similar images;

Algorithm 2. The search stage steps

Algorithm of adding new images to the existing knowledge base is as follows:

INPUT: New image

OUTPUT: Index for content based image search with new added image

Generate descriptors for new image;

Execute mean shift on new image descriptors;

foreach $newCluster \in newClusters$ **do**

    Add Word $(newCluster, imageID)$;

    Calculate $TF - IDF$ value;

end

Algorithm 3. Adding new image to existing index

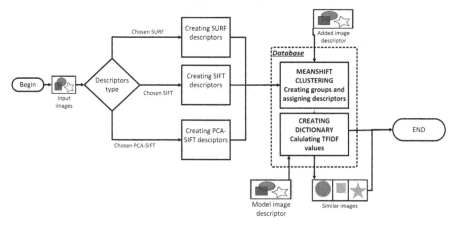

Fig. 1. Block diagram of the proposed algorithm

5.1 Estimation of h Parameter

In this section we present a novel, parameterless version of the mean shift algorithm. Mean shift is an algorithm that requires one input parameter h. Since this parameter depends on the number of generated groups. The small size of this parameter will generate a large number of groups with low coverage. On the other hand, a large range of the parameter will produce many groups which too much generalized data. The choice of this parameter is therefore very significant, because it determines the size of the database. We have developed a novel modification of mean shift consisting in estimation of the h parameter. The h parameter estimation algorithm is as follows:

1. Set the algorithm steps count K.
2. Calculate the distance $d_{i,j}$ between all points x_i and x_j where $i = 1, .., N, j = 1, .., N, i \neq j, N$ - the number of all the key points.
3. Determine the value of m parameter, that allow to increment h parameter by value m.

$$m = \frac{\max\limits_{\substack{i=1,...,N \\ j=1,...,N \\ i \neq j}} d_{i,j} - \min\limits_{\substack{i=1,...,N \\ j=1,...,N \\ i \neq j}} d_{i,j}}{K} \tag{7}$$

4. Algorithm step $n = 1$
5. Calculate h, depending on the step of the algorithm n

$$h = \min_{\substack{i=1,\ldots,N \\ j=1,\ldots,N \\ i \neq j}} d_{i,j} + m \cdot n \qquad (8)$$

6. Calculation of the derivative of $g_{ij}(x)$ the kernel $k(x)$ for the cartesian product of all the key points i, j, and $i \neq j$. This stage is designed to accelerate the calculations performed during the execution of algorithm.

$$g_{ij} = g\left(\| \frac{x_j - x_i}{h} \|^2\right) \qquad (9)$$

7. Density gradient (3) is transformed to the form:

$$\nabla \widehat{f}(x_j) = \frac{2c_{k,d}}{nh^{d+2}} \left[\sum_{i=1}^{n} g_{ij}(x_j)\right] \left[\frac{\sum_{i=1}^{n} x_i g_{ij}(x_j)}{\sum_{i=1}^{n} g(x_j)} - x_j\right] \qquad (10)$$

for $j = 1, .., N$.
8. Execute the mean shift algorithm. Save the clusters and cluster number generated by the algorithm.
9. Increase algorithm step $n = n + 1$,
10. While $h \leq max\, d_{ij}$, repeat steps 5-9,
11. The estimated parameter h is calculated by the formula (8) where n for such n that $G_n = \underset{i=1,\ldots,K}{\mathrm{med}}\, G_i$.

6   Experimental Results

Experiments were carried out in .NET environment on our own software written C#. Research includes experiments on various objects with background. Test images were taken from [22]. We chose images with different types of objects such as: dinosaurs (100 images), drinks (100 images), landscapes (150 images), castles (230 images), meals (100 images), cars (100 images), cards (100 images), doors (100 images), dogs (100 images), flowers (100 images), mountains (100 images). All 1280 pictures generated a total of 99353 key points that were reduced to 18611 clusters. In experiments we used 90% of each class for index creating and 10% as query images. Our simulations prove effectiveness of the algorithm. Percentage effectiveness is presented in Tab.1. Fig. 2 presents exemplary results of the experiments. In top, left corner we presented a query image (red frame) (Q). The rest of images are retrieved by the the proposed algorithm. In tab. 1 we present the percentage effectiveness of our algorithm for estimated parameter $h = 288$ and fixed parameter h from 100 to 400. Fig. 3 shows a chart of average percentage of correctly found images from all categories versus the h parameter value. As can be seen, the parameter $h = 288$, obtained as a result of our algorithm, provides search results that are near to the optimum found for $h = 250$.

Table 1. Correctly found images

h parameter	dinosaurs	drinks	landscapes	castles	meals	cars	cards	doors	dogs	flowers	mountains
estimated $h = 288$	75%	92%	63%	61%	78%	77%	68%	95%	78%	97%	67%
fixed $h = 100$	7%	0%	0%	2%	1%	0%	1%	0%	1%	0%	0%
fixed $h = 150$	9%	1%	0%	3%	1%	2%	5%	10%	3%	4%	0%
fixed $h = 200$	55%	48%	32%	21%	46%	60%	37%	60%	56%	51%	43%
fixed $h = 250$	78%	90%	71%	59%	79%	80%	66%	93%	83%	95%	68%
fixed $h = 300$	68%	90%	59%	56%	70%	72%	63%	88%	75%	92%	66%
fixed $h = 350$	45%	41%	37%	17%	20%	20%	8%	33%	50%	46%	29%
fixed $h = 400$	37%	40%	35%	11%	18%	14%	5%	30%	47%	32%	24%

The differences in efficiency retrieval on image content is caused by image background. Some key points are located not only on objects, but also in background. Thus, the descriptors of these key points may be assigned by wrong cluster. If we have a similar background in an image class (e.g. dinosaurs) then most of images are correctly found. Otherwise the algorithm can retrieve images from a wrong class.

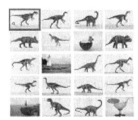

Fig. 2. Twenty example images from the experiment. Due to lack of space we had to narrow the example to 19 input (learning) images for each class and one query image. Top left image is a query image and the rest is 19 images retrieved by the algorithm. We can observe that 15 of them are retrieved correctly.

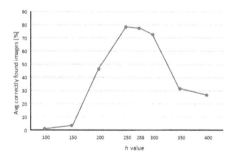

Fig. 3. Average percentage of correctly found images from all categories versus h parameter

7 Final Remarks

The presented algorithm is a contribution to fast content-based image search. Simulations verified the correctness of the algorithm. During initial simulations we discovered a method disadvantage, which is associated with the mean shift algorithm. Mean shift requires h parameter which is the key points search radius (window). Thus, the algorithm requires an experimental value of this parameter which resulted in the lack of automation during the indexation phase. Presented, improved version of the mean shift algorithm allows for automatic operation of the application. Our paper is a part of a larger system that allows to search and identify specific classes of objects that can be located in different places and in different number of images. In the future we will try to develop a method that removes background from input images. Thus, it will allow to resolve the problem of retrieving images from different classes.

Acknowledgments. The work presented in this paper was supported by a grant from Switzerland through the Swiss Contribution to the enlarged European Union.

References

1. Chang, Y., Wang, Y., Chen, C., Ricanek, K.: Improved image-based automatic gender classification by feature selection. Journal of Artificial Intelligence and Soft Computing Research, 241 (2011)
2. Cheng, Y.: Mean shift, mode seeking, and clustering. IEEE Transactions on Pattern Analysis and Machine Intelligence 17(8), 790–799 (1995)
3. Comaniciu, D., Meer, P.: Mean shift: A robust approach toward feature space analysis. IEEE Transactions on Pattern Analysis and Machine Intelligence 24(5), 603–619 (2002)
4. Cpałka, K.: On evolutionary designing and learning of flexible neuro-fuzzy structures for nonlinear classification. Nonlinear Analysis: Theory, Methods & Applications 71(12), e1659–e1672 (2009)
5. Derpanis, K.G.: Mean shift clustering. Lecture Notes, http://www.cse.yorku.ca/kosta/CompVis_Notes/mean_shift.pdf (2005)
6. Evans, C.: Notes on the opensurf library. University of Bristol, Tech. Rep. CSTR-09-001 (January 2009)
7. Gabryel, M., Korytkowski, M., Scherer, R., Rutkowski, L.: Object detection by simple fuzzy classifiers generated by boosting. In: Rutkowski, L., Korytkowski, M., Scherer, R., Tadeusiewicz, R., Zadeh, L.A., Zurada, J.M. (eds.) ICAISC 2013, Part I. LNCS, vol. 7894, pp. 540–547. Springer, Heidelberg (2013)
8. Gabryel, M., Nowicki, R.K., Woźniak, M., Kempa, W.M.: Genetic cost optimization of the $GI/M/1/N$ finite-buffer queue with a single vacation policy. In: Rutkowski, L., Korytkowski, M., Scherer, R., Tadeusiewicz, R., Zadeh, L.A., Zurada, J.M. (eds.) ICAISC 2013, Part II. LNCS, vol. 7895, pp. 12–23. Springer, Heidelberg (2013)
9. Gabryel, M., Woźniak, M., Nowicki, R.K.: Creating learning sets for control systems using an evolutionary method. In: Rutkowski, L., Korytkowski, M., Scherer, R., Tadeusiewicz, R., Zadeh, L.A., Zurada, J.M. (eds.) SIDE 2012 and EC 2012. LNCS, vol. 7269, pp. 206–213. Springer, Heidelberg (2012)
10. Górecki, P., Sopyła, K., Drozda, P.: Ranking by K-means voting algorithm for similar image retrieval. In: Rutkowski, L., Korytkowski, M., Scherer, R., Tadeusiewicz, R., Zadeh, L.A., Zurada, J.M. (eds.) ICAISC 2012, Part I. LNCS, vol. 7267, pp. 509–517. Springer, Heidelberg (2012)

11. Hare, J.S., Samangooei, S., Lewis, P.H.: Efficient clustering and quantisation of sift features: Exploiting characteristics of the sift descriptor and interest region detectors under image inversion. In: Proceedings of the 1st ACM International Conference on Multimedia Retrieval, p. 2. ACM (2011)
12. Lew, M.S., Sebe, N., Djeraba, C., Jain, R.: Content-based multimedia information retrieval: State of the art and challenges. ACM Transactions on Multimedia Computing, Communications, and Applications (TOMCCAP) 2(1), 1–19 (2006)
13. Lowe, D.G.: Object recognition from local scale-invariant features. In: The Proceedings of the Seventh IEEE International Conference on Computer Vision, vol. 2, pp. 1150–1157. IEEE (1999)
14. Lowe, D.G.: Distinctive image features from scale-invariant keypoints. International Journal of Computer Vision 60(2), 91–110 (2004)
15. Nister, D., Stewenius, H.: Scalable recognition with a vocabulary tree. In: 2006 IEEE Computer Society Conference on Computer Vision and Pattern Recognition, vol. 2, pp. 2161–2168. IEEE (2006)
16. O'Hara, S., Draper, B.A.: Introduction to the bag of features paradigm for image classification and retrieval. arXiv preprint arXiv:1101.3354 (2011)
17. Przybył, A., Cpałka, K.: A new method to construct of interpretable models of dynamic systems. In: Rutkowski, L., Korytkowski, M., Scherer, R., Tadeusiewicz, R., Zadeh, L.A., Zurada, J.M. (eds.) ICAISC 2012, Part II. LNCS, vol. 7268, pp. 697–705. Springer, Heidelberg (2012)
18. Ramos, J.: Using tf-idf to determine word relevance in document queries. In: Proceedings of the First Instructional Conference on Machine Learning (2003)
19. Salton, G., Buckley, C.: Term-weighting approaches in automatic text retrieval. Information Processing & Management 24(5), 513–523 (1988)
20. Sivic, J., Zisserman, A.: Video google: A text retrieval approach to object matching in videos. In: Proceedings of the Ninth IEEE International Conference on Computer Vision, pp. 1470–1477. IEEE (2003)
21. Veltkamp, R.C., Tanase, M.: Content-based image retrieval systems: A survey. Rapport no UU-CS-2000-34 (2000)
22. Wang, J.Z., Li, J., Wiederhold, G.: Simplicity: Semantics-sensitive integrated matching for picture libraries. IEEE Transactions on Pattern Analysis and Machine Intelligence 23(9), 947–963 (2001)

On Facial Expressions and Emotions RGB-D Database

Mariusz Szwoch

Gdańsk University of Technology,
Department of Intelligent Interactive Systems, Poland
szwoch@eti.pg.gda.pl

Abstract. The goal of this paper is to present the idea of creating reference database of RGB-D video recordings for recognition of facial expressions and emotions. Two different formats of the recordings used for creation of two versions of the database are described and compared using different criteria. Examples of first applications using databases are also presented to evaluate their usefulness[1].

Keywords: facial expressions recognition, affective computing, multimodal databases, Microsoft Kinect, RGB-D data.

1 Introduction

Human emotions are important element of non verbal human communication, often changing our behavior regardless of incoming verbal communicates, or even common sense. In contrary, the human-computer interaction is based on some kind of formal protocol, probable ergonomic but not allowing for any derogations or alterations, such causing the effect of contacting with "soulless machine". Human emotions can have great impact on the way of using software, human learning or training results, and also overall experience received from using computers. Therefore, robust and accurate recognition of human emotional states can play an important role in many software systems in different application fields.

Affective computing is one of the emerging research areas on human behavior that develop emotion recognition, interpretation, and processing methods to create *affective* and *affect-aware* software to better adapt its behavior to user needs. Such applications can have a significant impact in many fields such as healthcare, education, entertainment, software engineering, e-learning, etc. [16,12,21]. Affect recognition methods can use different information channels, such as video [8], audio [22], standard input devices [19,10], physiological signals [2,18], depth information [3], and others. As single channel approach can be limited by different factors, combination of multiple types of inputs from different modalities, or

---

[1] The research leading to these results has received funding from the Polish-Norwegian Research Programme operated by the National Centre for Research and Development under the Norwegian Financial Mechanism 2009-2014 in the frame of Project Contract No Pol-Nor/210629/51/2013.

S. Kozielski et al. (Eds.): BDAS 2014, CCIS 424, pp. 384–394, 2014.

different features over the same modality, can significantly improve the system's classification abilities [16,8]. This fact is quite obvious as people naturally communicate in a multimodal way by combining language, tone, facial expression, gesture, head and body movements, and posture.

The most popular source of non-invasive and non-intrusive information is video camera as it allows for recognition of facial expressions, gestures, and body movements. Some algorithms concentrate only on facial expression recognition (FER) as they are typically used by human to analyze the affect of other people [4]. Unfortunately, algorithms using video camera are very sensitive to face illumination conditions, causing great problems with a dark or unevenly illuminated scene. One of possible solutions is to use additional channel with scene depth information. As depth sensors use additional light source (e.g. infrared) the information is generally insensitive to different ambient light conditions. Rapid development and increasing availability of relatively cheap RGB-D consumer sensors allow for creation of on-line systems for recognition of facial expressions and, going father, human emotions and moods.

As most reported approaches use supervised training to create an emotion classifier, using a proper training and testing datasets is of great importance. Unfortunately, many researchers use different and relatively small data sets, what limits estimation of their generalizability and real recognition effectiveness, and makes difficult to compare different approaches. Many researchers use also very narrow set of recognized classes of facial expressions or emotions, data can be acquired in optimized conditions, and experiment participants can be very homogeneous in their appearance. All of these factors do not allow using such approaches in different application fields.

Although there are many available databases with face images for face recognition [7] only few databases have been collected in order to support recognition of facial expressions and emotions. Most databases offer only 2D images or video recordings [13,20], but some databases contain 3D images acquired by the means of stereography or specialized 3D scanners [15,1,5]. 3D face information can significantly improve recognition efficiency especially in difficult lighting conditions.

In this paper, two approaches to represent RGB-D video streams of facial expressions and emotions from Microsoft Kinect sensor are considered. The first one is a FEEDB ver.1 database [17] based on Microsoft proprietary XED format. This database consists of 1650 recordings of 50 persons posing for 33 different facial expressions and emotions. The second version of FEEDB consists of 1550 recordings of 50 persons recorded in two separate video streams, separately for RGB and depth channels. Both versions of FEEDB differ in their usability and handling aspects. Also initial results of using FEEDB for face recognition and emotion recognition are presented.

2   Background

Depth sensors such as Microsoft Kinect have become very popular nowadays allowing for development of new algorithms for recognition of human pose, gestures, face, and facial expressions. Enriching traditional algorithms based on

RGB camera with additional information channel can lead to better localization of parts of human's body as well as improve recognition efficiency in difficult illumination conditions.

Generally, recordings of human facial expressions and emotions may be divided into two categories: containing posed and spontaneous session. Posing allows for recording extreme expressions and emotions hardly observed in real computer using situations like human cry, envy, and so on. Drawbacks of posed recording are human problems to act such emotions and their artificial character. On the other hand recording of spontaneous emotions is much harder and requires a lot of time to catch the desired emotion or expression. Moreover, it is almost impossible to record extreme emotions mentioned earlier. That is why databases of facial expressions and emotions should contain images or recordings of both types.

There are a lot of face databases publicly available for researchers to be able to directly compare their results [7]. Unfortunately, only several of them contain data that can be used for recognition of facial expressions and emotions. And only few of them contain additional 3D information.

One of the first attempt to develop a comprehensive database for facial expression analysis was Cohn-Kanade AU-Coded Facial Expression Database (CK), in which each sequence was labeled by desired emotion to be expressed as well as by observed facial movements using Facial Action Coding System (FACS) [6]. Unfortunately, sequences were not verified against the real facial expressions they contained and the emotion label referred to what expression was requested rather than what could actually had been performed. The improved version of this database, CK+, contains additional 593 sequences with more frames per sequence [13]. CK+ contains both posed as well as spontaneous facial expressions. In this version validated emotion labels are provided together with FACS coding. Additional feature are recognition results for facial feature tracking, action units and emotions.

The Japanese Female Facial Expression Database (JAFFE) contains 213 images of 7 facial expressions posed by 10 Japanese female models [14]. Face Video Database (FVDB) of the Max Planck Institute for Biological Cybernetics contains short video sequences of facial Action Units recorded simultaneously from six different viewpoints. GavabDB contains 549 three-dimensional images of facial surfaces of 61 persons posing four facial expressions with neutral face recorder from different points of view [15]. The Bosphorus Database contains of 4666 3D faces from 105 persons acquired with professional 3D digitizer [1]. High resolution of about 0.3 mm allowed preparing very detailed RGB-D images of different posed facial expressions and emotions. All images are FACS indexed including intensity and asymmetry codes for each AU that makes this database very useful for all types of 3D face processing tasks.

Natural Visible and Infrared Facial Expression database (USTC-NVIE) contains both spontaneous and posed expressions of more than 200 subjects, recorded simultaneously by a visible and an infrared thermal camera [20]. The University of Milano Bicocca 3D face database (UMB-DB) is a collection of 1473 3D and

2D images of 143 people posing four basic facial expressions and different occlusion gestures [5]. All acquisitions were made with a high resolution laser depth scanner.

And finally, the EURECOM Kinect Face Dataset (EKFD) consists of multimodal 2D and 3D facial images of 52 people posing for 9 different facial expressions and occlusion situation [9]. Images are cropped and several characteristic facial points are marked manually. Though the database can be a good starting point for many face recognition approaches a small number of expressions limits its usage in facial expression and emotion recognition areas.

Some aspects of the described face databases are compared against FEEDB databases in Tab. 1.

Table 1. Comparison of sample databases containing face images or videos

Database	Number of Items	Content Type	Number of Persons	Posed / Spontaneous	FACS
CK	486	Frame sequences	97	Y/Y	Y
CK+	593	Frame sequences	123	Y/Y	Y
JAFFE	213	Images	10	Y/N	N
GavabDB	549	3D images	61	Y/N	N
Bosphorus	4666	3D and 2D images	105	Y/N	Y
USTC-NVIE	6300	RGB and infrared images	215	Y/N	N
UMB-DB	1473	3D and 2D images	143	Y/N	N
EKFD	468	RGB-D images	52	Y/N	N
FEEDB 1	1650	RGB-D video (XED)	50	Y/N	Y
FEEDB 2	1550	RGB-D video and images	50	Y/Y	Y

3 Databases of Facial Expressions and Emotions

When creating or evaluating a database of facial expressions and emotions several important questions must be answered taking into account different objectives that should be fulfilled, at least in some extent, to be useful for classifier's training and validation. These aspects cover such topics as [17]:

1. *Information channels* – though, the optical channel is the most obvious choice for acquiring information about the human face and its features, additional information channels can also be considered [20], including voice, scene depth, physiological signals and others. Numerous research projects proved that multimodal approach gives in general better results, in terms of recognition efficiency [16].

2. *Set of recorded expressions* – Choosing a set of facial expressions and emotions to be represented in a database seems crucial from its usability point of view. Unfortunately, there are no commonly accepted set of expressions that could be used in all applications. In fact, the choice of interesting set of expressions depends on a field of application and on assumed model of emotions.

3. *Posed vs spontaneous expressions* – Most collections of facial expressions and emotion recordings contain posed expressions, which can provide valuable information allowing to collect expressions that are hardly observable in real life.

4. *Static images and recordings* – As facial expressions and emotions are very dynamic processes, storing series of frames brings additional temporal context allowing to reason about more complex emotions consisting of several action units. The drawbacks of storing sequences of frames are their size, synchronization problems with other information channels and some labeling problems.

5. *Description level* – In order to be useful, each database of emotions and facial expressions should be properly described or labeled. The description should contain both classification information as well as additional metadata. For indexing facial expressions, the Facial Action Coding System (FACS) was proposed, which is a human-observer based system using so called action units (AU) to describe even subtle changes in facial features [6]. Although many emotional models have been proposed, in most cases emotions are labeled according to accepted set of expressions. For posed emotions, labels can be differentiated into intended and expressed ones, which not always are the same [13]. Apart from AU indexing and labeling, images and recordings should also include additional meta-data such as some information on each person, acquisition parameters, and scene organization. Also additional information about the data content can be included such as location of characteristic facial points (landmarks).

6. *Data representativeness* – Representativeness is the key attribute of a training set in each supervised training process. In general, a database of facial expressions and emotions should contain a large number of recordings of people of different age, gender, race, and occupation to cover a great variety in face shape, skin texture and color, facial hair, possible glasses and jewelry as well as variety in way of expressing different emotions. Though it is possible to narrow any aspect of participants due to assumed field of application, such as children emotion, received results could not be generalized on other cases, such limiting usefulness of a database.

7. *Scene composition* – Face location and orientation in an image play an important role in face recognition as well as in recognition of facial expressions and emotions. In expressions recognition field it is often assumed that face stays oriented towards the camera. However, it may not be true especially when expressing natural emotions. In that case human pose often changes dynamically what should be reflected in a database. Other important factors, that can influence face location or expressions recognition, are scene illumination, background and the presence of other people.

8. *Image resolution* – The effective resolution of the face in an image is an important factor of the image quality and may influence on recognition efficiency. For three-dimensional representation depth resolution (Z axis) should correspond to scene resolution in an image plane (X and Y axis). It means that a face should be located at certain "optimal" distance from the camera.

All above aspects are discussed in more detail in [17].

In the next subsections two versions of database of facial expressions and emotions (FEEDB) are described and finally compared. The goal of their creation was to deliver a comprehensive and reliable dataset of recordings that would allow developing different classifiers of facial expressions and emotions in a human-computer interaction. These databases can be used both for training and testing different algorithm of face analysis.

Both versions of FEEDB contain color and scene depth recordings of adult IT students of both genders which seems to be quite representative in some applications such as video games. The recording was done in standard scenery of an IT laboratory with mixed natural and fluorescent illumination (Fig. 1). No special control on illumination or other people in the background were done. All the images were taken against an artificially lighted nonhomogeneous background (walls and ceiling). Participants sat in an upright, frontal position at different places allowing for differentiated background and lighting conditions (Fig. 1). Participants were instructed to begin and end each expression with a neutral pose, and to estimate a difficulty in expressing of particular emotion.

All recordings for the FEEDB database were made using Microsoft Kinect sensor. Color video channel consists of frames 640 × 480 pixels in size with 24-bit color values sampled at 30 fps. Also depth channel frames consists of subsequent frames of size 640 × 480 depth pixels sampled at the same frequency with a distance (depth) value represented by 13 higher bits of 16-bit word. Kinect sensor was placed just below the monitor, standing on the platform of about 20 cm above the desk level. All participants were asked to keep the optimal distance to the sensor that is about 60-80 cm.

3.1   Facial Expressions and Emotions Data Base Version 1

The first version of Facial Expressions and Emotions Data Base (FEEDB) consists of 1650 short recordings of 50 persons trying to express the following 33 posed facial expressions and emotions: *neutral, surprise* (neutral and positive), *amazement* (positive and negative), *smile* (natural, fake and wry), *laughter, cheerful whistling, pleasure, excitement, speaking, singing, sadness, weeping, pain, fear, affright, sleepy, yawning, doubt, boredom, irony, frustration, anger, concentration, determination, attention, effort, scorn, rejection,* and *threaten* [17]. Sample expressions from FEEDB ver.1 are presented in Fig.1.

All recordings in FEEDB ver.1 are stored in XED file format which combines both color and depth streams received from a Kinect sensor. The advantages of this format are: storing both streams in one file and capability of using XED files with the same interface as a Kinect sensor. Unfortunately there are also several drawbacks. Firstly, XED is Microsoft proprietary file format with no possibility of accessing file content without a Microsoft Kinect Studio (MKS) application. Secondary, using XED demands Kinect sensor attached to the system which is sometimes inconvenient. The last problems concern the MKS software which allows doing some basic editing operations but final file is lossy compressed with a compression ratio about 3:1 which makes such recordings sometimes useless

a)   b)

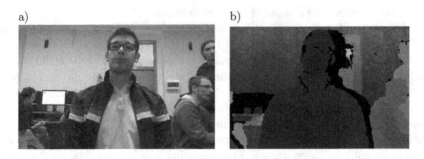

Fig. 1. Sample frames from FEEDB: a) video channel, b) depth channel

for further face processing. Unfortunately, there is currently no option available to switch off or reduce this compression.

3.2 Facial Expressions and Emotions Data Base Version 2

One of the goals of creation of the first version of FEEDB was to choose the final set of facial expressions and emotions which are relatively easy to express by most of users which might suggest their frequent and spontaneous use in some real life scenarios especially in human-computer interaction. The reported difficulties in expression of certain emotions have caused the change of the expressions set in the second version of FEEDB. So the new version contains the following emotions and facial expressions: *neutral, surprise, smile, laughter, anger, scorn, disgust, sadness, fear, effort, bored, shouting, whistling, chewing, speech, left eye blink, right eye blink, squinting eyes, wrinkling of the nose, determination, biting the upper lip, biting the lower lip, tumefaction of the mouth, raise the left corner of the mouth, raise the right corner of the mouth, biting the right inner cheek, biting the right inner cheek, wrinkle face, download eyebrows, raising eyebrows, assent, denying,* and *doubt.* Those emotions have been recorded for 50 persons resulting in 1550 recordings.

Additionally, the most interesting set of ten emotions had been selected to be expressed spontaneously. Special video materials had been prepared to provoke tested persons to naturally express the following expressions: *neutral, surprise, joy, anger, scorn, disgust, sadness, fear, concentration,* and *excitement.* Those expressions have been recorded for 25 participants of the experiment, giving 275 recordings.

All recordings have been carefully indexed with metadata about each participant (age, gender), the name of the emotion and its kind (spontaneous or posed), a list of action units (AU) found in each recording, three temporal indices for beginning, maximum and ending of the emotion, and a list of bitmap coordinates of 22 characteristic facial points for frames in three mentioned points of time (Fig.2a).

All recordings in FEEDB ver.2 are stored in AVI files, separately for color and depth channels. Specially developed recording software was used to manage

a)   b)

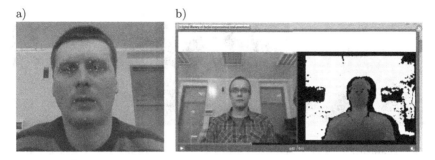

Fig. 2. Sample frames from FEEDB ver.2: a) fragment with marked characteristic facial points, b) in synchronous web player

the recording session and to save both data streams. The advantage of using AVI files is possibility of using them with no Kinect sensor attached to the system. Files can be batch processed, and read by any software. As example a special visualization application in Adobe Flash technology was developed for synchronous playback of both streams in a Web environment (Fig.2b).

Unfortunately, there are also some drawbacks of not using AVI files. As AVI is only a video container, different codecs can be used causing some compatibility problems. Additionally, as depth information is stored as 13-bit words there is a danger that the higher five bits would be compressed as belonging to separate channel of video stream. This can demand additional format conversion for depth channel, e.g. using look-up-tables (LUT) and careful choice of used codecs.

4   Experiments and Preliminary Results

FEEDB is still under development, but the set of over 3000 indexed and differentiated recordings provides sufficiently large and representative test-bed for development facial analysis algorithms that use additional information about the scene depth.

During FEEDB creation several interesting observations have been made on expressing different emotions by different people. First of all, many people have big problems in expressing certain expressions in a controlled way. It does not mean that they are totally unable to express them but they have serious limitations in doing it on demand. Additional polling of experiment attendees allowed indicating emotions and facial expressions easy and hard to express [17]. Analysis of these results showed that easy expressions contain mainly easy facial activities concerning more with facial expressions than emotions, such as speaking, singing or yawning. Expressions of medium difficulty level contain mainly emotions of a lower to an average arousal level such as surprise, laugher, boredom or moderate anger. Hard to express emotions cover mainly negative emotions (scorn, fear, frustration, threaten) or high arousal ones such as excitement, or weeping [17]. Interesting are also statements of experiment participants who noticed that

a)   b)   c)

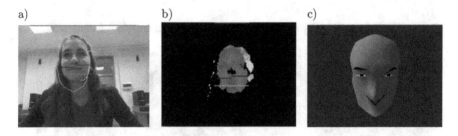

Fig. 3. Visualization of chosen stages of emotion recognition process: a) RGB image, b) depth image with supporting lines, c) 3D avatar reflecting recognized emotion (joy)

some emotions are so rare in everyday life that they were not aware how they expressed them, unless they had tried them earlier before a mirror like actors or actresses. This confirms a thesis that some expressions should be observed rather as spontaneous emotions than evoked ones.

The second observation is that there are serious individual differences in the way of expressing the same emotion by different people. This is true both for posed as well as natural emotions. It is evidently seen for spontaneous emotions recorded for FEEDB ver.2 where the same stimulation caused different user reactions. It indicates the need of creation of individual psychological profiles for each user.

Those observations can lead to some interesting conclusions. Firstly, individual differences in emotion expression can have psychological background. Secondly, emotion classification cannot average training data, but should use multimodal approach trying to cluster similar expressions.

The resolution of images from Kinect sensor seems to be sufficient for face analysis if user is within the close distance of about 60-80 cm from the sensor. In this range the image resolution is about 20-25 DPI in each direction [17]. Also the measurement error is only at about 1 mm.

FEEDB database has been used in two initial experiments. In the first experiment the recordings were used for face recognition. Three different approaches were used. Two approaches were based on Eigenfaces algorithm [11] from OpenCV library one for color, and the second for depth images. The third approach used also geometrical features for color images. The average recognition efficiency for each method was nearly the same at the level about 90% for ten tested persons. Additional tests with varying light conditions decreased the average recognition efficiency to 70% for side illumination, and to 50% with bottom or very strong illumination. In difficult lighting conditions the algorithm based on depth information was more efficient in most cases. It was especially evident in low light conditions, which indicate usefulness of depth sensors.

In the second experiment the database was used by application developed to recognize ten emotions basing only on depth channel (Fig.3). The set of emotions covers *neutral, joy, surprise neutral* and *positive, euphoria, fear, fright, anger,* and *scorn.* The application locates the face and its characteristic points, classifying then emotions on the base of recognized action units. The average efficiency of the recognition for 25 persons was 50%. Though, the results of experiments do not qualify them to be published yet experiments proved usefulness of FEEDB database and encourage for further research.

5   Conclusions and Future Works

Development of comprehensive and reliable database of facial expressions and emotions is a very difficult task that demands undertaking many project decisions and gathering a representative set of observation objects. Initial experiments have proved that both versions of FEEDB can be useful in research projects dealing with face recognition and analysis especially in human-computer interaction field. Moreover, additional depth channel allows not only for improving visual algorithms but can even be used alone in some applications.

FEEDB database will be further developed in different aspects. Some of the recordings will be exchanged as they contain not exactly proper or week expressions. Some part of new recordings will contain additional information channels, namely audio and several biometric sensors. The FEEDB database is available for other researchers in whole or in part as training or testing dataset.

References

1. Alyüz, N., Gökberk, B., Dibeklioğlu, H., Savran, A., Salah, A.A., Akarun, L., Sankur, B.: 3D face recognition benchmarks on the Bosphorus database with focus on facial expressions. In: Schouten, B., Juul, N.C., Drygajlo, A., Tistarelli, M. (eds.) BIOID 2008. LNCS, vol. 5372, pp. 57–66. Springer, Heidelberg (2008)
2. Bailenson, J., Pontikakis, E., Mauss, I., Gross, J., Jabon, M., Hutcherson, C., Nass, C., John, O.: Real-time classification of evoked emotions using facial feature tracking and physiological responses. International Journal of Human-Computer Studies 66, 303–317 (2008)
3. Burgin, W., Pantofaru, C., Smart, W.: Using depth information to improve face detection. In: Proc. of the 6th Int. Conf. on Human-Robot Interaction, pp. 119–120 (2011)
4. Castrillon-Santana, M., Deniz-Suarez, O., Anton-Canalis, L., Lorenzo-Navarro, J.: Face and facial feature detection evaluation - performance evaluation of public domain Haar detectors for face and facial feature detection. In: Third Int. Conf. on Computer Vision Theory and Applications, VISAPP 2008, pp. 167–172 (2008)
5. Colombo, A., Cusano, C., Schettini, R.: UMB-DB: A database of partially occluded 3D faces. In: Proc. of ICCV 2011 Workshops, pp. 2113–2119 (2011)
6. Ekman, P., Friesen, W.: Facial Action Coding System. Consulting Psychologist Press (1978)
7. Grgic, M., Delac, K.: Face Recognition Homepage (March 02, 2014), http://www.face-rec.org/databases/
8. Gunes, H., Piccardi, M.: Affect recognition from face and body: Early fusion vs. late fusion. In: Proc. of IEEE Int. Conf. on Systems, Man and Cybernetics, pp. 3437–3443 (2005)

9. Huynh, T., Min, R., Dugelay, J.-L.: An efficient LBP-based descriptor for facial depth images applied to gender recognition using RGB-D face data. In: Park, J.-I., Kim, J. (eds.) ACCV Workshops 2012, Part I. LNCS, vol. 7728, pp. 133–145. Springer, Heidelberg (2013)

10. Kolakowska, A.: A review of emotion recognition methods based on keystroke dynamics and mouse movements. In: Proc. of the 6th Int. Conf. on Human System Interaction, pp. 548–555 (2013)

11. Kshirsagar, V., Baviskar, M., Gaikwad, M.: Face recognition using eigenfaces. In: Proc. of the 3rd Int. Conf. on Computer Research and Development (ICCRD), pp. 302–306 (2011)

12. Landowska, A.: Affect-awareness framework for intelligent tutoring systems. In: Proc. of the 6th Int. Conf. on Human System Interaction, pp. 540–547 (2013)

13. Lucey, P., Cohn, J., Kanade, T., Saragih, J., Ambadar, Z., Matthews, I.: The extended Cohn-Kanade (CK+): A complete dataset for action unit and emotion-specified expression. In: IEEE Computer Society Conference on Computer Vision and Pattern Recognition (Workshops), pp. 94–101 (2010)

14. Lyons, M., Budynek, J., Akamatsu, S.: Automatic classification of single facial images. IEEE Trans. on Pattern Analysis and Machine Intelligence 21, 1357–1362 (1999)

15. Moreno, A., Sanchez, A.: GavabDB: A 3D Face Database. In: Proc. of the 2nd COST Workshop on Biometrics on the Internet: Fundamentals, Advances and Applications, pp. 77–82 (2004)

16. Picard, R.: Affective computing: From laughter to IEEE. IEEE Transactions on Affective Computing 1, 11–17 (2010)

17. Szwoch, M.: FEEDB: a multimodal database of facial expressions and emotions. In: Proc. of the 6th Int. Conf. on Human System Interaction, pp. 524–531 (2013)

18. Szwoch, W.: Using physiological signals for emotion recognition. In: Proc. of the 6th Int. Conf. on Human System Interaction, pp. 556–561 (2013)

19. Vizer, L., Zhou, L., Sears, A.: Automated stress detection using keystroke and linguistic features. Int. Journal of Human-Computer Studies 67, 870–886 (2009)

20. Wang, S., Liu, Z., Lv, S., Lv, Y., Wu, G., Peng, P., Chen, F., Wang, X.: A natural visible and infrared facial expression database for expression recognition and emotion inference. IEEE Transactions on Multimedia 12, 682–691 (2009)

21. Wrobel, M.: Emotions in the software development process. In: Proc. of the 6th Int. Conf. on Human System Interaction, pp. 518–523 (2013)

22. Zeng, Z., Pantic, M., Roisman, G., Huang, T.: A survey of affect recognition methods: Audio, visual, and spontaneous expressions. IEEE Transactions on Pattern Analysis and Machine Intelligence 31, 39–58 (2009)

The Optimal ChunkSize Pair Choosing Method for Dual Layered Deduplication Backup System

Mikito Ogata[1] and Norihisa Komoda[2]

[1] Hitachi Information & Telecommunication Engineering, Ltd., Japan
[2] Osaka University, Japan
mikito.ogata.mz@hitachi.com,
komoda@ist.osaka-u.ac.jp

Abstract. This paper proposes a multiple layers deduplication system for backup process in IT environment. The proposed system eliminates the duplication in target data by applying a series of plural ChunkSizes. The effectiveness of proposed system is evaluated by a simulator and a prototype machine on a linux server, comparing to a conventional single chunk system.The system realizes better deduplication capability with much less number of duplicated chunk reduction. This paper then proposes a choosing method to choose the optimal ChunkSizes to be set in layers. It maximizes the deduplication capability and minimizes the performance degradation caused by chunk reductions.

Keywords: deduplication, backup, archive, capacity optimization, enterprise storage.

1 Introduction

Due to an exponential increase of data, to take backup of the data to prepare the safe recovery in emergent cases becomes a burden for the system. Customers' key requirements for the backup operations are to reduce amount of stored data, processing time and resource utilization. Recently, a technology called "deduplication backup" is popular as an effective cure and implemented in many commercially available products. It divides target data into small segments in an adequate KB length and eliminates the duplication as much as possible and write only remaining parts into the backup storage. The size of division that leads the length of segments is called ChunkSize. Two objectives are especially important to implement the technology, one is to increase reduction capability and the other is to decrease processing time. Many improvements have been proposed[10,4,12,13,5,2,9,3,11]. However, because of a serious trade-off between the above objectives, it is not easy to improve both at the same time. Further, there seems to have been no study that tried to extend the layers of deduplication beyond one.

This paper proposes a new approach to aim at achieving both increasing reduction capability and decreasing processing time. The system has plural layers of deduplication algorithms and operates them in sequence on the data[6,8,7].

S. Kozielski et al. (Eds.): BDAS 2014, CCIS 424, pp. 395–404, 2014.
© Springer International Publishing Switzerland 2014

Our previous researches have clarified that the proposed system can increase the data reduction and decrease the processing time compared with conventional deduplication systems that have only one algorithm. Further, those have clarified the improvement depends on not only the combination of ChunkSizes in each layer but the target data characteristics.

Regarding practical requirements to adapt our system in production, it will be beneficial if there is a clear universal method to choose one ChunkSize in any environment. But there is no easy approach because the environments varies on the customers. This paper shows two methods to choose the optimal ChunkSizes that maximize deduplication capability and minimize the number of reduced chunks.

2   Layered Deduplication System

2.1   Conventional Single Layer Deduplication System

Any conventional deduplication system that is commercially available in production has one instance of mechanism for deduplication in the system. This is common even if the instance is implemented as a software application or a PBBA, Purpose Build Backup Appliance.

Here, an instance that eliminates the duplication, that is manipulates the deduplication, is called "Module", in which a suitable algorithm is implemented. A typical algorithm consists of chunking that is dividing the target data into plural small independent segments, then calculation of each identifier, checking the identification and storing only the remaining in the storage. Chunking means dividing the data into plural small segments as a unit of elimination. How much duplication is eliminated depends on the algorithm, especially the ChunkSize.

Generally, small ChunkSizes realize the finer segmentation and therefore the higher deduplication efficiency, and on the contrary, big ChunkSizes realize courser segmentation and therefore the lower deduplication efficiency. This correlation between ChunkSize and deduplication efficiency is a serious and non-linear trade-off. This makes it difficult to use smaller ChunkSizes as a measure to achieve higher reduction rate universally for overall environments.

2.2   Multiple Layers Deduplication System

Fig. 1 shows an example of our proposed "Multiple layers deduplication system" or "MLDS" as an abbreviation. The system has plural Modules. Each Module has an deduplication algorithm based on variable length chunking. A target data is deduplicated in series of the Modules, such as M_1, M_2, M_3 and so on. The order of Modules to manipulate the data is defined in advance so that any former Module has a larger ChunkSize than the later Modules. Target data is ingested into the system, where the first Module identifies the duplication in the data and eliminates it with the first algorithm, then the remaining data after the deduplication is ingested into the second Module. The second Module the identifies the duplication in the data and eliminates it with the second algorithm, and so forth. For convenience, we call a system which has only one Module "Single layer deduplication system" or "SLDS" as an abbreviation.

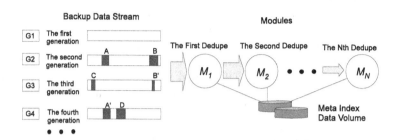

Fig. 1. Multiple layered deduplication system

As convenient notations hereafter, "Duplicated Parts" means the segments which are identical to at least one of previous segments, "Differed Parts" means not. "Duplication Rate" means the ratio of all the duplicate data to the whole data, that includes all Duplicated Parts and Differed Parts. "Deduplication Rate" means the ratio of eliminated data to the whole data. Each Module in the system, a Module in case of SLDS or plural Modules in case of MLDS, eliminates the Duplicated Parts according to their capacity of deduplication. The capacity depends on the implementation, such as algorithms, parameter settings and table structures, in the Module. The total deduplicated ratio by the system is called "System Deduplication Rate" and SDR as an abbreviation, and, the total reduced number of chunks is called "Number of Reduced Chunks" and NRC as an abbreviation. Further, we use the same sequence of number to identify the layer, that is the ChunkSize implemented in the first Module is called "the first ChunkSize", and the same manner is applied to each Module.

Fig. 2 shows an example of the manipulation in case of two layers system, called "Dual layered deduplication system" or "DLDS" as an abbreviation. The first Module divides the data, third generation, into four chunks, removes the Duplicatied Parts, B and C. Then A and D, which include Differed Parts, is concatenated as a data stream into the second layer. The second layer divides the data into seven chunks and remove a, c, e, f, g. Then, only b' and d' are added in the storage, the rest are managed in meta table link.

Our previous works analyzed the behavior of DLDS by changing their Chunk-Sizes and the data characteristics like the Duplication Rate and the average length of Differed Parts. Especially the following conclusions are reported[7].

- SDR of DLDS is the same or a slightly better comparing to SLDS of the same ChunkSize as the second ChunkSize of DLDS. The second ChunkSize is a dominant factor for the SDR.
- NRC of DLDS is notably lower than SLDS. In typical case, 60% bigger and in case of high Duplication Rate environment more than 90% bigger can be

achieved. The dominant factor to determine NRC is the first ChunkSize as a primary and the second ChunkSize as a complimentary.

These two aspects give the result that our DLDS provides the same SDR with much less NRC.

- The first ChunkSize of DLDS that realizes the minimum NRC , it is called "optimal first ChunkSize", depends on the target data characteristics. The ChunkSize is smaller for the data with lower Duplication Rate, becomes bigger for the data with higher Duplication Rate. In addition, the dependency is weakened under the comparison with the corresponding SLDS.

- The optimal first ChunkSize of DLDS and the improvement depends on the average length of Differed Parts. The size becomes bigger and the improvement bigger as the length increases. That indicates the improvement becomes bigger as the spacial locality in data increase.

Because all MLDSes have more than one ChunkSize simultaneously activated and each layer's behavior is affected by the ingested data characteristic, it is not obvious how to choose all the sizes in layers systematically in optimal way. This paper shows the choosing method under the various data characteristics to match predefined optimization policies.

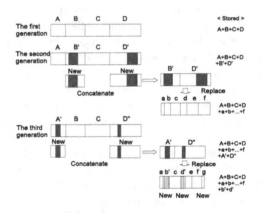

Fig. 2. Example of deduplication by dual layered deduplication system (DLDS)

3   Method of Choosing the Optimal ChunkSizes

3.1   Choosing Strategy

The optimal ChunkSizes are defined to guarantee one of the following policies.
(Case 1) To minimize NRC.
(Case 2) To maximize SDR and to minimize NRC with keeping under a predefined NRC lowest boundary.

Fig. 3 shows a example of NRC separation of the number of reduced chunks in each layer. The number of reduced chunks by the first layer decrease as the size increases. This is because large chunks can reduce coarsely. However, as the granularity decreases, the remaining Duplicated Parts that should be taken over in the second layer becomes bigger. This comes in larger consequent overhead in the following layer, and therefor additional number of reduced chunks are produced in the second layer. The optimal pair of the first ChunkSize and the second ChunkSize is decided as one that provides the minimum NRC in well organized way among all layers.

Fig. 4 shows a choosing strategy using simulation. The simulator predicts all the achievable SDRs and NRCs under assumed target data characteristics over all combinations of the first ChunkSize and the second ChunkSize. Then, a ChunkSize selector chooses a unique pair of sizes to match the objective.

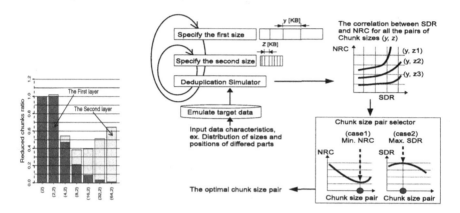

Fig. 3. Separation of NRC into each layer

Fig. 4. choosing strategy

3.2   Choice of the Optimal ChunkSizes

The indicators to be improved with high priority are deduplication capability and processing time. The KPI, Key Performance Indicator, to reflect the deduplication capability is naturally defined as SDR. On the other hand, processing time usually very much depends upon their system environment, like the hardware configuration, schedule of the operations and the traffic. Therefore, to keep the independency, NRC is adopted as a KPI to reflect processing time. This comes from the character that in a typical deduplication process, the processing time for Differed Parts is relatively lighter by using a common technique like bloom filter mechanism[1,13], however the processing time for Duplicated Parts is heavier to decide the uniqueness by tracking a big meta index table. SDR is recognized to be better if the value is bigger and NRC to be better if the value is smaller. From the practical implementation viewpoint, the range of each ChunkSize is defined as 2, 4, 8, 16, 32, 64KB.

(Case 1) This policy is applicable to the case that desire the reduction of NRC with keeping the same deduplication capability even if changing the system from existing SLDS to MLDS. As described in the previous section 2.2, MLDS provides the same or slightly better SDR with typically half NRC than SLDS and the dominant factor to define SDR is the last ChunkSize. The choice of the optimal ChunkSizes is first fix the first ChunkSize from the desired SDR, then fix the second ChunkSize to minimize NRC.

(Case 2) This policy is applicable to the case where the system requires the biggest boundary of NRC, which reflects the desirable backup time allowance, such as predefined backup windows. In that case, the system desires the maximum SDR under the condition of the requirement. The choice of the optimal ChunkSizes is, list all available pairs with keeping the NRC restriction, then pick up one of the maximum SDR.

4   Evaluation

4.1   Approaches

In order to evaluate the effectiveness of our proposed methods, a simulation and a prototype machine are configured. From practical usage, a simulation method is much more flexible and therefore affordable. In this paper, a prototype machine is complimentarily used as a validator for simulation. A simulator we used was reported in our previous paper[7]. The prototype is newly implemented in a Linux-based server, because no commercial machines with MLDS are available.

Tab. 1 shows the SDR and NRC obtained from simulator, prototype and their discrepancies. The discrepancies are under 5 %. Further, among the generations the covariance of SDR and NRC of both methods are also under 5 %.

Table 1. Difference of measured value to simulated one

ChunkSize pair	(y, 2)	(y, 4)	(y, 8)	(y, 16)	Average
Difference in SDR [%]	5.2	1.4	-3.4	-5.3	-0.5
Difference in NRC [%]	10.3	4.1	-1.8	0.8	3.4

4.2   Emulation of Target Data

In order to reflect practical environments, real sample data in a sequence of generations which are used in actual company were taken, and the difference among generations were analyses and the characteristics of each were abstracted. The average differences over generations are emulated as an average distribution of the length and the interval of Differed Parts laid between one generation and the following one. The distribution of length and interval are shown in Fig. 5. The left figure shows the distribution of the length of Differed Parts, right does the intervals. In both figures, x-axis means the length of bytes and y-axis means the probabilities. In this sample, the average Duplication Rate is 55 %.

For simulation, nine sequenced data that are generated like each of them follows the distribution are presumed as the difference between two generations. The simulator manipulates nine data as a series of difference. For the prototype, ten sequenced data are generated like each difference between two follows the distribution. The prototype manipulates ten data as a series of data.

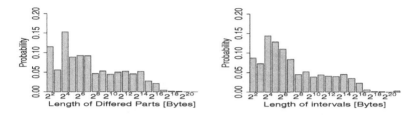

Fig. 5. Distribution of the length of Differed Parts and Duplicated Parts

4.3   ChunkSize Choosing to Minimize NRC

Fig. 6 shows the comparison of SDR predicted by simulation and the prototype. The graphs consist of three rows and four columns. The column means one of the cases of second ChunkSize equals to 2, 4, 8, 16. The left point in the box is the value in case of SLDS, the rest are the cases of DLDS, the points order follows the incremental sequence of the first ChunkSize. The solid line shows the result from the simulator, and the dotted line shows ones from the prototype.

The top row shows SDR, results in the fact DLDS has the same or slightly better SDR than the corresponding SLDS. The reason of better SDR is due to the additional potential of new boundary by the second chunking after the first chunking. As an example, after the first deduplication, two chunks both of which include any differed part are concatenated, in the case new space which comes from the attachment can be divided by the second deduplication, even each original space cannot be divided.   The middle row shows NRC in absolute number and the bottom does NRC in the relative ratio based on the number of SLDS. The bottom shows all cases of DLDS provide less NRC than SLDS, and further NRC in (16, 4) is the minimum, approx. 60 % less than SLDS.

In all cases, DLDS can deliver the same level of SDR with about a half NRC, in precise 40 % to 62 %. In addition, the heavy non-linear trade-off between SDR and NRC over the ChunkSizes is much weakened for DLDS than SLDS. This is shown in the series of is (8, 2)-(16, 4)-(32, 8)-(64, 16) comparing with (2)-(4)-(8)-(16). This means the system tuning is more stable for changing ChunkSizes and therefore safer from any unpredictable risks.

The optimal pair is (8, 2) for (2), (16, 4) for (4), (32, 8) for (8) and so on. The effectiveness of the optimal pair decreases as the second ChunkSize increases.

As a note, when the data is encrypted before backup process, the modification affects over the data, therefore the duplication behaves smaller. Even in this case, SDR is improved by DLDS, although the optimal chunk size become smaller and the corresponding SDR improvement become lower[8,7].

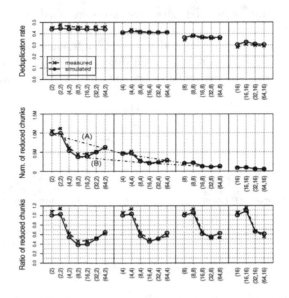

Fig. 6. Comparison of Deduplication Rate between simulation and measurement

4.4   ChunkSize Choosing to Maximize SDR

Fig. 7 shows the ChunkSizes in KiloByte that provide the maximum SDR under the NRC restrictions. x-axis means the NRC value boundaries. The solid line (S) shows sizes in case of SLDS, the solid line (D1) and (D2) show sizes in case of DLDS, (D1) is for the first ChunkSize and (D2) for the second. The dotted line means the correspondence of the equivalent SDR. The fact that DLDS can provide the equivalent SDR with less NRC means in the other way that DLDS can provide higher SDR with the condition of the same restriction of NRC. Tab. 2 shows the improvement of SDR under the assumption of NRC restrictions, the sizes are chosen to maximize SDR and at the same time minimize NRC under the restriction. Theoretically, any ChunkSize can be chosen in byte, here from practical reason, the ChunkSize is assumed to be one of 2, 4, 8, 16, 32, 64KB.

The choice of the optimal ChunkSize pair is first to fix a set of pairs that have NRC less than the restriction, then pick up one pair that has the maximal SDR. For example, in case of NRC restriction of 1000, (16) is the only one potential for SLDS, (64, 8), (32, 8), (16, 8), (64, 16), (32, 16) are possible pairs for Multiple layers. Among those, (32, 8) has the maximum SDR, therefore the optimal one. (32, 8) provides 20.6 % improvement. As NRC restriction becomes strict, the optimal ChunkSize becomes larger and the improvement bigger, up to 32 %.

Tab. 3 and Tab. 4 show the improvement of NRC and SDR, from the result of simulation method and from the prototype in sequence. The optimal pairs are the same. The improvements of estimated SDR are slightly bigger by simulation, the improvements difference of estimated NRC becomes bigger as the first ChunkSize becomes larger. It is because the actual lengthes of ChunkSize that are results of

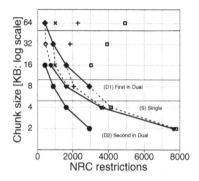

Fig. 7. The optimal ChunkSize pairs

Table 2. Improvement of SDR by the optimal ChunkSize

NRC restriction	Single Layer	Dual Layered	Improvement of SDR [%]
7643-	(2)	(8, 2)	0.0
3596-7643	(4)	(8, 2)	7.3
1618-3596	(8)	(16, 4)	12.5
681-1618	(16)	(32, 8)	20.6
-681	(32)	(64, 16)	32.1

chunking process are smaller in prototype than simulation. This is a consequence of additional chunking mechanisms in prototype[7].

Table 3. Improvement by Simulation

S D R	Single layer Dual layered	(2) 0.439 (8, 2) 0.440	(4) 0.411 (16, 4) 0.411	(8) 0.367 (32, 8) 0.368	(16) 0.305 (64, 16) 0.305
Increase [%]		0.23	0.00	0.27	0.00
N R C	Single layer Dual layered	(2) 978 (8, 2) 375	(4) 460 (16, 4) 209	(8) 207 (32, 8) 114	(16) 87 (64, 16) 53
Reduction [%]		-61.6	-54.6	-44.9	-39.1

Table 4. Improvement by Measurement

S D R	Single layer Dual layered	(2) 0.454 (8, 2) 0.459	(4) 0.404 (16, 4) 0.414	(8) 0.345 (32, 8) 0.357	(16) 0.280 (64, 16) 0.290
Increase [%]		1.10	2.48	3.48	3.57
N R C	Single layer Dual layered	(2) 1061 (8, 2) 451	(4) 485 (16, 4) 228	(8) 214 (32, 8) 107	(16) 92 (64, 16) 48
Reduction [%]		-57.5	-53.0	-50.0	-47.8

5   Summary

A new approach that improves deduplication capability and weakens performance penalty for data backup was proposed and evaluated. The proposed system, called Multiple Layers Deduplication System (MLDS) activates plural ChunkSizes simultaneously. Because the effectiveness of MLDS depends on the target data characteristics and setting ChunkSizes in each layer, we proposed two practical methods to optimize the ChunkSizes in all layers to maximize the effectiveness. The methods were practically evaluated in case of two layers by using a simulator and a prototype in comparison with conventional single deduplication system. As a result, the optimized DLDS achieved 40 to 62% reduction of chunk manipulation process and 0 to 32 % increase of deduplication capability.

References

1. Broder, A., Mitzenmacher, M.: Network applications of bloom filters: A survey. Internet Math. 1(4), 485–509 (2003)
2. Dubnicki, C., Grayz, C., et al.: HYDRAstor: a scalable secondary storage. In: Proc. of The 7th USENIX Conf. on File and Storage Technologies, FAST 2009, pp. 197–210 (2009)
3. EMC Corporation: EMC data domain boost software (2010)
4. Liu, C., Lu, Y., et al.: ADMAD: Application-Driven Metadata Aware Deduplication Archival Storage System. In: Proc. of Fifth IEEE International Workshop on Storage Network Architecture and Parallel I/Os (SNAPI 2008), pp. 29–35 (2008)
5. Meister, D., Brinkmann, A.: Multi-level comparison of data deduplication in a backup scenario. In: Proc. of The Israeli Experimental Systems Conf. (SYSTOR 2009), pp. 1–12 (2009)
6. Ogata, M., Komoda, N.: Improvement of performance and reduction of deduplication backup system using multiple layered architecture. In: Proc. of the First Asian Conf. on Information Systems (ACIS 2012), pp. 196–200 (2012)
7. Ogata, M., Komoda, N.: The assignment of chunk size associated with the target data characteristics in deduplication backup system. In: Proc. of the Second Asian Conf. on Information Systems (ACIS 2013), pp. 29–34 (2013)
8. Ogata, M., Komoda, N.: The parameter optimization in multiple layered deduplication system. In: Proc. of the 15th International Conf. on Enterprise Information Systems (ICEIS 2013)., vol. 2, pp. 117–124 (2013)
9. Quantum Corporation: Data deduplication background: A technical white paper (2009)
10. Tan, Y., Feng, D., Yan, Z., Zhou, G.: DAM: A data ownership-aware multi-layered de-duplication scheme. In: Proc. of 2010 Fifth IEEE International Conf. on Networking, Architecture and Storage (NAS), pp. 403–411 (2010)
11. Wallace, G., Douglis, F., Qian, H., Shilane, P., Smaldone, S., Chamness, M., Hsu, W.: Characteristics of backup workloads in production systems. In: Proc. of the 10th USENIX Conf. on File and Storage Technologies (FAST 2012), pp. 33–48 (2012)
12. Won, Y., Ban, J., Min, J., Hur, J., Oh, S., Lee, J.: Efficient index lookup for de-duplication backup system. In: IEEE International Symposium on Proc. of Modeling, Analysis and Simulation of Computers and Telecommunication Systems (MASCOTS 2008), pp. 1–3 (2008)
13. Zhu, B., Li, K., Patterson, H.: Avoiding the disk bottleneck in the data domain deduplication file system. In: Proc. of The 6th USENIX Conf. on File and Storage Technologies, FAST 2008, pp. 1–14 (2008)

Protection Tool for Distributed Denial of Services Attack

Łukasz Apiecionek[1], Jacek M. Czerniak[2], and Hubert Zarzycki[3]

[1] Casimir the Great University in Bydgoszcz,
Institute of Technology,
Bydgoszcz, Poland
lapiecionek@ukw.edu.pl
[2] Foundation for Development of Mechatronics in Bydgoszcz,
Bydgoszcz, Poland
jczerniak@mechatronika.org.pl
[3] Wroclaw School of Applied Informatics "Horyzonty",
Wrocław, Poland
hzarzycki@yahoo.com

Abstract. This article presents a potential tool for solving one of the biggest problems of security of network resources in computer networks. This problem are the Distributed Denial of Service attacks, which are able to block the computer networks. The authors introduce the nature of the problem and subsequently present a tool which provides a possible solution to deal with it. The proposed methods have already been tested in the range of their impact on network resources. The results presented in this article suggest that the method may be implemented as an efficient solution for some DDoS attacks.

Keywords: DDoS, IP Network, security, QoS.

1 Introduction

IT systems are nowadays omnipresent. The users need a fast access to information from every part of the network. Denial of Service attacks, or lately rather Distributed Denial of Service attacks (DDoS in short), have become a problem in the network systems operation, as by seizing the system resources in the computers forming the network until they stop working and thus by blocking services they cause network unavailability. A user who has already started working in the system loses the connection and cannot even log out of the system, which has to do it for him after the connection timeout is reached or when a broken connection is detected. DDoS attacks are an important issue for IT systems nowadays and they have to be eliminated. One of the solutions is the Enhanced Quality of Service method which could work on all routers in the network. This paper presents a possible way to protect network resources in case of DDoS. Chapter II describes how DDoS work on network traffic layer. Chapter III presents a possible method for network protection, followed by test results of the implementation

S. Kozielski et al. (Eds.): BDAS 2014, CCIS 424, pp. 405–414, 2014.

phase, compiled in chapter IV. In this chapter it is also demonstrated why this solution should work in a global network. Chapter V provides a conclusion and discussion over the developed method.

2   Distributed Denial of Service Attack Description

DDoS attacks are widely described in the literature [10,9]. These attacks can be performed on various system resources: TCP/IP sockets [9,11] or Domain Name System (DNS) servers. Regardless of the method, the main principle is to simulate so many correct user connections that their number exceeds the actual system performance and drives it to abnormal operation. Many authors [10,9,3,2] suggest dealing with the DDoS attacks by their global detection and emphasize the necessity of cooperation between network providers. The transmission of the attackers' packets is performed through the provider's network and if it cannot be blocked, it leads to data link saturation. Such saturation results in a lack of connection to the server [13]. Even applying some extra intrusion detection systems which use general purpose computing on graphics processing units [12] is not sufficient. In figures (Fig. 1, 2, 3, 4) the authors present how DDoS attacks affect the network. At first the network is stable (Fig. 1). There is lot of routers, computer machines and Web servers.

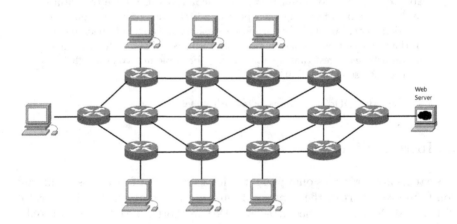

Fig. 1. A sample network

A hacker gets his chance to attack Web Server when he is able to control a sufficient amount of the machines (as presented in (Fig. 2). The computers under the hacker's control are marked with red color).

When the attack on Web Server is initiated, some network connections become saturated (presented in figure (Fig. 3 and 4) 4 with red dashed lines). During the attack more connections are affected, and finally, all connections to Web Servers are fully saturated and Web Server resources are insufficient to perform normally (Fig. 4).

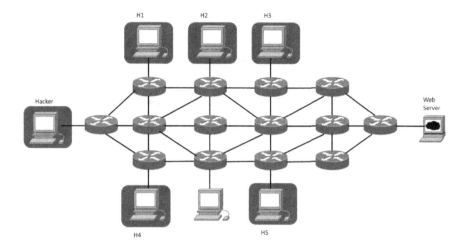

Fig. 2. A sample network with some machines, H1-5, under the control of the hacker

It is possible to deal with this problem and to prevent such situations by using some specialistic methods, such as Quality of Services, which could work on each router in the network and drop the unnecessary packets. This method is possible to implement and it is presented in the next chapter.

3   Enhanced QoS Method

3.1   Method Overview

Network bandwidth is managed using network bandwidth administration methods. They allow to set a given bitrate in the network for specific users or type of network traffic. In this manner a privileged traffic is selected. For the traffic not classified as privileged, the Fair Queue method is applied. This method allows fair bandwidth division between data sources. It is widely used on routers. It should be noted that when bandwidth is limited in the receiver's network, this method divides packets fairly, so that every data flow can reach the receiver. The presented in the own research [4] QoS method enhances the existing mechanisms for traffic control. It introduces a historybased "traffic activity fair queue", yet not basing on data source or category, but on connection history. In order to illustrate the mode of action of the presented solution, a model of DDoS attack was developed. It should be assumed that during normal network activity n data flows are transmitted to the receiver over a router. When an attack occurs, the number of data flows grows rapidly up to the maximum number supported by the network resources. Figure (Fig. 5) shows a hypothetical number of connections, which grows over time from the beginning of the attack on a particular network resource. The start of the attack was labeled in figure (Fig. 5) as DDoS. When the attack stops, the number of data flows decreases to an average number

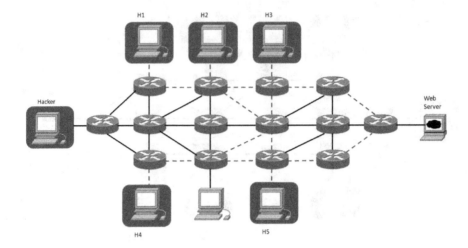

Fig. 3. A sample network during the hacker's DDoS attack on Web server

in the network. The number of data flows shown on the graph is a hypothetical value, just to illustrate how this mechanism works.

When the number of packets exceeds the given limits, in order to deal with the DDoS attack, QoS method has to remove them from the queue by using a wellknown mechanism called Random Early Detection (RED) with special condition. In a standard method, the RED mechanism is executed on incoming packets with data flows categorization. In the proposed Enhanced QoS method, the RED mechanism should not be executed on the data flows that had been transmitted earlier [8]. The key steps of the method are:

- keeping the data flows history from the moment before the attack and this period should not be neither too long, as otherwise it will not capture the appropriate traffic, nor too short, in order to protect the device from heavy load,
- detecting the moment of the attack, i.e. by monitoring the number of connections, and when this number exceeds the given limits, a potential DDoS attack signal is generated,
- after detecting the attack, setting the traffic filtering rules to privilege historical packets, i.e. by selecting them with the use of filtering rules on a router module,
- blocking all other malicious traffic, i.e. by using RED mechanism [5] or firewall rules [7,6].

Figure (Fig. 6) shows a general method scheme. The QoS Fair Queue method is used during usual operation of the device and transmits packets to the receiver. Correct connection history is gathered basing on the Fair Queue method. When a DDoS attack is detected, the packets are transmitted to a special data flow recognition module which compares them with the correct connection history

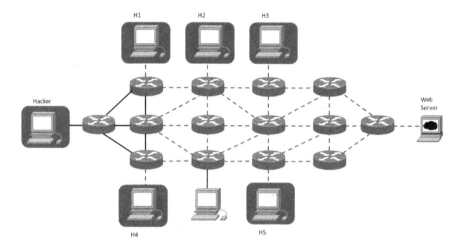

Fig. 4. A sample network after the hacker's blocking Web server using DDoS attack

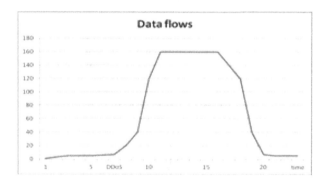

Fig. 5. IP packet flow over a router

database. After their categorization, the packets are either transmitted to the receiver or removed by the RED mechanism.

3.2   Implementation Results

The presented method is currently in the implementation phase. QoS Fair Queue methods and RED mechanisms are functional for this moment. Missing elements of the proposed method are yet to be implemented and all elements have to be integrated. The missing elements are:

- data flows history storage mechanism,
- a mechanism deciding if an attack on the system occurred,
- an additional data flows recognition mechanism.

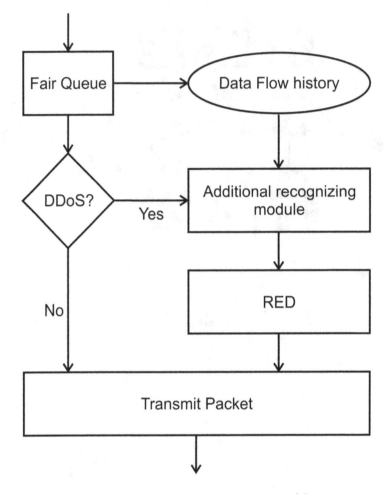

Fig. 6. Enhanced QoS method diagram [4]

Historical transmission data collection mechanism can be based on the information constantly received from the Fair Queue method, which monitors all the data flows. Detecting the moment of the attack is another step which can be achieved using simply detecting model by monitoring the number of transmitted packets or data flows in the Fair Queue method. The third element is constructing a mechanism for filtering the privileged traffic by its identification in the network history. This mechanism can be created in the Linux system by the IPTables firewall module [7,6]. A common problem with filtering the traffic is that it can reduce the device's throughput and in result, impair network operation. That is why this element was implemented in the first place. For this purpose, an IPTables module for Linux Debian system was developed, which has to detect a certain traffic based on a given signature: source IP address, sender and type of protocol. This module was prepared as an external Linux kernel module loaded on the user's demand. Then, a network performance test was made, using this module to filter packets. The bandwidth was tested with the use of IPerf software, version 2.0.5. Test environment consisted of:

- Router working on Linux Debian 2.6.32 system and IPtables 1.4.8 module, with an additional module,
- Sender host,
- Receiver host.

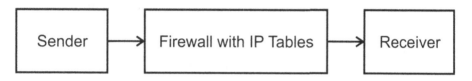

Fig. 7. Test network

The receiver host was working under Windows 7 64-bit system and was equipped with Intel i3 processor. Both the router and the receiver host were working in virtual VMWare environment, and were equipped with 512 MB of memory and 1 processor. These mchines were running on test data receiver host. The aim of this test was not to achieve large network bandwidth, but to check if, and in what extent adding another filtering rule would impact the router's bandwidth. The test conditions were as follows:

- Test period: 60 seconds,
- Test protocol: TCP.

The results of average transmission speed for four different conditions are listed in table (Tab. 1). Conditions A, B, C and D define the configuration of rules in IPTables module. Every condition was tested three times. In condition A, the module had no filtering module set. Condition B has one filtering condition:

- blocked all external traffic incoming to the firewall: INPUT deny any any.

Condition C consists of condition B with one more filtering rule implemented, executed on selected traffic while transmitting packets through a firewall:

- blocked all external traffic incoming to the firewall: INPUT deny any any.
- filtered traffic: FORWARD tcp on port 6000, with the use of the implemented module.

Condition D consists of condition C with blocking rule implemented, which was executed for other types of traffic:

- blocked all external traffic incoming to the firewall: INPUT deny any any.
- filtered traffic: FORWARD tcp on port 6000, with the use of the implemented module.

Table 1. Average speed test results

Test	Average results	
condition	Sender [kbit/s]	Receiver [kbit/s]
A	77.1	83.8
B	76.1	84.8
C	78.5	83.1
D	75.4	83.2

- blocked all external traffic forwarded through the firewall: FORWARD deny any any.

As it could be noticed, there is no significant impact on router throughput by filtering module. This is important, because it proves that this tool will have no impact on network throughput.

3.3   Proof of Concept

The analysis of the results suggests that adding an additional filtering rule does not cause a drop in packet transmission speed through the router or firewall. It confirmed the thesis that the proposed method will not result in much bigger load on resources of existing devices, and thus it will not cause slowing the network performance, but can help to fight the DDoS attacks. This method will not solve all DDoS attack problems, but it will enable the users to close their active connection when the attack starts. Moreover, some tests using Web browsers were made. Over 90% of users choose one of the following web browsers: Internet Explorer (12%), Mozilla Firefox (27%) and Google Chrome (55%) [1]. Thus, the tests were performed basing on those three most popular web browsers. The test procedure was to connect to the web server which did not exist and to check how the browser will send packets. Test condition:

- operating system Windows Vista,
- Internet Explorer version 9.0.8112.16421,
- Mozilla Firefox version 16.0.2,
- Google Chrome version 23.0.1271.64 m.

The results of debug packets from Wireshark program are presented in table (Tab. 1). As it could be noticed, Mozilla Firefox browser tries to make a connection 18 times, while Google Chrome and Internet Explorer stops after the 9th connection attempt.

The mode of action of the presented web page browsers enables the user to close the active connection and finish his work. During browsers tests, the authors recognized the fact, that web page browsers made automatic retransmissions and tried to connect to a web server more the once. Using the presented enhanced QoS method, it will be possible to transmit the packets to their destination. Depending on the user's browser, there are at least nine chances to transfer the appropriate data.

Table 2. Wireshark Web browsers connection test results

Number of packets	Delays before the next packet is send from the browser[seconds]		
	Mozilla Firefox	Internet Explorer	Google Chrome
1	0	0	0
2	0,254	0,001	0,001
3	2,996	2,995	0,25
4	3,246	2,995	2,996
5	8,997	8,995	2,996
6	9,247	8,995	3,246
7	20,996	20,995	8,997
8	21,239	23,991	8,997
9	21,246	29,992	9,248
10	23,995		
11	24,235		
12	24,245		
13	29,998		
14	30,238		
15	30,248		
16	42,231		
17	45,23		
18	51,233		

4   Conclusions

In this article, a new concept of eliminating some DDoS attacks was presented along with the evidence for the solution's practical application in common use. Implementation of this concept should based on cooperation between all internet providers. The prepared QoS method will allow to minimize the scale of the DDoS attacks. The best solution would be applying this method to all routers and firewalls in the network. Such algorithm does not require any new system resources of the network devices. Its implementation in the whole network and at all providers will result in reducing the number of attacking packets and as a result, deflect the attack. The mechanism of privileging the packets which were sent prior to the attack and come from the users operating correctly will result in their uninterrupted work in the network. Of course, it will not solve all the DDoS problems. If the attack lasts longer than the span of the stored history, this method will not be efficient. The same situation will occur when some new users try to connect to the network resources. However, in other cases it will provide some protection of the network and thus it will enable theusers to close their active session, opened prior to the DDoS attack. Fighting with the DDoS attacks is not easy, but it is essential to protect the society against the hackers' attacks, especially in the situations when network resources are really needed.

References

1. Browser statistics and trends,
   http://www.w3schools.com/browsers/browsers_stats.asp
2. Cert advisory ca-1996-01 udp port denial-of-service attack (september 1997),
   http://www.cert.org/advisories/CA-1996-01.html
3. Cert advisory ca-1996-21 tcp syn flooding and ip spoofing attacks (November 2000),
   http://www.cert.org/advisories/CA-1996-21.html
4. Apiecionek, L., Czerniak, J.: Qos solution for network resource protection. In: Proceedings of International Scientific Conference INFORMATICS 2013, Spiska Nova Ves, Slovakia (November 5-7, 2013)
5. Changwang, Z., Jianping, Y., Zhiping, C., Weifeng, C.: Rred: Robust red algorithm to counter low-rate denial-of-service attacks 14(5) (2010)
6. Chapman, B., Zwicky, E.: Building Internet Firewalls. O'Reilly "&" Associates, Inc. (1995) ISBN 1-56592-124-0
7. Cheswick, W., Bellovin, S.: Firewalls and Internet Security: Repelling the Wily Hacker. Addison-Wesley Publishing Company (1994) ISBN 0-201-63357
8. Kovác, D., Vince, T., Molnár, J., Kovácová, I.: Modern Internet Based Production Technology. In: New Trends in Technologies: Devices, Computer, Communication and Industrial Systems, pp. 145–164. SCIYO,
9. Moore, D., Voelker, G., Savage, S.: Interferring Internet Denial-of-Service Activity. No. NSN 7540-01-280-5500 (2006)
10. Rocky, K., Chang, C.: Defending against flooding-based distributed denial-of-service attacks: A tutorial. IEEE Communications Magazine, 42–51 (October 2002)
11. Schuba, C., Krsul, I., Huhn, M., Spafford, E., Sundaram, A.: Analysis of a denial of service attack on tcp, computer science technical reports. paper 1327 (1996),
    http://docs.lib.purdue.edu/cstech/1327
12. Vokorokos, L., Ennert, M., Hartinger, M., Raduovska, J.: A survey of parallel intrusion detection on graphical processors. In: Proceedings of International Scientific Conference INFORMATICS, spiska nova ves, slovakia (2013)
13. Wrzesień, M., Olejnik, Ł., Ryszawa, R.: Ds/ips: Detection and prevention systems of hacking the computer networks. Studies and Materials in Applied Computer Science 4(7), 16–21 (2012)

A Keystroke Dynamics Based Approach for Continuous Authentication

Dina El Menshawy, Hoda M.O. Mokhtar, and Osman Hegazy

Faculty of Computers and Information, Cairo University, Cairo, Egypt
{d.ezzat,h.mokhtar,o.hegazy}@fci-cu.edu.eg

Abstract. In this paper, we present the application of keystroke dynamics for continuous user authentication in desktop platform. We show the differences between static and continuous systems based on keystroke dynamics in terms of creating the template and authentication phase. The key factor in the continuous authentication system is monitoring the genuineness of the user during the whole session, and not only at log-in. Moreover, we propose a general approach for continuous authentication based on keystroke dynamics. In our experiments, we use the email application as a case study to present the effectiveness and efficiency of our proposed approach. Our main conclusion is that using keystroke dynamics can serve as a feasible and acceptable measure for continuous user authentication. The investigations have shown that it is feasible to authenticate users based on keystroke dynamics for continuous authentication systems.

Keywords: biometrics, keystroke, authentication, security, continuous.

1 Introduction

Computer security is an integral part of any computer activity. Nevertheless, ensuring security continues to be a challenging problem, especially when doing on-line transactions, so internet users should avoid disclosure of confidential information on the internet [3]. The main solutions to eliminate security threats are authentication and access control. Authentication is concerned with proving the identity of a person, while access control decides who is allowed to access the system and determine the privileges [16]. The increasing need for improving security systems led to more research in the application of biometrics in authentication systems. A *biometric* is a measurement of a biological characteristic such as fingerprint, iris pattern, retina image, face or hand geometry; or a behavioral characteristic such as voice, gait or signature [17].

In this paper, we propose an algorithm for continuous authentication based on keystroke dynamics by measuring the trust of the system in the genuineness of the current user. The rest of the paper is organized as follows: we present in section 2 an overview of biometrics with its different categories. Section 3 introduces the behaviorial biometrics in general with specific focus on keystroke dynamics. Section 4 presents the related work. Section 5 shows a comparison between static and continuous authentication. Section 6 illustrates the application

S. Kozielski et al. (Eds.): BDAS 2014, CCIS 424, pp. 415–424, 2014.
© Springer International Publishing Switzerland 2014

of keystroke dynamics in continuous authentication systems. Section 7 shows our proposed algorithm for continuous authentication. Finally, section 8 presents the conclusions and future work.

2   Biometrics: A Brief Overview

In general, there are three levels of computer security mechanisms: the first mechanism depends on something a person carries, such as an ID badge with a photograph, while the second scheme relies on something a person knows, such as a password. Finally, the third approach is related to a person's human attributes, such as fingerprint and/or signature. People have personal characteristics that uniquely identify them such as hand signature, fingerprint and voice. Those unique characteristics are known as biometrics. In general, biometrics is mainly divided into two categories, namely, physiological biometrics and behavioral biometrics. *Physiological biometrics* identifies a person based on his\her physiological characteristics such as eye retina, whereas *behavioral biometrics* relies on detecting the behavioral attributes of the user, such as keystroke dynamics and voice [11, 24]. Biometrics became popularly used as a security tool due to its universality and distinctiveness. A biometric based authentication system can be evaluated using either a genuine test or an impostor test, described as follows:

- The genuine test (or False Rejection Rate (FRR)): the percentage of valid inputs which are incorrectly rejected. It is denoted by the number of incorrectly rejected attempts divided by the total attempts of legitimate users who try to access the system.
- The impostor test (or False Acceptance Rate (FAR)): the percentage of invalid inputs which are incorrectly accepted. It is denoted by the number of incorrectly accepted attempts divided by the total attempts of impostors who try to access the system [11].

3   Behavioral Biometrics and Keystroke Dynamics

Behavioral biometrics, are a subset of biometrics that deals with a person's behavior. *Keystroke dynamics* means the pattern in which a user types characters, or numbers on a keyboard. Keystroke dynamics is used to define the person's identity as it resembles an individual's handwriting or signature. A user's keystroke rhythms are measured to generate a distinctive prototype of the user's typing patterns for use in authentication. A key advantage of using keystroke dynamics is that it can be captured constantly [25]. Moreover, no additional hardware is needed to collect keystroke data, the keyboard is enough [18]. In other words, keystroke dynamics is basically concerned with "how you type" rather than "what you type" [15,23]. The following figure depicts how a keystroke based authentication system works:

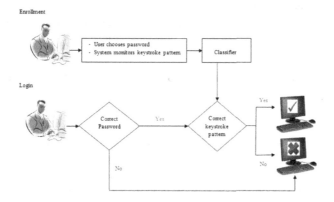

Fig. 1. Authentication Process (Adapted from [13])

4 Related Work

A huge number of researches were conducted to explore the utilization of keystroke dynamics for authentication in desktop platform. In [14, 19, 21], the authors collected keystroke data in the form of duration and latency when users enter a specific password, then different techniques were used to detect patterns in the typing rhythm. Also, in [4,12,22], the researchers utilized other features for authentication such as: total duration time of writing the password and release-press interval. In [10], the authors created a client side Javascript to record the time when a key was pressed, together with the ascii value of the key. In [5], the authors captured keystroke data by learning an event-length based Markov model from sequences of events. In [6], the authors combined keystroke and mouse dynamics to be used for users' authentication. The experiments showed that fusing both kinds of biometrics gave better authentication accuracy rather than utilizing only keystroke dynamics.

5 Static Authentication Versus Continuous Authentication

In a static authentication system, the identity of the user is verified at the start of the session while in continuous authentication, the identity of a user is monitored during the full session. This means that the genuineness of a computer user has to be checked from the moment he\she logs in (using static authentication) to the moment he\she logs out the computer [17]. Static keystroke analysis is executed on typing samples using predetermined text for users under observation while continuous keystroke analysis has no predefined text. The main drawback of static authentication systems is that whenever access is granted to a certain user, this will be valid throughout the whole session [1, 2]. On the other hand, continuous verification implies a continuous monitoring of issued keystrokes. Unfortunately, the verification algorithms, as well as the implementation process needed for continuous monitoring are complex [20].

6   Continuous Authentication Based on Keystroke Dynamics

The major difference between static and continuous keystroke dynamics is in the authentication phase [7]. In static keystroke dynamics, the complete typing of the fixed text of the input and template are compared. On the other side, this cannot be applied to continuous keystroke dynamics [2]. In the case of continuous authentication, the template has to include all letters and letters' combinations that are typed during the session but this will be very difficult. We need to restrict the template to a particular number of letters or letters combinations. A lot of work needs to be done to explore how keystroke dynamics can be employed for continuous authentication.

As a result, in this paper we proposed a new algorithm similar to the one implemented in [1] to handle the problem of huge templates. We proposed a general approach for continuous authentication based on keystroke dynamics. In the following section, we will present the proposed algorithm along with the experiments and results.

7   Proposed Algorithm for Continuous Authentication Based on Keystroke Dynamics

In this section, we will present the general approach for continuous authentication based on keystroke dynamics. Also, we present the applicability of the proposed algorithm in the email domain.

7.1   Template Generation

The first step is to create a template of the typing pattern of the user, as we mentioned earlier, creating a template with all typed keys and keys combinations typed by the user will be a time consuming process. Also, the template may not include all different key combinations, so even if the user behavior is monitored for few days or even a week, it does not have to include all possible key combinations. As a result, we need a solution for the problem of templates generation. We constructed a general approach for continuous authentication, but actually we applied it on an email application as authentication in emails based on only passwords has a number of weaknesses. We asked 15 computer users to monitor the most frequent words they type in emails. After combining all words, we noticed that the following words: dear, hi, hello, regards, thanks, meeting, time, AM and PM are the most common typed words in emails by different users. The idea behind choosing those words is that the typing pattern may be relatively stable because they are frequently typed by users.

7.2 Features Extraction

In a continuous authentication system, allowing or rejecting access to users is checked after each single keystroke. A user will not be locked out of the system based on only one key wrongly typed. The system will lock the user out, when he\she types in an incorrect manner over a longer period of time. In this paper, a little bit different keyboard features are used to explore various attributes which are not much used before, the features are:

- The difference between two press events (PP).
- The difference between two release events (RR).
- The difference between one press and one release events (PR).
- The difference between one release and one press events (RP).
- The total time taken to write a certain word [7].

The following figure illustrates the features when a user enters the word "hello":

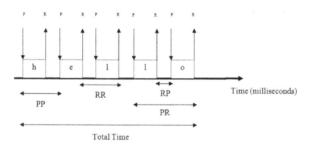

Fig. 2. Keystroke Features of Password Entry

The arrow direction pointing down, declares a key press, while the arrow pointing up denotes a key release. All timing features were measured in milliseconds.

7.3 Measuring Level of Trust

The system must determine a level of trust of the genuineness of the user based on the typing behavior. The system will only block access to a user if the trust level has dropped below a pre-determined threshold $T_{lockout}$. If a user gets locked out by the system, then a static authentication mechanism must be used to have access again [1]. The authors in [1] implemented the levels of trust through the use of a so-called penalty and reward function. In its simplest form, a user gets rewarded with an increased trust level if he follows his template, in other words, the user is following the pattern expected based on his pre-measured features. On the other hand, the user gets penalized if his typing rhyme deviates from what is expected. In this case, the user's trust level is decreased. Thus the proposed penalty-reward approach proceeds in 2 phases: a comparison (evaluation) phase, where typing rhythm is compared against template; and a scoring phase, where trust level is adjusted based on the first phase outcome.

To determine the degree of deviation from current typing pattern and template, a distance measure is usually employed. When the system determines the distance D between the current typing and the template, this value can be used in the second stage to update the trust level. At the beginning, the system should determine whether the trust level will go up or down, or in other words, if the distance value D will cause a a penalty or a reward. This can be done by comparing the value D to a threshold $T_{distance}$. If the distance value is below the threshold, then the user gets rewarded, otherwise he gets a penalty.

Now, how large the penalty or reward should be determined, i.e. how much the trust level C should decrease or increase. There are two different possibilities: a fixed change or a variable change. The easiest way to apply the penalty and reward is the first one. In the case of a fixed change, the trust level will change by a predetermined value. The level of trust was denoted by C and the value of C ranges from 0 which denotes no trust to 50 which declares complete trust. The trust level C will be initially set to 50 after a user has successfully completed a static authentication procedure. The same techniques as in static keystroke dynamics can be used in the first stage. We determine the distance D between the typed text and the template. Once we have a template, we need a distance metric and a penalty and reward function. Also, we need to determine the thresholds $T_{distance}$ and $T_{lockout}$. The easiest implementation of the fixed change for penalty and reward is to increase or decrease the value C by 1 in both cases. When the user is locked out of the system because the trust level dropped below the threshold, then he\she needs to use the system's static authentication mechanism to log in again [1].

7.4   Proposed Algorithm for Continuous Authentication in Email Application

At first, we asked 15 users to write some of the words previously listed. We chose only those words: dear, hi, hello, regards and thanks, as a start for our experiments. Each user was asked to write each word ten times. Some samples were used to construct the users' template while the rest were used to test the proposed algorithm. The template of each user was constructed, the template contains information about total duration time of each word. Moreover, it contains PP, RR, PR and RP values for each two consecutive keys in each word. For each of the four latencies: PP, RR, RP and PR, and total duration of each word typed by every user, the mean value of all samples of each feature was calculated. Also, the standard deviation was calculated for each feature.

This process was repeated for each user after writing every word. After that, the average standard deviation was calculated for all users after writing all words. Moreover, the same 15 users were allowed to write the word "fusion" (a word which is not so commonly typed by users). Also, this word was chosen as it contains keys from different locations on the keyboard. The same process of calculating the mean and standard deviation was implemented. After that, an average value for standard deviation was calculated for all users. It has been deduced that, the standard deviation of writing the word "fusion" is much higher than the average standard deviation of the list of words. The result of this difference may be because, users type frequent words in a more stable pattern than un-common typed words. As a result, keystroke dynamics of common typed words can be used for authentication in continuous systems and the proposed algorithm can be applied on email applications.

The proposed algorithm goes as follows:

Input: Template of words: dear, hi, hello, regards, thanks
Assume successful static authentication, so C = 50;
foreach *user* **do**
  foreach *word* **do**
    compute total time;
    foreach *two consecutive keys* **do**
      compute PP, RR, PR, RP;
      if $D<T_{distance}$ **then**
        increase C by 1;
      else
        decrease C by 1;
      end
    end
  end
  if $C>T_{lockout}$ **then**
    user continues access;
  else
    log out user;
  end
end

Algorithm 1. Proposed Algorithm for Continuous Authentication (Adapted from [1])

We assumed that users have to write all words of the template (although emails written by users don't have to include all those words, but we will have this assumption in the initial phase of our experiments). A vector of means and standard deviations were calculated for all features after writing all words by all users. The distance metric used to compute the distance D between the template and test input is the Euclidean distance. Standard deviation will be used to determine the threshold $T_{distance}$. Standard deviation measures the range of PP, RR, PR, RP and total duration that will be acceptable for each word typed by every user.

The decision of choosing the standard deviation is not easy because, if the standard deviation increases, the probability that an imposter or other users will overlap with the genuine user's profile increases, and as a result, FAR increases. On the other hand, the typing behavior of the genuine user will usually fall within this increased range of standard deviations, and as a result, FRR decreases [8,9]. The challenge is to tweak the value of standard deviation to have acceptable FAR and FRR error measures. Different values of standard deviations for all the features for all users were tried to reach a compromise between both error measures. Standard deviations of different features for all users varied from 1.5 to 3. This means that at the point of intersection (Equal Error Rate EER), this gives the best biometric system accuracy at standard deviation of 2. The following figure shows FAR and FRR error measures with varying the standard deviations:

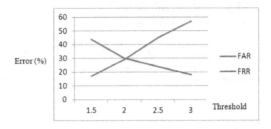

Fig. 3. FAR and FRR Error Measures

The other challenge of the algorithm is to determine the value of the threshold $T_{lockout}$. The same idea of choosing the value of the threshold $T_{lockout}$ is similar to that of choosing the values of the threshold $T_{distance}$. If the value of $T_{lockout}$ increases (becomes near to 50), the likelihood that a genuine user will be locked out increases as there will not be a wide range for typing discrepancy between the template and test inputs. As a result, FRR increases and the user will get frustrated from using the system. On the other hand, if the value of $T_{lockout}$ decreases (becomes near to 0), the likelihood that a malicious user will access the system increases as the system will allow for more typing discrepancy. As a result, FAR increases and the system will not be secure. After applying various values for $T_{lockout}$, it has been deduced that a value of 30 for the global threshold $T_{lockout}$ performs the best as only 4 users were logged out of the system.

8   Conclusion and Future Work

Keystroke analysis has proven to be a promising technique having achieved good results in continuous authentication systems. The main issue in continuous authentication system is monitoring the genuineness of the user throughout the whole session. We proposed an algorithm for continuous authentication based on keystroke dynamics and showed how it can be applied on an email application. In the future work, we can conduct more extensive experiments, by increasing the number of participants and the number of words. Also, we can explore other different ways of applying the penalty and reward function to keep track of the genuineness of the user during the whole session.

Acknowledgements. We would like to thank Romain Giot, Mohamad El-Abed and Christophe Rosenberger for making their GREYC-Keystroke software publicly available to help researchers to create keystroke dynamics databases.

References

1. Bours, P.: Continuous keystroke dynamics: A different perspective towards biometric evaluation. Journal of Information Security 17(1), 36–43 (2012)
2. Bours, P., Barghouthi, H.: Continuous authentication using biometric keystroke dynamics. In: Proceedings of the Norwegian Information Security Conference, Norweg (2009)
3. Chirillo, J., Blaul, S.: Implementing Biometric Security, 1st edn. Wiley Pub. (2003)
4. Choras, M., Mroczkowski, P.: Keystroke dynamics for biometrics. In: Proceedings of the 8th International Conference, Warsaw, Poland (2007)
5. Damiani, E., Gianini, G.: Navigation dynamics as a biometrics for authentication. In: Apolloni, B., Howlett, R.J., Jain, L. (eds.) KES 2007, Part II. LNCS (LNAI), vol. 4693, pp. 832–839. Springer, Heidelberg (2007)
6. El Menshawy, D., Mokhtar, H.M.O., Hegazy, O.: Enhanced authentication mechanisms for desktop platform and smart phones. International Journal of Advanced Computer Science and Applications 3(10), 100–106 (2012)
7. Giot, R., El-Abed, M., Rosenberger, C.: Greyc keystroke: a benchmark for keystroke dynamics biometric systems. In: Proceedings of the IEEE International Conference on Biometrics: Theory, Applications and Systems (BTAS), USA (2009)
8. Giot, R., El-Abed, M., Rosenberger, C.: Keystroke dynamics with low constraints svm based passphrase enrollment. In: Proceedings of the IEEE International Conference on Biometrics: Theory, Applications and Systems (BTAS), USA (2009)
9. Giot, R., El-Abed, M., Rosenberger, C.: Web-based benchmark for keystroke dynamics biometric systems: A statistical analysis. In: Proceedings of the 8th International Conference on Intelligent Information Hiding and Multimedia Signal Processing, Greece (2012)
10. Gunetti, D., Picardi, C., Ruffo, G.: Keystroke analysis of different languages: A case study. In: Famili, A.F., Kok, J.N., Peña, J.M., Siebes, A., Feelders, A. (eds.) IDA 2005. LNCS, vol. 3646, pp. 133–144. Springer, Heidelberg (2005)
11. Jain, A., Ross, A., Pankanti, S.: Biometrics: A tool for information security. IEEE Transactions on Information Foresnsics and Security 1(2), 125–143 (2006)
12. Joyce, R., Gupta, G.: Identity authentication based on keystroke latencies. Communications of the ACM 33(2), 168–176 (1990)
13. Kang, P., Hwang, S.-s., Cho, S.: Continual retraining of keystroke dynamics based authenticator. In: Lee, S.-W., Li, S.Z. (eds.) ICB 2007. LNCS, vol. 4642, pp. 1203–1211. Springer, Heidelberg (2007)
14. Kapczynski, A., Kasprowski, P., Kuzniacki, P.: User authentication based on behavioral patterns. International Scientific Journal of Computing 6(1), 75–79 (2007)
15. Karnan, M., Akila, M., Krishnaraj, N.: Biometric personal authentication using keystroke dynamics: A review. Journal of Applied Soft Computing 11(2), 2406–2418 (2011)
16. Kizza, J.M.: Ethical and Social Issues in the Information Age, Texts in Computer Science, 4th edn. Springer (2010)
17. Kumar, S., Sim, T., Janakiraman, R., Zhang, S.: Using continuous biometric verification to protect interactive login sessions. IEEE Transactions on Pattern Analysis and Machine Intelligence 29(4), 687–700 (2007)
18. Lau, E., Liu, X., Xiao, C., Yu, X.: Enhanced user authentication through keystroke biometrics. Final project report, Massachusetts Institute of Technology (December 2004)

19. Monrose, F., Rubin, A.: Keystroke dynamics as a biometric for authentication. Journal of Future Generation Computer Systems 16(4), 351–359 (2000)
20. Omote, K., Okamoto, E.: User identification system based on biometrics for keystroke. In: Varadharajan, V., Mu, Y. (eds.) ICICS 1999. LNCS, vol. 1726, pp. 216–229. Springer, Heidelberg (1999)
21. Shanmugapriya, V., Padmavathi, G.: Keystroke dynamics authentication using neural network approaches. In: Das, V.V., Vijaykumar, R. (eds.) ICT 2010. CCIS, vol. 101, pp. 686–690. Springer, Heidelberg (2010)
22. Stefan, D., Shu, X., Yao, D.: Robustness of keystroke dynamics based biometrics against synthetic forgeries. Journal of Computers and Security 31(1), 109–121 (2011)
23. Teh, P.S., Teoh, A.B.J., Tee, C., Ong, T.S.: Keystroke dynamics in password authentication enhancement. Journal of Expert Systems with Applications 37(12), 8079–8089 (2010)
24. Van Den Broek, E.L.: Beyond biometrics. In: Proceedings of the International Conference on Computational Science, The Netherlands (2010)
25. Yampolskiy, R.: Computer Analysis of Human Behavior, 1st edn. Springer (2011)

Map-Matching in a Real-Time Traffic Monitoring Service[*]

Piotr Szwed and Kamil Pekala

AGH University of Science and Technology, Poland
pszwed@agh.edu.pl, kamilkp@gmail.com

Abstract. We describe a prototype implementation of a real time traffic
monitoring service that uses GPS positioning information received from
moving vehicles to calculate average speed and travel time and assign
them to road segments. The primary factor for reliability of determined
parameters is the correct calculation of a vehicle location on a road segment, which is realized by a *map-matching algorithm*. We present an
a new incremental map-matching algorithm based on Hidden Markov
Model (HMM). A HMM state corresponds to a road segment and a sensor reading to an observation. The HMM model is updated on arrival of
new GPS data by alternating operations: expansion and contraction. In
the later step a part of determined trajectory is output. We present also
results of conducted experiments.

Keywords: ITS, GPS, map-matching, Hidden Markov Model, Viterbi.

1 Introduction

With the growing number of vehicles traveling on public roads, traffic congestion has become a serious problem in urban areas. It results in great variation of
travel times, has negative impact on planning in logistics, causes increased fuel
consumption and pollution. Real-time traffic monitoring is an important functionality of various Intelligent Transportation Systems (ITS) aiming at alleviate
this problem. Such systems can utilize traffic data originating from various types
of sensors being a part of road infrastructure: inductive loops, cameras and mi-
crophone arrays. Another source can be personal smartphone devices capable of
receiving positioning data and transferring them over cellular networks.

In this paper we describe a prototype implementation of a real time traffic
monitoring service developed within INSIGMA project [1]. The system uses GPS
positioning information received from moving vehicles and o calculate average
speed and travel time and assign them to road segments.

The primary factor for reliability of determined parameters is a correct calculation of a vehicle location on a road segment. This key task is realized by
a *map-matching algorithm*, which in the presence of uncertain sensor readings

[*] This work is supported by the European Regional Development Fund within IN-
SIGMA project no. POIG.01.01.02-00-062/09.

S. Kozielski et al. (Eds.): BDAS 2014, CCIS 424, pp. 425–434, 2014.
© Springer International Publishing Switzerland 2014

should select the most accurate link and the most likely vehicle position. The decision can be based on the current value obtained from a sensor or on the history comprising a number of past data.

The character of the developed system puts some requirements related to the map-matching algorithm applied. To give reliable results it should take into account road connectivity, what excludes such popular approaches, as point to curve matching. Moreover, to provide real-time operation the algorithm should be *incremental*, i.e. be capable of analyzing GPS trace on arrival of new data. This paper presents a new incremental map-matching algorithm based on Hidden Markov Model and discusses its application within the traffic monitoring system. In our approach a HMM state corresponds to a road segment and a sensor reading to an observation. The algorithm updates the HMM model on arrival of new GPS data by alternating operations: expansion (new states are added to the model) and contraction (dead ends are deleted, the graph root is moved forward along the detected path and a part of trajectory is output). We discuss results of initial experiments conducted for 20 GPS traces, which to test algorithm robustness, were modified by introduction of artificial noise and/or downsampled.

The paper is organized as follows: related works are reported in section 2. In the next section 3 the operational concept of the system is presented. It is followed by section 4, which introduces the HMM model used by the developed algorithm. Its description is provided in the next section 5. Conducted experiments are reported in section 6 and finally section 7 gives concluding remarks.

2   Related Works

Map-matching algorithms are core components of various Intelligent Transportation Systems. Their applications include personal navigation systems, traffic monitoring [18,8,4], vehicle tracking and fleet management [6].

More then thirty map-matching algorithms are surveyed by Quddus et. al in [13]. Authors divided them into four groups: geometric, topological, probabilistic and advanced. Algorithms employing geometric analysis take into account only shapes of road segments ignoring, how they are connected. The simplest approach consists in finding the closest map node (a segment endpoint) to the current GPS reading (point-to-point matching). Another option is to find the closest road segment (point-to-curve matching) [19] or to match pairs of points from the vehicle trajectory to the road segments [5].

Topological map-matching algorithms utilize information about connections between road segments. This removes leaps between map links that can be observed for algorithms based only on geometrical information [14].

Many positioning devices are capable of delivering a circular or elliptic confidence region associated with each position reading. The idea behind probabilistic algorithms is to select in the match mapping process only those road segments that intersects with the confidence region. If several candidates are found, only one of them with the highest probability is selected. Such approach was discussed in [11].

Advanced algorithms usually combine both topological and probabilistic information applying various techniques to assign road links to GPS readings: Kalman Filter, fuzzy rules [3] or particle filter [7].

Several map-matching algorithms are path-oriented, i.e. they maintain a set of candidate paths. In the algorithm developed by Marchal et al. [9] they were stored in a collection being sorted according to the path score based on distance to GPS trace. An idea of using a tree like structure representing a set of candidate paths was proposed by Wu et al. [20]. Both algorithms are *incremantal*, i.e. they update the path representation on arrival of new GPS reading.

Hidden Markov Model (HMM) [15] is a Markov process comprising a number of hidden (unobserved) states. Transition between states can occur with a certain probabilities. Each state is assigned with a set of observations. One of them is to be output as the state is reached. For a given state conditional probabilities of observations occurrence (*emission probabilities*) sum up to 1. A problem that can be elegantly formulated with HMM is the *decoding* problem: it consists in finding the most probable sequence of transitions between hidden states that would produce the given sequence of observations. Such sequence can be efficiently determined with the well-known Viterbi algorithm.

Application of HMM for map matching was discussed in [10,17]. In both papers hidden states correspond to projections of vehicle positions on road segments and observations to location data obtained mainly from GPS sensors. Transition probabilities are established based on links connectivity, whereas emission probabilities assume Gaussian distribution of GPS noise.

3   Operational Concept

The operational concept of the traffic monitoring system is shown in Fig. 1. The system receives (1) raw data from mobile terminals equipped with GPS receivers. They may buffer GPS readings in the internal memory and send a number of records in bulk feeds. The system may also accept data from a simulator.

1. Raw samples are *preprocessed*. This step includes trajectory smoothing with Kalman filter and interpolation of points between GPS readings.
2. Then cleansed and normalized samples (2) are submitted to a component implementing *map-matching* algorithm. The algorithm requires the map data. In the described implementation its source is OpenStreetMap (OSM) [12]. To accelerate computations, the map data are stored in the memory. At the output (3) trajectories of tracked vehicles are fed into the database.
3. Traffic parameters (4) are determined with periodically activated *Traffic parameters calculation* process and stored in the another database. The step involves also data aggregation based on values and timestamps. We calculate two parameters: average speed and traversal time for a road link.
4. The traffic parameters assigned to road links are to be used by other components of an ITS (not shown in the figure): route planning [16] or traffic control. For the testing purposes the system provides also *map visualization*.

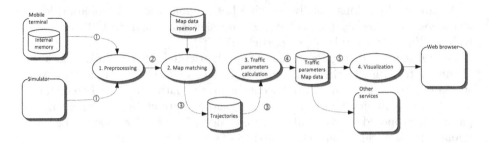

Fig. 1. Operational concept

4   Hidden Markov Model

In this section we discuss construction of the Hidden Markov Model that is the core concept used in the developed map-matching algorithm. Following the OSM structure we define a road network model as a directed graph $G = (V, E, I)$, where $V \in \mathbb{R}^2$ are graph nodes described by two coordinates: longitude and latitude, $E \in V \times V$ are straight road segments linking two nodes. Pairs of edges belonging to $I \subset E \times E$ can be used to specify forbidden maneuvers at junctions. Such information is available in OSM.

For a given edge $e = (v_b, v_e)$ and a point g, we define the projection $p(g, e)$ of g onto e as a point g' laying on e minimizing the distance, i.e.

$$p(e, g) = \underset{g' = v_b + t(v_e - v_b) \wedge t \in [0,1]}{\arg \min} d(g, g'), \tag{1}$$

where $d(g, g')$ is a distance between g and g' given by the *haversine* formula [2]. The projection point $p(e, s)$ calculated according to (1) can be either an orthogonal projection on a segment or one of its end points (see Fig. 2).

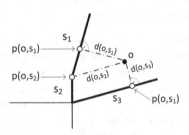

Fig. 2. Projections of a GPS point o and distances to road segments s_1, s_2 and s_3

A state in a Hidden Markov Model describe both a road segment and a projection of a GPS fix on the segment calculated according to formula (1). Thus, each state tuple $(e, p, i) \in Q$ has the following components: e - a road segment, p - a projection point and i a sequence number. In the assumed model observations O correspond to data obtained from GPS sensor, i.e. they are tuples (x, y, t), whose elements are longitude, latitude and time respectively.

Below we give the definition of Hidden Markov Model reflecting adaptation introduced to support the map matching problem.

Definition 1 (Hidden Markov Model). *Hidden Markov Model is a tuple* $\lambda = (Q, A, O, P_t, P_o, q_0)$, *where* Q *is a set of states*, $Q \subset E \times \mathbb{R}^2 \times \mathbb{N}$, $A \subset Q \times Q$ *is a set of arcs*, O *is a set of observations*, $O \subset \mathbb{R}^2 \times \mathbb{R}$, $P_t \colon A \to (0, 1]$ *is a function that assigns a probability to a transition between states*, $P_o \colon Q \times O \to [0, 1]$ *is an emission probability function satisfying* $\forall q \in Q \colon \sum_{o \in O} P_o(q, o) = 1$ *and* q_0 *is an initial (root) state.*

Two states $q_1 = (e_1, p_1, i_1)$ and $q_2 = (e_2, p_2, i_2)$, where $e_1 = (v_{11}, v_{12})$ and $e_2 = (v_{21}, v_{22})$, can be connected with an arc $a = (q_1, q_2)$, if $e_1 = e_2$ or there exists a path in a graph $\pi = v_{11}, \ldots v_{22}$ linking endpoints of road segments. Currently, in most cases we consider sequences of length 3, i.e. two consecutive segments having common endpoints. This assumption imposes the requirement that observations (locations obtained form a GPS sensor) should be dense enough to be assigned to consecutive segments. The interpolation conducted as a part of preprocessing (see Fig. 1) was introduced to satisfy this requirement.

To calculate the transition probability for an arc a linking states q_1 and q_2 a weight function $\theta(a) \colon A \to [0, 1]$ is used. Basically, it assigns 1 if q_1 and q_2 can be connected by a path, however if the possible path violates traffic rules or physical constraints (e.g. speed greater than 250 km/h) a small value (0.1) is used. Finally, the weights assigned to outgoing arcs for a given state q are normalized applying the formula (2) to give the probabilities.

$$P_t(a) = \frac{1}{Z_t} \theta(a),$$
$$\text{where } Z_t = \sum_{\substack{a_i \colon a_i = (q, q_i) \in A \\ a = (q, q_a)}} \theta(a_i). \tag{2}$$

Emission probability P_o is computed for a subset of states in HMM Q_H and an observation o. For a given HMM state $q = (e, p, i)$, where $p = (x_p, y_p)$ is the vehicle position, its GPS observations o can be distributed on XY plane around the point p. We have assumed 2-dimensional normal distribution given by (3).

$$P(x, y) = \frac{1}{D} e^{-k((x - x_p)^2 + (y - y_p)^2)}. \tag{3}$$

The D normalizing factor is given as $D = \int_{-\infty}^{\infty} \int_{-\infty}^{\infty} P(x, y) \, dx \, dy$. For k the value 0.01 was taken, what corresponds to noise giving translations of GPS readings by 10m. In such case $D \approx 314.0$. As the map data used in experiments used longitude and latitude coordinates, we applied, however, a modified version of (3), in which Euclidean distance was replaced by the haversine formula.

5 Map Matching Algorithm

The algorithm takes at input a sequence of GPS readings (observations) $\omega = (o_i \colon i = 1, n)$ and constructs a sequence of Hidden Markov Models $\Lambda = (\lambda_i \colon 0 = 1, n)$. Basically, it contains two stages: *initialization*, during which the first model

λ_1 is built and *processing* that is repeated for successive observations to give models $\lambda_2, \ldots, \lambda_n$. The processing stage is further decomposed into *expansion* and *contraction* .

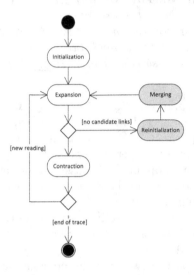

Fig. 3. Steps of the map-matching algorithm

Initialization. In this step a set of possible states (road segments), to which the initial vehicle position might be assigned is determined. The algorithm examines all road segments in a supplied part of the map and chooses only these, whose distance to the measured point if below a certain threshold r (e.g. 35 meters). At that point the construction of a HHM sequence, that can be perceived as a trajectory tree, begins. The tree root is set to a fictional state from which the vehicle might have moved to any of the states belonging to initial model λ_1.

Expansion. A new model λ_i is build for a given observation o_i. The GPS location embedded in o_i is projected on a set of road segments represented as states in λ_{i-1} with the timestamp $i-1$ as well as on segments connected to them according to the map model. The set of candidate segments is limited to those, for which the distance to GPS point is below the threshold r. New states and links are added to λ_i and probabilities are calculated as described in Section 4.

Contraction. The contraction stage has two goals: firstly orphan nodes without successors are removed, what keeps the detection model compact, secondly the HMM root is moved forward and a next part of the trajectory is output. Fig. 4 gives an example of HMM model being in fact a union of λ_4 and λ_5. The state numbering adopts the convention that q_{ij} is the j-th state added in the i-th step. States marked with the white color are to be removed during the contraction operation for the model λ_4.

The subgraph between q_{12} and q_4 is an interesting pattern that we call a *join*. Presence of a join in HMM indicates that during the map matching process vehicle positions were assigned to parallel roads that finally joined at a certain point. Hence, the algorithm faces the problem of selecting the most probable among at least two competing paths. In such situation the Viterbi algorithm is launched (yielding in the discussed example the path $(q_{12}, q_{23}, q_{33}, q4)$ and the HMM root moves to q_4.

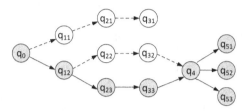

Fig. 4. Example of HMM, during the contraction step in 4th iteration all states filled with white color are to be deleted

An exceptional situation in the expansion phase occurs, if for a given observation o_i it is not possible to find candidate road links, on which the point would be projected. We may conclude then, that the map matching algorithm got lost. They may be several reasons of such situations. It may stem from noise that was not sufficiently removed by the Kalman filter. The other reason can be that observations are not dense enough to be matched to neighbor map segments. We handle this issue by performing *reinitialization* and obtaining a new model λ_{i0}. Depending on application the models λ_{i-1} and λ_{i0} can be merged. In the case of traffic parameters calculation some tracking errors can be accepted. Thus, λ_{i-1} model is processed with the Viterbi algorithm to get the most probable path and the whole matching process restarts from λ_{i0}.

The Viterbi algorithm is also used to find the most probable path in the last model λ_n (in the case, when the input sequence of observations is finite.)

6   Results

The algorithm was tested on the map of Kraków in Poland. The map originated from OpenStreetMap project [12]. The input dataset was represented by 20 GPS traces, which were recorded during several car trips throughout Kraków with EasyTrials GPS[1] software running on iPhone 5. The total length of traces used in experiments was 148.46km. Both input and map-matched trajectories were stored in GPX format that is supported by JOSM, the OpenStreetMap editor.

During the tests the collected data were fed in four forms: original, modified by artificially introduced random noise (magnitude between 0 and 20 meters added to each sample), half sampled (H-S) and half-sampled with the noise.

As a basic quality criterion we have taken the number of reinitializations for a given trace. Analyzing the traces manually we realized that forced reinitialization

---

[1] http://www.easytrailsgps.com/

Table 1. Test results

No	Length (km)	Samples	Original		Noise		H-S		H-S & Noise	
			RI	RIS	RI	RIS	RI	RIS	RI	RIS
1	9.45	256	0	0	3	0.012	3	0.012	3	0.012
2	8.26	248	3	0.012	10	0.04	2	0.008	1	0.004
3	7.78	261	2	0.008	5	0.019	1	0.004	2	0.008
4	7.67	267	0	0	6	0.022	3	0.011	5	0.019
5	9.67	233	3	0.013	0	0	0	0	0	0
6	6.40	209	0	0	0	0	0	0	0	0
7	5.59	108	0	0	0	0	0	0	1	0.009
8	9.03	259	0	0	10	0.039	0	0	3	0.012
9	7.05	216	1	0.005	1	0.005	1	0.005	0	0
10	7.94	248	2	0.008	4	0.016	1	0.004	0	0
11	7.19	190	0	0	1	0.005	1	0.005	10	0.053
12	11.22	273	1	0.004	7	0.026	1	0.004	2	0.007
13	4.19	118	0	0	0	0	1	0.008	1	0.008
14	5.96	192	2	0.01	2	0.01	0	0	2	0.01
15	9.03	271	0	0	2	0.007	0	0	0	0
16	6.45	242	1	0.004	3	0.012	2	0.008	3	0.012
17	7.95	228	4	0.018	7	0.031	3	0.013	3	0.013
18	7.41	192	1	0.005	3	0.016	2	0.01	2	0.01
19	7.06	283	4	0.014	7	0.025	2	0.007	6	0.021
20	3.16	188	0	0	2	0.011	0	0	1	0.005
Total	148.47	4482	24	0.005	73	0.016	23	0.005	45	0.010

Fig. 5. The map with marked average speed values. Legend: red [0,20); yellow [20,50); green [50,90); blue: [90,∞]. Assumed units: km/h.

in about 30% of cases results in lost of traffic information for a given road segment or in a bad assignment to a neighbor link.

The obtained values are gathered in Tab. 1. Each table row shows test results for a particular GPS trace. Subcolums marked with *RI* give number of algorithm reinitializations in the selected mode, *RIS* denotes average number of reinitalizations per sample. The best results were achieved by running the tests with the original input data. However, for applied half-sampling the number of reinitalizations was practically identical. This effect can be probably attributed to the interpolation. The worse indicator value was obtained for noisy data. Nevertheless, all obtained values are fairly good. In the normal mode the reinitialization occurred once per 6.18km and for 0.5% samples.

The algorithm implemented in C# language and published as a RESTfull web service was capable of processing 20 simultaneous feeds with 50 times speed-up, i.e. time intervals between subsequent send operations were 50 times smaller then differences between sample timestamps. This corresponds to 1000 mobile sensors feeding real-time data simultaneously.

For testing purposes we have also implemented a web based visualization of calculated traffic parameters. Example results (for unmodified dataset) is presented in Fig. 5.

7 Conclusions

This paper discusses a prototype real-time traffic monitoring system based on GPS positioning information originating from traveling vehicles. The input data are passed through Kalman filter, then normalized by interpolation and finally delivered to the component implementing map-matching algorithm, which determines vehicle trajectories obtained by matching GPS data with a road network stored in a digital map. In turn, the trajectories are a basis for calculation of traffic parameters.

We describe a map matching algorithm based on Hidden Markov Model. In each iteration it updates the HMM by expanding it with new states corresponding to road segments and contracting to output a next part of a vehicle trajectory. The structure of obtained HMM in most cases forms a tree similar to that proposed by Wu et al. [20]. However, our model accepts parallel roads. Compared to earlier works [10,17], our algorithm described in Section5 is incremental, i.e. it does not build a HMM model for a given GPS trace to be analyzed afterwards with the Vitrebi algorithm, but on each input updates the HMM model and, if possible, outputs next trajectory points.

References

1. INSIGMA project, `http://insigma.kt.agh.edu.pl` (last accessed: December 2013)
2. CodeCodexWiki: Calculate distance between two points on a globe, `http://www.codecodex.com/wiki/Calculate_Distance_Between_Two_Points_on_a_Globe` (online: last accessed: December 2013)
3. Fu, M., Li, J., Wang, M.: A hybrid map matching algorithm based on fuzzy comprehensive judgment. In: Proceedings of the 7th International IEEE Conference on Intelligent Transportation Systems, pp. 613–617 (2004)

4. Google Official Blog: The bright side of sitting in traffic: Crowdsourcing road congestion data, http://googleblog.blogspot.com/2009/08/bright-side-of-sitting-in-traffic.html (online: last accessed: December 2013)

5. Greenfeld, J.S.: Matching GPS observations to locations on a digital map. In: National Research Council (US). Transportation Research Board. Meeting (81st: 2002: Washington, DC). Preprint CD-ROM (2002)

6. Gurtam: Commercial GPS solutions for vehicle tracking and fleet management, http://gurtam.com/en/ (online: last accessed: December 2013)

7. Gustafsson, F., Gunnarsson, F., Bergman, N., Forssell, U., Jansson, J., Karlsson, R., Nordlund, P.J.: Particle filters for positioning, navigation, and tracking. IEEE Transactions on Signal Processing 50(2), 425–437 (2002)

8. INRIX: Inrix home page, http://www.inrix.com/default.asp (online: last accessed: December 2013)

9. Marchal, F., Hackney, J., Axhausen, K.: Efficient map-matching of large GPS data sets-tests on a speed monitoring experiment in Zurich. Arbeitsbericht Verkehrs-und Raumplanung 244 (2004)

10. Newson, P., Krumm, J.: Hidden Markov map matching through noise and sparseness. In: Proceedings of the 17th ACM SIGSPATIAL International Conference on Advances in Geographic Information Systems, pp. 336–343. ACM (2009)

11. Ochieng, W.Y., Quddus, M., Noland, R.B.: Map-matching in complex urban road networks. Revista Brasileira de Cartografia 2(55) (2009)

12. OpenStreetMap: OpenStreetMap Wiki (2013), http://wiki.openstreetmap.org/wiki/Main_Page (Online; accessed December 2013)

13. Quddus, M.A., Ochieng, W.Y., Noland, R.B.: Current map-matching algorithms for transport applications: State-of-the art and future research directions. Transportation Research Part C: Emerging Technologies 15(5), 312–328 (2007)

14. Quddus, M.A., Ochieng, W.Y., Zhao, L., Noland, R.B.: A general map matching algorithm for transport telematics applications. GPS Solutions 7(3), 157–167 (2003), http://dx.doi.org/10.1007/s10291-003-0069-z

15. Rabiner, L., Juang, B.: An introduction to hidden Markov models. IEEE ASSP Magazine 3(1), 4–16 (1986)

16. Szwed, P., Kadluczka, P., Chmiel, W., Glowacz, A., Sliwa, J.: Ontology based integration and decision support in the Insigma route planning subsystem. In: FedCSIS, pp. 141–148 (2012)

17. Thiagarajan, A., Ravindranath, L., LaCurts, K., Madden, S., Balakrishnan, H., Toledo, S., Eriksson, J.: Vtrack: accurate, energy-aware road traffic delay estimation using mobile phones. In: Proceedings of the 7th ACM Conference on Embedded Networked Sensor Systems, pp. 85–98. ACM (2009)

18. University of California, Berkeley: Mobile millenium project, http://traffic.berkeley.edu/ (online: last accessed: December 2013)

19. White, C.E., Bernstein, D., Kornhauser, A.L.: Some map matching algorithms for personal navigation assistants. Transportation Research Part C: Emerging Technologies 8(1), 91–108 (2000)

20. Wu, D., Zhu, T., Lv, W., Gao, X.: A heuristic map-matching algorithm by using vector-based recognition. In: International Multi-Conference on Computing in the Global Information Technology, ICCGI 2007, p. 18 (2007)

The Extended Structure
of Multi-Resolution Database

Krystian Kozioł[1], Michał Lupa[2], and Artur Krawczyk[3]

[1] AGH University of Science and Technology,
Faculty of Mining Surveying and Environmental Engineering,
Department of Geomatics,
al. A. Mickiewicza 30, 30-059 Kraków, Poland
krystian.koziol@agh.edu.pl
http://www.wggiis.agh.edu.pl
[2] AGH University of Science and Technology,
Faculty of Geology, Geophysics and Environment Protection,
Department of Geoinformatics and Applied Computer Science,
al. A. Mickiewicza 30, 30-059 Kraków, Poland
mlupa@agh.edu.pl
http://www.geoinf.agh.edu.pl
[3] AGH University of Science and Technology,
Faculty of Mining Surveying and Environmental Engineering,
Department of Mine Areas Protection, Geoinformatics and Mine Surveying,
al. A. Mickiewicza 30, 30-059 Kraków, Poland
artkraw@agh.edu.pl
http://www.wggiis.agh.edu.pl

Abstract. The aim of the paper is to show how to extend the database of topographic objects for the purposes of automatic data generalization. Generalization is performed basing on topographic databases, that exist in various scales (BDOT10K - scale 1:10 000, BDOO - general geographic objects database 1:100000). Under the law, these databases are fundamental source of information about the spatial location and characteristics of topographic objects in Poland. BDOT is a multi-resolution database, which is fed by the objects with varying degrees of details and accuracy, depending on the data source. It is also assumed that the BDOT will be a source of data for the editorial of standard cartographic studies for various scales. Objects for a given scale are obtained by the generalization processes, which allow to generate less detailed data (for example, in a scale of 1:50 000) from the reference dataset (scale 1:10 000). The authors have proposed the structure of MRDB system, which assumes the existence of Web Generalization Services (WGS). These services can provide a remote access to simplification algorithms and the data generalization "on the fly" [5, 4]. It can significantly reduce the process of data producing through automation of the manual work, resulting in a significant optimization of the database at the level of its power. Therefore, we can observe the optimization of the work with databases, already at the level of their feeding.

Keywords: spatial databases, generalization, MRDB, WGS, topographic database, GIS.

S. Kozielski et al. (Eds.): BDAS 2014, CCIS 424, pp. 435–443, 2014.
© Springer International Publishing Switzerland 2014

1   Introduction

The basic functionality of contemporary IT systems, especially Geographic Information Systems (GIS), is the processing of spatial data stored mainly in databases, which often have the ability to process certain selected geometric objects [1, 22, 11]. All of this led to the fact, that the data previously cataloged in analog form - using paper maps, were subjected to the process of digitization, and this required the development of suitable methods for archiving data. Unfortunately, these solutions have a lot of drawbacks, in which the most difficult to accept are the problems with low efficiency of data operations as well as the constraints on data processing and thus the generalization of data.

The first one, as shown in the work [2, 3, 17], can be solved using optimization methods for SQL queries. Unfortunately, in the case of the data, in which the geometry of the objects is different depending on the adopted scale, this approach has the limitations in the form of queries, which can be created only between the tables with same level of detail (e.g.1:5000) [17, 7]. Furthermore, these databases are the source for the editorial process of cartographic maps, according to the agreed model of DCM (Digital Cartographic Model) and WMS/WFS (Web Map Service/Web Feature Service) servers, sharing maps and spatial data in the network. This approach to pure spatial databases would be highly inefficient, due to the need of processing high detail objects. However, the perception varies depending on the scale and some maps do not require the high-resolution data. For example, the objects in the case of small-scale maps are mostly distorted or invisible and some of them change their structure (e.g. from surface to point) [10]. Therefore, the generalization requires an appropriate structure of the database. The structure of database, in which successive data representations are created in the process of reference data generalization, can be called as a multi-resolution. Moreover, the correct process of data generalization is extremely complicated, due to the many aspects of visualization and various topographic data contexts. For this reason, the existing solutions were dominated by a huge amounts of decisions that needed to be taken by the expert (cartographic database administrator) as well as low level of unambiguity (lack of processes automation). The work on the automation of the generalization processes is currently one of the main research problems in the processing of spatial data [16, 18].

2   The Database of Topographic Objects in the SDI Structure

European Parliament directive 2007/2/EC of 14 March 2007 establishes a consistent spatial information infrastructure (ESDI - European Spatial Data Infrastructure) for the European Union and the EFTA. In addition, INSPIRE implies the harmonization of systems operating at different levels of particulars, which means that the multi-resolution database will be the source for databases with less details. Implementation of SDI defines the requirements with respect to topographic objects databases, concerning the harmonization and interoperability of the data. Currently, topographic objects are recorded and stored at three levels of details: BDOT500 (LoD1), BDOT10K (LoD2) and BDOO [19–21]. The database of topographic objects (BDOT) should be a conceptually consistent and nationwide system for collecting, storing, viewing, sharing and managing topographic data at three levels of generalization. Among the competencies of BDOT, it also should include data management, financing and organizing the maintenance of this resource in timeliness. The repository of BDOT at the appropriate level of detail should satisfy the expectations of local and national administration. Furthermore, under the law, a topographic objects database serves as the primary source of information about the characteristics and location of spatial objects. However, the characteristics of the data collected at three levels of detail are different and thus it is difficult to provide a unified model of these databases. The topographic objects databases, at levels from 1: 500 to 1: 5000 (BDOT500) have a registration character, at the level of 1:10 000, BDOT10K is a base topographic reference for standard cartographic studies in terms of medium scales, while the general geographic objects database (BDOO) is responsible for the small scale data.

The topographic database structure will vary according to the level of detail, which is referred by the scale. Therefore, the differences are visible also at the level of the base feeding. For example, data sources of BDOT10K (scale 1:10 000) include records collected in the state of geodesy and cartography, that is responsible for: register of land and buildings, national register of borders and surface units of territorial divisions of the country, the national register of geographical names, register of the villages, streets and addresses, aerial and satellite imaging. However, in the field of building classes, built-up area and territorial development, the main source of the data will be BDOT500 (scale 1:500 or lower). Due to the large amounts of data used in creating and updating BDOT10k, there are used a multiple sources of reference data. In addition, the supplement and verifying of the data from registers are obtained by terrain interview.

3   The Concept of the Generalization in MRDB

The hierarchical structure of database forces the feed system of generalized data to lead them to the appropriate level of detais. Therefore, it has become necessary to develop algorithms for automatic generalization of the data, as well as feeding and updating databases with lower details "on the fly". Previously proposed models of BDOT do not have all the necessary information which are required to properly perform the process of generalization. Consequently, it is necessary to supplement the multi-resolution database with topographic information from other data schemas. This allows to synthesize geographic information into a single coherent database schema. Synthesis embrace not only the geometry, but also the construction and editing data dictionaries of the attributes. Supplemented model of BDOT allows to generate derivative spatial objects. Spatial objects assume the ready-made base tuples form, which contain information about the object geometry and assigned attributes.

As a result of the research presented in this paper, the authors decided to build an automatic generalization system using geo-portal software type. Its concept provides for the feed and updating derivatives and less detailed databases in an automatic way. Undoubtedly, one of the key functionalities of the system will be the remote access to the spatial processing algorithms as network processing services (Web Processing Services - WPS). WPS services are defined in OGC standard as a Web Generalization Services (WGS), because of its generalization nature [5, 4].

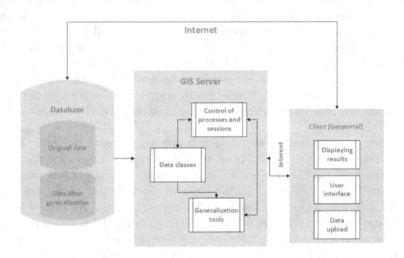

Fig. 1. Reduction of a point on a line

The diagram above (Fig. 1.) illustrates the designed MRDB system with a three-tier architecture. Data layer is a spatial database, which is responsible for the validation of input data, the completeness and correctness of topological objects sent by the user (geoportal). The database allows also to store and share information, obtained as a result of generalization processes, as well as to supply databases with less details. The logic of the system is located in GIS server side, which provides the appropriate generalization tools and algorithms as WGS services. GIS server also exercises protection over the processes and sessions. The presentation tier is a web service (geo-portal), responsible for the management and visualization of the dynamic maps (before and after generalization).

4   The Schemes of Data Acquisition

The primary data source for the generalization system should be the objects BDOT500 database. However, most of these objects requires to be extended with the attributes from other sources. A very important feature of the system architecture are the detailed classes of objects cast, belonging to the basic map,

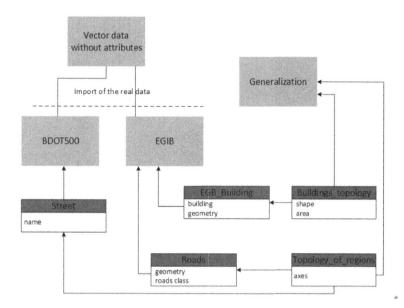

Fig. 2. Illustration of data flow for selected classes of MRDB in the generalization process

on the generalized MRDB model (Fig. 2.). A generalized MRDB model was completely subordinated to the geometry.

The diagram above, concerning the "edges", consists of the geometry and attributes originated from the processing of database BDOT500. The road parcel defined in the BDOT500 model is joined to the road parcel from EGiB (The Lands and Buildings Register) model. Integration of spatial data in MRDB is based on two major types of schemas: "topology of buildings" and "topology of regions", which are essential in the process of generalization. This schema element is temporary and dynamic, but, nevertheless, it can be the source of attribute values in derivative database or the subject of generalization tools [8, 9, 15].

Currently, the database schema structure of BDOT500 does not provide for one important type of objects, which are the edges of linear objects (according to theory of graphs). This problem can be solved by the skeletonization (Fig. 3.). The most popular algorithms of skeletonization are the methods of computational geometry, such as: Delaunay triangulation, Voronoi diagrams, Straight Skeletons, Medial Axis Transform [14, 12, 13, 6, 23].

One of the key operations of generalization is the data segmentation. The segmentation is achieved by construction of regions, basing on the linear objects. This process is a great illustration of the integration of two dato one database structure for generalization. The segmentation consists of following steps:

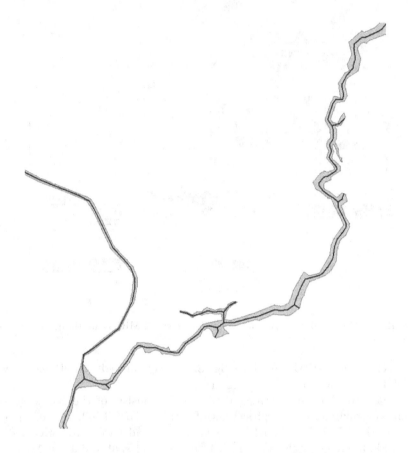

Fig. 3. Examples of the river polygon skeletonization

- generation of the elementary regions,
- classification of the streets into four groups, according to their attribute - class, which is determined by the category of its management (for example, class 1 - national roads),
- the construction of the regions, basing on a prior classification.

As a result of this process, the database is fed by five new classes of objects - elementary regions and four classes of other, component regions (Fig. 5.).

Creation of the regions is a process, in which the edges of the different classes roads are combined in a linear closed chain. The start and end nodes of the roads are joined in the way to form a closed figure, which will be a region of the planar graph. To perform this process properly, the planar graph should have a correct topology for the network objects (roads, railways, rivers). What is more, the performed classification of the structural regions, formed from the road network, will allow the automatic elimination of the less important elements (regions classified below). Nevertheless, it is important to clarify the decisive factor in

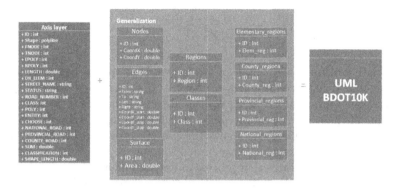

Fig. 4. Structure of the topological database in relation to the structure of communication network edges

Fig. 5. An example of building objects clustering at different levels of the network hierarchy

the removal of the considered region. Elimination of the regions removes its all internal objects.

5 Conclusions

Solution is based on a segmentation of objects from spatial database is not a new issue, and it has been applied in practice. However, there are no tools for applying segmentation on the geographical objects. It is due to the necessity of using value of their descriptive attributes which often differ between databases. Standardization of BDOT base allows the use of predefined attributes in segmentation of processed regions. Automatic grouping of objects is one of the key components in the generalization process of data for small scales.

As a result, costly, repeated acquisition of the same data for the individual scales of BDOT can be avoided. The lack of topology in geographical data of BDOT standard is still the main concern. The topology is a key element in the process of linear segmentation. It is impossible create consistent levels of hierarchy of extracted data regions without correct topology. The authors also defined a structure of database which allows to process spatial data in the process of segmentation.

The proposed solution allows, (in the future) the segmentation of geographic features to be used as a service network processing WPS (Web Processing Service). This application, due to its nature, is called the network service generalization WGS (Web Generalization Service) also.

Acknowledgments. This work was financed by the National Science Center (Poland), under the agreement No. N N526 064340, as a part of the project: "Automatic data feeding into database and actualisation of data concerning buildings in the multiresolution topographic database".

References

1. OpenGIS. Implementation Specfication for Geographic information - Simple feature access - SQL option, version 1.1 (November 22, 2005) http://www.opengeospatial.org/standards/sfs
2. Bajerski, P.: Optimization of geofield queries. In: Proceedings of the 2008 1st International Conference on Information Technology, pp. 1–4 (2008)
3. Bajerski, P., Kozielski, S.: Computational model for efficient processing of geofield queries. In: Cyran, K.A., Kozielski, S., Peters, J.F., Stańczyk, U., Wakulicz-Deja, A. (eds.) Man-Machine Interactions. AISC, vol. 59, pp. 573–583. Springer, Heidelberg (2009)
4. Bloch, M., Harrower, M.: A map generalization web service. In: Proceedings of AutoCarto, pp. 1–11 (2006)
5. Burghardt, D., Neun, M., Weibel, R.: Automated processing for map generalization using web services. Journal Geoinformatica 13(4), 425–452 (2009)

6. Christensen, A.H.J.: Line generalization by waterlining and medial-axis transformation. successes and issues in an implementation of perkal's proposal. Cartographic Journal 37(1), 19–28 (2000)
7. Chrobak, T., Keller, S., Koziol, K., Szostak, M., Zukowska, M.: Podstawy Cyfrowej Generalizacji Kartograficznej. Wydawnictwa AGH (2007)
8. Chrobak, T., Koziol, K.: Digital cartographic generalization of buildings layer in creating data of the topographical database. Archives of Photogrammetry, Cartography and Remote Sensing 19, 59–69 (2009)
9. Chrobak, T., Koziol, K., Krawczyk, A., Lupa, M.: Koncepcja architektury systemu generalizacji obiektów przestrzennych na przykadzie zabudowy. Roczniki Geomatyki 10(7), 7–14 (2012)
10. Chrobak, T., Koziol, K., Krawczyk, A., Lupa, M., Szombara, S.: Automatyzacja procesu generalizacji dla wielorozdzielczej bazy danych. Wydawnictwa AGH (2013)
11. DRAFT ISO/IEC 13249-3:1999: Information technology - Database languages - SQL Multi-media and Application Packages - Part 3: Spatial International Organization for Standardization, version 68 (April 04, 2003), http://jtc1sc32.org/doc/N1101-1150/32N1107-WD13249-3--spatial.pdf
12. Gold, C., Snoeyink, J.: A one-step crust and skeleton extraction algorithm. Algorithmica 30(2), 144–163 (2001)
13. Haunert, J., Sester, M.: Area collapse and road centerlines based on straight skeletons. GeoInformatica 12(2), 169–191 (2008)
14. Jones, C., Bundy, G., Ware, J.: Map generalization with a triangulated data structure. Cartography and Geographic Information Science 22(4), 317–331 (1999)
15. Koziol, K.: Eliminacja obiektów liniowych z zastosowaniem regionów strukturalnych na przykadzie sieci drogowej. Roczniki Geomatyki 4(3), 109–117 (2006)
16. Koziol, K.: Operatory generalizacji warstwy zabudowy. Roczniki Geomatyki 10(7), 45–57 (2012)
17. Lupa, M., Piórkowski, A.: Spatial query optimization based on transformation of constraints. In: Gruca, A., Czachórski, T., Kozielski, S. (eds.) Man-Machine Interactions 3. AISC, vol. 242, pp. 627–636. Springer, Heidelberg (2014)
18. Mackaness, W., Ruas, A., Sarjakoski, L.: Generalisation of geographic information: cartographic modelling and applications. Elsevier Science (2011)
19. Minister Administracji i Cyfryzacji: Rozporzadzenie z dnia 12 lutego 2013 r. w sprawie bazy danych geodezyjnej ewidencji sieci uzbrojenia terenu, bazy danych obiektów topograficznych oraz mapy zasadniczej, dz.u. 2013 nr 0 poz. 383
20. Minister Spraw Wewnetrznych i Administracji: Rozporzadzenie z dnia 17 listopada, w sprawie bazy danych obiektów topograficznych oraz bazy danych obiektów ogolnogeograficznych, a takze standardowych opracowan kartograficznych, DzU nr 279, poz. 1642
21. Minister Spraw Wewnetrznych i Administracji: Rozporzadzenie z dnia 20 pazdziernika 2010 w sprawie ewidencji zbiorów i usług danych przestrzennych objetych infrastruktura informacji przestrzennej, DzU nr 201, poz. 1333
22. Piorkowski, A.: Mysql spatial and postgis - implementations of spatial data standards. Electronic Journal of Polish Agricultural Universities 14(1), 1–8 (2011)
23. Szombara, S.: Application of elementary triangle in collapse operator of digital cartographic generalisation process. In: 4th Doctoral Seminar on Geodesy and Cartography, Olsztyn, Poland, June 9-10, Wydawnictwo Uniwersytetu Warmińsko-Mazurskiego, Heweliusza (2012)

Platform for Storing and Searching Different Formats of Spatial Data

Przemysław Kulesza and Michał Wójcik

Maritime Institute in Gdańsk, Długi Targ 41/42, 80-830 Gdańsk, Poland
{przemyslaw.kulesza,michal.wojcik}@im.gda.pl

Abstract. The BalticBottomBase (BBB) project aims at sharing geospatial data on the South Baltic Sea environment among researchers from different research units. The platform, which is developed as a part of the project, contains modules for storing, processing, serving and searching of geospatial data. One of the main challenges is to provide a unified way to store, describe and present different types of geospatial data stored using different formats. The platform is developed as a distributed infrastructure to better utilise available resources and uses open-source solutions in order to achieve cost efficiency.

The focus of this paper is set on examining part of BBB system used to store and serve Geographic Information System (GIS) data. The paper presents different configurations of BBB system components including storing and map serving tools. Properties (including performance and maintainability) of each components configuration have been considered and the best combination of components have been chosen to fulfil BBB project requirements. These requirements include performance of serving GIS data to end users and ease of maintenance. All the tests have been carried out in real life scenarios.

Keywords: BalticBottomBase, Geographic Information System, storage, database, GIS, map server, searching.

1   Introduction

Maritime Institute in Gdańsk manages a large amount of South Baltic data which has been gathered from different sources and represents different types of data, e.g. oceanographic, bathymetric, geological, sonar, spatial planning and magnetic field data. The data are stored in various formats like raster and vector files as well as raw data fetched from measurement devices or specialized software databases. The main goal of the BalticBottomBase (BBB) project (http://balticbottombase.eu/info) is to make the data more readily available for other institutions (e.g. scientific community, local and state authority, maritime safety agencies and community as a whole) in a convenient way and in compliance with geographic data sharing standards.

The paper presents a distributed platform for GIS (Geographic Information System) which allows serving, storing and searching data. BBB system uses four

S. Kozielski et al. (Eds.): BDAS 2014, CCIS 424, pp. 444–453, 2014.

different types of GIS data which have been described in this paper. Individual functions of the system like storing and serving spatial data are performed by independent components. The components need to cooperate in order to make GIS data available to the end user in a standardized form. Performance of the system as a whole does not depend only on the performance of each component alone, but also on a particular combination. Therefore a more system wide performance measurement is necessary. The paper presents performance tests of several combinations of components to select the most appropriate configuration of storing and serving components. Configurations have been reviewed from performance and BBB project requirements perspective. Chosen components have been implemented in BBB system which is currently used in commercial projects and tested by Maritime Institute in Gdańsk and cooperating research units.

2   Known GIS Storage and Dissemination Systems

There are many open source and commercial GIS storage and serving (generating map images and disseminating it over internet protocols) systems available. Map servers can cooperate with different GIS storage methods what allows for a selection of appropriate solutions for the defined issue. These products differ in such aspects as:

- support for Open Geospatial Consortium (OGC) [14] standards,
- number of supported data types,
- reliability of provided services,
- performance of processing and publishing (registering data in map server) data,
- portability between different operating systems,
- cooperation with other GIS systems.

We have described some of the well known systems both for storage and serving of GIS data.

2.1   Database Storage Systems

One of the two most popular storage systems for spatial data, PostGIS, is based on an open source object-relational database system PostgreSQL [23]. PostGIS provides support for both vector and raster data and allows for spatial queries to be executed in native SQL. Moreover it supports all the objects and functions specified in the OGC standards.

Another popular spatial data storage system is Oracle Spatial, which is a commercial alternative to PostGIS. Oracle Spatial is an Oracle Database option, which adds support for spatial data (vector and raster). It takes benefits from database technology developed by Oracle [21]. The solution can be integrated with Oracle Data Mining (ODM) applications to provide spatial analysis, which allows to detect spatial correlations (impact of one spatial object on another) or collocation (associations between spatial objects in relation to non-spatial attributes). Oracle Spatial has useful functionalities like Routing Engine (computes routes, distances, times between locations) which can be used in the area of logistics or transportation.

MSSQL (Microsoft SQL Server databases) [12] is an example of widely used storage system with built-in support for spatial data. It allows for storing vector geometry and geography data which complies with the OGC standards. SQL Server contains functions used to make spatial queries and spatial indexes. Disadvantage of this product is lack of raster data type. It stores raster as a regular Binary Large Object (BLOB) without any metadata allowing for spatial analysis. SQL Server can not be installed on another operating system than Microsoft Windows.

Raster support is available in both PostGIS and Oracle Spatial. PostGIS provides a `Raster` type to store all data and metadata information about rasters in one table [22]. Oracle Spatial stores metadata and information about retrieving raster data in a separate table that contains column `SDO_GEORASTER`. The table for data storage is called Raster Data Table and includes a `SDO_RASTER` type which contains BLOB (Binary Large Object) `RASTERBLOCK` [20] columns.

PostGIS and Oracle database supports storing of GIS vector and raster data. Oracle database is an example of a commercial product which provides vendor support, contains a large number of additional functionalities (e.g. Routing Engine, integration with Oracle Data Mining) and is based on a mature database engine. PostGIS is a reliable and a more popular solution as a data source in open-source and commercial GIS products than Oracle Spatial.

2.2   Map Servers

Most of the GIS data is presented in a form of a map. The data can consume a large volume of storage which makes them problematic to be served directly through the network. In order to deal with these problems, GIS data serving applications are used. Their role is to serve particular map (or only its part selected using geographic coordinates) displayed in a human readable manner. The most popular solutions in this case are MapServer and GeoServer platforms [27,26,8,10,2].

MapServer is a mature, standalone platform for publishing GIS data. It is based on the technology called Common Gateway Interface (CGI) [25,9]. This allows MapServer to be compatible with a large group of web servers supporting CGI scripts. Configuration of a single map is stored in a text configuration file (*mapfile*) and modified manually or by mapscript API for which bindings in multiple programming languages are available.

GeoServer [7] uses Java web technologies, which allow deployment in many servlet containers like Apache Tomcat [1] and application servers like Glassfish [18]. The main advantage of GeoServer is the ease of installation and use. All administration activities can be done with a web or RESTful interface [19]. Support for new formats can be easily added without the need for server code recompilation.

Both MapServer and GeoServer are open source projects that expose interfaces in accordance with the Web Map Service (WMS) [16] and the Web Feature Service (WFS) [15] standards.

ArcGIS for Server [5] is an example of a commercial product which has capabilities for serving GIS data through web services [19]. ArcGIS for Server services can be integrated with a number of clients (e.g. desktop or mobile) using a large range of OGC standards. Administration tasks can be preformed analogously to GeoServer with the use of REST interfaces. ArcGIS for Server can run on 64-bit systems with the ability to scale horizontally (adding new servers) in order to reach an additional boost of performance.

3   Sources of GIS Data

Measuring instruments in the Maritime Institute in Gdańsk provide data in different formats. There are four main categories of data types which have been taken into consideration:

- geospatial vector data — describes geographical features using geometrical shapes (points, lines and polygons) linked to attribute values (e.g. name, unit, administrative area code), stored using Esri shapefile [4] format,
- geospatial raster data — describes measurement using non scalable (usually extremely high-resolution) raster images with embedded georeferencing information (e.g. map projection, coordinate system, ellipsoids, datums, etc.), the most popular format is GeoTIFF [11,24],
- profile data — describes subsequent soil layers below the bottom of the sea measured in a straight line along the water surface,
- sample data — describes subsequent soil layers below the bottom of the sea measured using single deep sea probe.

Rasters and shapefiles (vector data) are gathered from a file system. Profiles and samples can be fetched from specialized geological and geophysical software which stores their data in Microsoft Access [13] and MSSQL. Profile data consists of two parts: raw data (proprietary binary format of a Survey Engine Seismic tool [3]) and soil layers interpretation (stored in MSSQL database). The sample data are available in GeODin [6] Microsoft Access databases.

Raster data collected from measurement devices can vary in size according to their resolution and area of measurement. The data usually consists of one band (layer of pixel values) and its size ranges from a dozen of megabytes to a couple of gigabytes (3 GB) per single set of data. The size of a single shapefile data is much smaller than the size of raster data and usually ranges from a couple of kilobytes to one megabyte.

4   Storing GIS Data

PostGIS and PostgreSQL have been chosen as a underlying database because it is very mature and reliable solution used by other open-source libraries and tools. In addition the PostGIS database can register raster in one of two modes available for **Raster** type:

- by storing raster binary data directly in database (in-db),
- by storing file system path to the raster data in the database (out-db).

A raster layer can be stored and published in three ways:

- by storing the concrete raster type in the database as binary data (in-db) and publish it as a PostGIS connection,
- by storing the concrete raster type in the database as a path to a file in a file system (out-db) and publish it as a PostGIS connection,
- by storing raster as a file in a file system and publish its path directly.

Shapefile layer can be stored and published in two ways:

- by storing a shapefile in a database as a table and publish it as a PostGIS connection,
- by storing a shapefile in a file system and publish its path.

Profile and sample data, as presented by the BBB system, consist of two main parts. First part is responsible for storing location about data and is represented as a point or a line (spatial metadata). Second part contains detailed information about soil layers. Spatial metadata for both of types of data can be defined in two ways:

- by storing location data in a database as a geometry type,
- by storing location data in a shapefile in a file system and publish its path.

These methods differ in performance and complexity of maintenance. All of the methods of storing GIS data have been analyzed from the performance perspective.

5   Serving Performance Tests

Several possibilities for serving GIS data were considered. Based on project requirements the choice of technologies was narrowed to the following set of candidate technologies: GeoServer, MapServer. Both map servers allow for publishing data directly from file system or database.

Tests were performed to examine the performance of:

- processing a single request send by a single user;
- processing a single request send concurrently by a group of 20 users.

Each measured result is an average of 300 consecutive requests. Measurements were made by jMeter load testing tool. Tests were performed on the Ubuntu 12.10 system with Core i7-2630QM processor, 8GB memory and 5400 rpm hard drive. Map servers has been configured to serve GIS data without using built-in cache mechanisms.

Raster tests include a representative one band raster dataset composed of:

- a small raster (21 MB) in 739x1941 resolution,
- a medium raster (91 MB) in 7490x4055 resolution,
- a large raster (216 MB) in 19704x10960 resolution.

The first test takes into account the performance of serving raster data stored in the database and the file system. Unfortunately the out-db method of storing raster data in a database was not measured due to errors in PostGIS 2.0 [17]. In the current stable version 2.1 of PostGIS this issue has been fixed, but the fix is not currently compatible with PostgreSQL versions supported by the Red Hat Enterprise Linux operating system on which the release product is deployed.

Fig. 1a shows a huge difference in performance for raster data stored in a database and a file system. In the case of a large raster, retrieving it from the file system is approximately 660 times faster. Time of retrieving raster data from database increases with data size. On the other hand serving time of rasters stored in the file system does not seem to depend on the file size, at least in the measured size range. As a result of significant differences in performance and being prone to errors, registering raster in database as a `Raster` type was not considered in the subsequent tests.

Fig. 1b and Fig. 1c show performance of retrieving different kinds of raster data published in MapServer and GeoServer. In both cases it can be seen that better performance can be achieved with the use of MapServer.

Shapefile (vector) data used in tests has been divided into two sets which represent typical data used in the BBB system. First set contains four shapefiles with different size while second differs in number of geographic objects. Each shapefile in set consists of polygons.

Shapefile samples with a different size:

- sample 1 – shapefile with the size of   7 kB containing 1 geographic objects,
- sample 2 – shapefile with the size of   13 kB containing 1 geographic objects,
- sample 3 – shapefile with the size of   26 kB containing 1 geographic objects,
- sample 4 – shapefile with the size of   51 kB containing 1 geographic objects.

Shapefiles samples with different number of small polygon objects:

- sample 1 – shapefile with the size of   30 kB containing 100 geographic objects,
- sample 2 – shapefile with the size of   60 kB containing 200 geographic objects,
- sample 3 – shapefile with the size of   139 kB containing 400 geographic objects,
- sample 4 – shapefile with the size of   240 kB containing 800 geographic objects.

Fig. 1 shows that in all cases storing shapefiles in a file system is the most efficient way in both MapServer and GeoServer. MapServer is clearly slower in all performed tests and the most in shapefiles stored in a database. Fig. 1e and 1g points that different number of geographic objects have a negligible impact on time of serving data unlike one large object (Fig. 1d, Fig. 1f). There is also a huge difference in case of MapServer while serving shapefile from database and file system. In GeoServer this difference is much smaller.

Based on raster performance results MapServer takes lead over GeoServer what is clearly shown in Fig. 1c. Solution based on storing shapefiles as files is more efficient but requires attribute redundancy. The same shapefile attributes should be stored in a database and in a shapefile what is troublesome from the maintenance perspective. This means that all shapefiles should be stored in a database.

Despite more complex data management MapServer has been chosen as a map serving tool for serving the data in the BBB project. The greatest impact on this decision has the fact that rasters cover approximately 80% of all data which will be stored in BBB system and MapServer is more efficient in serving this kind of data.

6   The BBB System Architecture

The choice of the architecture for the BBB project and services was based on the test results presented beforehand, ease of data handling and project specific conditions like a searching engine. Architecture based on storing the GIS data in a database is less complicated in maintenance in comparison to storing all the data directly on a file system. If the data are stored on a file system then there is an added complexity resulting from the need for information redundancy: keeping path to data on disk in a searchable way in a database and additionally in a map server layer configurations. Storing the data in a database is more consistent because all GIS data is kept in one place. Additionally, the method is more convenient, because of the possibility of usage of spatial analysis functions in a database. Managing and analysing the GIS data located in a file system requires external tools or libraries as opposed to database which already contains functions that operate on spatial data.

Specific solutions for all categories of data are described below:

– Rasters are stored as files in a file system, which is the most suitable solution from the performance and reliability perspective. Current GIS tools are error prone when handling rasters stored in a database, what has been mentioned together with performance tests.
– Shapefiles are registered in a database as tables. The BBB project requires access to shapefile attributes using the web interface. The easiest method is based on loading shapefile file to a database as a single table with geometry and non-geometry attributes as columns.
– Sample location and description data is fetched from GIS application GeODin which stores it in Microsoft Access database. In the BBB system this data is stored directly in a database as a geometry type because soil layers are simply represented as subsequent information about depth and layer.
– Profile location data is available from marine survey tool Survey Engine Seismic. Location of this data is stored in a MSSQL database and further loaded to the BBB database as a geometry type. Data describing profile is

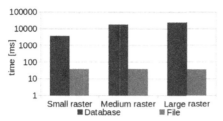

(a) Performance of retrieving raster data
stored in database and file system

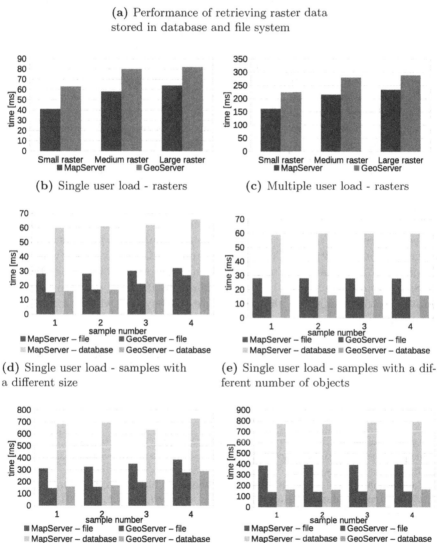

(b) Single user load - rasters

(c) Multiple user load - rasters

(d) Single user load - samples with
a different size

(e) Single user load - samples with a different number of objects

(f) Multiple user load - samples with
a different size

(g) Multiple user load - samples with a
different number of objects

Fig. 1. Comparison of performance in publishing raster and shapefile data

complex and is usually represented as tens of thousands of points. Storing this data in a database for a large number of profiles causes performance issues. To avoid this problem non geographic data of a profile is stored as a file.

7   Conclusions

It can be seen that the platform for storing, serving GIS objects can be successfully built in a cost efficient way with the usage of free software without proprietary solutions. Individual components used to store and serve GIS data can be combined in different configurations. Performed tests show difference in performance of each configuration of storing methods and map servers. Configuration focused on performance of serving GIS data and ease of maintenance various types of data has been chosen.

Presented tests compare different ways of publishing raster and vector GIS data using two popular map serving solutions: GeoServer and MapServer accompanied by PostgreSQL database server enhanced with PostGIS. Performance from the end user's point of view depends on the exact combination of these components. Performed tests show which solution is best in the context of BBB project case. When it comes to the performance of serving data to the end user and ease of maintenance MapServer and PostGIS are the best configuration.

The presented configuration of map server and storage method were implemented and tested in a production environment. Commercial projects conducted by The Maritime Institute in Gdańsk are currently using BBB system where benefits of chosen configuration are already seen. Performance of serving GIS data has been favourably accepted by users in Maritime Institute in Gdańsk and co-operating research units. In the near future BBB system will be available for external users.

Acknowledgements. Project co-financed by the European Regional Development Fund under the Operational Programme Innovative Economy.

References

1. Apache Software Foundation: Apache Tomcat online documentation, http://tomcat.apache.org/tomcat-7.0-doc/introduction.html (access: May 18, 2013)
2. Bandyophadyay, M., Pratap Singh, M., Singh, V.: Integrated visualization of distributed spatial databases an open source web-gis approach. In: 2012 1st International Conference on Recent Advances in Information Technology (RAIT), pp. 619–621 (2012)
3. Coda Octopus Products Ltd.: Survey Engine Seismic+, http://www.codaoctopus.com/products/survey-engine-seismic (access: November 28, 2013)
4. Environmental Systems Research Institute, Inc.: ESRI Shapefile Technical Description (1998)

5. ESRI: ArcGIS for Server, `http://www.esri.com/software/arcgis/arcgisserver` (access: December 02, 2013)
6. Fugro: GeODin Software, `http://www.geodin.com/` (access: November 28, 2013)
7. The GeoServer Team: GeoServer online documentation, `http://docs.geoserver.org/` (access: May 18, 2013)
8. Huang, Z., Xu, Z.: A method of using geoserver to publish economy geographical information. In: 2011 International Conference on Control, Automation and Systems Engineering (CASE), pp. 1–4 (2011)
9. Kropla, B.: Beginning MapServer, Open Source GIS Development. Apress (2005)
10. Lin, B., Zhang, C., Wang, X., Fan, J.: Design of electric webgis sharing platform based on openscales. In: 2011 2nd International Conference on Artificial Intelligence, Management Science and Electronic Commerce (AIMSEC), pp. 2955–2958 (2011)
11. Mahammad, S.S., Ramakrishnan, R.: Geotiff - a standard image file format for gis applications. In: Conference Proceedings of Map India 2003(2003)
12. Microsoft: Microsoft SQL Server 2012 (2012), `http://technet.microsoft.com/en-us/library/bb933790.aspx` (access: January 28, 2014)
13. Microsoft Corporation: Microsoft Access, `http://office.microsoft.com/en-us/access/` (access: November 28, 2013)
14. Open Geospatial Consortium: Standards and Supporting Documents, `http://www.opengeospatial.org/standards` (access: May 18, 2013)
15. Open Geospatial Consortium: Web Feature Service standard, `http://www.opengeospatial.org/standards/wfs` (access: May 18, 2013)
16. Open Geospatial Consortium: Web Map Service standard, `http://www.opengeospatial.org/standards/wms` (access: May 18, 2013)
17. Open Source Geospatial Foundation: ST_AsBinary semantic discrepancy defect, `http://trac.osgeo.org/postgis/ticket/2217` (access: May 18, 2013)
18. Oracle: GlassFish Open Source Edition documentation, `https://glassfish.java.net/docs/index.html` (access: May 18, 2013)
19. Oracle: RESTful Web Services, `http://docs.oracle.com/javaee/6/tutorial/doc/gijqy.html` (access: May 18, 2013)
20. Oracle: Oracle Spatial 11g GeoRaster, An Oracle Technical White Paper (2007)
21. Oracle: Advanced Spatial Data Management for Enterprise Applications, An Oracle White paper (2010)
22. PostGIS Raster Team: PostGIS Raster documentation (2007), `http://trac.osgeo.org/postgis/wiki/WKTRaster` (access: May 18, 2013)
23. The PostgreSQL Global Development Group: PostgreSQL online documentation, `http://www.postgresql.org/docs/` (access: May 18, 2013)
24. Ritter, N., Ruth, M.: GeoTIFF Specification, Revision 1.0 (1995)
25. Team, T.M.: MapServer online documentation. Open Source Geospatial Foundation, `http://mapserver.org/documentation.html` (access: May 18, 2013)
26. Wang, J., Su, T.: Design and implementation of a webgis-based marine geophysical information sharing platform. In: 2010 Second IITA International Conference on Geoscience and Remote Sensing (IITA-GRS), vol. 2, pp. 393–396 (2010)
27. Wang, W., Luo, H.: Design and construction of environmental monitoring data public system based on webgis platform. In: 2010 18th International Conference on Geoinformatics, pp. 1–5 (2010)

The Impact of Geometrical Objects Generalization on the Query Execution Efficiency in Spatial Databases

Michał Lupa and Adam Piórkowski

AGH University of Science and Technology,
Faculty of Geology, Geophysics and Environment Protection,
Department of Geoinformatics and Applied Computer Science,
al. A. Mickiewicza 30, 30-059 Kraków, Poland
{mlupa,pioro}@agh.edu.pl
http://www.geoinf.agh.edu.pl

Abstract. This paper presents the problem of geometrical objects generalization in relation to the geospatial queries efficiency. Authors in their research focused on a qualitative assessment of data, which are derived from the lossy generalization processes. Furthermore, they also attempted to determine the critical level of profitability of lossy generalization. There was also examined a correlation between the reduction of data details by removing vertices and a geospatial query execution speed.

Keywords: spatial databases, object generalization, query optimization.

1 Introduction

Spatial databases have a key role in contemporary information systems, which are based on the processing of geometric objects. Considerable technological progress in the field of computer graphics has enabled the simultaneous development of digital cartography, where reality was mapped using vector objects. The next step was the creation of Geographic Information Systems, in short - GIS. Unfortunately, in most cases, these systems are used only as a storage for spatial data, and the processing of these data mostly takes place in specialized programs, outside of database management systems (DBMS). Data, previously cataloged in analog form - using paper maps, were subjected to the digitalization processes, which required the development of suitable methods for data storage. The development of spatial data storage and processing methods has resulted in the SQL language extensions, namely, the first standard by OGC [2,1], adding basic operations on points and shapes, and then the second, which is a separate section of the standard SQL/MM (ISO) concerning spatial data (SQL/MM - Spatial) [10]. All these solutions have made that today spatial databases are used in many branches of industry, in particular, where the appropriate spatial analysis are a key task.

S. Kozielski et al. (Eds.): BDAS 2014, CCIS 424, pp. 454–464, 2014.

Therefore, we can observe a significant number of the spatial database uses, which rely on performing rapid analysis and generating results in the shortest time (analysis in crisis management, logistics). Unfortunately, as it is shown in [21,4], query execution times are often burdened with significant overhead of time. Moreover, solutions to this problem are difficult to obtain, and thus - are not widely known. Among the studies of increasing the efficiency of queries we observe a tendency in which the authors are focused on the SQL syntax. One of them is the work [17], associated with algebraic transformations, concerning query selectivity [3]. The authors [4,6,5] have drawn attention to the possibility of using Peano algebra to the query decomposition, whereas in the case of [16,20], the authors have proposed rule-based query optimization, at the stage of its development. A special case is a spatial database and geospatial queries, whose effectiveness often depends only on the volume of geometric data [22]. Therefore, it is worth to draw attention to the problem of geometric objects generalization, where the reduction of redundant or less important points have a considerable impact on increasing the efficiency of queries [7,8,14].

2 Methods of Reducing Vertices in Objects

2.1 Lossless Reduction

One of the methods of spatial data generalization is a lossless reduction. This method involves the removal of redundant and unnecessary points, in such a way that this does not change the shape of the figures or loss of any information relevant to the analysis. An example of lossless reduction is described below.

The analyzed object is a closed convex shape. Sequence of three points (A, B, C) can be simplified without any consequences for the changes in the shape by rejecting the central (B) and allowing the extremes (A, C) (Fig. 1.), if the points (A, B, C) lie on a straight line, and the middle point (B) is not a node of another sequence (Fig. 2).

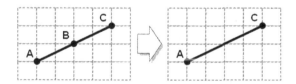

Fig. 1. Reduction of a point on a line

2.2 Lossy Reduction

The case of lossy reduction embrace the situation where the vertices (points) do not lie on a straight line. The removal of point B causes a loss of information about the shape of the figure (Fig. 3). In some cases, the results of the measurements (or data obtained in the process of automatic digitalization) are affected

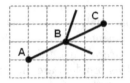

Fig. 2. A node

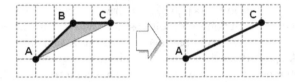

Fig. 3. Lossy reduction

by some errors, for this reason, some points may not severely affect on the shape of the objects. Similarly, in the case of the figures with large number of vertices, the reduction of some of them does not cause large changes in shape. Therefore, the vertex rejection (reduction) should be done only under certain criteria, which will allow to minimize the loss of information [12].

The Criteria of Lossy Reduction. Lossy reduction has to be carried with a criteria that will organize an algorithms, as well as to minimize errors, which could be accompanied with generalization, if this would not be in any way controlled [13]. The process of generalization could then be done as:

- an independent point procedure - the removal of point without checking its importance in the shape, sample algorithms of removing the n-th point [25], or random elimination [24];
- local processing procedure - consists of considering the direct neighborhood of the point (two adjacent vertices), which will be discussed further;
- procedures for the extended local processing, where the context of the study includes more than three points [15,19,9,18].

The most common methods of generalization, in which each point is analyzed separately, are the local processing procedures. The exemplary methods are the algorithms proposed by Jenks [11]. These methods involve the removal of the point which is a vertex of the angle whose measure does not exceed a certain value (in the range of half full angle). The second approach is to set the distance from the vertex of the straight line joining the adjacent points. In both cases, this may apply to removing the large-surface triangle, despite the fact, that the distance (or angle) were insignificant (Fig. 4). This criterion is questionable in the case of triangles with large shoulders, due to its influence on the shape.

The criterion that the authors have proposed during the studies described in this paper is a criterion based on the minimization of the loss of the surface object area. The vertex will be removed if the loss of the area after reduction is minimal. In case where the changes are slight (e.g. a given percent of the surface of the object), the reduction is acceptable, and the loss of data precision (such as the surface and the probability of incorrect assessment of the relationship between objects) is under control.

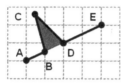

Fig. 4. The problem of vertex reduction with forms of acute angle

3   The Vertex Reduction for Convex Figures

The vertices reduction in convex figures is the lossy reduction, which runs basing on algorithm of minimization of the surface loss. The authors proposed also the division of the objects using the criterion of convexity, due to differences in the behavior of objects during generalization. They will be described in more detail in the next section of the article.

3.1   The Adaptive Algorithm

Most of the algorithms implementing the removal of vertices are based on the assumption of a constant threshold of data loss (for example, the maximum angle, distance from the point to the line, the triangle). The authors suggest an adaptive approach - a figure is tested by removing the vertices in the order implementing the increasing loss of data quality. This approach is certainly much more time consuming and computationally complex, but the process of generalization is realized only once.

The adaptive algorithm allows for a single, local analysis of each of the vertices of the figure and, based on the adopted criterion, determines its removal. In the next iterations the i-th vertex is selected, for which the surface area of a triangle formed between two adjacent vertices (i-1, i+1) is the smallest. The vertex is reduced, and then there is a new object edge between the vertices i-1 and i+1 created. A similar algorithm was proposed by Visvalingam [26]. Below there is the adaptive algorithm, presented as a pseudo-code form:

```
min := start_value
while vertices_count > 3
{
    for i = 0 to N-1 do
    {
        triangle := triangle(i-1, i, i+1)
        if area(triangle) < min then
        {
            min := area(triangle)
            vertice_to_del = i
        }
    }
    if(stop_criteria == true)
        break
    delete vertice_to_del
}
```

The algorithm can be stopped by testing a sample stop conditions:

- the total change of shape exceeded the predetermined threshold, 15% of the area,
- next removed points reduce the surface area of the figure by more than predetermined percentage of the area assigned to one point removed (cost of point reduction in quality is greater than the gain of the acceleration process).

3.2   The Results

The most important stage of the research was to develop a measure that would allow to capture adequately the moment, in which further reduction of vertices becomes meaningless. Due to the significant reduction of the figure area, the spatial query time gain would be disproportionate to the loss of information about the geometry of the analyzed object. The authors decided to demonstrate experimentally the moment of profitability reduction, by consulting the loss of the area in relation to the reducing query execution times.

The aim of the first part of the study was to analyze the loss of a convex object area in subsequent iterations. For this purpose, the authors made a test object, consisting of 24 vertices, which then was reduced by the procedure described in section 3.1. The input object (Fig. 5) and the results (Fig. 6) are shown in figures. All geospatial queries, which operate on geometric objects in real applications, process large amounts of data. The analysis of each spatial object is associated with reading of its binary representation - Well Known Binary (WKB). Therefore, the time of this processing can be proportionate to the number of object vertices. For example, the calculation of the area of any polygon is performed by dividing the figure for the component triangles, which area is calculated and then they are added up. Therefore, it is worth to check, what effect the generalization has on the query execution times.

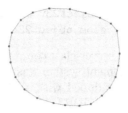

Fig. 5. The input object - visualization using ArcGIS software

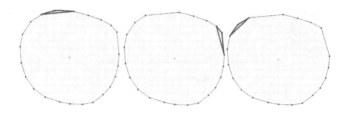

Fig. 6. The next three triangles (iterations) of the algorithm

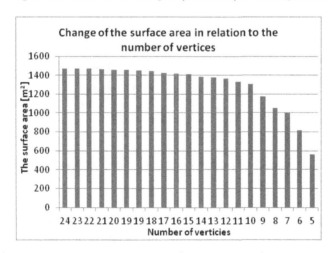

Fig. 7. Change of the surface area in relations to the number of vertices

The test data was the objects with different area and various numbers of vertices (from 24 to 4). For the purpose of the tests, there was created a tables containing a thousand rows (objects). In order to verify the time execution gain, the authors proposed two following geospatial queries.

- Z1 - The area of sum of objects:

  `SELECT Area(ST_Union(object.geom)) FROM object;`

- Z2 - The difference of object pairs, by 24, 23 vertices, 22, 21, vertices, 20, 19, etc.:

```
SELECT * FROM object24v, object23v WHERE
ST_Difference(object24v.geom,object23v.geom);
```

Both queries were performed in server-class computer running on Windows 2008 RT. The used database management system was PostgreSQL 9.0 with PostGIS 2.0 extension. The server was equipped with an Intel Core i7-2600 CPU@3.40 GHz processor and 16 GB RAM DDR3 333 MHz memory. Execution speed of each query was verified experimentally ten times. The results of the query Z1 execution times [ms] are presented in graph (Fig. 8). The graph shows the minimum query execution times, obtained during tests 1-10, which were taken as representative, due to the small scatter of values. Summary of query Z2 results

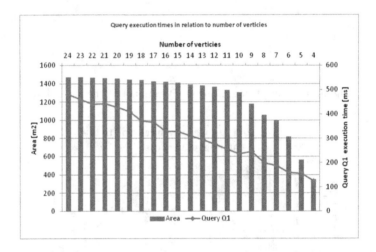

Fig. 8. Query Z1 execution times (line) and shape area reduction (bars) in relation to number of vertices

is shown in the graph (Fig. 9) The graph shows the minimum times, obtained in the tests from 1 to 5. By observing the results it is easy to see that the processing time is linearly related to the number of vertices of the shape. The situation is different with the quality of the data after generalization. Reduction of points at first does not change much the quality of shape (here defined as the reduction of surface area), however, beyond a certain moment, we have to deal with considerable loss of quality. The task ahead is to develop a multi-criteria optimization, which ensures a compromise between acceleration of query execution times, and the loss of data quality.

4   The Vertex Reduction for Concave Figures

Adaptation of the minimum loss of surface algorithm in the case of the concave shapes causes significant fluctuations in the total area of the object (next

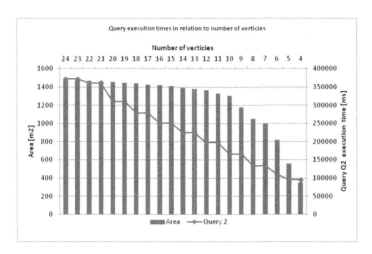

Fig. 9. Query Z2 execution times (line) and shape area reduction (bars) in relation to number of vertices (pairs of the objects). Visualization using ArcGIS software.

Fig. 10. The input concave object

iterations of algorithm provides excess or insufficiency of object geometry with respect to the geometry of the initial size). This situation is illustrated by tests performed on an concave object with 47 vertices (Fig. 10). Variation of the area, typical to concave figures is given in the graph bellow (Fig. 11). This property cause that the adaptation algorithm described in section 3.1 does not bring the expected results, both in terms of detail reduction and increased efficiency of query execution. All of this lead to inability to determine the measure of profitability of lossy reduction. This kind of figures is most common, real or generated [23]. The problem of the concavity of the figures can be solved by the application of minimum convex hulls, through which there will be made an conversion from concave to convex figures. This method will also allow for an assessment of the degree of differentiation of the edge of the figure, which determines the level of growth or loss of the area, by comparing the original figures with its convex hulls. This assessment may serve as a parameter decision-making in the context of the selection methods of simplifying the process of multi-criteria optimization.

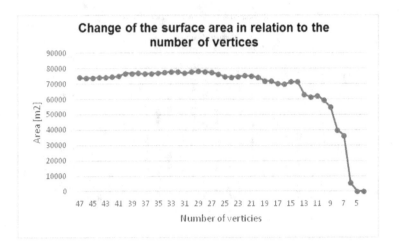

Fig. 11. Change of the surface area in relations to the number of vertices (concave objects)

5   Conclusions and Further Work

This paper describes methods of lossless and lossy generalization of geometric objects with respect to the efficiency of geospatial queries. The authors proposed the algorithm for vertices reduction, which is based on the minimization of the loss of generalization object area. Experimentally, there has been shown an acceptable level of loss of information in relation to the processing time gain, which can be a measure of the profitability of lossy generalization. There was also presented the problem of concave shapes, which can be solved based on the method of minimum convex hulls. Another task is the construction of new measures of cost-effectiveness for concave figures which might properly evaluate the quality of the shape of the generalized figures. The solution to this problem may be a part of the decision-making adaptive algorithm, in which the using of the methods and criteria for the optimization will depend on the concavity (or convexity) of the object. Moreover, the authors intend to develop a method of generalization, where the simplification of the figure will be based on ranking list of vertices. Rank will be formed basing on the loss of the surface area, after removing of vertex and angle measurement, which will be created by the adjacent arms. Further study of the authors will also focus on the development of appropriate methods of multi-criteria optimization. The criteria of the query execution time in relation to data loss will be extended with a recognition at a predetermined scale.

Acknowledgments. This work was financed by the AGH - University of Science and Technology, Faculty of Geology, Geophysics and Environmental Protection, Department of Geoinformatics and Applied Computer Science as a part of statutory project.

References

1. OGC - The Open Geospatial Consortium, http://www.opengeospatial.org/
2. OpenGIS Implementation Specfication for Geographic information - Simple feature access - SQL option, http://www.opengeospatial.org/standards/sfs
3. Augustyn, D.R.: The method of query selectivity estimation for selection conditions based on sum of sub-independent attributes. In: Gruca, D.A., Czachorski, T., Kozielski, S. (eds.) Man-Machine Interactions 3. AISC, vol. 242, pp. 601–609. Springer International Publishing, Switzerland (2014), http://dx.doi.org/10.1007/978-3-319-02309-0_65
4. Bajerski, P.: Optimization of geofield queries. In: Proceedings of the 1st IEEE International Conference on Information Technology, pp. 1–4 (2008)
5. Bajerski, P.: How to efficiently generate pnr representation of a qualitative geofield. In: Cyran, K., Kozielski, S., Peters, J., Stanczyk, U., Wakulicz-Deja, A. (eds.) Man-Machine Interactions. AISC, vol. 59, pp. 595–603. Springer, Heidelberg (2009), http://dx.doi.org/10.1007/978-3-642-00563-3_62
6. Bajerski, P., Kozielski, S.: Computational model for efficient processing of geofield queries. In: Cyran, K.A., Kozielski, S., Peters, J., Stanczyk, U., Wakulicz-Deja, A. (eds.) Man-Machine Interactions. AISC, vol. 59, pp. 573–583. Springer, Heidelberg (2009), http://dx.doi.org/10.1007/978-3-642-00563-3_60
7. Chrobak, T.: Podstawy cyfrowej generalizacji kartograficznej. AGH Uczelniane Wydawnictwa Naukowo-Dydaktyczne (2007), http://books.google.pl/books?id=QfeVGAAACAAJ
8. Chrobak, T., Koziol, K., Krawczyk, A., Lupa, M., Szombara, S.: Automatyzacja procesu generalizacji dla wielorozdzielczej bazy danych. AGH Uczelniane Wydawnictwa Naukowo-Dydaktyczne (2013)
9. Douglas, D.H., Peucker, T.K.: Algorithms for the reduction of the number of points required to represent a digitized line or its caricature. The Canadian Cartographer 10(2), 112–122 (1973)
10. International Organization For Standardization: ISO/IEC 13249-3:1999, Information technology - Database languages - SQL Multi-media and Application Packages - Part 3: Spatial (2000)
11. Jenks, G.F.: Lines, computers, and human frailties. Annals of the Association of American Geographers 71(1), 1–10 (1981)
12. Koziol, K.: Comparison of selected simplification algorithms on the example of a representative test area. Annals of Geomatics 9, 49–57 (2011)
13. Koziol, K.: Generalisation operators of buildings layer. Annals of Geomatics 10, 45–57 (2012)
14. Kozioł, K., Szombara, S.: New method of creation data for natural objects in mrdb based on new simplification algorithm. In: 26th International Cartographic Conference. International Cartographic Association (2013)
15. Lang, T.: Rules for robot draughtsman. The Geographical Magazine 42, 50–51 (1969)
16. Lupa, M., Piorkowski, A.: Rule-based query optimizations in spatial databases. Studia Informatica 33, 105–115 (2012)
17. Lupa, M., Piorkowski, A.: Spatial query optimization based on transformation of constraints. In: Gruca, D.A., Czachorski, T., Kozielski, S. (eds.) Man-Machine Interactions 3. AISC, vol. 242, pp. 621–629. Springer International Publishing, Switzerland (2014), http://dx.doi.org/10.1007/978-3-319-02309-0_67

18. McMaster, R.B.: Automated line generalization. Cartographica 24(2), 74–111 (1987)
19. Opheim, H.: Fast data reduction of a digitized curve. Geo-Processing (2), 33–40 (1982)
20. Papadias, D., Mamoulis, N., Theodoridis, Y.: Constraint-based processing of multiway spatial joins. Algorithmica 30(2), 188–215 (2001), http://dblp.uni-trier.de/db/journals/algorithmica/algorithmica30.html#PapadiasMT01
21. Park, H.H., Lee, Y.J., Chung, C.W.: Spatial query optimization utilizing early separated filter and refinement strategy. Information Systems 25(1), 1–22 (2000)
22. Piorkowski, A., Krawczyk, A.: The problem of object generalization and query optimization in spatial databases. Studia Informatica 32, 119–129 (2011)
23. Pluciennik, T., Pluciennik-Psota, E.: Using graph database in spatial data generation. In: Gruca, D.A., Czachorski, T., Kozielski, S. (eds.) Man-Machine Interactions 3. AISC, vol. 242, pp. 643–650. Springer International Publishing, Switzerland (2014), http://dx.doi.org/10.1007/978-3-319-02309-0_69
24. Robinson, A., Sales, R., Morrison, J., Ostrowski, W.: Podstawy kartografii. Państwowe Wydawnictwo Naukowe (1988), http://books.google.pl/books?id=nQE1AAAACAAJ
25. Tobler, W.: Numerical Map Generalization: And, Notes on the Analysis of Geographical Distributions. Discussion paper series, Department of Geography, University of Michigan (1966), http://books.google.pl/books?id=MbUUnQEACAAJ
26. Visvalingam, M., Whyatt, J.D.: Line generalisation by repeated elimination of points. The Cartographic Journal, 46–51 (June 1993)

Path Features in Spatial Data Generation

Tomasz Płuciennik and Ewa Płuciennik

Silesian Technical University, Institute of Computer Science, Akademicka 16, 44-100
Gliwice, Poland
{Tomasz.Pluciennik,Ewa.Pluciennik}@polsl.pl

Abstract. The following paper is a continuation of the topic of spatial
data generation. The idea is to be able to create customisable sets of
spatial layers for further use in simulation and testing of GIS systems'
behaviour. Since the test data for a specific GIS implementation is not
always available, a special generator is being developed and this article
will focus on the latest results, which is a generation of synthetic geometries that could represent e.g. rail tracks or rivers related to already
existing map features. The process is supported by a graph database and
its mechanisms. The other goal of the article is to give a current status
update and to group all tasks and ideas for the future research.

Keywords: GIS (Geographic Information Systems), graph database,
spatial information.

1   Introduction

Geographic Information Systems are designed to maintain and process information concerning objects (features) on the earth's globe [12]. Features can be
represented as geometrical objects or raster data. They can also have additional
attributes attached. During the development process of a GIS system the target
data is not always provided up-front or it is very limited. The publicly available
data may not satisfy some requirements (especially in terms of accuracy). Therefore an idea of the spatial data generator was born. The main requirement is
a possibility to produce different types of data with configurable characteristics
and comprehensive sets of attributes. Another problem has been recently identified, namely simulation of data based on a limited number of information (e.g.
data provided from Global Navigation Satellite System-based measurements),
so that the generator could become an integral element of a GIS system. This
could help with urban planning or simulation of geologic or industrial processes.

The topic of the spatial information generator was already presented in [14–16]
and it is still explored. In short, there is a set of supported datasets (layers) that
can be produced. They are: roads, parcels, buildings (2D and 3D), trees and simple
3D terrain. Additional layers can supplement the program via implementation of
plugins cooperating together to make layers coherent. All data is placed on a chosen
area of the globe. The main goal is to expand upon existing software and present
possibilities for new data types.

S. Kozielski et al. (Eds.): BDAS 2014, CCIS 424, pp. 465–471, 2014.

In this paper only a part of the generation process will be described. It provides a possibility to create rail tracks and rivers. The algorithm uses previously generated layers and additional data structures. The next chapter contains a short description of what is available as an input for the line features generation. In the following chapters the new algorithm is described in more detail, example results are presented and conclusions for the future are drawn. Let us note that some data to large extent still remains simplified in both existing and newly developed layers.

2   Background

The spatial data generator is being developed in Java 7 and it uses GeoTools 10.2 [1] GIS library, JTS (Java Topology Suite) 1.13 [5] spatial geometry library and Neo4j [18, 13, 6] graph database with Neo4j Spatial [18, 3] extension (version 0.11 neo4j 1.9) providing e.g. spatial indexes, spatial searches, mapping of geographic layers, integration with GeoTools etc. Output from the program is stored in ESRI Shapefile [8] files and it is presented here using uDig 1.4.0 [4]. Preview of the graph database is done using Neoclipse 1.9.1 [2].

Rail tracks and rivers can be represented as curves in the simplest scenario. Of course perfect curves (described by i.e. polynomials) are not used. The geometries are stored as linestrings [5] – lists of points which are then connected in order (though creation of such features is supported by i.e. B-splines or Bezier curves). For now, it is assumed that neither rail tracks nor rivers are connected with each other.

Two other layers are de facto required to be able to build upon them the linestring layers: roads and parcels. Roads constitute a base for the parcels generation and some parcels are later marked as roads. Roads are also linestrings creating a mesh, like the very simple one presented in Fig. 1a. Parcels are generated as polygons placed where the roads are and in empty spaces between them (Fig. 1b). This algorithm is complicated (refer to [16]) and it will not be described here. The important thing is that both layers are stored in a graph database. In case of roads this is simple and commonly done (e.g. in [9]). Parcels, on the other hand, require more complicated hierarchical graph [17] with each descending level containing a view of the layer in more detail. Using a graph database in this particular solution greatly improves the generation process. It can describe complicated relations between multiple objects like in e.g. [7], as well as complicated objects themselves as in [11].

The part of the graph that is interesting here (general view in Fig. 2) is its lowest level containing nodes with all of the final parcel geometries with their attributes (especially parcel types). These nodes are connected with *IS_NEIGHBOUR* (custom) relationship when the corresponding geometries' edges touch each other. The relationships contain information about a common point of the two geometries. Here it is a centroid point (geometric centre of a polygon) of the intersection [5] (linestring) of the two geometries.

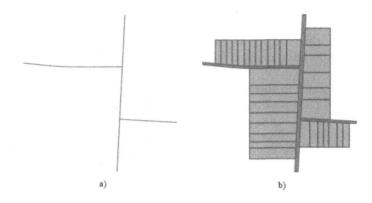

Fig. 1. a) Example road structure. b) Road structure with matching parcels.

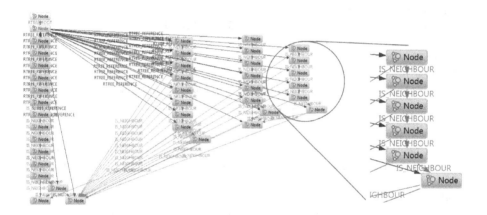

Fig. 2. General view of the graph for parcel features

3   Line Features Algorithm

The new layers – rails and rives – can be imagined as paths traversing through the graph. This can be related to route finding and navigation, as graphs are widely used to represent possible paths in e.g. [17, 10, 19]. Since the generation process for the both layers is essentially the same and, as stated before, no new features will intersect with each other, the search will be limited to finding a single path. The algorithm simply traverses through the parcels' graph and returns the found path in respect to some additional conditions. It can be easily extended to take into account the existing paths, based on prerequisites that will be presented (e.g. not using the same parcel twice, making sure crossing of rails/rivers is believable). Furthermore, let us assume that the path goes through the whole map i.e. theoretical ends are somewhere outside of it. The path feature have to fulfil specific conditions:

- no self intersections (each parcel on a path is unique),
- passing only through road-type and not yet used (empty) parcels (empty parcels on a path will be marked as e.g. rail track-type and any unused parcels will be later marked for placing other features on them like e.g. buildings),
- not crossing roads in succession,
- crossing road parcels in (usually) perpendicular manner.

At the beginning of the algorithm two parcels have to be chosen – start and end of the path – on the edges of the map, preferably (but not necessarily) far from each other. The following code snippet presents the general structure of the traversal [6] definition:

```
TraversalDescription description = Traversal.traversal()
  .depthFirst()
  .uniqueness(Uniqueness.NODE_PATH)
  .relationships(RelationshipTypes.IS_NEIGHBOUR)
  .evaluator(new Evaluator() {
    public Evaluation evaluate(Path path) {...}
  })
  .evaluator(Evaluators.includeWhereEndNodeIs(endNode));
Traverser traverser = description.traverse(startNode);
```

The description of the traversal is fully customisable. The *depthFirst()* method makes it favour moving deeper through the graph than checking the immediate relationships first, which increases the possibility of quickly reaching the end parcel. Uniqueness setting *NODE_PATH* ensures that paths will not have duplicated nodes. Only *IS_NEIGHBOUR* relationships are traversed. The traversal is done between chosen *startNode* and *endNode*. The *traverser* object provides the found paths containing ordered nodes. Only one path will be used as a result.

Custom evaluators can be added to the traversal. Evaluators are called after every jump between nodes and decide whether to continue through the current path or exclude the jump. Here the evaluator is taking care of several conditions. First, it makes sure that only road and empty parcels will be returned in the path:

```
ParcelType type = getParcelType(path.endNode());

if (type != ParcelType.ROAD && type != ParcelType.EMPTY) {
  return Evaluation.EXCLUDE_AND_PRUNE;
}
```

Then, when the path is longer (path length is the number of relationships), undesirable situation is traversing through two roads consecutively:

```
if (path.length() > 0) {
  ParcelType type0 = getParcelType(nextToLastNode);
```

```
ParcelType type1 = getParcelType(lastNode);

if (type0 == ParcelType.ROAD && type1 == ParcelType.ROAD) {
    return Evaluation.EXCLUDE_AND_PRUNE;
  }
}
```

During the jump over a road the distance of the chosen common points of it and its two neighbour parcels is checked (common points are taken from the relationships). This distance should not exceed much the width of the road. Additionally, parcels should be big enough so that the path will always cross to the other side of the road. This condition is implemented as follows:

```
if (path.length() > 1) {
  ParcelType type0 = getParcelType(secondFromTheEndNode);
  ParcelType type1 = getParcelType(nextToLastNode);
  ParcelType type2 = getParcelType(lastNode);

  if (type0 == ParcelType.EMPTY && type1 == ParcelType.ROAD
      && type2 == ParcelType.EMPTY) {
    Coordinate c1 = ...; // read from the first relationship
    Coordinate c2 = ...; // read from the first relationship
    double road_width = ...; // read from the roads graph

    // 45 degrees max
    if (c1.distance(c2) > Math.sqrt(2.0) * road_width) {
      return Evaluation.EXCLUDE_AND_PRUNE;
    }
  }
}
```

Finally, if none of the above conditions excluded the current path, the traversal can continue:

```
return Evaluation.INCLUDE_AND_CONTINUE;
```

With more complicated neighbour geometries (e.g. concave polygons), the points stored in the relationships may not be exactly common to the neighbours but in the worst case scenario a path simply will not be found there.

Non-road parcels on the path are now marked as e.g. rail-type and will not be used in further traversals. The path geometry is finally build based on the points taken from e.g. the path parcels centroids or the relationship properties. Additional smoothing can be also applied.

4   Results

The rail track or river found by traversing the graph from Fig. 2 is presented in Fig. 3. Fig. 4 presents a fragment of a more complicated example with a significantly bigger structure (covering 100 km^2). Though geometries are still crude

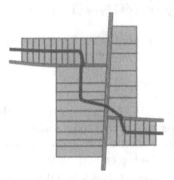

Fig. 3. Example of a simple path

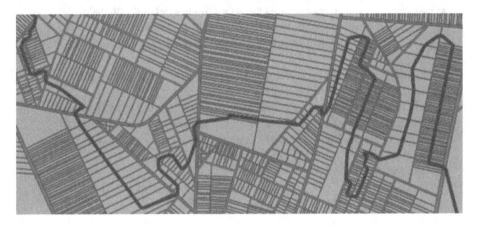

Fig. 4. Path in a complex structure

and simplifications are assumed, there is a potential to create realistic maps. Changing the fashion the parcels are generated and letting the user customise the characteristics of the paths might be useful. Furthermore, one can imagine extending the traversal to include outgoing/intersecting paths and more possibilities of choosing where paths end.

5   Conclusion

Planned development of the generator involves adding a possibility to define attributes for any layer (name, type, way of value generation, value limitations etc.). Other necessary work includes further improvement of how geometries look and integration with GNSS. This should lead to more lifelike structures with fully-fledged rail tracks (i.e. merging tracks, blind tracks) and rivers flowing into other rivers or lakes. All this could be also modified by the terrain shape or vice versa, force specific terrain features. The data from GNSS receivers could be helpful as reference points upon which to conduct the generation.

All this have to be extendable by any new data layers that may be required in the future. There is a strict relation between the layers, as they must overlap each other correctly. The new generator plug-ins can use every data structure the existing ones provide. This means evolution of the graph database to encompass more data types and their relations as it is impossible to store everything in simpler or inadequate data structures e.g. flat files or relational databases. Using a graph database means also a possibility to add layers to existing datasets. Research on spatial data generation is therefore still ongoing.

References

1. GeoTools, `http://docs.geotools.org/stable/userguide/` (access December 10, 2013)
2. Neoclipse, `https://github.com/neo4j/neoclipse` (access December 10, 2013)
3. NeoSpatial, `https://github.com/neo4j/spatial` (access December 10, 2013)
4. uDig, `http://udig.refractions.net/` (access December 10, 2013)
5. JTS Topology Suite Technical Specifications, Version 1.4 (2003)
6. The Neo4j Manual, v1.9.5 (2013)
7. Bakalov, P., Hoel, E.G., Heng, W.L., Tsotras, V.J.: Editing and versioning dynamic network models. In: Proceedings of the 16th ACM SIGSPATIAL International Symposium on Advances in Geographic Information Systems, ACM-GIS 2008, Irvine, California, USA, November 5-7 (2008)
8. ESRI Shapefile Technical Description: An ESRI White Paper (1998)
9. Gerke, M., Butenuth, M., Heipke, C., Willrich, F.: Graph supported automated verification of road databases using aerial imagery. In: 2nd International Symposium on Spatial Data Quality, pp. 412–430 (2003)
10. Jain, S., Seufert, S., Bedathur, S.J.: Antourage: Mining distance-constrained trips from flickr. In: Proceedings of the 19th International Conference on World Wide Web, WWW 2010, Raleigh, North Carolina, USA, April 26-30, pp. 1121–1122 (2010)
11. Li, S.: A layered graph representation for complex regions. CoRR abs/0909.0109 (2009)
12. Longley, P., Goodchild, M., Maguire, D., Rhind, D.: Geographic Information Systems and Science. John Wiley & Sons Ltd. (2005)
13. Pater, J., Vukotic, A.: Neo4j in Action. Manning Publications (2012)
14. Płuciennik, T.: Generating roads structure of a virtual city for gis systems testing. Studia Informatica 32(2B(97)), 43–54 (2011)
15. Płuciennik, T.: Virtual city generation for gis systems testing. Studia Informatica 33(2B(97)), 117–129 (2012)
16. Płuciennik, T., Płuciennik-Psota, E.: Using graph database in spatial data generation. In: Gruca, A., Czachórski, T., Kozielski, S. (eds.) Man-Machine Interactions 3. AISC, vol. 242, pp. 649–658. Springer, Heidelberg (2014)
17. Stoffel, E.P., Schoder, K., Ohlbach, H.J.: Applying hierarchical graphs to pedestrian indoor navigation. In: Proceedings of the 16th ACM SIGSPATIAL International Symposium on Advances in Geographic Information Systems, ACM-GIS 2008, Irvine, California, USA, November 5-7 (2008)
18. Webber, J., Robinson, I., Eifrem, E.: Graph Databases. O'Reilly Vlg. Gmbh&Co., O'Reilly & Assoc. Inc. (2013)
19. Westphal, M., Renz, J.: Evaluating and minimizing ambiguities in qualitative route instructions. In: Proceedings of the 19th ACM SIGSPATIAL International Symposium on Advances in Geographic Information Systems, ACM-GIS 2011, Chicago, IL, USA, November 1-4, pp. 171–180 (2011)

Importance of Some Topics of Data Management in Cloud-Based Maritime Fleet Management Software

Jolanta Joszczuk-Januszewska

Gdynia Maritime University,
al. Jana Pawla II 3
81345 Gdynia, Poland
jolajj@am.gdynia.pl

Abstract. Fleet management refers to the management of ships while at sea also. In terms of maritime fleet management, there is important function the software must possess - data management. This also means transfer and replicate data between the office and the individual ships of the fleet. This ensures that the respective databases on shore and on board remain synchronized and that all involved personnel are working with the same information. Cloud-based maritime fleet management software performs exceptionally well on these and more requirements. The aim of this paper is to underline some important topics in the area of data management in cloud-based maritime fleet management software. Above all are explain: role of software as a service (SaaS) which is considered be part of the nomenclature of cloud computing (CC) in data quality, and role of Key Performance Indicators (KPI) in data management process.

Keywords: Maritime Fleet Management Software, Cloud Computing, Data Management.

1 Introduction

Fleet management is a term used to describe the management of any/all aspects relating to a company's vehicle. Fleet vehicles can be defined as vehicles over which a business has some degree of influence in their selection and operation.

Fleet management refers to the management of ships while at sea also.

Maritime fleet management software enables people to accomplish a series of specific tasks in the management of aspects like crewing, maintenance, and day-to-day operations.

In terms of maritime fleet management, there is important function the software must possess - data management. This also means transfer and replicate data between the office and the individual ships of the fleet. This ensures that the respective databases on shore and on board remain synchronized and that all involved personnel are working with the same information.

Ship owners and ship managers are interested in further optimising their cost structure, in simplifying processes and automating their information flows.

S. Kozielski et al. (Eds.): BDAS 2014, CCIS 424, pp. 472–481, 2014.
© Springer International Publishing Switzerland 2014

That's where cloud computing (CC) comes in. Cloud-based maritime fleet management software performs exceptionally well on these and more requirements. CC is not a new technology, it's a new way of delivering computing resources.

Cloud computing is a type of computing that relies on sharing computing resources rather than having local servers or personal devices to handle applications [4,1].

The aim of this paper is to underline some important topics in the area of data management in cloud-based maritime fleet management software. Above all are explain:

– role of software as a service (SaaS) which is considered be part of the nomenclature of CC in data quality, and
– role of Key Performance Indicators (KPI) in data management process.

2   SaaS in Maritime Fleet Management

Growth and popularity surround CC within the business world, but shipping industry is far more complex than the average office set up [2,3].

There are many definitions available to describe the term CC. One simple definition refers to CC as the delivery of computing service without owning an own infrasrtucture.

Other entity such as U. S. National Institute of Standards and Technology (NIST) defines cloud computing as a model for enabling ubiquitous, convenient, on demand network access to a shared pool of configurable computing resources (e.g. networks, servers, storage, applications, and services) that can be rapidly provisioned and released with minimal management effort or service provider interaction. This cloud model is composed of five essential characteristics: on demand self-service, broad network access, resource pooling, rapid elasticity, measured service [8].

NIST has explained that CC has also service model - SaaS through which applications are provided in the cloud.

This indicates CC as a concept that works for shipping. This is exactly what we need in fleet management when managing ship to shore data synchronisation [12,11].

The situation today means that ever more complex systems aboard the ships must be operated by fewer crew members. But even with small crews, ships must be kept in top condition so as to avoid any unnecessary downtime or costly repairs.

That's where CC comes in. Cloud-based software may be the answer to those seeking to reduce operational costs and minimize capital employed.

We all know, employing smart tools and state-of the-art software can improve efficiency. This is particularly important for technical management software such as planned maintenance, procurement and crewing. Such software must ensure all relevant people - aboard or ashore - are provided with the right level of information whenever they need it.

For example, crew members must be able to plan their work in detail (what to do, when and how to do it) and report back to shore; staff ashore needs the correct information (performance data, running hours, purchase requests etc.) at the click of a button; and management must be able to compare, analyze and benchmark data across the fleet. The information must be available on-time and in one place to ensure data integrity. Cloud-based software performs exceptionally well on these and more requirements.

Let's look at the underlying technology of cloud-based software such as SaaS. With SaaS, we're not talking about a new technology. The SaaS concept is based on the idea to provide, support and run software via the Internet. So companies starting to employ SaaS will not have to change technology. They will just make use of a new way of accessing computing.

The popularity of SaaS is steadily increasing because it reduces costs and simplifies deployment. With the introduction of SaaS, shipping companies can optimise capital as well as operational costs, and relieve their organisation of tasks that do not belong to their core competencies [12,11].

With SaaS, the software provider can support many customers with a single version of a product. This approach allows customers to scale as fast as needed without replacing information technology (IT) infrastructure or adding IT staff. A SaaS application means there is one central server and one central database. All employees share the same database, all employees have access to the same current information (within their access rights).

For SaaS applications to work efficiently, access to the Internet is essential. For use aboard ships, the SaaS application is managed offline, and data transfers as well as software updates are conducted via regular synchronization.

In terms of technical fleet management, there are two distinctive features the software must possess: high quality data management and offline availability.

In terms of data management, the SaaS provider manages - on behalf of all customers - the data and documents that are shared by all users of the application, for instance product specifications.

In terms of ownership of the data, customer specific or vessel specific data belongs to the customer and not to the system provider, and cannot be accessed or viewed by any other user of the system. Customers, on the other hand, can access and download their data in any industry standard format.

Having offline availability means that office staff can access the central database in real-time through a secure internet connection - no matter when and where they are located.

Crews aboard the vessels work offline by accessing the database onboard, which is mirrored to the central server's database ashore. Regular synchronisation as part of the standard sync schedule of each vessel ensures that both ship and shore work with the same up-to-date information.

2.1   SaaS versus on Premise Solutions

There are some distinct differences between on premise solutions and SaaS solutions which are descrbed in [9].

With SaaS, the software is neither installed nor operated on the company's IT landscape, but is offered as a hosted service. Employees use the application via a web-based frontend, or open an offline software client which regularly synchronizes with the central database.

The latter is particularly important in the maritime industry, i.e. for the ships which have no continuous internet connection. Customers access the software on the central server by using their existing computers. The data remains on the central server; no need for customers to worry about security, backups, implementing software upgrades and other IT related tasks.

2.2 Importance of Security

Despite the advantages of SaaS, some reservations exist. Issues such as security, data sovereignty and inflexibility in terms of customization of the software tend to be mentioned. When it comes to extremely sensitive data, particular attention must be paid to data security and legal issues. This applies to both traditional as well as SaaS applications.

In terms of data sovereignty, some people may worry about losing physical control over their data. Experience shows, however, that security issues tend to be dealt with more professionally at a dedicated SaaS application provider than would be the case with in-house solutions. Cloud providers have a vital interest in safe IT environments. Security problems would soon mean the end of a SaaS company.

3   Importance of Mespas R5 Software Solution

Today, MESPAS is the world's leading fleet management software provider based on CC and the fastest growing SaaS company in the maritime industry [9].

The mespas R5 is a software solution covering all major functionalities of modern maritime fleet management: asset management (including planned maintenance), procurement, and crew management. But there's more: with mespas R5 can be offered system based on cloud computing in the maritime industry, allowing SaaS.

Software and data are provided via the Internet, as is the server infrastructure. With MESPAS, the user can minimize the costs for data management and IT, regardless of the fleet's size or growth rate.

All users need standard IT hardware to access both application and data. There are two types of data available within the MESPAS database: master data and customer specific business data. Master data can be accessed by all relevant users, business data is accessible only by eligible users.

At the heart of the system lies the state-of-the-art database architecture and central server infrastructure. The future-proof technology allows the user to focus on core business rather than on IT issues or data implementation.

3.1   The Concept

The mespas R5 software is installed both on board the vessels as well as on the office staff's personal computer. The members of the office staff access the central database, containing the entire fleet's information, in real-time through a secure internet connection - no matter when and where they are located.

Crews aboard the vessels work offline by accessing the database on board, which is mirrored to the central server's database ashore. Regular synchronization of data as well as information ensures that both ship and shore work with the same up-to-date information.

Suppliers will receive requests for quotation as well as purchase orders via the same mespas R5 platform. Working with only one central database is the basis for efficient information flow and offers transparency, e.g. on the status of purchase requests or purchase orders.

One of benefit is improve efficiency by complete and high quality data. Importance of mespas cloud solution involved from comparison to traditinal/in-house installed software, in terms of data management, is given in the Tab. 1.

3.2   Asset Management

With mespas R5 Asset Management the users achieve cost-efficient operations and a reliable performance of their fleet. Crews can efficiently plan and execute the technical maintenance of their vessels, and office staff are given an accurate, immediate overview of past, current and future tasks across the fleet.

Asset Management helps ensure that all aspects of fleet management are in compliance with regulations and requirements. Planned maintenance takes place on time, and vessels are kept on schedule and in top condition. The solution significantly reduces the life-cycle costs of equipment and machinery. Mespas Asset Management can be used as a standalone solution or combined with mespas R5 Procurement for better stock control and more efficient operations.

Mespas R5 Asset Management is a user friendly and smart planned maintenance software. It is linked to the MESPAS central database, giving users the added benefit of working with a comprehensive and up-to-date set of data.

Data quality is of great importance in order to run reports, to compare and benchmark. The MESPAS database is based on original equipment manufacturer (OEM) information. It contains most commonly used parts and products, and is continuously growing. With MESPAS, there is no need for tedious and error prone data entry and data management, saving a great deal of time and money.

3.3   Easy Installation, Usage and Synchronization

Managing the IT side of a technical fleet management system on board of a vessel can be somewhat of a challenge. Installing a server, maintaining it, and implementing stringent back-up procedures, all these are tasks which require profound IT knowledge.

MESPAS accounts for the rough environment and the limited IT knowledge on board of vessels by introducing the mespas Cube. The mespas Cube is an offshore server (small industrial design PC) with no moving parts, compact and handy as an external hard disk. It comes pre-configured with the vessel specific database and the mespas R5 software.

The crew simply connects the mespas Cube to the network on board. Everything is pre-installed, all configuration will be done automatically. Also, synchronization between ship and shore can easily be automated.

Table 1. MESPAS cloud solution versus in-house installed software, in terms of data management [9]

MESPAS cloud solution		Traditional / in-house installed software
	CONCEPT	
Available 24/7 via secure internet connection; server-farm that meets highest standards: server redundancy, power backup, data backup, internet connection, fire protection, environmental control	Availability, security and disaster recovery	Local solution features significant lower physical and digital security. Higher disaster recovery costs
	DATA	
Data specialists implement master data according to stringent rules and industry requirements. All data are entered once only - no data duplication, ensuring highest data quality and comparability	Data quality	Company must define and enforce own standards and regulations. No industry comparisons; comparability restricted to own company; often not even vessel-to-vessel comparisons available
	OPERATIONS	
No client-based staff needed to manage synchronization; synchronization can be automated. Assured reliability of synchronization process and data quality	Synchronization ship-shore	Reliability of synchronization process not guaranteed; typically time-consuming process
Sophisticated reporting tool available to general real-time reports (standard or customized) for single vessels or across fleet	Reporting	Very limited reporting; comparisons restricted to data on same server; manual intervention and processing before data can be used for reporting

3.4   Real-Time Key Performance Indicators across Fleet

In fleet management operations, a huge amount of data is recorded and documented every day - ashore and especially on board the vessels. This information is required by internal management as well as external stakeholders such as ship owners, port authorities, classification societies, customers, and so on.

The mespas Reporting Engine was developed with the needs of these stakeholders in mind. The web-based mespas Reporting Engine provides the required data in every detail or as Key Performance Indicators (KPI). The user can run reports on demand as well as schedule regular reports. These will be generated automatically at the preferred intervals and conveniently delivered by e-mail.

Are created reports on topics such as:
− Performance of critical equipment;
− Maintenance activities;
− Running hour;
− Benchmarking of engine performance;
− Deficiencies, incidents, and non-conformities;
− Purchasing volume analysis;
− Inventory levels and trends;
− Budgeting and controlling;
− Certificates' overview.

4   Importance of Solution Proposed by iFleet Systems Company

iFleet Systems Company is a specialist provider of integrated maritime systems and service solutions for the commercial shipping industry. The primary focus is to enhance fleet management and operations by providing cost effective and reliable vessel management systems using unique capabilities and ship to shore data synchronization [10].

Using the latest proven technologies such as cloud computing this solution can develop, manage and integrate ship IT systems and fleet management software on board vessels and shore based operations.

4.1   Business Intelligence Reporting

Once an organisation has analysed its mission, identified all its stakeholders, and defined its goals, it needs a way to measure progress toward those goals. Key Performance Indicators (KPIs) and Key Risk Indicators (KRIs) are these measurements and iFleet Systems will allow the user to warehouse clean data to build KPIs and KRIs that are reliable and reflect the organisation's goals. iFleet systems can develop and design KPIs and KRIs which can be delivered to senior management via a wide array of media applications.

Once collated the data is used to produce KPIs and KRIs that allow the fleet to be monitored.

4.2   Cloud Computing as Hosted Services

iFleet Systems Company has developed a suite of specialist IT data management systems for the maritime industry. He helps companies manage the flow of information between the ships and offices, improving data handling, document management and reporting systems.

His success is to provide bespoke systems that fit the customer's needs rather than a packaged solution that contains hidden on going additional expense.

The cloud hosted solutions are built using a secure development lifecycle that ensures security and privacy is incorporated into services by design, from software development to service operations. This approach results in 5 different layers of security - data, application, host, network and physical.

The following list identifies the essential services:

- **Parts Management.** Fleet Broadband is a maritime global Satellite Internet, Telephony, SMS Texting and ISDN Network for ocean going vessels using portable domed terminal antennas. Using Fleet Broadband company provides access to a secure hosted system in data centre of company which means that he can avoid the pitfalls caused by having programs and databases on board which generally are updating issues, patches, corruption, backup failure, hardware failure and vessel visits.
- **Planned Maintenance.** Based on the same access as Parts Management system means maintenance tasks can be added shore side that show on the vessel either vessel wide or on a user level ensuring that tasks are not missed. Planned tasks are setup with workflows that alert users to tasks coming up and tasks that have not been completed. Manuals for system maintenance can all be stored on the system and made available for download by the vessel. The Parts and Maintenance modules complement each other and are able to provide full reporting utilities including KRIs and KPIs.
- **Vessel Performance.** Data is collated from the vessel that the users fleet and produce reports that show up-to-date and accurate information that can be shown in the fleet management system or integrated into the current systems.
- **Vessel Auditing.** Data is collated from auditors who inspect the fleet and by using a bespoke iPad application are able to report on the condition of each vessel; these reports are very wide ranging and cover all aspects of the vessel from the general appearance down to engine parts. Once collated the data is used to produce KPIs and KRIs that allow the fleet to be monitored.
- **Document Management.** DM system provides document synchronization from shore to vessel and back again using a variety of file types including the industry standard Microsoft Office suite of products ensuring the vessels are kept up to date at all times. This system is idea for Quality Management and day to day operations. He designs networks that are robust and secure and bring the vessel into the corporate network by using the latest technologies for data synchronization. He specialises in virtualisation that reduces hardware costs and downtime for vessels.

5   Importance of Marineopsys Software

Marineopsys with cloud-based complete fleet management software suite is a SaaS enterprise fleet management software for the maritime industry [7]. Marineopsys automates maintenance, supply chain, chartering, operations, crewing, accounting and reporting - in a single, integrated and powerful business management software solution. This solution is built on two main platforms: SaaS and Cloudsuite framework.

5.1 KPI Module

The KPI module helps to track various Performance Indicators, KPI and Shipping Performance Indicators (SPI) as per industry standards. This module is integrated with the entire system and calculates performance on a quarterly basis based on actual operational data. The users can drill down from the SPI level to the lowest data point which caused the SPI.

Some of the main features of the KPI module include:
- Vessel specific quarterly Performance Indicators,
- Vessel specific quarterly Key Performance Indicators,
- Vessel specific quarterly Shipping Performance Indicators,
- Drill down to lowest event which triggered a performance update.

5.2 Replication Management

The replication manager module helps to replicate data between shore and remote assets. The replication can be configured to run as per the clients requirements.

Some of the main features of the replication module include:
- Scheduled replication between shore and remote assets,
- On demand replication,
- Conflict resolution.

6 Conclusions

Cloud computing plays now and will play an increasingly important role in the maritime fleet management software in the future [5,6]. Data management is a critical step in maritime fleet management.

A great part of this paper was be devoted to presentation of three commercial cloud-based maritime fleet management software systems: mespas R5, solution proposed by iFleet Systems company, marineopsys.

Based on examples this paper showed the importance of some topics of data management (e.g. data security management, data quality management, data replication, data analysis) in cloud-based maritime fleet management software solutions. Above all:
- SaaS has important role in high data quality, and
- KPI are effective measurements of timeliness, quality and compliance in data management process.

For example, with MESPAS and mespas R5, data across the fleet is stored in one central location, ensuring a high integrity of recorded, processed and stored data. The mespas Reporting Engine securely retrieves the information via the web, and produces comparable and meaningful reports at various levels of details. These dynamic KPI reports give the users an immediate overview across the fleet and help make decisions based on accurate facts.

References

1. Babcock, C.: Management Strategies for the Cloud Revolution. McGraw-Hill, New York (2010)
2. Business Software Alliance: BSA Global Cloud Computing Scorecard (2013)
3. Cisco Global Cloud Index: Forecast and Methodology, 2010-2015 (2011)
4. Franclin Jr., C., Chee, B.J.: Cloud Computing. CRC Press Inc. Taylor & Francis Group, Boca Raton (2010)
5. Joszczuk–Januszewska, J.: The Benefits of Cloud Computing in the Maritime Transport. In: Mikulski, J. (ed.) TST 2012. CCIS, vol. 329, pp. 258–266. Springer, Heidelberg (2012)
6. Joszczuk–Januszewska, J.: Importance of Cloud-Based Maritime Fleet Management Software. In: Mikulski, J. (ed.) TST 2013. CCIS, vol. 395, pp. 450–458. Springer, Heidelberg (2013)
7. Marineopsys: http://www.marineopsys.com
8. Mell, P., Grance, T.: The NIST Definition of Cloud Computing. National Institute of Standards and Technology, Information Technology Laboratory (2011)
9. MESPAS: http://www.mespas.com
10. iFleet Systems: http://www.ifleetsystems.com
11. Thoma, C.: Fleet management on the basis of cloud software. Ship & Offshore (3), 76–77 (2011)
12. Thoma, C.: Software as a Service - revolutionizing fleet management. Digital Ship 12(1), 40–42 (2011)

The Models of Determination of Spin Values from Experimental Properties of Nuclei on the Base of Other Experimental Properties

Andrzej Krajka, Zdzisław Łojewski, and Robert Mitura

Maria Curie-Skłodowska University,
Pl. Marii Curie-Skłodowskiej 5,
20-031 Lublin, Poland
akrajka@gmail.com

Abstract. In December 2012 the group of physicists from Atomic Mass Data Center, located at Centre de Spectrométrie Nucléaire et de Spectrométrie de Masse (CSNSM), Orsay, France, shared the databases of atomic nuclei (cf. [1]) containing a lot of experimental data characterizing the nucleus, for example, nuclear ground-state masses and radii, magnetic moments, half-lives, spins and parities of excited and ground-state, their decay modes and the relative intensities of these decays, the deformations and many others. Today there are about 2830 nuclei that have been observed. It is estimated that the total number of nuclei which may be obtained by experimenters is between 6000 and 7000. It is evident that many properties of atomic nuclei are so far unknown and we usually define them using a variety of theoretical methods, based on those properties of nuclei which are already known. One of the most important properties is the spin of the nucleus in the ground state, specifying also the possible excited states. The main aim of the presented paper is to use the latest experimental data to check the existing theoretical methods of the prediction of unknown spins of nuclei and proposals for new estimates based on the methods of data mining and artificial intelligence (cf. [4–6, 8, 9, 11]). We compare the different models of the prediction of spin value with the help of the R programistic language, well fitted to statistical and data mining modelling. The properties of atomic nuclei have been collected in the MySQL database. In order to integrate the R language with MySQL we use the package RMySQL which contains the interface to communicate with the MySQL.

Keywords: nucleons, spin, energy, atomic mass, data mining, R language, MySQL database, regression trees, neural networks.

1 Introduction

Nuclear Physics started a little bit more than 100 years ago. Since then, scientists have accumulated a huge amount of data for a large number of nuclides. Today there are about 2830 nuclei that have been observed. It is estimated that the

S. Kozielski et al. (Eds.): BDAS 2014, CCIS 424, pp. 482–491, 2014.

total number of nuclei which may be obtained by experimenters is from 6000 to 7000 [3, 13].

Thus, today there exists an enormous number of experimental data on the atomic nucleus that need to be interpreted, sorted, treated in a homogeneous way, while keeping traceability of the conditions under which they were obtained.

There are a lot of experimental data characterizing the atomic nucleus, for example, nuclear ground-state masses and radii, magnetic moments, half-lives, spins and parities of excited and ground-state, their decay modes and the relative intensities of these decays, the deformations and many others.

In paper [1, 2], the so-called AME2012 and NUBASE2012 data presents the results of the most recent studies that contain full information on nuclear experimental data (in ASCII format). The tables present the values and their uncertainties for all experimentally known nuclei as a function of the mass number A, the number of protons Z or neutrons N.

In [2] section 2.4 there is mentioned the method used for computing unknown spin properties, the TNN methods (trends in neighbouring nuclides). The base of this method is the knowledge of spin values of "similar" nuclei. From the information science viewpoint this is a variant of k-neighbourhood method, but the neighbourhood is treated in a particular way: for N-odd and Z-even there are (N-2,Z) and (N+2,Z) elements, for N-even and Z-odd there are (N,Z-2) and (N,Z+2) elements whereas for N-odd and Z-odd we have the set {(N-2,Z-2), (N-2,Z),(N-2,Z+2), (N,Z-2), (N,Z+2), (N+2,Z-2), (N+2,Z), (N+2,Z+2)}. We check this method (henceforth referred to as TNN) and compare it with other data mining methods.

2   The Tools

We prepare the data for the database. The Relational Database Management System (RDBMS) is included in the Oracle service among Oracle Business Intelligence. For us the best module is Oracle R Enterprise (cf. Fig. 1). However, we did not make use of this solution because on the Oracle website it only the image disk for the virtual machine (Virtual Box) is shared. The image has the size of about 25GB and it is necessary to have a very effective and fast computer. Generally, a virtual machine works very slowly.

A free software tool available in the public domain is MySQL database. MySQL supports the RDBMS functionality and the same time it is not much different from commercial tools. Nubase 2012 files on which our research is based are imported to the MySQL database (Tab. 1) afterwards by the package RMySQL is created communication with R language, and our research is carried out in this environment. The R environment is a programming language for statistical computing and graphics. The R environment is a free software, and it is often used in data analysis and data mining. Surveys of data miners show how R's popularity has increased substantially in recent years.

We prepare the data for the MySQL database. Further we install the R platform ([10, 14]) and interface between the R language and the MySQL database

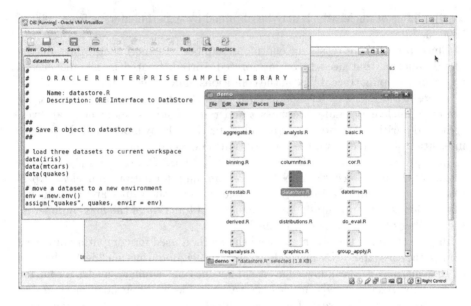

Fig. 1. View sample of the Oracle R Enterprises library

Table 1. View of a fragment of table in MySQL

N	Z	AMASS	BETA_DECAY_ENERGY	MASS_EXCESS	J
22	17	38968008.15	3441.973	-29800.218	1.5
24	19	42960734.71	1833.529	-36575.379	1.5
26	21	46952402.49	599.632	-44336.798	3.5
28	23	50943956.80	-752.623	-52203.902	3.5
30	25	54938043.94	-231.090	-57711.701	2.5

named RMySQL. For details cf. [7], [12]. Later the communication runs mainly in the SQL language.

3   The Data

In the MySQL database we denote the experimental data, the so-called "weak" argument experimental (cf. [2] p.1170) data, the computed values according to the notation from paper [2]. The primary key of our base were neutrons (N) and protons (Z) values.

The introduced variables in a natural way label the four classes of nuclei (even-even, odd-even, even-odd and odd-odd). In addition, the variables labelled as NMAGIC and ZMAGIC indicate the interval where N and Z belong, between the "so-called" magic numbers intervals:

$(0, 2], (2, 8], (8, 20], (20, 40], (40, 52], (52, 82], (82, 114], (114, 126], (126, 152], (152, \infty)$,
however, we restrict our considerations only to the nuclei with $N \geq 16, Z \geq 16$.
Thus for example if $N = 32$ then $NMAGIC = 4, NPARITY = 0$.

The S_n, S_{2n}, S_p, S_{2p} denote the energy of separation of the last neutron, the last two neutrons, the last proton and the last two protons, respectively. The attribute MASS denotes the atomic mass of nuclei, Q_α is the energy α decay for given nuclei, BETA_DECAY_ENERGY, BINDING_ENERGY, MASS_EXCESS are naturally named. Because the neural networks are, in reality, linear combinations of inputs (except for the action of activation functions) in sequential analysis we try different functions of attributes. The best results (statistically significant) are obtained with the following (added in further analysis to the data) variables $S_p/Z, S_{2p}/Z, S_n/N, S_{2n}/N$. From a physical point of view they are binding energies of the odd nucleon, which is responsible for the spin of atomic nuclei.

Furthermore, on the base of the decay mode we construct the graphs (pointers in tables of database) of sequenced decay nuclei. The paths of these sequences end with stable nuclei or nuclei with internal transition or spontaneous fission. The variable JPREV indicates the value of spin of preceding in this path nuclei and RPREV indicates the type of decay ($\alpha, \beta-, \beta+$, proton emission ...) which gives the considered nuclei. We suppose that the values of the previous, in the decay path, spin and the type of decay of the previous nuclei are significant for the prediction of spin values. We try to consider all the paths of decay, by far analogy to the ARIMA methods (for categorial value - spin), but it seems that only the last value JPREV is significant for the spin evaluation. This is a correct conclusion from a physical point of view, because only the direct ancestor of the nascent nucleus can affect its spin. All the analyses are made in three cases:

- for all data(AL) - named later ALL,
- for N-even and Z-even case and allows cases together - named NEE,
- for four cases (N, Z): even-even(EE), even-odd(EO), odd-even(OE), odd-odd(OO) - named POZ.

Obviously, the case (N, Z) even-even is trivial (in this group the spin is always 0), but because most data mining methods work better in the larger set of observations, thus we do not allow this case.

The analysis of the most significant predictors for spin evaluation is presented in Tab. 2. All values presented in Tab. 2 have the probability of being significant less than 0.00001. We see that the previous spin value in the decay path (JPREV) plays the important role, but mainly in the N-even and Z-odd case. The type of decay of the previous nuclei (RPREV) is not important. In the case of N-odd the great role is played by parameters characterizing N as $S_n, S_{2n}, N, NMAGIC$ whereas in the case of Z-odd the parameters connected with Z as Z, S_p, S_{2p}. In summary, as expected, the greatest influences on the spin of the formed nucleus are mostly associated with the size of the binding energy of the odd nucleon and spin of his direct ancestor.

Table 2. The best predictors for spin evaluation

AL		OE		EO		EE	
variable	χ^2	variable	χ^2	variable	χ^2	variable	χ^2
JPREV	1112	S_n	270	Z	541	S_{2p}	499
ZPARITY	1079	S_{2n}	259	BINDING		N	341
NPARITY	1071	MASS	252	_ENERGY	382	MASS	320
Q_{da}	493	S_n/N	251	MASS	381	S_p	301
S_p	465	S_{2n}/N	218	N	350	JPREV	227
DECAY_TYPE	457	N	215	MASS_EXCESS	284	NMAGIC	202
N	455	Z	180	ZMAGIC	240	Z	185
Z	419	S_p	171	S_p	215	S_{2p}/Z	163
MASS	412	$JPREV$	161	S_{2p}	181	BINDING	
NMAGIC	380	S_{2p}/Z	152	JPREV	179	_ENERGY	161
S_n	343	NMAGIC	111	NMAGIC	177	MASS_EXCESS	158
BINDING		ZMAGIC	106	S_p/Z	173	S_n/N	126
_ENERGY	341	DECAY_TYPE	53	BETA_DECAY		ZMAGIC	60
MASS_EXCESS	338			_ENERGY	159		
BETA_DECAY				S_{2p}/Z	157		
_ENERGY	332			DECAY_TYPE	123		
ZMAGIC	320						

4   The Neural Networks

All the methods considered in this paper are compared by the coefficient of percent of positively classified spin values (ACC). Although the method POZ works separately on the subsets (N,Z) - even-odd, odd-even, odd-odd, in this case we add the value 0 for the case (N,Z) - even-even and adjust all four groups with the one ACC coefficient. We proceed similarly with the NEE methods.

Among the investigated neural networks the best one was the multilayer perceptron. The ACC coefficient for different neural networks is summarized in Tab. 3.

We see that the difference between the partition with respect to parity of N and Z in the ACC coefficient is small, but for further analysis we take the method NEUPOZ only.

5   Classification Tree

The finally produced trees constitute the compromise between the complexity (number of nodes, leaves and levels) and quality (number of wrongly classified items). We show two trees produced by us: TREE87 with 87 leaves and TREE155 with 155 leaves. In those trees the case N-even and Z-even is immediately identified by building the trees procedure. In the case of N-odd and Z-odd the great role is played by NMAGIC, ZMAGIC and JPREV, in the area N-odd, Z-even, JPREV and atomic MASS and in the area N-even, Z-odd ZMAGIC, energy of proton separation S_p and MASS_EXCESS.

Table 3. The best obtained neural networks

Method	EE	EO	OE	OO	AL	Percent evaluated	Type - Input - Output
NEUALL	100	58.1	61.8	62.0	**70.5**	44.4	MLP 86-13-15 (Tanh/Softmax)
NEUNEE	100	61.1	65.6	62.7	**72.4**	56.3	-; MLP 54-8-15 (Linear/Logist)
NEUPOZ	100	64.6	69.2	69.9	**76.0**	56.2	-; MLP 50-12-6 (Logist/Softmax)
							MLP 38-12-6 (Logist/Softmax);
							MLP 44-11-9 (Tanh/Softmax)

The significant role of variables marked as NMAGIC and ZMAGIC, which label the so-called 'magic numbers', should be noted. From the point of view of the structure of the atomic nucleus, the 'magic number' is nothing but the number of nucleons (protons or neutrons), at which it fills up completely the next nucleon shell. It is clear that the transition to the next shell must be related to changes in the spin of the nucleus. The remaining variables, as previously described, are related to the characteristics of the last odd particle and the spin of the ancestor.

So, the classification tree model well classify the importance of existing characteristics, affecting the spin nucleus.

6   The Quality of Models

We considered also other methods of classification apart from neural networks and classification trees like naive Bayes methods (the result is not good, because most of the attributes do not have normal distribution), k-neighbourhood (this method is not tally with the physical facts), Multivariate Adaptive Regression Splines (MARS), the previously mentioned TNN method (trends in the neighbouring nuclide), Support Vector Machine method (SVM), as well as neural networks and classification trees presented in the previous sections. The best of them, obtained by us, are summarized in Tab. 4.

We conclude that TNN is 20% worse than the SVM method, 25% worse than the Neural network method and TREE87 method and 33% worse than the TREE155 method. It is recommended to replace the classical TNN method by neural networks and/or classification trees to decrease the probability of wrong classification by 25-33%.

It is interesting to analyse wrongly classified nuclei in the NEUPOZ, TREE87 and TREE155 methods. They are shown in Figs. 2-3. The well-classified nuclei are light green, the wrong ones are dark red, the magical numbers are denoted by a blue line. Although we expected some relation between the badly-classified nuclei and the magical number, this relation is not so obvious, but we see it in the areas where there are no experimentally computed spin values. It is interesting that for the neural networks method you can see the good predicting spins of heavy and superheavy regions of nuclei, where as we know the spins

Table 4. ACC of different methods

Model	Data	N-Z					Percent
		e-e	o-e	e-o	o-o	all	evaluated
k-neighbourhood	NEE	100.0	44.3	43.3	45.6	**58.4**	31.8
naive Bayes	NEE	100.0	52.2	43.8	40.3	**59.1**	56.3
TNN	ALL	100.0	46.1	50.0	45.2	**60.4**	52.7
MARS	NEE	100.0	54.4	57.7	55.8	**67.1**	56.3
SVM	NEE	100.0	59.9	65.5	64.8	**72.6**	56.3
NEU	POZ	100.0	64.6	69.2	69.9	**76.0**	56.2
TREE87	–	100.0	66.6	71.3	67.5	**76.4**	44.4
TREE155	–	100.0	70.0	78.2	72.8	**80.3**	44.4

Table 5. Wrongly classified cases

Ele-ment	N	Z	NEU/TREE	True value	TNN	Ele-ment	N	Z	NEU/TREE	True value	TNN
Rh	53	45	5	**2**	6	Rh	54	45	9/2	**1/2**	9/2
Sb	61	51	5	**3**	4	Te	63	52	5/2	**7/2**	3/2
Br	47	35	2	**5**	3	Pr	71	59	5	**6**	4
Ge	45	32	1/2	**7/2**	1/2	Nd	75	60	5/2	**9/2**	5/2
Er	89	68	5/2	**3/2**	5/2	Gd	89	64	5/2	**3/2**	5/2
Ho	93	67	1	**5**	2	Re	107	75	1	**7**	3
Hf	105	72	1/2	**7/2**	7/2	Ir	115	77	1	**4**	2
Pa	143	91	1	**4**	2	Np	146	93	3/2	**5/2**	3/2

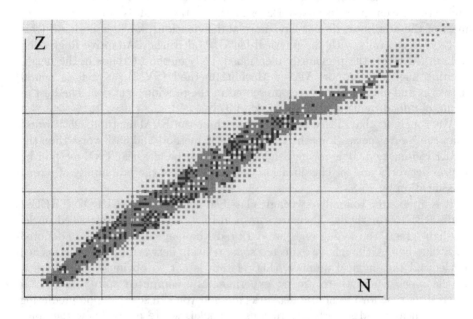

Fig. 2. NEUPOZ classification errors

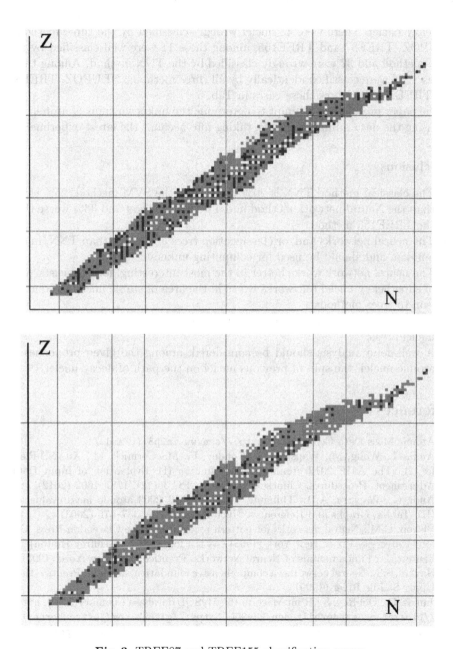

Fig. 3. TREE87 and TREE155 classification errors

are very important but difficult to reproduce, but worse work in the area of stable nuclei than trees methods. Hence, we believe that our calculations will be useful in planning experiments related to obtaining unknown heavy and superheavy nuclei. There were 48 nuclei wrongly classified by the three methods: NEUPOZ, TREE87 and TREE155, among these 11 were well classified by the TNN method and 37 were wrongly classified by the TNN method. Among these 48 cases, 16 were classified identically by all three methods: NEUPOZ, TREE87 and TREE155. We show these cases in Tab. 5.

The paper presents an attempt to determine the unknown spins of atomic nuclei using the data mining methods, taking into account the latest experimental data.

Conclusions

* The classical method TNN is 20% worse than the SVM method, 25% worse than the Neural network method and TREE87 method and 33% worse than the TREE155 method.
* The neural networks and/or classification trees are better than TNN, in our opinion, and should be used for computing unknown spins.
* The neural network works better in the most interesting, for physicists, area of superheavy nuclei but works worse in the area of stable nuclei in comparison to trees methods.
* In the TREE155 method the errors have almost uniform distribution for all nuclei areas.
* A well-done analysis should be considered, among the given properties of atomic nuclei, the spin of previous nuclei on the path of decay nuclei.

References

1. Atomic Mass Data Center page, http://csnwww.in2p3.fr/amdc/
2. Audi, G., Wang, M., Wapstra, A., Kondev, F., MacCormick, M., Xu, X., Pfeiffer, B.: The AME 2012 atomic mass evaluation (I). Evaluation of Input Data, Adjustment, Procedures. Chinese Physics C, CPC 36(12), 1287–1602 (2012)
3. Audi, G., Wapstra, A.H., Thibault, C.: The AME 2003 atomic mass evaluation (II). Tables, Graphs and References. Nucl. Phys. A 729, 337–676 (2003)
4. Bishop, C.M.: Neural networks for pattern recognition. The Clarendon Press, Oxford University Press, New York (1995) (with a foreword by Geoffrey Hinton)
5. Fausett, L.: Fundamentals of Neural Networks. Prentice Hall, New York (1994)
6. Haykin, S.S.: Neural networks: a comprehensive foundation, 2nd edn. Prentice-Hall, Upper Saddle River (1999)
7. James, D., DebRoy, S.: R interface to the MySQL database (January 2012), http://biostat.mc.vanderbilt.edu/RMySQL, https://github.com/jeffreyhorner/RMySQ , http://cran.r-project.org/web/packages/RMySQL/RMySQL.pdf
8. Nisbet, R., Elder, J., Miner, G.: Handbook of Statistical Analysis and Data Mining Applications. Academic Press, Elsevier Inc., Burlington, MA (2009)

9. Patterson, D.: Artificial Neural Networks. Prentice-Hall, Singapore (1996)
10. The home page of R environment, http://cran.r-project.org
11. Ripley, B.D.: Pattern recognition and neural networks. Cambridge University Press, Cambridge (1996)
12. Using MySQL with R (2009), http://oz.berkeley.edu/classes/s133/Db3.html
13. Wapstra, A.H., Audi, G., Thibault, C.: The AME 2003 atomic mass evaluation (I). Evaluation of input data, adjustment procedures. Nucl. Phys. A 729(1), 129–336 (2003)
14. Zoonekynd, V.: Statistics with R, http://zoonek2.free.fr/UNIX/48_R/all.html

Developing Lean Architecture Governance at a Software Developing Company Applying ArchiMate Motivation and Business Layers

Jan Werewka[1,2], Krzysztof Jamróz[2], and Dariusz Pitulej[1,2]

[1] AGH University of Science and Technology,
Department of Applied Computer Science,
al. A. Mickiewicza 30, 30-059 Kraków, Poland
{jan.werewka,d.pitulej}@agh.edu.pl
http://www.agh.edu.pl
[2] ATSI S.A. (Advanced Technology Systems International)
ul. Krakowska 386, 30-080 Zabierzów, Poland
info@atsisa.com
http://www.atsisa.com

Abstract. The work presents a lean approach of architecture development at a software development company. In this paper, a meta-model of the ORRCA methodology of IT architecture development is described. The methodology approach is used to support the architecture development and maintenance at a software development company by proposing a solution including processes, repositories, organizational structures, etc. The meta-model considered here is based on the motivation layer and business layers described in the ArchiMate notation. For a practical model verification, all six ArchiMate standard viewpoints for modeling ORRCA motivational aspects are used. The proposed meta-model applies to the development of software systems, for which bridging to business goals is essential.

Keywords: software architecture, enterprise architecture, architecture governance, lean approach, ArchiMate, meta-model.

1 Introduction

In growing organizations, whose main area of operation is software development, three main issues can be identified:

- Increasing complexity of developed systems over time. The effort required to develop new systems or to add functionalities to existing ones is increasing.
- Problematic integration of new systems with existing ones.
- Poor business alignment of developed systems. Organizations have been facing growing obstacles in aligning these increasingly costly IT systems to business needs of customers.

S. Kozielski et al. (Eds.): BDAS 2014, CCIS 424, pp. 492–503, 2014.

It is believed that a critical point has been reached, at which the complexity of IT systems is growing exponentially, while the chances of delivering a real value to customers are decreasing. This is the reason why organizations look for new means to successfully manage the complexity of existing systems and to correctly plan the development of new systems. The key to success is the introduction of the architecture governance that includes all the important factors influencing the development of a lean system's architecture. Architecture governance [4] is the practice and orientation by which (...) architectures are managed and controlled at an enterprise-wide level. The main task of architecture governance is implementing a system of controls over the creation and monitoring of all architectural components and activities. It is possible to base the introduction of architecture governance on a general-purpose architecture framework, like TOGAF. However, tailoring such universal tool that could be used in any industry, may be laborious. This problem is especially visible at a Software Development Enterprise, which is very specific because of its strong emphasis on IT systems. Therefore, the article describes an approach already adapted to exclusive needs of the software industry. In the proposed solution, the following core EA governance activities are distinguished:

- Governance strategy.
- Enterprise architecture landscape modeling.
- Next generation enterprise landscape evolving.
- Enterprise architecture modeling.
- Architecture evaluation and architecture decisions.
- Architect competency development and assessment.

The proposed solution uses ArchiMate which offers [5] an integrated architectural approach that describes and visualizes the different architecture domains and their underlying relations and dependencies. ArchiMate provides a graphical language for the representation of enterprise architectures consisting of motivation, business, application, technology and also implementation and migration layers (Fig. 1).

2   Related Works

The problem of aligning the architecture with business is considered broadly. There are different solutions proposed, but the Business Model Canvas approach [11,9] seems to gain popularity.

Developing, deploying and maintaining the enterprise software is a major challenge for both the organization developing and deploying the software and the one which is to use it. To ensure that these organizations cooperate effectively, it is necessary to build broader and deeper relationships that go beyond the simple rules of cooperation between the client and the contractor. In [14], a SMESDaD (Synergetic Methodology for Enterprise Software Development and Deployment) is proposed that concerns the operation and cooperation of two organizations (enterprises). One of these organizations is an IT company supplying software

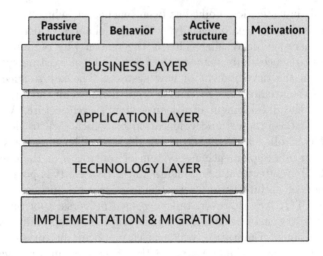

Fig. 1. ArchiMate layers

(SDE - Software Delivering Enterprise) for the main business line of the second company operating on the market (MOE - Market Operating Enterprise). One of the important challenges for the SDE is to make proper architectural decisions regarding the software development. To ensure this, a proper environment at the SDE must be developed. The examples of how to apply enterprise prescriptive methodologies at the SDE, such as TOGAF [4] and Zachman Framework, are known and described in the literature, for example [1]. Likewise, the issue of creating and maintaining the software architecture is also widely explored and described in existing publications [2,15] and [1].

The analysis of solutions for enterprise architecture models usually focuses on the most popular architectural models: the Zachman framework, TOGAF (The Open Group Architecture Framework), FEAF (Federal Enterprise Architecture Framework), The Gartner Methodology, and DoDAF (Department of Defense Architecture Framework). As a result of analyzing the methods of describing enterprise architectures, it was assumed that TOGAF [4] is a sufficient tool for organizing or adapting the method of the enterprise architecture development, which can also be used with other structures. TOGAF aims at providing a practical, easily-accessible, standardized, industrial method of designing the enterprise architecture.

Each area of architecture has its own concepts for modeling and visualizing the internal cohesion. These specific models and visualizations simplify the communication, considerations and analyses of the field. The communication aspects are important in the architecture development process. The communication, modeling and analysis on the enterprise level is described in detail in [8]. The assessment of the architectural decision using the ontology is presented in [18]. For the field of software engineering, a Knowledge View concept [3] and its application in software engineering was introduced. This work aimed at bringing knowledge engineering and Semantic Web technologies closer to software engineers and programmers.

Communication of the solutions must be universal enough to cope with the task of describing the enterprise and software architectures. ISO/IEC/IEEE 42010 defines architecture description (AD) standards and specifies requirements regarding architecture descriptions [6]. Architecture description languages (ADL) are a form of expression to be used in architecture descriptions. The standard avoids choosing a particular ADL, but defines minimum requirements on an ADL. To conform to the standard, an ADL must specify: the identification of concerns, the identification of stakeholders having those concerns, the types of models implemented by the ADL that frame these concerns, any architecture viewpoints and correspondence rules. Examples of ADLs include: 1st generation languages, such as Wright, Rapide, or Acme and 2nd generation languages, such as AADL, ArchiMate, SysML, UML (and profiles) and the viewpoint languages of RM-ODP (in ISO 10746).

ArchiMate [5] is an architecture description language (ADL) that is mostly used for modeling the enterprise architecture and is a very convenient tool for communicating solutions among business and technology staff. ArchiMate has some predefined views suitable for different stakeholders. There are some proposals to use the notation to create new viewpoints. In the paper [12], a Model Driven Engineering (MDE) framework to support the EA modeling activity is proposed, basing on the ArchiMate language. The definition of the ArchiMate language was accompanied by the assumption that in order to build an expressive business model, it is necessary to use relations linking completely different fields: from business motivations to business processes, services and the infrastructure. ArchiMate goes where UML does not - it defines a meta-model that can be used to create and portray relations between elements of various layers. The ties and limitations specified in the meta-model support the analysis, coherence checking, identifying and tracing links as well as the management.

3   ORRCA - The Proposed Solution

Basing on the solutions, standards and own experience, an ORRCA (Open Robust and Reference Component's Architecture) prescriptive methodology is proposed. The best solution to create a correct architecture governance at a software development company is the transition from both customer related requirements and enterprise architecture related requirements to the final software architecture. Our proposal constitutes an attempt to create a meta-model that fills the gap and transit from the enterprise architecture to software architecture for the SDE. This gap is filled by creating the meta-model for the motivation and business layer. The main task is to identify key requirements, goals and principles that are connected to the enterprise architecture and are crucial for the software architecture. Knowing these areas is enough to model motivation and business layers.

The best way to perform these actions is to use the existing notation, in our proposition - ArchiMate, which is enough to describe different levels of abstraction suitable for modeling enterprise and software architectures. ArchiMate provides a complete set of concepts for describing relationships between architecture fields, at the same time producing a simple and uniform structure. The basic assumption of the language is that services play a key role in relations between domains. The ArchiMate notation also provides a meta-model to describe the motivation and business layer that is adapted by us and used to create the ORRCA meta-model.

4   The ORRCA Motivation Layer

The motivation viewpoint [5] allows the designer or analyst to model the motivation aspect without focusing on certain elements within this aspect. For example, a viewpoint can be used to present a complete or partial overview of the motivation aspect by relating stakeholders, their primary goals, the principles that are applied, and the main requirements on services, processes, applications, and objects. The motivation can be considered as an IT architecture manifesto proposed for small and medium-sized software development companies.

A driver element is defined within the ArchiMate motivation layer as something that motivates the change at an organization. Typical reasons for a change originate from what a company wants to achieve: to reduce development costs, to be more competitive, and to increase customer satisfaction. ORRCA defines the following set of drivers:

1. Software development companies must reconcile the interest in delivering new systems using modern technologies with delivering functionalities that already exist in legacy systems.
2. Too high redundancy in existing systems causes high development and maintenance costs.
3. Existing monolithic solutions are difficult to be integrated and cause problems with reference to software reusability and scalability.
4. Inability to integrate existing solutions influences the time and costs of systems under development.
5. Technical debt causes problems with the software quality and maintenance.
6. Wrong architectural decisions influence negatively the time of requirement realization.
7. Omitting of an evolutionary (agile) development causes the increase in costs.
8. Building architecture solutions without considering the domain (industry branch) knowledge results in a lack of market usability when it comes to the developed software products.

An assessment concept defined in ArchiMate is an outcome of an analysis of a driver. It is common for enterprises to undertake an assessment of these drivers using the SWOT analysis.

The notion of a goal in ArchiMate refers to the end state that a stakeholder intends to achieve. ORRCA defines the following set of goals:

– Architecture development goals:
  • Build systems for an easy and effective integration
  • Prepare legacy/existing systems to be integrated through the common integration platform
  • Refactor existing systems to achieve a proper grain level
  • Establish processes to ensure a proper design of the architecture
  • Minimize the technical debt in critical parts of existing systems
– Architecture processes improvement goals:
  • Establish processes to ensure the analysis of existing solutions before a new development
  • Establish procedures to properly identify architecture decisions
  • Create and maintain a service repository
  • Establish procedures to continuously manage the technical debt
  • Use an iterative approach to evolve the architecture of existing products
  • Improve effectiveness of the technical support for existing products
  • Use market knowledge to deliver better products
– Architecture competence development goals:
  • Develop competences of software architects
  • Continuously increase the technical skills of development teams
  • Involve to projects people with high domain knowledge and experience
  • Establish processes that ensure the use of agile techniques in the development of the software architecture
  • Include an analysis and development of the software architecture into agile development processes

A principle is an element of an ArchiMate motivation layer defining [5] a normative property of all systems in a given context, or the way in which they are realized. ORRCA defines the following set of principles:

– lean architecture,
– architecture governance,
– components reusability,
– integration readiness,
– portability,
– scalability,
– data as an asset.

The base principle in ORRCA is the lean architecture based on the ideas of Taiichi Ohno, the originator of the Toyota Production System known as Lean Manufacturing. Eliminating wastes is a fundamental lean principle from which all other principles have originated. The principles and practices of Lean Manufacturing are adapted by Lean Software Development (LSD) [17,10,13].

The principles and goal contribution viewpoints in ArchiMate allow [5] an analyst or a designer to model principles that are relevant to the design problem at hand, including the goals that motivate these principles. Fig. 2 describes relationships between principles and their goals in ORRCA. The principles may

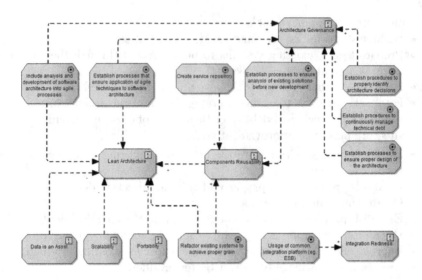

Fig. 2. Goal contribution viewpoint

influence goals positively or negatively. The resulting views can be used to analyze the impact that the goals have on one another or to detect conflicts among stakeholder goals.

A requirement in the ArchiMate motivation layer is defined [5] as a statement of need that must be realized by a system. In this respect, requirements represent the "means" to realize goals. ORRCA defines the following set of requirements:

- The architect (taking architectural decisions) should have appropriate competences
- The architecture development should undergo well defined governance
- Transparent development (such as in the Open Source development)
- Building a common solution architecture
- Building the architecture landscape for an industry branch
- Defining preferable technology stacks
- Taking in regard future solutions (for example, Open Group Platform 3)
- Architectural documentation
- Architecture models (for example, different viewpoints in ArchiMate)
- Architecture development (TOGAF)
- Usage of component technologies and solutions
- Collecting operational critical data: usability data, user behavior, load and traffic, abnormal system behavior

ArchiMate [5] defines a constraint as a restriction on the way in which a system is realized. In contrast to a requirement, a constraint does not prescribe some intended functionality of the system to be realized, but imposes a restriction on the way in which the system may be realized. ORRCA regards architecture tradeoffs and defines the following set of constraints:

- The architecture governance is responsible for defining what architecture decision is and what design decision is.
- It is not feasible to:
  - maintain the simplicity of usage with increasing reusability (maximized reuse and minimized use)) [7];
  - make a big architecture up front (BAUF) [7];
  - design an infinitely flexible system) [7];
  - ignore the existing code base and domain knowledge;
  - create a totally closed solution;
  - infinitely fulfill quality attributes, such as security, scalability and availability.
- Coarse grained modules are easier to use, but fine grained modules are more reusable) [7].
- Lightweight modules are more reusable, but heavyweight modules are easier to use) [7].
- The gain of using architecture for the project should be higher than the cost of using it.

5   Dependencies between Motivation and Business Layers in ORRCA

The business layer offers external customers products and services that are realized at the organization by business processes performed by business actors. The business layer describes how an organization performs in conformance with the motivation layer.

There is a limitation regarding the mapping of motivation to business elements. Fig. 3 shows the mapping of motivation to the business layer derived from the ArchiMate description. In the business layer, active entities (for example, business actors or business roles) perform actions related to business processes or functions (capabilities). The stakeholders of the motivation layer are assigned to business actors of the business layer.

ArchiMate defines the meaning) [5] as the knowledge or expertise present in a business object or its representation, given a particular context. Typical examples of meaning descriptions are definitions, ontologies, paraphrases, subject descriptions, and tables of content. The knowledge considered here may concern, for example, the knowledge of stakeholders (experience, the knowledge base) or the description of organization processes.

The value in ArchiMate is defined as a relative worth, utility, or importance of a business service or product. An important element of the ArchiMate business layer is a product which is a coherent collection of services, accompanied by contracts, which is offered as a whole to internal or external customers. A product can contain the following architecture services for new systems: selecting the best architecture under given constraints, delivering suitable architecture tools and models, delivering competency and coaching, effective development (reducing development costs, being more competitive, increasing customer satisfaction).

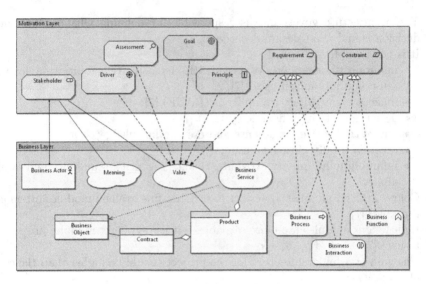

Fig. 3. Mapping of the motivation layer to the business layer

The requirements realization viewpoint in ArchiMate allows the designer to model the realization of requirements by the core elements, such as business actors, business services, business processes, application services, application components, etc.

In the ORRCA requirements realization viewpoint model, the following service groups are distinguished (Fig. 4): architecture services for new systems, architecture services for legacy systems, architecture consulting services, and the support of architecture development for the industry branch (domain).

6   Verification of the Solution

To verify the described approach, it is required to ensure that all development activities of an SDE are properly aligned with its business goals. It can be done by adjusting the SDE's internal process and introducing the mechanisms that support software architecture governance within the enterprise. It is crucial to ensure that all architecture decisions are made in accordance with the strategy of the enterprise. Software architecture reviews are a good example of tools that can be used for this purpose. Introducing an Architecture Conformance Review) [1] at the beginning of each project provides a better alignment of software project activities with long-term goals and the strategy of the SDE. It is important to verify the software architecture at the very early stage of the project, because the costs of implementing changes and corrective actions are relatively low at this stage. Another important architecture governance activity is the Post-Implementation Review) [1]. It verifies whether the software is implemented according to the design. If any non-compliance is found, corrective

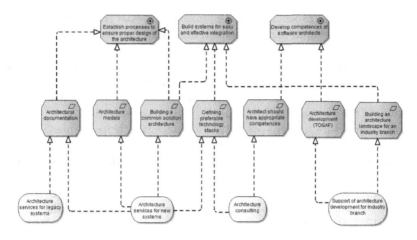

Fig. 4. Realization of requirements

actions can be planned. In addition, the Post-Implementation Review should contain lessons learned that aid in a continuous improvement of architecture governance processes. Of course, it is not sufficient to control the software architecture only at the beginning and at the end of the project. Every change of the scope and the resulting architecture modifications should be verified in order to check whether they are compliant with the long-term strategy goals of the SDE (Architecture Conformance Review). Performing Post-Implementation Reviews periodically in the course of the project is also a good practice. It allows for tracking non-compliance sooner, which can significantly reduce costs of introducing and implementing corrective actions.

One of the key elements of the software architecture review is the verification in terms of adhering to the architecture principles defined by the senior management of the SDE. Principles are general rules that are a basis for all architecture decisions that aim at ensuring that they are in line with the long-term strategy of the enterprise. Architecture principles are described in more detail in enterprise architecture frameworks: TOGAF [4] or PEAF [16]. In an SDE, there is usually a conflict between realization of short-term goals specified by a customer within a single project, and satisfying long-term strategy defined by the senior management of the SDE. Verifying the software architecture according to architecture principles helps to find a good balance between both short-term and long-term goals.

7   Conclusions

It is very important to consider the ORRCA methodology to successfully develop the software and move from the enterprise architecture to software architecture. To describe the ORRCA motivation, all six ArchiMate standard viewpoints for modeling motivational aspects are used: stakeholder, goal realization, goal contribution, principles, requirements realization and motivation viewpoint.

The presented ORRCA meta-model can be used by an SDE to create its own architecture governance model facilitating the development of the system's architecture for its products. An architecture governance model based on ORRCA includes all important factors that should be taken into consideration during the creation of the system's architecture for a given organization. A successfully developed architecture governance for the organization helps in managing the overall current IT landscape complexity and allows for better mapping of customer's business needs to requirements developed in the implemented systems. The key to market success for any SDE is not only knowing and fulfilling customer's requirements but also taking into account its own constraints, strategic goals and existing capabilities located in a properly described IT landscape. According to the gathered ORRCA modeling experience, the ArchiMate language and its meta-model is a sufficient tool for modeling and describing the architecture of existing and future systems at an SDE. A properly described motivation and business layer is helpful in developing systems that better fit customer business needs, are cheaper to create, easier to be integrated with current and future solutions, and require shorter time of software development and delivery.

The authors know from their own involvement in various projects that in many cases, large software systems can be successfully developed only provided that the involved organizations strictly cooperate, and this cooperation is synergic in nature. This can be achieved by a properly developed architecture governance process.

References

1. Bente, S., Bombosch, U., Langade, S.: Collaborative Enterprise Architecture: Enriching EA with Lean, Agile, and Enterprise 2.0 Practices. Morgan Kaufmann (2012)
2. Eeles, P., Cripps, P.: The Process of Software Architecting. Addison Wesley Professional (2010)
3. Goczyła, K., Piotrowski, P.: Application of knowledge views. Studia Informatica 31(2A), 77–88 (2010)
4. The Open Group: TOGAF Version 9.1, p. 692 (2009-2011)
5. The Open Group: ArchiMate 2.0 Specification, p. 183 (2009-2012), http://pubs.opengroup.org/architecture/archimate2-doc/toc.html
6. ISO/IEC/IEEE: Systems and software engineering – architecture description. ISO/IEC/IEEE 42010:2011(E) (Revision of ISO/IEC 42010:2007 and IEEE Std 1471-2000, pp. 1–46 (January 2011)
7. Knoernschild, K.: Java Application Architecture: Modularity Patterns with Examples Using OSGi. Robert C. Martin Series. Pearson Education (2012)
8. Lankhorst, M.: Enterprise Architecture at Work: Modelling, Communication and Analysis. Enterprise engineering series. Springer (2009)
9. Meertens, L., Iacob, M., Jonkers, H., Quartel, D., Nieuwenhuis, L., van Sinderen, M.: Mapping the business model canvas to archimate. In: Proceedings of the 27th Annual ACM Symposium on Applied Computing, pp. 1694–1701. ACM (March 2012), http://doc.utwente.nl/82858/

10. Morien, R.: Agile management and the toyota way for software project management. In: 3rd International Conference on Industrial Informatics, pp. 516–522 (2005)
11. Osterwalder, A.: The Business Model Ontology: a proposition in a design science approach. Dissertation, Université de Lausanne, Ecole des Hautes Etudes Commerciales (2004)
12. Pena, C., Villalobos, J.: An mde approach to design enterprise architecture viewpoints. In: Seventh IEEE International Conference on E-Commerce Technology (CEC 2005), pp. 80–87 (2010)
13. Poppendieck, M., Poppendieck, T.: Implementing Lean Software Development: From Concept to Cash. A Kent Beck signature book, Addison Wesley Professional (2007)
14. Rogus, G., Skrzyński, P., Szwed, P., Turek, M., Werewka, J.: SMESDaD – synergetyczna metodyka rozwijania i wdrażania oprogramowania korporacyjnego. Pomiary, Automatyka, Robotyka R 15(12), 196–209 (2011)
15. Rozanski, N., Woods, E.: Software Systems Architecture: Working with Stakeholders Using Viewpoints and Perspectives. Addison-Wesley (2011)
16. Smith, K., Graves, T.: An Introduction to Peaf: Pragmatic Enterprise Architecture Framework. Pragmatic EA, Limited (2011)
17. Steindl, C.: Lean software development, ibm corporation, p. 65. IT Architect and IT Specialist Institute Central Region in Herrenberg (2005)
18. Szwed, P., Skrzynski, P., Rogus, G., Werewka, J.: Ontology of architectural decisions supporting atam based assessment of soa architectures. In: Ganzha, M., Maciaszek, L., Paprzycki, M. (eds.) Proceedings of the 2013 Federated Conference on Computer Science and Information Systems, pp. 287–290. IEEE (2013)

Preconditions for Processing Electronic Medical Records

Kazimierz Frączkowski, Zygmunt Mazur, and Hanna Mazur

Wroclaw University of Technology, Faculty of Computer Science and Management,
Institute of Computer Science
Wyb. Wyspiańskiego 27, 50-370 Wroclaw, Poland
{kazimierz.fraczkowski,zygmunt.mazur,hanna.mazur}@pwr.wroc.pl

Abstract. Due to the abundance and diversity of document types (both external and internal ones), the process of switching to electronic medical records requires a number of organizational, technical and legislative adjustments. The development of those processes and the current stage of works are described in this article. The first step in selecting documents that will become subject to electronic storage and processing in the first place was a survey in which respondents of nine Regional Project consortiums chose medical records most frequently used in business processes conducted by service providers. Based on the Pareto principle, 20% of medical records used in 80% of medical services provided were identified.

Keywords: electronic medical records, metadata.

1 Introduction

Computerization of the public administration and healthcare is to bring notable economic and social benefits. The Communication from the Commission to the European Parliament, the Council, the European Economic and Social Committee and the Committee of the Regions of 6 December 2012 on e-Health Action Plan 2012-2020 – *Innovative healthcare for the 21st century*, points to the necessity to eliminate barriers to effective use of IT, one of which is lack of interoperability in this area. According to the Commission, interoperability on the following four levels – legal, organizational, semantic and technical – preconditions obtaining the added value from investments in e-Health, forecast in studies and pilot projects. Studies conducted in 2008 by Gartner in six countries (the Czech Republic, France, Spain, the Netherlands, Sweden and Great Britain) showed, among others, that the introduction of an electronic prescription transfer system would prevent writing five million incorrect prescriptions to outpatients a year, while electronic systems supporting management of medical orders and taking clinical decisions would eliminate side effects among 100 thousand inpatients a year [15]. In 2012 in Poland doctors issued 23 thousand incorrect prescriptions amounting to PLN 1.6 million in total. At that time 13366 pharmacies and 1145 dispensaries were authorized to issue prescriptions reimbursed by the National Health Fund [19].

S. Kozielski et al. (Eds.): BDAS 2014, CCIS 424, pp. 504–514, 2014.
© Springer International Publishing Switzerland 2014

Access to data on medical services and events to authorized persons at any time and place would facilitate quicker and cheaper diagnosis. Current use of IT in healthcare does not meet existing needs or take advantage of available possibilities. There are a number of reasons behind this situation, among others, complex issues related to the interoperability of IT systems and the security of medical data. What also matters is effective search and sharing of medical data, which requires solving the problem with creation and application of information on data, that is metadata. These issues have been explored for years now by theoreticians and practitioners who have been developing healthcare IT systems [18]. The aim is to standardize the description of data [11] in such a way to allow for unambiguous and secure identification of a patient, for reading of the source of the data and the description containing values of features used in the description of clinical values. In 2011 in Poland doctors wrote more than 800 million prescriptions, around 19 million sick notes, while the remaining medical documents were estimated to amount to ca. 2 billion. Creation of an effective information system that would meet current needs requires modifications and new solutions in a number of areas. Apart from IT solutions, what also matters are legal regulations, in particular the Regulation of the Minister of Health of 25 June 2013 on the statistical system in healthcare and the Regulation (EC) No 1338/2008 of 16 December 2008 of the European Parliament and of the Council on Community statistics on public health and health and safety at work [8].

The problem of introducing the Electronic Medical Record is being dealt with by world leading economic powers. A report on the EMR market shows that the USA, Australia, Canada, France, Germany, Japan, Scandinavian countries, Spain and Great Britain, India and China have not met all the needs in this scope so far. The level of implementation is not satisfactory while IT companies estimate the value of the world market of EMR creation and storage at 600 trillion $ [6]. It is estimated that US expenditures on EMR have increased from 7.4 billion dollars in 2010 to 9.8 billion in 2013. As the cloud computing technology is expanding, ever more often we hear about the need to increase the efficiency of EMR processing and its security. Schweitzer in his work [10] points to the cost efficiency of such solutions stressing at the same time insufficiency of legal regulations in this area. EMR implementation models are dominated by opinions based on research and pilot project results that recommend gradual and selective choice of medical records and application of tools that facilitate decision-making processes that increase the safety of patient's health when a doctor or a nurse have access to the EDM [12].

The co-author of this article while working as a Coordinator of Information Project in Healthcare from 2009 to 2013 took some actions aimed at developing an effective method of switching from traditional medical records to the EMR. Given the scale of such a project (more than 100 thousand service providers and ca. 350 thousand medical staff) as well as the above-mentioned number of medical records created and processed (prescriptions, referrals, certificates, etc.), it was necessary to develop the method and obtain the consent of leading stakeholders for its implementation. Based on the Pareto principle, it was assumed that there are 20% of medical records used in 80% of business processes conducted by medical service providers that need to be identified. In 2013 the Project Coordinator Bureau, the Centre for Healthcare Information Systems, made a survey in which service providers identified medical records most often created, shared and processed in the healthcare system in Poland [1].

2   Main Objectives and Principles of P1 Project

The main objective of the project called "Electronic Platform for Storage, Analysis and Sharing of Digital Resources Regarding Medical Events" (marked as P1) is to build an electronic platform for public healthcare services, which will help public administration bodies and entrepreneurs (among others, healthcare centers, pharmacies, medical practices) gather, analyze and share digital resources regarding medical events in compliance with the Act on the healthcare information system. The main advantage of the system is quick and convenient access to data gathered in registers in the P1 system concerning medical events in Poland. It will improve the quality of service, healthcare planning, electronic accounting and efficient management of emergencies. The P1 project meets all the legal regulations listed on the project website [20].

An important requirement in the P1 project co-financed with the European Regional Development Fund as part of the Innovative Economy Operational Programme 2007-2013 is the credibility of data on medical events and services and the interoperability of healthcare information systems. The interoperability of two or more systems should be understood as a possibility to exchange information between them "understandingly", which requires development and implementation of standards that will guarantee one interpretation – understanding of data – by different systems.

The P1 system will be used by patients, service providers, pharmacists, medical and administrative personnel. So far, as part of the P1 system, prototypes of the following two systems have been created – Online Patient Account and e-Prescription. In 2011 the e-Prescription system was first tested in Leszno, while the Online Patient Account (pol. *Internetowe Konto Pacjenta* – IKP) in Kraków. The next stage of the P1 project is the integration of the two systems with the implementation of the rules of semantic and technological interoperability developed by the Center for Healthcare Information Systems based on the HL7 CDA standard in other locations in 2013.

From the perspective of medical data exchange, the main product of the P1 project is portal – an electronic platform that participates in exchange of medical records, which offers, among others, mechanisms of authorization and security of medical information supported by the Abuse Detection System. The system is to monitor data and electronic transaction effectively in order to detect potential abuses and irregularities.

3   Electronic Medical Records – A Survey

In accordance with the Act on the healthcare information system [3], electronic medical records are an electronic document that contains data on healthcare services that have been provided, are being provided or planned, which allows the patient to take advantage of healthcare services. Electronic medical records should, among others, be available in XML and PDF, allow for exporting all the data as XML so that it will be possible to read them in some other ICT system, allow for adding other documents created in other forms (e.g. paper version, X-rays, etc.) by digital mapping of the documents, which next will be transported to the ICT system, at the same time ensuring clarity, accessibility and coherence of the entire documentation. The above expectations must be clarified in the resolution of the Minister of Health due to additional requirements imposed on electronic medical records. For example, they must contain a predefined set of metadata that describe the medical document (e.g. a scanned paper document), ensure clear and reliable identification of the type of the document, must be ascribed a unique (at least in the country) ID and have a structure that can be processed in information systems [16,17].

Electronic medical records are produced as a result of provision of a health service defined as an action "aimed at maintaining, recuperating, restoring and improving health and other medical actions arising from treatment or regulations" [4]. A medical event, on the other hand, is "an action initiating a business process in healthcare, stages or the outcome of this action, which are to be recorded, transferred, analyzed and stored in the healthcare information system" [9].

Current works are aimed at defining the scale and types of medical records, expanding the notion of the electronic medical record and specifying stages of its wide-ranging implementation.

Types of documents most frequently exchanged between healthcare entities, their quantity and related business processes were defined following the collection, distribution and analysis of results of the survey conducted among service providers – entities participating in the E-Health Regional Projects [7]. 122 questionnaire forms were analyzed (sent from 9 consortiums participating in Regional Projects), and it turned out that most common referrals are those to: laboratory test, hospital, specialist or diagnostic clinic, medical rehabilitation, systemic rehabilitation, health-resort, healthcare and curative institution or healthcare and nursing institution. The information serves as a guideline as to which medical records are to be created (as the most important) in electronic form in the first place. Medical data named by the respondents as the most frequently processed in their systems are orders (prescriptions, demands and referrals), data on medical events – that result from consulting a doctor and are to be made available on the IKP, critical data and next verification of authorizations in the eWUŚ system and reporting to the National Health Fund [7].

According to Art. 56 of the Act on the healthcare information system, a medical record created after 31 July 2014 must be kept in an electronic form. In projects such as epSOS (Smart Open Services for European Patients – Open eHealth Initiative for a European), TRANSFoRm (Translational Research and Patient Safety in Europe), in which more than ten leading EU universities participate as part of 7PR, electronic medical records are used as a source of information about GERD, a chronic disease. Results of the works on metadata that describe information about this disease are the basis for building tools used to find medical information about patients from five EU countries, including Poland, in order to build a system supporting medical decisions based on facts and improve patients' health safety [2].

4   Classification Model; Medical Records Subject to Implementation in the First Stage

Clinical data are usually kept in paper form, as scanned documents (i.e. in electronic form but not as electronic medical records), images, photos and voice records. A number of service providers have developed their own standards and medical information systems.

A standard applied in medical circles (mainly in the USA) for electronic information exchange is Health Level Seven (HL7), which supports communication among different medical systems.

In an ICT system, apart from the electronic form of a scanned document, we also need additional information allowing for data processing in IT systems. An electronic medical record must comply with HL7 CDA R2 (Clinical Document Architecture Release 2).

A CDA document is made up of a heading and a body. Each medical record is ascribed a classifier, e.g. the type of the document (e.g. a consultation) and perspectives (a set of multidimensional properties) included in the heading of the document, described for example as follows: persistence – the period of time when a clinical document is not changed, stewardship – data on the subject who created the document and is responsible for it, potential for authentication, wholeness, human readability (Fig. 1). The classification is to serve for presenting the history of client-healthcare institution contact.

The P1 system contains not only data on medical services but also on documents produced while providing the services. Each document should be ascribed an unambiguous ID – a classification code compliant with resort codes ascribed on the registration of healthcare entities as shown in Fig. 2 [21].

In September 2013 the Center for Healthcare Information Systems made available materials containing a set of business and validation rules for electronic medical documents, definitions of the structures of such documents, requirements concerning data and glossaries to be used to classify data in the document [13]. The business rules included:

- "A document MUST comply with HL7 CDA R2".
- "If a patient is not one year old yet and has not been ascribed a PESEL number or an ID other than the ID in the patient base used by the system in which the document was created, then the document MUST have at least

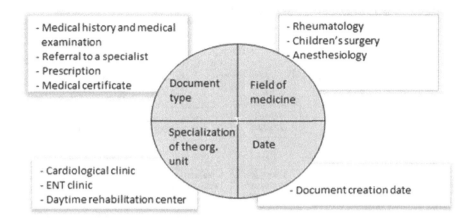

Fig. 1. Classification of electronic medical records – perspectives (Own study based on [21])

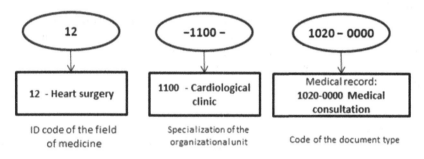

Fig. 2. Classification codes for medical records (Own study based on [21])

one ID of the patient's mother and the date of the patient's birth, and in the case of a multiple pregnancy, also the ID of the newborn baby in the multiple pregnancy".

Document [13] contains business rules that apply to records selected in the survey conducted. It is necessary to write down business and validation rules also for other medical documents so that standardization covers at least those documents or their fragments being source of data for messages sent to P1. Fig. 3 shows a structure of the Referral. There are different objects that describe a referral to a health resort, to an institutional care, for long-term nursing, to a mental hospital, to the Voivodship Medical Board of the Ministry of Interior.

Another document created as part of standardization of electronic medical records (which will allow for adequate interoperability) is called "Model for transport of data on medical events and the index of electronic medical records stored in P1". The model is to help us define the format and scope of information sent. Given the variety and wide scope of information on medical events and medical

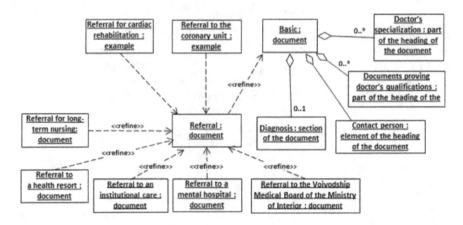

Fig. 3. Object Referral (Own study based on [13])

records, we assume that the System for Storage of Medical Data and Medical Events (SGDM-ZM) will store and process sets of information about medical events and medical records and separately an outline of a medical event or medical records, which are not complete information/documents [14].

The recommended structure of the medical records index looks as follows:

Identification part:

- Service provider (rpwdl.csioz.gov.pl)
- Patient (PESEL)
- Organizational section (part VIII ID code determining the specialization of organizational sections)
- Field of medicine (ID code of the specialization)

Content:

- Service code (code of the healthcare function related to the code of the statistical unit)
- Healthcare employee
- Medical procedure (ICD-9)
- Diagnosis (ICD-10)
- Related cause (ICD-10)

Related information:

- Type of information
- Location of the records (ID of the document instance)

The assumption made for the P1 platform that it will be supported by the PESEL base (there will be no own patient base created) will make it difficult to group documents with no PESEL number provided.

5   Metadata-Based Storage and Search for Medical Information

Metadata, often called "data on data" or "information on information", help users find relevant information in the cyberspace and learn about interrelations between the result and other information. Traditional cataloguing methods do not work in the World Wide Web – the rapid growth in the quantity of data prevents such an option. Metadata, on the other hand, allow for such organization and management of information present in the WWW system that will support our search. In the P1 project, which is to service so many stakeholders (around 38 million citizens), process documents (more than 700 million prescriptions a year), data on medical events (more than a billion a year) and cover different entities related to the types of medical records we face the problem with their storage and effective sharing. The P1 project entails the development of a portal participating in exchange of medical data in which metadata are stored centrally. According to applicable regulations set out by the Inspector General for the Protection of Personal Data and the Resolution of the Minister of Health [5], service providers are obliged to store medical data. Achievement of the objectives of the P1 project requires central storage of metadata on data stored in the system. With relevant records the P1 system will participate in sharing of medical data by collecting information about record type, the place it was created and the entity that created it.

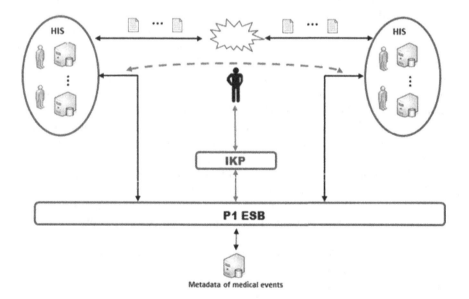

Fig. 4. Model architecture for sharing electronic medical information with central record of metadata on "medical events" (Own study)

Information on electronic medical records shown in Fig. 4, such as prescriptions, sick notes, referrals and others will be sent to the P1 system using for this purpose the ESB (*Enterprise Service Bus*) form Hospital Information System (HIS) and stored there centrally, in compliance with HL7 CDA. Information on other medical records (metadata), on existence of such documents, who created them, etc., will be available in the Online Patient Account portal (IKP).

Medical records of a patient, e.g. test results or a discharge note, can be exchanged between service providers (who created such information and store it in an electronic form) only upon the patient's consent. The P1 platform is a trustworthy participant in authorization of the exchange.

6   Conclusion

The practice of keeping only paper version of medical records is to last until 31 July 2014. The Center for Healthcare Information Systems is intensively working on the P1 and P2 systems which are to support medical entities in creating, storing and sharing electronic medical records.

Types of medical records subject to electronic storage and processing, their standard electronic form, mechanisms for generating sets of metadata in order to mark and classify them as well as the pace of implementing the agreed changes depend on work performed by thousands of stakeholders. What will matters are also the results of research carried out as part of projects such as epSOS (aimed at developing mechanisms for data exchange between European countries that apply to, among others, prescriptions), TRANSFoRm and others being part of the new Horizon 2020 program in which one of the priorities is personalized healthcare and digital security.

An important objective of the P1 project is to store electronic medical records on local service providers' servers, while metadata – centrally. This concept arises, among others, from the estimated of the quantity and capacity of produced medical documents mentioned in the Introduction, but also existing legal regulations, standards and trends observed in the process of development of similar systems in the European Union. Fact-based medicine being one of challenges set in the "Horizon 2020" program requires exploration of huge quantities of data. Metadata management and related issues such as metadata standardization, quality of information, validation, authentication are the foundation for creating big databases, individual elements will be covered by ISO/IEC 11179 [11]. Results of research on metadata show the direction for further research and new IT projects still focused on this issue as distributed data processing using the semantic web as well as cloud processing rely on effective representation of metadata.

Further works on electronic medical records will probably aim at formulation of recommendations for other electronic documents based on HL7 CDA, including discharge notes and other external documents related to consultation or test results. In order to standardize certain documents it may be necessary to form specialization teams responsible for standardization of types and content of medical documents. IT specialists, on the other hand, will need to deal with marking descriptive information in digital representation generated automatically. Such metadata could be obtained with no extra costs. Thousands of service providers and a number of applications developed in recent years must be adapted to the requirements of interoperability related to integration with the P1 central system.

References

1. http://www.csioz.gov.pl/wydarzenieDetail.php?id=6
2. Act of 27 August 2004 on provision of health services financed with public funds
3. Act of 28 April 2011 on the healthcare information system
4. Act of 6 November 2008 on patient' s rights and the Commissioner for Patient' s Rights
5. Announcement of the Minister of Health of 4 July 2013 on announcement of a consolidated text of the Regulation of the Minister of Health on the scope of necessary information stored by service providers, detailed method of recording the information and sharing it with entities obliged to finance the services with public funds (December 5, 2013)
6. Overview of International EMR/EHR Markets Results from a Survey of Leading Health Care Companies, http://www.accenture.com/SiteCollectionDocuments/ PDF/Accenture_EMR_Markets_Whitepaper_vfinal.pdf (August 2010)
7. Preliminary summary of the second edition of the study
8. Regulation (EC) No. 1338/2008 of 16 December 2008 of the European Parliament and of the Council on Community statistics on public health and health and safety at work. Journal of Laws. L 354 of 31.12.2008, http://europa.eu/ legislation_summaries/employment_and_social_policy/ health_hygiene_safety_at_work/em0013_en.htm
9. Regulation of the Minister of Health of 28 March 2013 on requirements regarding the Medical Information System
10. Standard ISO/IEC 11179, Information Technology – Metadata registries (MDR), http://metadata-standards.org/11179
11. TRANSFoRm, http://www.transformproject.eu
12. Collins, S., Bakken, S., Vawdry, D., Colera, E., Currie, L.: Model development for EHR interdisciplinary information exchange of ICU common goals. International Journal of Medical Informatics 80(8) (2011), ncbi.nlm.nih.gov/pmc/ articles/PMC3044780
13. CSIOZ: Business and validation rules defining the structure of medical documents (ePrescription, eReferral and eOrder) processed on the P1 platform. Warsaw (2013)
14. CSIOZ: Model for transport of data on medical events and the index of electronic medical records stored in P1. Warsaw (2013)
15. Eurobarometer 70: Public option in the european union (December 2008), http://ec.europa.eu/public_opinion/archives/eb/eb70/eb70_first_en.pdf
16. Frączkowski, K., Sikorski, L.: Electronic medical records – part. I. National Review of Medicine 4 (2012)
17. Frączkowski, K., Sikorski, L.: Electronic medical records – part. II. National Review of Medicine 5 (2012)
18. Moehrke, J.: Healthcare metadata (May 14, 2012), healthcaresecprivacy. blogspot.nl/2012/05/healthcare-metadata.html

19. National Health Fund: 23 thousand incorrect prescriptions amounting to pln 1.6 million (April 28, 2013), http://wiadomosci.gazeta.pl/wiadomosci/1,114871, 13820932,NFZ__23_tys__blednie_wypisanych_recept_o_wartosci.html
20. P1 project: Electronic Platform for Storage, Analysis and Sharing of Digital Resources Regarding Medical Services, http://www.p1.csioz.gov.pl
21. P1 project: Approach to classification of medical records P.1. Recommendations for classification of electronic medical records. Warsaw (2012)

Concept of Database Architecture Dedicated to Data Fusion Based Condition Monitoring Systems

Marek Fidali and Wojciech Jamrozik

Institue of Fundamentals of Machinery Design, Silesian University of Technology,
Konarskiego 18a, 44-100 Gliwice, Poland
{marek.fidali,wojciech.jamrozik}@polsl.pl
http://ipkm.polsl.pl

Abstract. In the paper a concept of object-relational database dedicated to monitoring and diagnostics systems is presented. Important feature of proposed solution is the treatment of database as an integral sub-system dedicated to collect, manage and share data required for configuration and correct operation of the diagnostic system. It was assumed that diagnostic system has a modular structure, what determined the database architecture. Each module of the monitoring system and all operations performed in the system are reflected in the database by instances of designed objects connected with relations (references). In order to limit amount of stored data the database will collect only primary data like e.g. raw signals, configuration constants and meta data (rules of data management and parameters of data processing implemented in the diagnostic system). The concept of proposed database was implemented in limited form in commercial database system Oracle 12c using object-oriented features introduced in the SQL:1999 dialect.

Keywords: monitoring systems, diagnostics, data fusion, object relational database.

1 Introduction

In the domain of machinery and industrial process diagnostics different kinds of one- and multidimensional data coming from heterogeneous sources are utilized [10,15]. Owing to this fact more and more often data fusion algorithms are applied. Data fusion is a domain of science dealing with synergistic combination of data from multiple sources (e.g. sensors) in order to obtain new data. More reliable and accurate information can be extracted from the fused data than it could be acquired by processing data from single sources (sensors) separately. Data fusion techniques integrate knowledge from different domains of science, like control theory, signal processing, artificial intelligence, probability, statistics, etc. [18,13,12]. Three categories of data fusion methods can be distinguished:

- Data (signal) level fusion - combines the raw data from multiple sources of signals into a single one.

S. Kozielski et al. (Eds.): BDAS 2014, CCIS 424, pp. 515–526, 2014.
© Springer International Publishing Switzerland 2014

- Feature level fusion - at this level various features extracted from source data are merged together.
- Decision level fusion - combines results from multiple algorithms to yield a final fused decision.

It should be noticed that results of data fusion could not be considered as a fundamental and the only valuable source of diagnostic information, but in supportive manner especially in estimation of general conditions of diagnosed objects or processes. Data fusion could be performed in on-line and off-line modes as well as in static and dynamic manner. Regardless of the applied data fusion algorithm the question arises: how to represent, store and manage data used for fusion and being result of fusion. A partial answer could be found in several interesting papers. In [16] a brief survey of requirements and application of information fusion from the federated database point of view is presented. A novel, unified and flexible database approach to implementing systems dealing with fusion of spatial and temporal data was presented in [2]. In this paper authors analysed the requirements and designed the data model used to implement the airspace command and control system. In [1] a top-down view of key algorithm and database management system requirements associated with advanced data fusion application was presented. Issues related to the design of a conceptual model and architecture for a medical database that models and stores multi-modal and multimedia medical datasets and take into consideration data fusion technology was presented in [7]. All related works take into consideration object-oriented database types. In this article authors present a concept of a database dedicated for monitoring and diagnostics systems. In domain of technical diagnostics different database solutions are applied beginning from databases dedicated for single objects and maintained by dedicated software through commercial database systems and finishing on unified solutions [5]. While there are different interesting solutions of diagnostic databases [3], authors do not know any database realization that deal with data generated by fusion algorithms. Approach presented in this paper is based on object-relational database architecture. Important and original feature of proposed solution is the treatment of the database as an integral sub-system dedicated to collect, manage and share data required for configuration and correct work of the diagnostic system.

2   Model of Diagnosed Object

Elaboration of diagnostic system database utilizing data fusion algorithms require assumption of a model of diagnosed object. For this purpose one can adopt approach taken from systems theory where object is isolated from environment and treated as real system represented in form of "black box", what allows omission of its internal structure [6]. Interactions between the object and environment are proceeded only through its inputs and outputs. Such model should be treated as a dynamic system characterised in time domain, where values of output signals in determined time points depends not only on values of input signals in the same moment but also on values taken in preceded moments of time. It means that internal features of the system could be described by condition (state) parameters (Fig. 1). Knowing condition parameters and input signals allows us to determine output signals. For a certain object and time period, model can be defined in following way:

$$\mathbf{Y} = \mathbf{X} \times \mathbf{C} \tag{1}$$

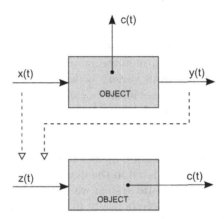

Fig. 1. Object model and way of its transformation [6]

For purposes of object diagnostics identification of changes of machine conditions on the basis of input and output signals is important. Thus model (1) of object could be defined as following relation:

$$\mathbf{M} \subset \mathbf{X} \times \mathbf{Y} \times \mathbf{C} \tag{2}$$

where:
$\mathbf{M}$ – object model,
$\mathbf{X}$ – set of input signals,
$\mathbf{Y}$ – set of output signals,
$\mathbf{C}$ – set of condition (state) parameters,

Such machine model allows to merge input and output signals (Fig. 1) in order to define space of interaction signals:

$$\mathbf{D} \subset \mathbf{Z} \times \mathbf{C} \tag{3}$$

where

$$\mathbf{Z} = \mathbf{X} \times \mathbf{Y} \tag{4}$$

and:
$\mathbf{D}$ – machine diagnostic model,
$\mathbf{Z}$ – set of interaction signal values.

Assuming introduced models (2), (3) object can be considered as a source of interactions which parameters could be observed in a form of signals. Object condition is changing in time what causes changes of observed signals features. According to above statements it can be claimed that on the basis of signal analysis and/or data fusion results it is possible to identify and forecast changes of machine condition.

3   Structure of Monitoring System Utilizing Data Fusion

The task of monitoring and diagnostic systems is to elaborate and provide to operator information about the condition of a object/process to check the correctness of its operation. Modern monitoring and diagnosing systems work online and gather information from variety of sources, which may be different types of sensors (e.g. temperature, vibration, cameras, etc.). To extract diagnostic information signals from raw data, operations like processing, analysis and recognition are carried out. After any of these operations diagnostic data can be fused. In Fig. 2 general structure of a monitoring and diagnosing systems using data fusion was presented. Application of data fusion at each stage of diagnostic decision making process is considered in the diagram.

In the monitoring and diagnostic system we have to deal with the one– and multidimensional data, what is symbolically marked in Fig. 2 by use of different style of arrows. Data can flow on two levels. The first one could be called the primary level (level 0) and is associated with data streams considered independently from each other, while the second level (level 1) refers to aggregated data created after fusion operations. For each of data transfer layers one can distinguish three types of operations that can be performed: processing, analysis and recognition. Depending on the system purpose nodes and data flow paths can be added or removed from the general system structure presented in Fig. 2. Depending on assumed system structure the data stream could flow through different paths. The common goal of all operations on data streams is diagnosing and/or predicting of object conditions, in order to provide feedback to the control system. At each stage of the flow, data are transformed what affects their dimensionality and representation. It can be assumed that the amount of data at level 1 will be less than at the primary level due to the properties of fusion operations.

4   Database Concept for Diagnostic Systems Using Data Fusion Procedures

When designing a database for monitoring and diagnosing system of machines and industrial processes it is necessary to take into account the function of the database in the system and which data is collected. One can assume different degrees of detail of stored data and various degrees of database integration with the monitoring and diagnostic system. The authors assumed that database should

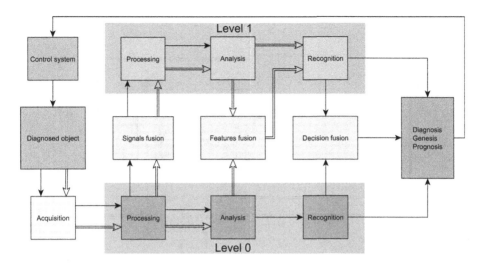

Fig. 2. General presentation of fundamental operations and data flow in monitoring and diagnostic system utilizing data fusion procedures

be an integral part of a system. It is not only responsible for data storing but also provides intermediary role in the supply and control of data flow between system and operator and vice versa (Fig. 3).

It is assumed that the following four groups of data are collected: raw data, meta data, general data, decision and control data. Raw data should be understood as the values obtained directly from the acquisition system. This data after acquisition should not be subjected to any transformation which can significantly change its structure and values.

Meta data are values of parameters required for the implementation of all operations performed on the data by procedures implemented in the diagnostic system. The meta data also provide information about sequence of operations what ensures traceability of data flow paths.

General data is any kind of data related to description of diagnosed object and its technical parameters such as technical data, nominal process parameters, configuration parameters, etc.

Decision data are diagnoses (decisions) developed by the diagnostic system and group of responsible experts. In this category one can classify also a control data necessary for regulation of process parameters.

Proposed approach of database allows its integration with the structure of diagnostic system in such a way that every change of parameters values of processing, analysis and recognition procedures must be firstly saved in the database and then transferred to the system. Such approach makes the database something more that only a storage module. The database becomes an active part of system configuration manager. Storage of a raw data gives opportunity to use the system in off-line mode for testing of procedures under new parameter values. Having raw and meta data a system operator is able to precisely in-

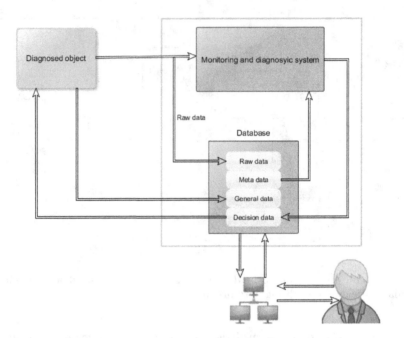

Fig. 3. Place and function of the database in monitoring and diagnostic system

vestigate the way in which the data were transformed and how diagnosis was formulated. Thus there is an opportunity to transfer the raw and meta data into diagnostic system procedures so one can improve and compare results of different diagnosing algorithms. Additional advantage of storing raw and meta data is possibility of fast implementation of data mining and machine learning procedures [11]. Taking into consideration concept of database structure from authors point of view the most convenient architecture to realise the database can be object-relational. The choice is driven by the fact, that complex data types are suitable to describe the system structure. Using object-relational mapping, classes can be reflected in a relational database structure [14]. Despite the fact that many applications have been developed allowing this kind of mapping, object-relational impedance mismatch is still a real problem. Object-relational database is not suffering this difficulties. Additionally it allows modelling of all introduced data types and create relations between them simultaneously assuring data integrity. It is a good compromise between the advantage of relational databases useful when there are large volumes of simple transactional data and object oriented approach suitable for data schemes with complex relations (especially many to many object relationship) [17,19,4]. Object-relational approach is also well suited for proposed database, while breaking complex information out into simple data takes time and is labour intensive (code must be written).

5   Example of Implementation of the Database Concept to Welding Monitoring System

The concept of database were partially implemented to welding process monitoring system developed by authors during earlier research [8,9]. In this system several types of data were acquired such as: one dimensional data including nominal welding parameters, material properties, time vector, values of on-line acquired signals of welding current, arc voltage, welding speed, shield gas flow rate, wire feeding speed and two dimensional signals like sequences of infrared thermograms and visible light images of welding arc, welding pool and welded joint area. The welding monitoring system has implemented algorithms which perform data fusion on the signal and decision level. On the signal level additional virtual sensors are obtained by calculation of welding linear energy and resistance. The decision level fusion based on the belief functions theory is used to combine outputs from several k-Nearest Neighbour classifiers in order to make one, final diagnostic decision.

In order to show the database concept from the object-oriented perspective a class diagram was elaborated. It allows the translation of real world problem to the database schema. Simplified form of database class diagram of monitoring and diagnosing system of welding process is presented in Fig. 4. In presented approach, a common UML modelling software (UML MagicDraw) was used to design the structure of objects stored in the database. As it can be seen system operator has access to the information about the observed objects, measuring devices, acquired data and the data flow paths during the diagnostic decision making process. Because of choose of object-oriented approach in the designing process some original data types and tables were created in order to map the real world objects into database structure. In the presented example only class attributes are shown, whereas user defined methods are omitted in this first conceptual approach.

According to the proposed concept the data stored in database can be assigned to one of four categories. Raw data is described by object type hierarchy. The super-type Data_objtyp refers to meta class *Data*. In proposed model Data_objtyp object type can be extended to create more specialized objects for certain data types of different dimensionality Data1D_objtyp , and Data2D_objtyp . 1D data is fully stored in the database, while in the case of 2D data only path to original images or thermograms and thumbnails (BLOB type) are stored in the database. For the sake of clarity in forthcoming description some definitions were simplified while other were completely omitted. User types described above are defined in following manner:

```
CREATE TYPE Device_objtyp AS OBJECT (
  deviceID NUMERIC,
  ...);
CREATE TYPE Data_objtyp AS OBJECT (
  acq_time TIMESTAMP,
  acq_device Device_objtyp,
  ...) NOT FINAL, NOT INSTANTIABLE;
```

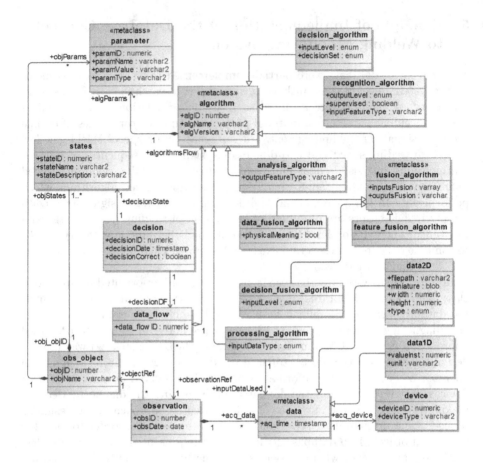

Fig. 4. Simplified conceptual model of diagnostic system database

```
CREATE TYPE Data1D_objtyp UNDER Data_objtyp (
  valueInst NUMERIC,
  unit VARCHAR2(10)
  ...);
CREATE TYPE DataList_ntabtyp AS TABLE OF Data_objtyp;
```

Information about each observation of object or process (certain time period or operation) is gathered in the `Observation_objtab` table. It contains reference to table with information about monitored object (being a foreign key in form of reference to table `Obs_object_objtab` , that is a typed table of objects representing the *obs_object* class in form of `obs_object_objtype`) and complete acquired data in the form of nested table `acq_data_ntab` with `DataList_ntabtyp` where each row of the table is of `Data_objtyp` type. Code used to generate this table is presented below.

```
CREATE TYPE obs_object_objtype AS OBJECT (
  objID NUMERIC,
  ... );

CREATE TYPE Observation_objtyp AS OBJECT (
  obsID NUMBER,
  obsDate DATE,
  objectRef REF obs_object_objtype,
  acq_data_ntab DataList_ntabtyp
  );

CREATE TABLE Obs_object_objtab OF obs_object_objtype (
  objID PRIMARY KEY,
  ... )
  OBJECT IDENTIFIER IS PRIMARY KEY;

CREATE TABLE Observation_objtab OF Observation_objtyp (
  PRIMARY KEY (obsID),
  FOREIGN KEY (objectRef) REFERENCES Obs_object_objtab)
  OBJECT IDENTIFIER IS PRIMARY KEY
  NESTED TABLE acq_data_ntab STORE AS Poacq_data_ntab (
  (PRIMARY KEY(NESTED_TABLE_ID, acq_time))
  ORGANIZATION INDEX COMPRESS);
```

One of the key objects for meta data description is `Algorithm_objtyp` object type. It is a super-type for group of object types considering the specificity of the system meta class for object `Fusion_Algorithm_objtyp` that inherits properties of `Algorithm_objtyp`.

```
CREATE TYPE Algorithm_objtyp AS OBJECT(...) NOT FINAL, NOT INSTANTIABLE;
CREATE TYPE Processing_Algorithm_objtyp UNDER Algorithm_objtyp(...);
CREATE TYPE Fusion_Algorithm_objtyp UNDER Algorithm_objtyp (...) NOT
    FINAL, NOT INSTANTIABLE;
CREATE TYPE Data_Fusion_Algorithm_objtyp UNDER
    Fusion_Algorithm_objtyp(...);
(...)
```

The way in which acquired data is transformed and diagnostic decision is reached is represented by `Data_flow_objtyp`. It contains reference to a specific observation of certain object or process. Each observation is the source of diagnostic signals for decision making process. Each diagnosis can be reached using different analytical or heuristic techniques, starting from different signal processing algorithms and ending with various decision making methods. As a nested table `AlgorithmsFlow_ntabtyp` of `Algorithm_objtyp` objects the sequence of applied algorithms (and their parameters) leading to certain decision is gathered. The implementation of `AlgorithmsFlow` as a separate table was avoided. It gives more flexibility and allows power of SQL (including sub-queries) and more options for access of individual records.

The acquired data (1D and 2D signals) being an entrance point to the processing path is referenced in the `Processing_Algorithm_objtyp`.

Beside tables mentioned above, user can store data in following tables:

- `Device_objtab` with information about acquisition devices, sensors, cameras. etc.,
- `States_objtab` containing possible object/process states among which diagnosis can be made, with and additional attribute being reference to the row in `Obs_object_objtab` object table,
- `Algorithms_objtab` an object table of `Algorithm_objtyp` objects.

The LabVIEW Database Connectivity Toolkit was used to connect to database and perform all needed database operations. Because Database Connectivity Toolkit offers complete SQL capabilities, comprehensive querying as well as data insertion, modification and deletion can be performed. Additionally simple querying is possible without SQL statements using only functions provided in toolbox. Nevertheless text based querying could not be omitted, because of non common database scheme.

Object identifiers (OIDs) and references (REFs) are used for relation creation. It is a convenient way to create complex relational joints, when attribute-based identifiers are unavailable. This approach offers a possibility to define simple queries with high clarity, as in the following example, where data acquired during certain observation of an object is listed:

```
SELECT C.obsID, C.objectRef.objID, {TREAT(VALUE(D) AS
    Data1D_objtyp).valueInst}
  FROM Observation_objtab C, TABLE(C.acq_data_ntab) D
  WHERE C.obsID = 23;
```

Basic querying is already implemented in the diagnosing system, so there is no need to formulate SQL queries by system user. Despite this there is still lack of several useful functions, allowing advanced data search and custom data projection form e.g. external data processing software.

6   Summary

In the article the general concept of database dedicated to monitoring and diagnostic system utilised data fusion procedures was presented. Considered idea spread the database functionality not only to data storage but also to supporting and supervising of a decision making during diagnosing of objects and process. The database concept was implemented in basic way to welding process monitoring and diagnostic system. Architecture of the database is object-relational what from one side allow to model all real life objects, procedures and data flows and from the other side could be implemented by use of available and proven commercial database solution. References in the database are useful to model relations between the signal features and conditions of diagnosed object or process. Relations also are useful to defining connections between diagnostic and data fusion procedures. Proposed solution closely integrate a management application layers of the database and diagnostic system where database has a leading function in controlling of data sharing and flow. Next step in applying of the database into the diagnostic system will be elaboration of methods extending the possibilities of database, by placing specific operations connected with keeping data integrity but also data formatting and performing simple arithmetic operations on stored data in form of user defined types. Additionally it will increase the efficiency of the whole application by transferring part of LabVIEW software tasks into the object-relational database management system.

Acknowledgments. Scientific work financed from resources assigned to statutory activity of Institute of Fundamentals of Machinery Design, Silesian University of Technology at Gliwice.

References

1. Antony, R.: Database support to data fusion automation. Proceedings of the IEEE 85(1), 39–53 (1997)
2. Aschenbrenner, T., Brodsky, A., Kornatzky, Y.: Constraint database approach to spatio-temporal data fusion and sensor management (1995)
3. Batko, W., Borkowski, B., Głocki, K.: Application of database systems in machine diagnostics monitoring. Maintenance and Reliability 1, 7–10 (2008)
4. Bütüner, H.: Advantages of object-oriented over relational databases on real-life applications. Research Journal of Economics, Business and ICT 5 (2012)
5. Cholewa, W.: Bazy danych w diagnostyce technicznej. In: Proc. Midzynarodowy Kongres Diagnostyki Technicznej, Diagnostyka, Warszawa, September 19-22 (2000) (in Polish)
6. Cholewa, W., Kaźmierczak, J.: Data processing and reasoning in technical diagnostics. WNT, Warszawa (1995)
7. Coleman, J., Wloka, M., Haley, M.: Data fusion for the multimedia medical database, http://cseweb.ucsd.edu/~goguen/courses/275f00/mmmdb.html (Online; accessed January 30, 2014)
8. Fidali, M., Bzymek, A., Timofiejczuk, A., Jamrozik, W., Czupryski, A.: Welding process evaluation on the basis of video and thermal images. Welding International 26(7), 523–530 (2012)
9. Fidali, M., Jamrozik, W.: Diagnostic method of welding process based on fused infrared and vision images. Infrared Physics and Technology 61, 241–253 (2013)
10. Jardine, A.K., Lin, D., Banjevic, D.: A review on machinery diagnostics and prognostics implementing condition-based maintenance. Mechanical Systems and Signal Processing 20(7), 1483–1510 (2006)
11. Kasprowski, P.: Choosing a persistent storage for data mining task. Studia Informatica 33(2B), 509–520 (2012)
12. Kuncheva, L.: Combining Pattern Classifiers. Methods and Algorithms. Wiley (2004)
13. Mandic, D.P., et al.: Data fusion for modern engineering applications: An overview. In: Duch, W., Kacprzyk, J., Oja, E., Zadrożny, S. (eds.) ICANN 2005. LNCS, vol. 3697, pp. 715–721. Springer, Heidelberg (2005), http://dx.doi.org/10.1007/11550907_114
14. Murakami, T., Amagasa, T., Kitagawa, H.: Dbpowder: A flexible object-relational mapping framework based on a conceptual model. In: Computer Software and Applications Conference (COMPSAC), pp. 589–598 (2013)

15. Nowicki, A.N.: Infrared Thermography Handbook, Applications, vol. 2, British Institute of Non-Destructive Testing (2004)
16. Sattler, K.U., Saake, G.: Supporting information fusion with federated database technologies. In: Proc. 2nd Int. Workshop on Engineering Federated Information Systems, EFIS 1999, pp. 179–184 (1999)
17. Saxena, V., Pratap, A.: Performance comparison between relational and object-oriented databases. International Journal of Computer Applications 71(22), 6–9 (2013)
18. Stathaki, T.: Image fusion: Algorithms and Applications. Academic Press (2008)
19. Tuzinkiewicz, L.: Analysis and comparison of relational and non-relational databases. Studia Informatica 34(2B), 323–337 (2013)

Applying NoSQL Databases for Operationalizing Clinical Data Mining Models

Marcin Mazurek

Military University of Technology
Warsaw 00-908, Kaliskiego 2, Poland
marcin.mazurek@wat.edu.pl

Abstract. Access to data mining models built in clinical data systems is limited to relatively small groups of researches, while they should be available in real-time to clinicians in order to deliver the results at the point where it is most useful. At the same time, complexity of data processing grows as volume of available data exponentially rises and includes unstructured data. Clinical decision support systems based on relational and multidimensional technology lack capabilities of processing all available data because of its volume and format. On the other hand, NoSQL repositories offer great flexibility and speed in terms of data processing, but requires programming skills. A proposed solution presented in this paper is to combine both of the technologies in a single analytical system. Dual view of the data gathered in the repository allows to use data-mining tools, while Big Data technology delivers necessary data. Key-value style of querying a database enables efficient retrieval of input data for analytical models. Online loading processes guarantee that data is available for analysis immediately after it is produced either by physicians or medical equipment. Finally, this architecture can be successfully moved to the cloud.

Keywords: clinical decision support system, big data, architecture.

1 Introduction

Hospitals collects more and more data about patients, both structured and unstructured. It can be than used to make better clinical decision, as doctor can rely not only on their own knowledge, but also take advantage of others experience [11]. The main benefit from clinical decision support systems is that general treatment guidelines can be customized to cases based on evidence. In order to make data available, data warehouses are built. Then, upon the relational structures exposed by the data warehouse predictive models are constructed. This shifts standard medical practice from ad-hoc and subjective decision making to evidence-based treatment. As more data is taken into consideration model has higher accuracy and eventually outperforms physician prediction [9]. What is unique in data mining in medicine is the need for robust operationalization of the models for staff engaged in treatment. Operationalizations is the process of execution of constructed models for operational data. In medicine, operational

S. Kozielski et al. (Eds.): BDAS 2014, CCIS 424, pp. 527–536, 2014.
© Springer International Publishing Switzerland 2014

data are data describing patient. The real breakthrough in the treatment process can be achieved, when during patient examination, doctors would be able to score individual case against some of the published models or just find the most similar cases. With traditional data warehouse technology patients record has to be loaded to data warehouse and then become available for data analysis. As an alternative, doctors can manually input the patient record into the system and then run selected models on the data. None of this approaches is satisfactory. Cyclical Extract Transform Load (ETL) processes cause latency. Inserting data to analytical system by doctors is time-consuming and redundant operation, since the information anyway has to be earlier written to Hospital Information Systems (HIS).

The aim is to make use of unstructured data and facilitate execution of analytical models on data provided by physicians in real time. Before Big Data technology has changed the architectural landscape of decision support systems, such models where either moved to transactional systems as a calculation procedures (e.g credit scoring procedure) or their outcomes where shifted back as additional measures (e.g. churn probability). This causes latency which makes it useless in medicine applications. To boost usage of computer-aided decision making, such models has to be available online and give immediate answers.

To overcome these flaws, already existing data warehouse solutions might be supplemented with NoSQL repositories. They can be loaded online from Hospital Information Systems and directly from laboratory equipment. With their flexible data model it is easier to query database for attributes, which are specified as inputs data in predictive models. Although primary concern in this article is efficient operationalization of predictive data models, they also offer capabilities of efficient processing unstructural data, which is very common in medicine (scans, clinical notes and so on).

The proposed architecture is derived from evaluation of real-world implementation of data warehouse in cardiology clinic. The scope of project covered implementation of data warehouse based on relational and dimensional repository. The whole data warehouse solution has brought value for management processes, but the utilization of system as a clinical decision support system has been very limited. Accompanying data mining tools allow for building rather descriptive then predictive analysis. This is because usage of the model is restricted to historical records which has been already loaded to data warehouse. To be truly useful in clinical decision support, data mining model should be applied to new data. Models should be operationally deployed by moving them to operational systems and mobile devices.Another solution is real-time transfer of operational data to the analytical repository. The first choice, while being a common practise in business, is usually a lengthy process. Based on experiences from use cases of relational data warehouse I believe that the latter option is better, particularly in situation where input data for models are gigabyte images or video sequences.

The rest of the paper is organized as follows. The first section describes architectural principles and describes how different users will use the system. Next section introduces nonrelational repositories. Then the concept of architecture is discussed.

2   Usage Scenario and Design Principles

The target group of the system are varied. Access to advanced analytics will benefit doctors, researchers, patients and system of education. We distinguish different usage scenarios for the system:

1. Knowledge extraction from big volumes of both structured and unstructured data.
   (a) Actors: Data scientists team.
   (b) Input: All available data about patients (including unstructured data like medical images), drugs, treatment procedures and clinical paths. Patients data should undergo process of de-identification.
   (c) Output: Tables with extracted features, algorithms, visualisation of patterns.
   (d) Tools: Statistical packages and data analysis languages like Python[1] or R[2].
2. Predictive modelling.
   (a) Actors: Medical researchers.
   (b) Input: Relational repository with anonymised health records.
   (c) Output: Various classifiers (e.g. decision trees, regression models), prediction models. clusters, similarity measures, association rules.
   (d) Tools: Data mining tools and statistical packages.
3. Operational: using models to make better decision during patients examination.
   (a) Actors: Physicians.
   (b) Input: Models and data describing particular case.
   (c) Outcome: Prediction.
   (d) Tools: Mobile and desktop custom application.
4. Education.
   (a) Actors: Students and academical staff.
   (b) Input: Models and data describing hypothetical case.
   (c) Outcome: Prediction for hypothetical case, aggregated statistics for selected population.
   (d) Tools: Desktop custom application.

Usage scenarios described above can be used to verify completeness of the vision. The main principles of architecture work are as follows.

Single Point of Data Entry. All data that is manually inserted to information systems by doctors is inserted once. Currently, doctors fills in required forms to enter data into Hospital Information Systems (HIS). They should not be forced to repeat this step to use decision support models. Once the data is loaded, they should be automatically available in analytical repository.

---

[1] http://www.python.org/
[2] http://www.r-project.org/

Online Loading. The latency between entering the data about patient case and the moment, when they are available as input for analytical models should not exceed single minutes. This is due to the limited amount of time doctors spend on examining each patient.

Ease of Models Dissemination. Models should be easily accessible to all doctor. Using it as a supplementary tool in their practise does should not enforce additional effort. Real-time learning from clinical data and delivery of results to point of meeting patient and doctor may become a methodology for transforming healthcare [12]. This also imposes requirements on end user devices and appplication interfaces.

Capability to Process all Available Data. In addition to clinical data stored in EHR, there are others sources of data which may contribute to better understanding of relations between patients and treatment outcome. Examples are pharmaceutical R&D databases and social networks that can be mined for patients activities and preferences.

Providing User with Proper Tools and Interfaces. This is particularly important principle, as doctors are not supposed to be specialist in information technology. Each group of users mentioned earlier has different functional requirements. Data scientists with programming background should be provided with more powerful tools than physicians and access data in their native formats.

3   NoSQL Databases

NoSQL databases cover a range of data storage solutions, which differs significantly from each other. However, they have common properties which make them fit the proposed solution:

1. They consume data coming in any of digital forms.
2. Data is loaded real-time. Unlike to ETL processes in data warehouse, cleaning and transformation is deferred. Batch processing is directed by analytical goals.
3. Distributed storage and processing based on Map Reduce paradigm allows for processing huge volumes of data. This architecture is very scalable.
4. Flexible output data model. Key-value databases allows for efficient retrieval of features. Fig 1. shows representation of simplified laboratory procedure outcomes.

Idea of applying NoSQL database to medical data storage is widely discussed in bibliography. As a key driver of change, architects point out capability of processing constantly growing volumes of medical data, often unstructured [6]. While this remains unquestionable, NoSQL databases can also fill the gap between medical data mining researchers and practitioners. With means of key-value databases interfaces, preparation of input data for analytical models is

```
 1  {
 2  Patient:{
 3              ID: "20123312344" ,
 4              "DOB - Date of birth" : new Date ("Oct 23, 1967"),
 5              "Patient sex" : "Female"
 6              } ,
 7  "Date of procedure": new Date ("Jun 23, 2005"),
 8  "Complete blood count":
 9              {
10              "RBC - Red blood cell count": 5.20,
11              "WBC - White blood cell count": 8.1,
12              "Haematocrit - PCV level": 37.5,
13              "Erythrocyte volume":90,
14              "MCH - Mean corpuscular hemoglobin": 30
15              }
16  }
```

Fig. 1. Procedure outcomes represented in key-value document data model

much easier than in relational structures. This task can be automated - it is sufficient for doctor to choose case identification and model. The rest is performed by system.

Among examples of open-source solutions that can be used are MongoDB [2] integrated with Hadoop [1]. Big data technology, which NoSQL databases are part of, grows very dynamically so selection of platform should not be done in much advance. Besides comparing the database management system functionality and query language, special consideration is required when moving solution to the cloud [3].

4   System Architecture

The key concept of architecture is enabling dual view of the data: relational and key-value. Data from these repositories partly overlaps: attributes and measures describing history of treatment is available from relational and nonrelational repository. The latter stores also semistructured or unstructured data, from which features are extracted. The structures of the repositories are linked by common definitions stored in metadata repository. The logical components of the system are presented on Fig. 2.

Data Sources. Electronic Health Records (EHR) stored in Hospital Information Systems are primary source of data for decision algorithms. The problem is that they cover only a subset of digitalized information about patient. The remaining sources are medical images, laboratory tests outcomes, health monitoring equipment, remote sensors, pharmaceutical databases.Additional potentially valuable data may be gained from social networks. The Big Data technology with NoSQL repositories is trying to give an answer how we can process such volume of unstructured data constantly flowing into the system.

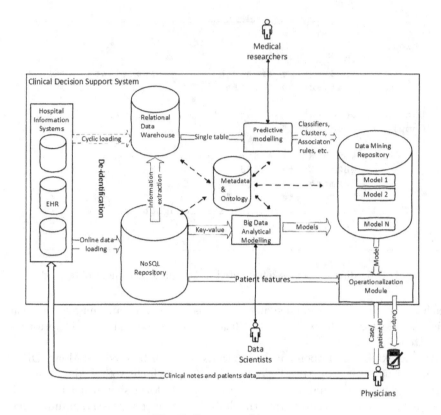

Fig. 2. System architecture

From operational point of view a most important source of data is a physician examining the patient. Data from clinical notes is input in decision process. These notes includes patient's complaint, symptoms, social circumstances, etc. Currently, doctors are obliged to enter some part of it into HIS. Much have been done to set standards of treatment description and patients state. There are international dictionaries like ICD-10. Because this data is used solely for statistical analysis and operational management its quality and scope is limited, and so is the possibility to treat them as complete source of knowledge about patient. This situation may change with implementation of decision support tools, that will be fueled by this data.

NoSQL Repository. NoSQL repository is loaded online with content stored in HIS and all other sources of data mentioned earlier. The loading infrastructure has to guarantee that raw data will become available in the system within minutes from being produced. The next step in the process, which is executed in batch is feature extraction. These features extend set of attributes and metrics available through relational repository. Identification of essential features and algorithms of its extraction are job of data scientist team. This is closed-loop process as available information scope constantly spreads. Data from this repository is indexed with concepts from knowledge base.

NoSQL database systems may also provide user with relational view of the stored data, and thus relational repository described below may be actually a part of Big Data ecosystem. This makes cyclically executed ETL processes redundant in the architecture. However, in many medical centers there is data warehouse already deployed with set of customized application.

Relational Data Warehouse and Data Mining Tools. Relational data warehouse allows for immediate answers for ad-hoc queries about either features of individual patients or some population statistics. The repository is loaded with features extracted form unstructured content processed in Big Data part of the system.

Relational data warehouse delivers data to data mining tools used by medical researchers. Data mining tools are used to perform all kind of exploratory analysis and predictive modelling. Various techniques and algorithms are usually delivered to the researchers as black-boxes modules run in graphical user interfaces. The tools support researchers on all stages of the process: data sampling and transformation, modelling, assessment. Ease of use allows to engage wide range of medical experts to build and verify models. They do not have to posses knowledge of data processing techniques and programming, like data scientist teams.

Relational data mining requires that all input data for data mining tasks is in attribute-value form. A training examples have to be stored in single-table, which requires rather extensive preprocessing of relational data model. This includes:

1. Calculation of derived measures (ratios, frequencies of occurrence, flags, levels according to norms).
2. Flattening of structures like time-series (repeated lab-test).
3. Variable transformation: bining, removal of outliers etc.
4. Handling multiple-instance learning problem - patient may have more than one diagnosis.

Some of transformations are used solely for model building, some has to be performed also on input vector. To perform tasks mentioned above, medical researchers should be supported, as transformations like these requires at least familiarity with logical structures of data warehouse. This is a bottleneck of data mining system in medicine. In proposed architectures, those transformation are performed in nonrelational repositories.

Repository of Models. The content of analytical repository is made of calculation procedures and algorithms constructed either on relational repository with data mining tools or directly on NoSQL database. Content include but is not limited to:

1. Predictive models, which come in variety of forms: regression equations, neural networks, decision trees, Bayesian networks, complex classifiers.
2. Similarity measures between clinical cases, clusters and their parameters.
3. Association rules.
4. Visualization of patterns.

Each model is accompanied with set of metadata, describing its purpose, set of inputs, accuracy measures. The latter is constantly evaluated in order to withdraw models that performs poorly on new cases. Models may be ranked based on their performance.

Metadata and Ontology. Metadata allows for semantical linking data from relational and non-relational repository. This is accompanied by ontology module which is responsible for providing semantics to the diverse medical information stored in repositories [8]. It includes the ontologies used to define and classify different types of health care data. Ontology concepts serve as keys for non-relational repositories and enables users to communicate with the system with domain terms. Fig 1 is an example of representing laboratory test outcomes with keys which are taken from SNOMED CT ontology [7]. For sake of clarity instead of concept keys the names were presented.

Operationalization Module. This component of the system is responsible for querying database for all relevant attributes required by the predictive models and concerning a particular medical case, which is being evaluated. An examining doctor provide ID of the case. Then a calculation procedure has to be executed with parameter values retrieved from repository. The outcome is presented online.

Implementation of this module will rely on scoring engine executing predictive models. Scoring engine will be decoupled from modelling tools by means of using standardized form of models like Predictive Modelling Markup Language (PMML) [4]. It allows for using any of of data mining tools as long as they provide possibility of export models to PMML. The key point is mapping data dictionary part of PMML model with operational data stored in repository. The assumption is that both data dictionary elements in model and keys in NoSQL database refer to common ontology concepts.

De-identification of patient's data and possibility to query particular case by doctors at the same time can be achieved by reversible coding of id attributes [5]. This form of de-identification goes with mapping tables, which maps artificial keys, which are used in repository with real ones. Nevertheless, security architecture of such system remains a challenge, because medical data often carry an unremovable identification footprint.

5   Conclusions

In this paper I presented a concept of NoSQL repository in architecture of clinical decision support systems. It allows for a shift from monitoring to prediction and simulation [6]. Data warehouses will begin to play substantial role not only as a tool for better health management system, but also will support doctors in their practise. Doctors point out that even visualizing patient's case against population and comparing case to most similar cases would be helpful, especially for less-experienced ones, who begin their carrier.

Key characteristics of the presented architecture is real-time loading of all available data to non-relational repository, from which processed data is published. Data is exposed in relational structures and key-values stores. The first model is best for data-mining tools, while the latter simplifies querying databases for input attributes of predictive models.

This architecture might be easily scaled to cover data from more than single medical institution. The quality of prediction rises with bigger set of evidence. Finally, already built classifiers might be published in 'Analytics-As-a-Service' model in cloud, and it is not technological difficulties which are worst to overcome. Except for data privacy matters, the concept of common repository and data sharing itself seems to be troublesome. Those concerns should be opposed to better outcomes of treatment procedures at possibly lower costs, thanks to rejecting the paths that somewhere proved to be unsuccessful. The portals like PatientsLikeMe[3] proves that evidence-sharing may be valuable tool in decision making even by non-professionals.

The feasibility of the architecture is still to be proved by set of prototypes. Some of them are already implemented: scoring engine based on Python written Augustus [10] and another one operating in Amazon Cloud based on ADAPA [13]. Future work will concentrate on structures and efficiency of loading data to NoSQL database from operational data sources.

References

1. Hadoop (February 14, 2014), http://hadoop.apache.org
2. Mongo DB (February 14, 2014), http://www.mongodb.org
3. Bajerski, P., Augustyn, D.R., Bach, M., Brzeski, R., Duszeko, A., Aleksandra, W.: Databases vs. cloud computing. Studia Informatica 33(2A), 9–25 (2012)
4. Data Mining Group: PMML 4.1. Specification (February 14, 2014), http://www.dmg.org/v4-1/GeneralStructure.html/
5. Emam, K.E.: Guide to the De-Identification of Personal Health Information. CRC Press (2013)
6. Groves, P., Kayyali, B., Knott, D., Van Kuiken, S.: The 'big data' revolution in healthcare. Tech. rep., McKinsey & Company (January 2013)
7. International Health Terminology Standards Development Organisation (IHTDSDO): SNOMED CT (December 13, 2013), http://www.ihtsdo.org/
8. Khan, A., Doucette, J., Jin, C., Fu, L., Cohen, R.: An ontological approach to data mining for emergency medicine. In: 2011 Northeast Decision Sciences Institute Conference Proceedings 40th Annual Meeting, Montreal, Quebec, Canada, pp. 578–594 (April 2011)

---

[3] www.patientslikeme.com

9. Oberije, C.: Mathematical models out-perform doctors in predicting cancer patientsŕesponses to treatment (April 2013), http://www.sciencedaily.com/releases/2013/04/130420110651.htm (retrieved December 13, 2013)
10. Open Data: Augustus. PMML model producer and consumer. Scoring engine (February 14, 2014), https://code.google.com/p/augustus/
11. Savage, N.: Better medicine through machine learning. Commun. ACM 55(1), 17–19 (2012), http://doi.acm.org/10.1145/2063176.2063182
12. Yu, S., Rao, B.: Introduction to the special section on clinical data mining. SIGKDD Explor. Newsl. 14(1), 1–3 (2012), http://doi.acm.org/10.1145/2408736.2408738
13. Zementis: ADAPA Scoring Engine (February 14, 2014), http://www.zementis.com/adapa.htm/

Database Application in Visualization of Process Data

Karolina Nurzyńska[1,2], Sebastian Iwaszenko[1], and Tomasz Choroba[1]

[1] Central Mining Institute
Plac Gwarków 1
40-160 Katowice
Poland
[2] Institute of Informatics
Silesian University of Technology
ul. Akademicka 16
44-100 Gliwice
Poland
karolina.nurzynska@polsl.pl

Abstract. An information system for visualization of Underground Coal Gasification (UCG) has been developed. The main goal for the system is to provide means for better understanding and control of the process. The system uses mainly data generated during UCG process numerical simulation. Apart from that, the system is capable of storing and visualizing measurement georadar data, gathered during in-situ processes. Development software consists of three modules: the data acquisition, data preprocessing, and data visualization. All modules communicate through a database, the central point in the system architecture. Data from modeling or measurements are stored in database in raw format. Then the preprocessing module converts them for visualization purposes. The visualization module shows preprocessed data to the user using 3D graphics. The visualization is dynamic (visualized data changes in time) and allows observing several properties of the process - gases concentration, gasification cavern development, and so on. The software was developed using .NET framework and XNA Game Studio 4.0 library.

1 Introduction

The search for new innovative sources of energy is one of the most important challenges for a modern world. The renewable energy sources, though very promising, have to be supported by conventional ones. Nevertheless, there is a constant pressure to reduce environmental impact of energy sector. The global warming causes emission of greenhouse gases, such as CO_2 or CH_4, to be a main point of interest. There is no surprise that law regulations, especially in UE, put severe restrictions on amount of greenhouse gases, which may be emitted to the atmosphere by industry.

Struggling against global warmth is particularly difficult for countries, which depend on coal as a basic fuel. It is economically infeasible to switch to alternative sources of energy, on the other hand, the environment protection is a

S. Kozielski et al. (Eds.): BDAS 2014, CCIS 424, pp. 537–546, 2014.

fundamental matter. One of the attempts to solve this stalemate is to convert coal into more environmentally friendly fuel. Coal gasification, and specifically Underground Coal Gasification (UCG) is one of the promising technologies. It allows converting coal in-situ into gases, which can be used as fuel or as a substrate for chemical processing. It is possible to obtain high hydrogen concentration in resulting gas, much more friendly to environment than coal itself. However, obtaining gas of high hydrogen concentration and of constant parameters during the process, requires appropriate process controlling and monitoring. Usage of modelling tools is one of the methods for determining appropriate process conditions. Supported with online measurement of process development and especially so called 'cavern' formation plays important role in process control.

The article presents development of software tools helping in UCG process understanding and control. 3D visualizing of modelled process data along with georadar gathered measurements allow observation in easy way how UCG performs. In section 2 the overview of coal gasification process is presented. Section 3 presents system architecture and construction. Section 4 describes input data and section 5 describes the database solution used. Finally, the conclusions are drawn in section 6.

2   Underground Coal Gasification

The process of underground coal gasification is seen nowadays as an interesting option for coal excavation. Especially, that it is the only option to exploit the shallow seams which mining with conventional methods is unprofitable from economical point of view. Moreover, it is believed that this process may diminish the influence of greenhouse gases, which result from fossils combustion. Because the harmful elements stay underground, it is believed to have less enviromental impact than other ways of coal usage. Not to mention, the potential possibility of carbon dioxide sequestration in cavities resulting from the process.

The general idea of the underground coal gasification process is to convert coal into gas in situ. The objective is to obtain a syngas of high quality. That means with high concentration of flamable gases and low concentrations of inerts. Because the gas composition depends strongly on the process conditions, it is crucial to be able to control the temperature of the process which corresponds to the reactions taking place. It is also important to be able to control the displacement of the reaction zone in the seam.

Figure 1 presents the schema of the process. As mentioned before the process takes places in the coal seam. There are two wells connecting the gasified seam with the surface. The injection well is used for process initiation and then the gasification agent is introduced throught this place. Resulting from the gasification process syngas is produced and collected by the receiving well. There are in situ experiments described in the literature [2], but also many laboratory or semi-technical experiments took place [1,3,4,6].

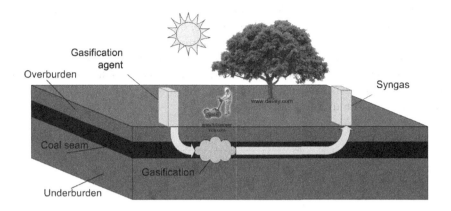

Fig. 1. Underground coal gasification schema

3   System Overview

In the process of underground coal gasification (UCG) an important role plays monitoring of the whole phenomena during this process, in order to determine the optimal operating conditions and to obtain a gas of the required characteristics. Monitoring of the coal gasification process and defining the parameters affecting the progress of physicochemical process provide the basis for monitoring and controlling of the entire process.

Due to the surface nature of most of the chemical reactions occurring in the process of coal gasification the particulary important role in gasification process plays the size and shape of the cavity. This area is limited by the space formed in the place of coal consumed in the process. Additionally, an area separating the char layer as a result of pyrolysis reaction from seam coal and the surrounding rocks can be highlighted. Knowledge of these surfaces, for example, their shape, location in space, and size, is one of the most important parameters allowing to control the gasification process and to describe the process using models.

The presented application enables the visualization of the UCG process in given environment and with assumed parameters of the gasification process. The data for presentation might be acquired with georadar during the experimental process or is estimated by the dedicated mathematical model. In order to facilitate the visualization process and to make the visualization tool independent from the data type, all information is stored in the relational database. This approach enables not only easy processing of input data for visualization but also allows to build a functional tool which improves the understanding of the process. Due to the specific system application, following functionality should be available:

– visualization of the UCG process,
– prediction of the UCG process basing on the mathematical model,
– prediction of the course of ongoing UCG process.

The presented system for UCG process visualization consists of three independent modules: the data acquisition module, data preprocessing module, and finally visualization module. The separate modules communicate with each other only through the data gathered in the database as it is presented on Fig. 2. It was assumed that no other possibility of communication should be available to assure full independence of each module from the others, what would enable easy addition of novel modules, if neccessary.

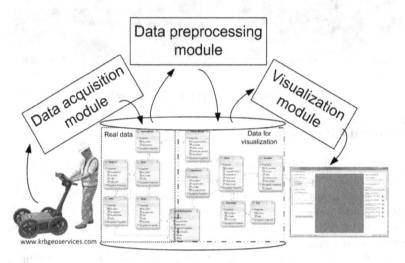

Fig. 2. Module communication with data

The data acquisition module is responsible for storing the process information provided by georadar images describing the changes during the experimental process or text files resulting from the mathematical model processing. Nevertheless the origin of the data, it is incorporated in the same data tables, which describe the shape of the cavity in chosen points of time. In case of the mathematical model, there are additionally stored information about values of many parameters describing the process, for instance, various gas concentrations, temperature, flow velocity, etc.

The task of the data preprocessing module is to generate a visual objects and assure its proper visualization on the basis of given data. This consists of generating visual objects considering the chosen parameters for visualization and assuring proper interpolation between consequtive frames. Some sophisticated algorithms for data aproximation are exploited in order to cut off the redundant data and in consequence allow to create the visualization with less data but conveying the same information. In general this module reads the real data from the database and generates the so called visualization data.

It is worth to mention that the process of UCG slowly varies within time, therefore it is possible to generate a detailed visualisation of whole process presented in much shorter time. However, the data preprocessing module enables the possibility to choose the timespan of visualization by the user. The information about the prepared visualization, which means the description of chosen data, each frame object description is stored also in the database, which allows for multiple visualization of the same data. Additionally, this approach saves time, as the preparation of data for visualization is time consuming.

Finally, the visualization module reads the visualization data describing a single simulation and prepares its visualization. It means that for each described in database object an adequate visualization technique is applied. The generated models, depicting the cavity shape in case of surface visualization or distribution of the process parameters over the gasification zone when the point cloud visualization tooks place, change according to the prepared drawing scenario in consequtive frames. This module also supports the user interaction which enables simulation rewinding, stopping, and repeating. The possibility to move the camera in order to observe the proces from any point of view is supported. Additionally, in each point of the simulation it is possible to generate an arbitrary cross-section throught the visualized volumen to observe the distribution of parameters in this section.

4   Input Data

The system is prepared to store data from different sources. It results from the possible operation mode as well as existing data sources. Uptil now, the data consists of radar readouts recorded during the experiments and data generated by the mathematical model of UCG process with ANSYS Fluent software.

During the experiment of UCG process the only data possible to gather for the system are radar profiles, which describe the cavity growth in time. They are recorded every few days as it is sufficient to notice the slow change of the cavity shape. One measurement consists of over a dozen radar cross-sections of the reaction zone. The exact number should be constant during one experiment, because each cross-section should have corresponding cross-section in each period of experiment. That is necessary in order to assure proper data for visualization needs. However, it may vary between experiments. Generally, the number of cross-sections should fit the size of the reaction zone. Fig. 3 presents an exemplary radar profiles.

There are many approaches to describe the UCG process mathematically [5,8,7]. In our design the gasification canal is build from a set of circular ribs, which the cavity surface is stretched on. The radius of each rib changes according to the chemical reactions, which take place in its neighbourhood. The model probes process values in the three-dimensional space described by automatically generated mesh. Therefore, for each mesh node there are stored information about the coal amount, generated gases (e.g. carbon dioxide, carbon monoxide, hydrogen, etc.), current temperature, flow velocity, and flow direction.

The model is implemented in ANSYS Fluent software, therefore, in order to access the data additional user defined functions must have been implemented, which allow for storing the details of the simulation into the text files. The text

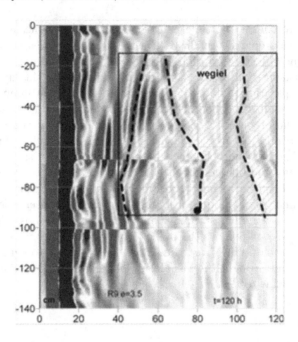

Fig. 3. Exemplary data recorded by the radar. The surface is on the left side. The black dashed line marks the probable surfaces between coal and gas.

files constitute the input data for presented visualization program. In comparison with experimental data, this simulation gives more detailed description of the cavity growth. However, the cavity surface in this case is not described strightforwardly. It needs processing of the data to find its position. Next, it is possible to define the frequency of simulation state saving. This functionality provides the option for adjusting the model data acquisition with real experiment conveyed in similar conditions.

5   Database

UCGVision application uses Microsoft SQL Server 2008 R2 as an information storage database server. The collected data, from both external sources (metrology information and simulation of physical models) as well as any information resulting from the internal processing (as graphical models), are stored in designed for this purpose database. The communication between the database layer and application layer corresponds to the functionality of a dedicated accessor class, based on the Entity Framework version 5.0. The whole solution provides a convenient, though not without problems, access to data through a unified interface.

The database structure was designed in accordance with the rules, which are inherent to the relational model, characterized by the principles of standardization. All tables meet the conditions of the third normal form, which, in the case of this project also fulfills the needs of the higher normal forms, concerning the multivalued dependency. Due to the fact, that the Entity Framework requires that each of the database tables have a simple primary key (to clearly identify a record), the mechanism of this key implementation each time was adopted (despite the fact, that in several cases, the apropriate candidate key existed). There are two distinguished thematically blocks in the database structure as it is presented on Fig. 4. One deals with real data concerning the physical model description (tables: PhysicalModels, Meshes, Snapshots, Values, Ranges, and ModelledParameters) and the other describes the data processed in application to create the graphical models (tables: GraphicalModels, Objects, ObjectParams, FrameObjects, Datum).

Preparation of 3D visualization involves several operations and requires a set of data describing what should be visualized and how should it be done. The fundamental information of the visualization is stored in table GraphicalModels. The table form the core around which other information is gatherd. It was assumed that one data set obtained from numerical model and from georadar measurements can be potentially visualized in several different ways. Because of that, the PhysicalModels is related to GraphicalModels with one to many relation. Each GraphicalModels row is further related to at least one row in Objects table. Each row in Objects table represents an Object - a set of physical model properties, which are visualized in the same way. The Object is aware of visualized parameters characteristic through the table ObjectParams representing n:m relation with modelled parameters (ModelledParameters data table). The relation identifies which of the modeled parameters are visualized by the Object. The relation is also used for interpretation of binary data stored in Datum table. It is assumed, that field 'data' in that table holds binary large object interpreted as sequence of 3 spatial coordinates followed by visualized parameters values. Identification of parameters is done using field 'number' in ObiectParams' table row. The Datum table is not related directly with Objects. Instead, relation is set up by FrameObjects table. The FrameObjects table is responsible to match parameters values with time. Each Object is capable of visualizing parameters' values as a function of time. It is achieved by use of 'frame' idea. Frame represents the visualized parameters values snapshot. As user is interested in visualization of parameter values in certain moment in time since simulation beginning, the system calculates appropriate frame number and shows data related to that frame (by means of FrameObjects table). It is also possible to animate parameters' values in time during visualization simply presenting data related to consecutive sequence of frames. The data are visualized using Visualizer object. The table Visualizes is a dictionary and a list of graphical visualizers exploited to represent objects. Detailed structure of database dedicated to visual representation related information storage is depicted in the Fig. 5.

As it was previously mentioned, the Entity Framework in version 5.0 acts as a contextual layer to support database. There is an interesting fact to point out, that the DLL library of this version is detected within the Visual Studio 2010 (using .NET Framework 4.0) as version 4.4, which was officially never

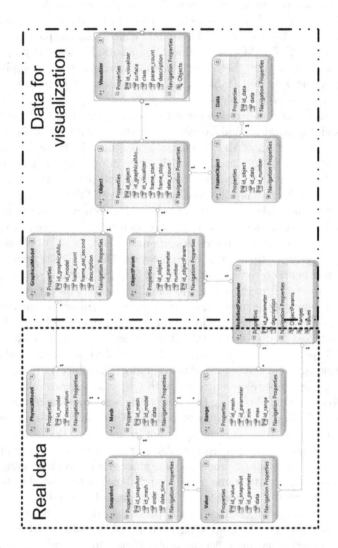

Fig. 4. The structure of database with marked relations between tables

distributed (Microsoft has vever released this version of the Framework). The big advantage of the Entity Framework is the simplicity of configuration, and especialy the automatic mapping of entity objects. Unfortunately, there is also a single, fundamental flaw. In the case of transmission of large amounts of data into the database, which takes place during the import of binary data files specifying physical models or when the graphical models are generated, the time overhead associated with the performance of database operations is significant, and influences software performance.

The accessor class is a container of contextual layer, which is implemented basing on the singleton design pattern. This solution provides a uniqueness facility to manage data access (via a collection of Entity Framework objects) while delivering a consistent interface. For the most part, it consists of a number of methods responsible for adding objects (entities) in the database, its collection, and disposal. Because the data management implementation was realized as a separate project belonging to the UCGVision solution, the convenience of data access is almost similar from each source code level. It is achieved due to the fact, that to achieve the communication with a database it is suficient that each created project is supplied with reference to the DLL library of the accessor and adequate for it namespace.

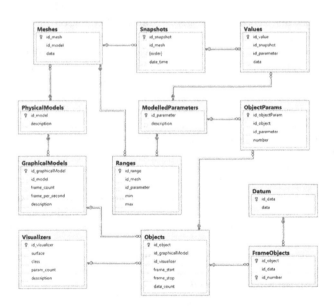

Fig. 5. The structure of database with marked relations between tables

6 Conclusions

In the article the construction of software visualizing UCG process is described. Using of possibilities offered by 3D graphics allow showing the dynamic data in user friendly way. The different requirements posed by input data format (real data) and their graphic representation (data for visualization) were reconciled by database structure. Specification of three main modules and a workflow surrounding them allowed for clear separation of system functionality, increasing its cohesion and improving manageability. Using of XNA made implementing advanced 3D graphics much easier than struggling with DirectX. Entity Framework proved its usefulness in data access layer implementation, though special attention have to be paid when high efficiency is necessary.

Acknowledgments. The research presented in this work is funded by the National Research and Development Centre in project titled: "Visualization of the development" of cavities in the process of underground gasification of coal". Agreement no LIDER/09/30/L-2/10/NCBiR/2011.

References

1. Kostur, K., Kacur, J.: The monitoring and control of underground coal gasification in laboratory conditions. Acta Montanistica Slovaca 13(1), 111–117 (2008)
2. Rauk, J.: Podziemne zgazowanie węgla w Stanach Zjednoczonych - Doświadczenie "Hanna II". Wiadomości Górnicze 11(78), 312–316 (1978)
3. Stańczyk, K., Dubiński, J., Cybulski, K., Wiatowski, M., Świądrowski, J., Kapusta, K., Rogut, J., Smoliński, A., Krause, E., Grabowski, J.: Podziemne zgazowanie węgla - doświadczenia światowe i eksperymenty prowadzone w KD Barbara. Polityka Energetyczna 13(2), 423–433 (2010)
4. Stańczyk, K., Howaniec, N., Smoliński, A., Świądrowski, J., Kapusta, K., Wiatowiski, M., Grabowski, J., Rogut, J.: Gasification of lignite and hard coal with air and oxygen enriched air in a pilot scale ex situ reactor for underground gasification. Fuel 90(5), 1953–1962 (2011)
5. Wachowicz, J., Janoszek, T., Iwaszenko, S.: Model tests of the coal gasification process. Arch. Min. Sci. 55, 245–258 (2010)
6. Wiatowski, M., Stańczyk, K., Świądrowski, J., Kapusta, K., Cybulski, K., Krause, E., Grabowski, J., Rogut, J., Howaniec, N., Smoliński, A.: Semi-technical underground coal gasification (UCG) using the shaft method in experimental mine "Barbara". Fuel 99, 170–179 (2012)
7. Yang, L.H.: Numerical study on the underground coal gasification for inclined seams. Environmental and Energy Engineering 51, 3059–3071 (2005)
8. Yang, L.H., Liu, S.: Numerical simulation of heat and mass transfer in the process of underground coal gasification. Numerical Heat Transfer 44, 537–557 (2003)

A Data Model
for Heterogeneous Data Integration Architecture

Michał Chromiak[1] and Krzysztof Stencel[2]

[1] Institute of Informatics, Maria Curie-Skłodowska University, Lublin, Poland
mchromiak@umcs.pl
[2] Institute of Informatics, University of Warsaw, Warsaw, Poland
stencel@mimuw.edu.pl

Abstract. Modern approaches to data analysis often require an intense integration of data from multiple data sources. The gap between utilized data models and schemata of pulled data require a significant effort to unify and deliver a clean view of an integrated data grid. This paper includes a discussion of a data model that challenges the most severe problems of data integration. The discussion includes most common issues in the area of data integration like horizontal, vertical and mixed fragmentation, replication or integrated data storage sparsity. The proposed data model is aimed to be used as a integral part of a polyglot data linkup entity as a central part of integrated environment.

Keywords: database architectures, heterogeneous integration, fragmentation, replication, heterogeneous databases, database integration, XML and semi-structured data designing, formal models and database usage, metadata management algorithms, data exploration and knowledge bases, data extraction and data integration.

1 Introduction

The integration of data sources is a main goal of many research projects. The present integration techniques focus around solving the problems with integration of DBMS's. Modern integration software design and development aims to provides efficient, reliable, convenient, multi-user safe access storage to massive amounts of persistent data. Mostly integration is based on XML [1,3,8]. Due to the widespread of interest in the databases utilization in numerous areas, their mechanisms has become diversified considering multiple application flavors. This even includes medicine [7] or real-time systems [2]. Thus, the integration of database systems seems to be a crucial and very complex task to accomplish. Nowadays, when the SaaS software delivery model is becoming more popular and endorsed it is often adopted as a cloud data integration platform [5,4].

The process of database integration can be constrained to general areas formed by progressive level of complexity and abstraction not bounded to their specific use. The database integration issues focus around basic problems like:

S. Kozielski et al. (Eds.): BDAS 2014, CCIS 424, pp. 547–556, 2014.
© Springer International Publishing Switzerland 2014

- Scalability (not only for local grids but also involving distribution) and homogeneous database environment (i.e. based on one vendor's solutions), covered by the DBMS vendors across most of the leading enterprise solutions such as Oracle, MS SQL Server, MySQL, PostgreSQL, etc.
- Grids. Multiple vendor originated databases. Such heterogeneous nature is imposed by environmental conditions. In that conditions we can consider the data sources as heterogeneous. This is how we refer to heterogeneity.
- Data model mismatch between the integrated data sources. In this case, the most well known and reported issue is the problem of impedance mismatch [9]. This issue has been addressed by numerous solutions namely in form of an object-relational mappers (ORM). Under this circumstances the integration procedure must consider the translation between the data models, despite the fact that it becomes a source of system overhead and may cause integration bottlenecks. This situation has been widely accepted as a indispensable condition of a heterogeneous DBMS integration. In this paper we want to propose an improvement for this process of heterogeneous integration in an efficient way.

The paper is organized as follows. Section 2 summarizes the general architecture assumptions. Section 3 will contain detailed description of the idea based on example, section 4 concludes.

2   Architecture - Meaning of Access Object

The goal of the presented architecture is to present data storage as a service. The effort here is to prepare the architecture to store notion of physical data existence focused on dealing with metadata instead the data itself. Unlike many present solutions performing physical gathering of data we aimed to store only access metadata of a specific resource. To fulfill such assumption we need to introduce some polyglot interface. Considering many data sources with different data schemes we are after customizing those data to form their arbitrary view, that can be accessed by client. The integrated grid will provide data that can (not obligatorily) conform some common scheme. All of scheme-dependent or unrelated data can be referred as data scene that can be freely used for data composition in form of linkup entity[1]. This is what we will discuss in this paper. Due to size limits for this paper more details and examples on polyglot linkup data model design, that is a major part of heterogeneous integration, please refer to [6].

To handle the heterogeneous integration problem there is a need for providing some sort of unified design. For this purpose we introduce reference object that plays the role of interoperable data access object (iDAO). Such object is responsible for storing all the metadata necessary to localize the data stored physically in the grid. Thus, integrator's client, when requesting for a specific

---

[1] Otherwise referred as cuboid due to its intuition emanation in geometric shape.

data in the grid, prior to reach actual data, would have to reach for its iDAO stored in linkup entity.

Let us discuss the design for an interoperable data object (iDAO). The iDAO as a structure describing data in a grid would require notion of a integrated data repository ID. However, as each repository can contain more then one schema, the ID of a repository must be accompanied with schema name including the described data. Therefore the ID can be constructed as follows `<DB_ID:SchemaID:EntityID>` which can be exposed as `DB1:HR:Employees`. The additional *DB1* id is required due to the fact that each repository can also contain many stores containing same schema with the same entity name. This way we have unambiguous repository, store, schema and entity metadata access information.

Second set of information that is required for data identification in integrated grid is a sequence of access data. Such sequence would provide the grid with potential multiple access methods for the same data. This is being done thanks to access profiles and brings the following benefits:

- gaining references to replications of a resource from the integrated grid
- enables defining arbitrary access protocols
- covers the mixed fragmentation issue

2.1   Distribution Integration

The basic representation of iDAO is presented as a structure below:

```
AccessObject {
    int  entity;  //  0 - object represents entire DB entity
    string repo_ID;
    sequence<ContactProfile> profiles;
}
```

In case when access object describe entity object (i.e. entity field flag equals 0) the *ContactProfile* sequence is responsible for covering the horizontal fragmentation of an entity. Each contact profile then describes following records of this entity i.e. each contact profile contain all the metadata to reach records from integrated environment. In case when access object is not describing an entity (values different than 0), then the each contact profile is describing a single record or in case of vertical fragmentation only a part of record in form of one (cell) or more (tuple) attributes. Moreover, under such circumstances the flag *entity* of following records or their parts would store order in which the fragmentation must fulfill the entity definition (see section 3 for example).

Let us elaborate more on *ContactProfile*.

```
typedef enum SourceID
    SELF,  Postgres, MS_SQL, MySQL, Oracle_11g, mongoDB
}
ContactProfile {
```

```
        bool vert_f;
        sourceID src_ID;
        sequence<ContactDetails> objectREFs;
}
```

Each *ContactProfile* from the sequence of the object profile indicates not only the source but also the fragmentation of the requested actual data. The fact of potential vertical fragmentation is represented with Boolean field. The datasource characteristics are responsible if the data are:

- SELF – Data store within the requesting client (if the client is a part of an integrated environment); we assume no vertical fragmentation at client store
- heterogeneous – The data present in integrated environment but heterogeneous in relation to data requesting client (native query from datasource wrapper; heterogeneous query engine)
- homogeneous – The data is present on a remote data source that is homogeneous in relation to data requesting client

The *vert_f* flag in contact profile brings a very important information. Depending on this flag, the sequence of contact details will adapt different form conforming vertical fragmentation. The *ContactDetails* itself is responsible for assembling integrated data into one full [2] record of the integrated entity. Because the object key is included in structure of contact profile in case of many contact profiles it would have to be repeated many times in the definition structure. Therefore, we provide a design that overcome this issue and guaranties storing of a object key only once with many different contact details. The details are depict in listing below:

```
ContactDetails{
    CommunicationConf {
    bool customSPEC  TRUE; // false  use some default settings for
            // the profile protocol configuration options used
                // to load balancing or integrity update check
        bool verify   FALSE;
        string date last_update;
        [...]
    };
...
    Details {
        string host;
        unsigned short port;
        CommunicationConf protocolSPEC;
        ObjectBody object;
        sequence<AccessDetails> objectView;
    };
};
```

The contact details of an object is a complex structure. The details substructure contains basic contact data such as host address, port number where

---

[2] Regarding record definition in entity.

the datasource hosting the requested object is available. Additionally the access method to each object with specific profile contact details contains specification for a specific access protocol. In the specification from the listing the *protocol-SPEC* field is responsible for the choice of a protocol and its configuration for a specific contact details. The type of the field for object access protocol definition contains a parameter (*customSPEC*) accepting boolean values. The TRUE value represent the fact that the protocol setting from *protocolSPEC* are ought to be used. Otherwise the access for an object will be commenced using the default protocol settings defined for the entire integrated environment or its part.

Additionally hash check for object (see below) is defined in *protocolSPEC* in field: *verify*. It defines whether the object content is ought to be validated against the present[3] object state in repository. The hash value form repository is provided by the repository wrapper using the protocol for particular profile. The most important value in the contact details structure is the object field that stores the actual native access methods for stored data.

```
ContactDetails{
    ...
    typedef enum nativeFastAccessMethod{
        PK, OID[, ... ]
    };
    typedef unsigned long BRI;
    FAM {
        nativeFastAccessMethod nFAM_f; //flag
            // for FAM ie 'PK'for RDBMS or 'OID'for ODB
        BRI  BRI_val;
        string  accessMethod; // query for RDBMS
    }
    ObjectBody {
        sequence<binary>  objectValHash;  //object value hash
        FAM  accessMethod;
        sequence<string>attributes;
    }
    Details {
        ...
        ObjectBody object;
        sequence<AccessDetails> objectView;
    };
};
```

It includes two important fields. First of the fields is *objectValHash* which is the value of the hash[4] for described object value i.e. cell, tuple, record (row). The hash value could have significant meaning in case of indexing such an object. It can be used as a index key value for given object where the non-key value would be the fast native access method (FAM). It is FAM that is the second important

---

[3] This value is not useful for fast access methods. However, it can be crucial for optimization methods like index.

[4] Is assures not only that the values are unambiguous, it also allow easy verification if the value has changed since last check.

field: *accessMethod*. As the FAM will differ along different/heterogeneous data-sources the FAM type contains the best record id (BRI) as the fastest and most effective method to reach each data within particular query engine and data model. E.g. the primary key in relational databases or the object id in object databases. What is more the attributes field in *ObjectBody* contains information about the entity's attributes that are being integrated.

The last field in details of an object is a sequence of access parameters – *AccessDetails*.

```
ContactDetails{
    ...
    typedef enum compID { //data source, security configs etc.
        REPLICA, [SOURCE_CONF, SECURE_CONF, ...]
};
    AccessDetails {
        compID  flag;
        sequence<binary>  property;
    };
};
...
Details {
    string              host;
    unsigned short      port;
    CommunicationConf   protocolSPEC;
    ObjectBody          object;
    sequence<AccessDetails> objectView;
};
```

Each parameter is defined using flag describing the type of a parameter and binary data. The binary data are serialized and represent any structures or objects that define access parameters. It is an element that can be expanded with arbitrary parameters. Data stored in binary form in 'property' field will be interpreted accordingly to 'flag' while deserializing. Without any modifications this property is by default responsible for resource replication handling. In that case with the REPLICA flag binary sequence would be a sequence of triples: <host, port, AccessObject>, storing information about other sources of a described data (i.e. cell, tuple, record/row).

This way *AccessObject* contains full information about integrated piece of entity.

3   Example of Integration

Let us present an example of how the proposed solution would act in details. At first we need to define a description for exemplary *Employees* entity. We assume that the actual data are originated from seven sources. Additionally the sources integration include mixed fragmentation. Moreover, we assume that data sources DB2, DB4 and DB6 represent the same data model as the client, therefore the client native queries can simply be passed directly to a data source.

The access object below will describe an entity with five records storing metadata for data from seven sources:

```
<i_dao, 0, DB1:HR:Employees,//entity
    <i_dao, 1, DB1:HR:Employees, ...>//record 1
    <i_dao, 2, DB1:HR:Employees, ...>//record 2
    <i_dao, 4, DB1:HR:Employees, ...>//record 3
    <i_dao, 6, DB1:HR:Employees, ...>//record 4
    <i_dao, 7, DB1:HR:Employees, ...>//record 5
>
```

Please note that each of the access object is not describing entity and therefore it can only cover the vertical fragmentation as it is with records two and three. In such case record's objects would contain one or more subobjects depending of fragmentation flavor. The subobjects would be included in contact profile. To fully describe entity in distributed environment we assume that it has replications for part of its data. To limit the listings size we will focus only on first record replication. It will have replication represented by three vertical plains.

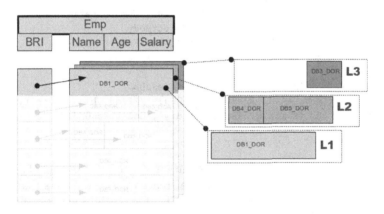

Fig. 1. Figure shows assumed fragmentation and replication using cuboid data representation

As shown in the Fig.1. we have the first record entirely present in DB1 regarding first plain L1. Therefore it will contain all the contact details for adequate record from DB1 database. Next plain L2 contains homogeneous data source DB4 with the attribute of NAME whereas the age and salary are originated from heterogeneous source DB5. Additionally the salary can also be accessed from L3 consisting of only DB3 reference.

Overall the entire replication definition and records integration in particular contact profile is as follows:

```
<i_dao, 1, DB1:HR:Employees, // record 1
    <FALSE, Postgres, ... > // L1: 1st. lvl plain = ContactProfile 1
    <TRUE,  MS_SQL,   ... > // L2
    <TRUE,  MySQL,    ... > // L3
>
```

The first record in L1 plain is not fragmented vertically (FALSE) but its replications in L2 and L3 are replicated (TRUE) independently. The presented example describes mixed fragmentation present in replicated parts of entity. Let us focus on detailed description of such variant. To described this kind of data state we will use a sequence of contact details. In this case it will cover the vertical fragmentation of each replication plain for the record one. Below we enclose all three contact profiles focusing on information contained in contact details. We assume here that the protocol is default and standard and that the databases have their IP address differing with only last digit. All databases work same port 4444: DB1=x.x.x.1:4444, DB2=x.x.x.2:4444 etc.

L1 vertical plain:

```
<FALSE,    Postgres,    // FALSE  no fragmentation; ContactDetails
    < x.x.x.1, 4444,// Details
        < TRUE, ... >              // TRUE  default CommunicationConf
        < '!@##$\%_HASH_\%$#@!',    // ObjectBody
        < PK, '1', 'SELECT * WHERE PK = 1;'> //FAM
        < name, age, salary >              // attributes
    >
    < vREPLICA,      {x.x.x.4, 4444, NameAccessOBJ}
            {x.x.x.5, 4444, AgeSalAccessOBJ} >//flag, bin. property
    < pREPLICA,      {x.x.x.5, 4444, 'salary', SalAccessOBJ} >
    >
>
```

In this plain we have the entire representation of a horizontal plain of a cuboid (i.e. layer). Its description contains exact reference to access objects that covers vertical fragmentation of a record in L2 (vREPLICA). Additionally pREPLICA (partial replica) indicates, that in L3 we have only partial representation of a record definition containing salary data.

L2 vertical plain:

```
<TRUE, MS_SQL,        // TRUE  fragmentation; ContactDetails
    < x.x.x.4, 4444,          // Details
        < TRUE, ... >          // TRUE  default CommunicationConf
        < '!@##$\%_HASH_\%$#@!', // ObjectBody
            < PK, '1', 'SELECT * WHERE PK = 1;'> //FAM; ie. NAME
        < name >//attributes
    >
    < SOURCE_CONF, {'additionalUSER', 'pass'} > //flag, bin. Property
    >
    < x.x.x.5, 4444, // Details
        < TRUE, ... > // TRUE  default CommunicationConf
        < '!@##$\%_HASH_\%$#@!', // ObjectBody
```

```
      < PK, '1', 'SELECT age, salary WHERE PK = 1;'>  //FAM;ie.AGE,
                                                    // SALARY
        < age, salary >                            //attributes
  >
    < SOURCE_CONF, {'additionalUSER', 'pass'} >//flag, bin. Property
  >
>
```

In Layer 2 we include access object to name attribute. Additionally in contact details we have extra access configuration to access DB4 which in this case is additional user and password. Second position in FAM, part of contact details sequence, contains list of attributes that can be obtained. It indicates potential existence of name attribute at DB5 which however, has not been designed to be integrated during the integration configuration process.

L3 vertical plain:

```
<TRUE, MySQL,    // TRUE  fragmentation; ContactDetails
    < x.x.x.3, 4444,          // Details
    < TRUE, ... >           // TRUE  default CommunicationConf
    < '!@##$\%_HASH_\%$#@!', // ObjectBody
        < PK, '1', 'SELECT * WHERE PK = 1;'>    //FAM ie. SALARY
        < salary >                  //attributes
    >
    < SECURE_CONF, {...} >        //flag, bin. Property
  >
>
```

This plain describes access to replication of salary attribute. Additionally some safety configuration stored in SECURE_CONF is required to reach to data about salaries.

In first record layer there was no fragmentation but its replications were accessible from within layer L2 and L3. However, in record two, the vertical fragmentation is present right in its L1. What needs to be noted is that in record one information about all of its replications from following layers is stored in L1. Yet, in record two, due to its vertical fragmentation, all of the replication reference data are stored in first section of contact details in L1. Each and every next record can be represented following this method.

4   Summary and Future Work

Current research in the field of database integration is focused around detailed areas without general approach that would be a complete (i.e. independent, extensible by design) and tested concept. Whereas the proposed solution brings the flexibility and abstraction layer enabling unrestricted design for networking and schema customization.

In this paper we have presented a solution that reduces the overhead from intermodel translations in a heterogeneous and potentially distributed database grid. In general, most approaches consider wrapper that plays the role of a mediating actor for each client query to the wrapped resource. The presented architecture however, is focused on overhead reduction. We eliminate this situation. This gain comes with the benefits of the presented architecture design and the general algorithm for direct and native requests to each integrated data source. The presented solution will not reject the existing achievements. It actually embraces what has already been done in field of managing low level storage.

In future papers we will discuss the grid architecture that incorporate the cuboid integrator idea. Furthermore, we will discuss adopting reliable optimization techniques in terms of the presented architecture.

References

1. Basu, J., Chanchani, N.: Heterogeneous xml-based data integration. XML Journal (2004)
2. Chang, F., Zhu, L., Liu, J., Yuan, J., Deng, X.: A universal heterogeneous data integration standard and parse algorithm in real-time database. In: Lu, W., Cai, G., Xing, W.L.W. (eds.) Proceedings of the 2012 International Conference on Information Technology and Software Engineering. LNEE, vol. 211, pp. 709–720. Springer, Heidelberg (2013)
3. Frank, T.: Xml-based heterogeneous database integration for data warehouse creation. In: PACIS 2005 Proceedings (2005)
4. Goyal, P.: Enterprise usability of cloud computing environments: Issues and challenges 2010. In: 19th IEEE International Workshops on Enabling Technologies Infrastructures for Collaborative Enterprises, pp. 54–59 (2010)
5. Liu, G.L., Yang, C.H.: A method of data integration based on cloud. Applied Mechanics and Materials, 433–435 (2013)
6. Chromiak, M., Stencel, K.: The linkup data structure for heterogeneous data integration platform. In: Kim, T.-H., Lee, Y.-H., Fang, W.-C. (eds.) FGIT 2012. LNCS, vol. 7709, pp. 263–274. Springer, Heidelberg (2012)
7. Planey, C.R., Butte, A.J.: Database integration of 4923 publicly-available samples of breast cancer molecular and clinical data, pp. 138–142 (2013)
8. Shiang, W.J., Ho, M.Y.: An interactive tool based on xml technology for data exchange between heterogeneous erp systems. Journal of CIIE 22(4), 273–278 (2005)
9. Subieta, K.: Impedance mismatch (2008), http://www.sbql.pl/Topics/ImpedanceMismatch.html

Applying Web 2.0 Concepts to Creating Energy Planning Portal

Artur Opalinski and Jerzy Buriak

Gdansk University of Technology, 80-233 Gdansk, Poland
`Artur.Opalinski@pg.gda.pl`
`http://ely.pg.gda.pl/~{}aopal`

Abstract. Community authorities in Europe are tasked to create Advanced Local Energy Plans (ALEP), which encompasses collecting local data on current energy use and generation, as well as generating future development scenarios. Both the data and the development scenarios must be publicly accessible as a basis for energy-related decisions taken by residents, local companies and other local institutions. The data must partially be collected voluntarily at least in the part regarding residential generation and use. On the other hand to offer value the data should be possibly complete. Therefore the need arises to solicit residential input and use of the data. This paper presents a Web2.0 based approach to creating a social environment where residents are willing to exchange energy-related information.

Keywords: Web2.0, social network, Advanced Local Energy Planning.

1 Introduction

Mitigation of energy dependence is one of the goals of European Union. This could be obtain by renewable energy sources and energy effectiveness of conventional fuel and energy use. European Commission indicated national targets up to 2020 for share of renewable energy in energy balance, total CO_2 emission limits, and finally energy effectiveness increase. Targets for Poland are additionally defined in the Energy Policy up to 2030 [30] and in the Energy Low which implements directives of European Commission.

These targets cannot be obtained without local action plans. The actions plans should be result of modern planning process. The modern energy planning is known as Advanced Local Energy Planning (ALEP), and described by best practice documents and guidebooks [7,31,15,12]. The adjective "advanced" is justified by enabling and allowing interaction between the planners and the key actors on the local energy market. ALEP consists of the following seven steps:

1. Analysis and formulation of initial conditions.
2. Establishing a detailed description of the present system.
3. Assess local, EU and global goals on sustainable development.
4. Identifying and assessing key technologies which can bridge to a future sustainable system.

S. Kozielski et al. (Eds.): BDAS 2014, CCIS 424, pp. 557–568, 2014.

5. Identifying key actors in the region (to ensure correct decisions and competitive markets).
6. Formulating and analyzing pathways towards a more sustainable energy system.
7. Establish pathway (with respect to technologies, markets, institutions).

In step 2 current description of the energy system is captured, which involves collecting data from local energy companies, as well as ordinary users. Primary energy sources, fuels and energy conversion, distribution networks, final consumers, energy flows, etc. are to be identified in this step.

In step 6 above, current trends in energy consumption and production and current trends in technology development needs to be defined, to create so called business as usual scenario (BAU). BAU is one of possible pathways, which are understood as energy scenarios to achieve goals. More than one development pathway should be formulated for a community or a region. The other scenarios should be more oriented towards sustainability, e.g. renewable energy development scenario with introduction of innovative technologies or other low CO_2 emission scenarios [7].

In step 7 above, the pathways are transformed into action plans. The energy action plans have to include the point of view of all players, including ordinary energy consumers, and not only local authorities and local energy companies. Engagement of ordinary consumers is necessary in planning process because their decisions are crucial to realize the plans and achieving the goals. Significant acceleration of development of small-scale distributed energy generation is currently expected in Europe, which leads to creating a so called prosumer. A prosumer is a consumer who also generates energy, and who partially uses the energy for own consumption while also selling overproduction to power system [4,1].

Ordinary energy consumers and prosumers are not prepared to communicate on the field of energy. The ordinary consumers are additionally generally not very interested in local energy planning process. Therefore dedicated mechanisms have to be developed and applied to engage them in the planning process. An innovative way to embrace the ordinary energy consumers and prosumers in the ALEP process is by Web 2.0 mechanisms.

Web 2.0 is sometimes defined by the tools which are predominantly used. The most traditional tools include wikis, blogs [35,27], and online forums [9]. Newer tools are in the form of Rich Internet Applications (RIAs) [26], characterized by thin clients, which are browser-based and store their state on managed servers. Content aggregation capabilities are greatly expanded by moving from portal-based technologies to so-called web mashups [26,25]. Web 2.0 services often mentioned are Facebook, YouTube, Twitter, Flickr or Wikipedia [18,22]. Not the tools alone however, but user-centric paradigm and user collaboration are the most distinguishing features of Web 2.0 [25,22]. It represents the evolution of the web from a source of information to a platform. Users are no longer only consumers, they become contributors to the collective knowledge accessible on the web. The hesitations whether Web 2.0 brings new qualities indeed [25], and dubbing Web 2.0 a hype, seem to be over as real, new problems arise with its pervasive use, like performance management of new applications [27,26], tagging a growing amount of data [24], content extraction [37], making social recommendations [33], or creation of new services based on it in telecommunication [17], network management [10], language learning [11] and other forms of education, marketing [28], or citizen participation in digital society [23,32].

The goal of this paper is to select and employ Web 2.0 elements in the ALEP portal to attract users, in order to convince them to provide their energy-related data and knowledge. The thesis of this paper is that employing ALEP-specific Web 2.0 elements in the portal attract users to the portal.

2   State of the Art

The basic international document describing the ALEP procedure [7] states in section 3.2.6 that the monitoring phase requires that a data collection and reporting process be organized and suggests in section 4.2.1 that the data collection is to be done in a traditional manner. But in reality, traditional surveys are ineffective and costly when a large number of consumers has to be involved. Section 4.5.2 of this document refers to use databases in municipal administrations and statistical offices, and suggests that other information can be obtained from statistical, market research, or other already existing surveys, or from studies performed for comparable areas. While such abundance of ready to use information may exist in Sweden where the author of this part of the guidelines is based, and it may perhaps to some extent exist in other participating countries (Germany, Italy, The Netherlands), it is not available in other European countries, including Poland.

In Poland planners typically use energy-related data acquired from energy companies. Data about municipal buildings comes from from Municipal Offices, and Statistical Yearbook of Voivodships. More detailed data can also be obtained from Statistical Yearbook related to the voivodship of the region or community. The data sets created in this way most often need additional work to be included in ALEP. Therefore various auxiliary methods of data collection are used. One example is electronic or paper survey, but low quality and small amount of data are the weaknesses of this method. The response rate and data quality are low because there is no legal basis to oblige citizen to provide the requested data.

As he law obligates energy companies to share energy data, engagement of experts from energy companies is an option to acquire data of possibly high quality. Unfortunately, exiting regulations do not define pricing for this data, making this option risky and costly.

The task of completing a database from the data possessed by ordinary consumers is nearly impossible for a single expert or even a team of planners when working in a traditional way. Additional mechanisms and tools for data collection are therefore used. An example tool supporting data collection is SEC-BENCH Internet service [5,3]. Towns and municipalities can conduct benchmarking of indicators of energy consumption, emissions, costs of modernization, failures, equipment life, performance, and comfort through its Web portal. Comparisons are done relative to a certain standard e.g. standard energy demand (as defined in the Regulation on Technical Conditions to be met by Buildings [2]) or other industry standards, such as Best Available Technology (BAT). SEC-BENCH service is not dedicated to energy planning and can therefore only serve as an example of benchmarking of buildings and energy supply technologies.

Regional Energy Analysing Model (REAM) [21] is a multi-scenario tool for local and regional energy planning that is capable of analyzing the energy system in a municipality, region, selected sub-sector or another geographic area. This tool calculates the solution on the least-cost basis. REAM users can choose the level of detail for ease of use, e.g. if they are insufficiently experienced to perform more complex analyzes. While REAM is an advanced tool for energy planning, it needs to be installed on personal computers and does not offer a fully functional Web interface.

An application is therefore needed to implements both the idea of Web data collection like REAM and energy planning capabilities of REAM. Web data collection from ordinary users requires to find methods to convince ordinary users to participate and to share their data or knowledge. While individual users are to provide pieces of portal content, its value is created by consolidating data from many users so the influence of portal creator remains paramount. The portal creator has to introduce the framework to attract users to build the aggregate value of the portal. The value of the data and knowledge does not depend only on their potential use, but also on their completeness and correctness. Unfortunately, the methodology to create Web 2.0 sites remains unknown, which this paper is trying to overcome.

Technology Acceptance Model 2 (TAM2) proposed in [34] has been applied to Web 2.0 sites by [39], and the influence of different factors on intention to use a site has been defined by statistical processing of survey results. The most important factor (path coefficient: 0.51) turned out to be perceived usefulness of the site. The perceived usefulness in turn is determined by the following factors, by descending path coefficients:

- tangibility of the results using the system (result demonstrability), with path coefficient of 0.41,
- individuals perception regarding the degree to which the target system is applicable the task at hand (job relevance), with path coefficient of 0.21,
- the degree to which one perceives the effectiveness of a system (output quality), with path coefficient of 0.18,
- attitudes and belief of others in groups to which one belongs to (subjective norm), with path coefficient of 0.11,
- the extent to which use of an innovation is perceived as enhancement of ones status (image), with path coefficient of -0.22.

The influence of the perceived ease of use is more complex, as it influences both intention to use and perceived usefulness. The study is not constructive: it does not explain what constitutes the most influencing factors in any given scenario, especially not for collecting voluntary data or other knowledge, as needed for ALEP.

Knowledge sharing is something considered the key problem of knowledge management research [16,38]. But as [8] notes, this does not seem to apply to social networking sites, where voluntary knowledge is abundant, even if of unequal quality. [8] proposes six basic knowledge-sharing capabilities recognized on eleven Web 2.0 sites selected for case studies. These capabilities include profiles, groups, connections, category, commentary and ratings. That paper does not try to rate their relative strengths or interdependencies. It seems to assume that individual contributions are attributable to social group members, which should not take place in ALEP data sets due to privacy reasons.

Widely used Technology Acceptance Model (TAM) [19] has been researched by [36] on the influence of emotion and motivation on user behavior in Web 2.0 systems. Authors state that perceived usefulness of the Web 2.0 sites differ, depending on utilitarian or hedonistic goals of users visiting the site, with utilitarian goals having stronger influence. Authors confirm the previously known results that both positive and negative emotions influence perceived usefulness of the site, with positive emotions having stronger influence than the negative ones. These results are not constructive as they still leave open the question how to influence positive and negative emotions. Data collection for ALEP is by its nature utilitarian.

When employing Web 2.0 to marketing activities by commercial companies, users are provided incentives to drive their community involvement. According to survey conducted in [28], most of the researched companies did provide small monetary rewards or prizes, such as gift certificates. Many users valued company-branded gifts, that could not be purchased by the public, over a monetary reward. Peer recognition was also deemed important, when contributor recognition system was attributing points to those who posted content. It has been discovered that moderator role can also ensure that the community is active by encouraging member participation in blogs, wikis and forums, pointing out community events such as podcasts, training sessions, or conferences, and sharing electronic articles and links to relevant websites.

3   Solution

According to [34], the three factors most influencing users intention to use a web portal are: result demonstrability, job relevance, output quality. Therefore these factors have been reflected when constructing the ALEP Web 2.0 portal.

The main purpose of the Web 2.0 portal is to support the planning process at the local and regional level. This objective can be seen as a scientific issue. On the other side, regional strategies are important for implementation of the national energy policy and for fulfillment of the international obligations in the field of energy and air quality. These obligations apply all EU member countries.

Achievement of the main objective is only possible through reaching certain additional, portal specific goals:

- attracting to the portal users willing to share their energy systems and energy consumption data,

- involving all the key players in the local energy market in the planning process conducted with the Web 2.0 portal.

Only fulfillment of both the above goals allows to build a strategy covering entire local energy market, including verification of the strategy based on individual case data and statistical methods.

The data seeming attractive from the point of view of the individual energy consumers and owners of energy devices, include data on fuels and modern technologies. Access to this data enables users to perform some simulations of their cost of supply of fuel and energy, to lower their maintenance costs. But the data on fuels and technologies alone may not be sufficient to encourage users to participate in the ALEP process, particularly during the calm periods out of public consultations and out of planning time for the supply of fuel and energy.

To attract users willing to share data about their buildings and energy demands, information referring to the situation of other users may be very valuable. Such information includes type of technology applied for heating and hot water, annual energy or fuel consumption, chosen technology for modernization. Hence, a supplementation of the data is necessary in order to ignite the attractiveness of the system, particularly in the first period of its usage. For this reason, example cases corresponding to popular solutions of heating and domestic hot water systems have been entered into the system. Also, descriptions of sample technologies of energy supply of different objects were provided: electric heating, gas, oil and coal boiler and heat from district heating. It is also planned to provide information regarding examples of distributed electricity sources like photovoltaic or small wind turbine, which is important for prosumers.

In addition the application presents the characteristics of fuels, like heat value and emissions, rates of fuel and energy tariffs, and modern technologies. The attractiveness of the portal is improved by links to other Web services, which can be maintained and ranked by portal users. The thematic links lead to sites related to: heat pumps, solar collectors, biogas plants, small wind turbines, photovoltaic panels. The portal contains a table of tariffs. In particular it contains sets of tariffs for: natural gas, district heating, electricity, prices of solid fuels, prices of liquid fuels. Prices of solid and liquid fuels have strong regional and local character and the portal gives a possibility to present and compare local prices of wood, coal, fuel oil, briquettes, pellets and other biomass.

4   Verification

The Web 2.0 portal has been implemented as ALEP-PL application [13] using ASP.NET Web Application template of Microsoft Visual Studio 2010 environment and database on MS SQL Server 2008 R2 [20]. The business logic of the application is implemented with C# classes [29]. The ASP.NET schema of application users, roles and permissions was integrated with ALEP-PL database [20].

The fitness of the portal framework to attract users and to motivate them to share their energy-related data and knowledge has been verified by means of an Internet survey. The survey was available through one of the portals tabs to all of its users.

Table 1. Analysis of content used and valued by users, by content category

Sub-category	Resp.	Main cat.	Abs. resp.	Mean resp.	Rel.resp.
1 legislation on thermal insulation of buildings, renewable energy sources (RES) and high-efficiency cogeneration	61.54%	Tradit.	134.62%	67.31%	0.32
2 lower heat value of fuels	73.08%				
3 fuel and energy prices	80.77%				
4 energy consumption in similar buildings	80.77%				
5 technical and economic data of energy technologies	42.31%	Web 2.0	276.92%	69.23%	0.65
6 advertisements of energy technologies producers and energy companies	26.92%				
7 funding opportunities of investment in renewable energy	46.15%				
8 others	11,54%	N/A	11.54%	11.54%	0.03

The survey [6] consists of six questions; questions 1-3 are used to filter out responses of accidental users or users who do not yet know site capabilities. Questions 4 is used to recognize users who are only interested in static content added by portal creator, i.e. non-Web 2.0 content. Question 5 is used to measure the relative values of static and Web 2.0 content for the user. Open question 6 exists to verify that the participants understand the goals of the ALEP site and the survey, and to solicit their free comments on the initiative.

A total of 26 valid responses have been obtained, after filtering-out not compliant responses by means of questions 1-3. As the site provides both traditional, relatively static Web content provided by the portal creator, as well as Web 2.0 content built by users, their relative importance to the users has been rated. The following content categories have been assigned:

− Traditional content:
  • legislation on thermal insulation of buildings, renewable energy sources (RES) and high-efficiency cogeneration,
  • lower heat value of fuels.

Table 2. Analysis of content value comparison, by content sub-category. Column with caption 1 denotes attaching the highest value to the sub-category; column with caption 8 denotes attaching the lowest value to the sub-category. ´

Sub-category	Final value	1[%]	2[%]	3[%]	4[%]	5[%]	6[%]	7[%]	8[%]
1 legislation on thermal insulation of buildings, renewable energy sources (RES) and high-efficiency cogeneration	38.46	15.38	23.08	11.54	15.38	23.08	0	11.54	0
2 lower heat value of fuels	42.30	23.08	15.38	11.54	19.23	15.38	0	3.85	3.85
3 fuel and energy prices	38.46	11.54	23.08	15.38	23.08	15.38	0	3.85	7.69
4 energy consumption in similar buildings	50.00	30.77	15.38	19.23	15.38	3.85	15.38	0.0	0
5 technical and economic data of energy technologies	-26.92	0	3.85	19.23	3.85	23.08	26.92	23.08	0
6 advertisements of energy technologies producers and energy companies	-34.61	3.85	7.69	11.54	11.54	7.69	30.77	15.38	11.54
7 funding opportunities of investment in renewable energy	-23.08	7.69	7.69	11.54	11.54	11.54	19.23	26.92	3.85
8 others	-76.92	7.69	3.85	0	0	0	0	15.38	73.08

- Web 2.0 content:
  - fuel and energy prices,
  - energy consumption in similar buildings,
  - technical and economic data of energy technologies,
  - advertisements of energy technologies producers and energy companies,
  - funding opportunities of investment in renewable energy.

The responses to question 4 on the content used and valued by users are presented in Tab. 1. Responses of users valuing Web 2.0 content amounted to 276,92% (more than one answer was allowed, so results above 100% are correct), which is twice as much in relation to the responses attributed to the traditional content. This relation is partially dues to the number of sub-categories in this Main category, as the mean percentage of responses in both main categories is very similar. Nevertheless it is to say that Web 2.0 content constitutes for users an important value of the portal, as related to both other main categories possible. The responses to question 5, which requested the participants to compare and order the value of each content sub-category, are presented in Tab. 2. It gives a closer look on how users value each of the content sub-categories. The final value in the table has been computed by substracting the sum of percentages of responses which attached higher value to a sub category, from the sum of percentages of responses which attached lower value to a sub-category. Placing an item on one of the top three places on the list was regarded attaching higher value to this item. Placing an item on one of the bottom three places on the list was regarded attaching lower value to this item. Placing an item in the middle of the list, i.e. on position 4 or 5, was considered undecided response and was not taken into further consideration.

The items top ranked by users are items number 4, 2, 1 and 3. Two of these top items, including the top ranked item 4 belong to the Web 2.0 category. Besides the catch-all item 7 (others), advertisements (item 6) ranked lowest. This item belongs to the Web 2.0 category, but probably users connotations on advertisements is low. The consolidated energy-use related data could be accessed only after sharing own energy-related data, therefore the responses on the value of data are treated as confirmation that the users perceived value of accessing them is justified by his/her cost of sharing own data.

5   Conclusion

The issue of selecting the right mechanisms and elements in Web 2.0 portals is not sufficiently researched. There is a significant gap between the high-level ideas, and the drivers to select practical elements of a Web 2.0 site. In the particular case of ALEP, the elements have been thoroughly considered and selected, based on the general TAM2 taxonomy, on the specific needs of the ALEP process, and on the attributes of casual users. The goal was to attract ordinary energy consumers by convincing them to the usefulness of the site.

Among the main values attracting users to the ALEP portal turn out to be the purposely designed Web 2.0 portal elements, i.e.: fuel and energy prices, and energy consumption in similar buildings.

The most attractive of these turned out to be the data on building energy use, provided by individual contributors. This kind of data is unavailable by traditional means, as it can only be delivered by the individual users. Therefore Web 2.0 technology seems appropriate for collecting user data for the ALEP process.

The portal in its Web 2.0 form may be further extended to support creation of producer groups and societies, devoted to promotion of energy technologies. Another feature of value for a broad category of portal users may be technology benchmarks based on investments costs against incomes, costs and maintenance issues.

The portal should contain additional code to assist users, especially ordinary users, in entering complex technical and economic data. The users may have difficulty in decision which units should be applied to enter a value. Particularly gas tariffs are cumbersome and troublesome. In this tariff the price of power is per cubic meters per hour for hour ((currency unit)/(m3/h for h)). The perception of these units as an expression of power is difficult even for engineers, not to mention ordinary consumers of energy. Therefore, the program must assist users in entering data on fuel and ordered power. Currently, the software assists users by automatically determining the approximate value of ordered power according to the information about the type of gas tariff and about gas consumption. Conversion formulas are included in the methods of handling events caused by the data inserts and data updates. The system also employs auxiliary and dictionary tables populated with values, which are used to create input option lists.

These tables are but the first attempts to facilitate the communication process and to contribute to data quality improvement. [14] presents results of analysis of the errors made by users during entering data into ALEP-PL. One of the result is observation that the service could be further simplified. In addition to the above mentioned problem of units interpretation of technical and economic data, another problem is the complex, professional terminology. The vocabulary of energy sector widely used in the ALEP-PL portal may be incomprehensible for ordinary users and some labels and link descriptions have to be changed to more common names and titles.

References

1. Directive 2012/27/EU of The European Parliament and of The Council on energy efficiency, amending Directives 2009/125/EC and 2010/30/EU and repealing Directives 2004/8/EC and 2006/32/EC (2012)
2. Polish Ministry Technical Code for Buildings of 12 April 2002 (2012), in Polish: Tekst ujednolicony rozporzadzenia Ministra Infrastruktury w sprawie warunkow technicznych, jakim powinny odpowiadac budynki i ich usytuowanie z dnia 12 kwietnia 2002 r
3. 5th SEC-BENCH Newsletter (2013), http://www.sec-bench.eu/docs/5th_SEC-BENCH_Newsletter.pdf (DOA: 15.12.2013I)
4. Polish Energy Law Act of 26 July 2013 (2013), in Polish: Ustawa z dnia 26 lipca 2013 r. o zmianie ustawy-Prawo energetyczne oraz niektorych innych ustaw
5. SEC-BENCH Project Report (2013), grant agreement no: EIE/07/067/SI2.4662, http://www.sec-bench.eu/docs/DELIVWP6/WP6_D6.6_-_Publishable_Technical_Brochure.pdf (DOA: 15.12.2013)
6. The verification survey, accessible over ALEP-PL portal, used to assess site Web 2.0 elements valuable for users, available online in Polish (2013) , https://www.surveymonkey.com/s/2LC6ZBG (DOA:15.12.2013)
7. Advanced Local Energy Planning (ALEP): A Guidebook, Annex 33 of Energy Conservation in Buildings and Community Systems Programme. Reinhard Jank, International Energy Agency (2000)
8. Allen, J.: How web 2.0 communities solve the knowledge sharing problem (2008)
9. Baxter, G., Connolly, T., Stansfield, M., Tsvetkova, N.: Introducing web 2.0 in education: A structured approach adopting a web 2.0 implementation framework. In: Proc. of 7th International Conference on Next Generation Web Services Practices (NweSP), pp. 499–504 (2011)
10. Bezerra, R., dos Santos, C., Bertholdo, L., Granville, L., Tarouco, L.: On the feasibility of web 2.0 technologies for network management: A mashup-based approach, pp. 487–494 (2010)

11. Boruta, S., Chang, V., Gutl, C., Edwards, A.: Foreign language learning environment built on web 2.0 technologies, pp. 82–88 (2011)
12. Bucko, P., Buriak, J., Renski, A.: Total Costs Comparison for heat consumers on Local Competitive Energy Carrier Market. Rynek Energii 4(53), 44–51 (2004) (in Polish)
13. Buriak, J.: Computer-aided local energy planning using alep-pl software 21(1), 7–23 (2013)
14. Buriak, J.: Disruption of the communication process on the example of alep-pl web application (2013)
15. Buriak, J., Jaskolski, M.: Energy roadmaps for the city of gdansk. Acta Energetica 2, 4–19 (2009)
16. Cabrera, A., Cabrera, E.: Knowledge-sharing dilemmas. Organization Studies 23(5), 687–710 (2002)
17. Carlin, J., Trinugroho, Y.: A flexible platform for provisioning telco services in web 2.0 environments, pp. 61–66 (2010)
18. Danyaro, K., Jaafar, J., De Lara, R., Downe, A.: An evaluation of the usage of Web 2.0 among tertiary level students in Malaysia. In: Proc. of International Symposium in Information Technology (ITSim), vol. 1 (2010)
19. Davis, F., Bagozzi, R., Warshaw, P.: User acceptance of computer technology: A comparison of two theoretical models. Management Science 35, 982–1003 (1989)
20. Evjen, B., Hanselman, S., Rader, D.: Professional ASP.NET 4 in C# and VB. Helion (2011), in Polish: Moch, W., Walczak, T.: ASP.NET 4 z wykorzystaniem C# i VB
21. Johnsson, J.: Models in local energy planning - the ream model as an example (2013), http://www.pepesec.eu/cms/wp-content/uploads/2010/01/Advanced-local-energi-planning_John-Johnsson_profu.pdf (DOA: 15.12.2013)
22. Knights, M.: Web 2.0. IET Communications Engineer 5, 30–35 (2007)
23. de Kool, D., van Wamelen, J.: Web 2.0: A new basis for e-government? (2008)
24. Kumar, S., Inbarani, H.: Web 2.0 social bookmark selection for tag clustering, pp. 510–516 (2013)
25. Murugesan, S.: Understanding web 2.0. IEEE IT Professional 9, 34–41 (2007)
26. Nagpurkar, P., Horn, W., Gopalakrishnan, U., Dubey, N., Jann, J., Pattnaik, P.: Workload characterization of selected JEE-based Web 2.0 applications. In: Proc. IEEE International Symposium on of Workload Characterization (IISWC), pp. 109–118 (2008)
27. Ohara, M., Nagpurkar, P., Ueda, Y., Ishizaki, K.: The data-centricity of web 2.0 workloads and its impact on server performance, pp. 133–142 (2009)
28. Parise, S., Guinan, P.: Marketing using web 2.0, pp. 281–287 (2008)
29. Powers, L., Snell, M.: MS Visual Studio 2008 Unleashed (2011), in Polish: Walczak, T. Microsoft Visual Studio 2010, Ksiega eksperta
30. Rada Ministrw RP: Polish national energy policy up to 2030 of 23 september 2010 (in Polish: Polityka energetyczna polski do roku 2030) (2010)
31. Stenlund, N., Martensson, A.: Municipal energy-planning and development of local energy-systems. Applied Energy 76, 179–187 (2003)
32. Tsui, H.D., Lee, C.Y., Yao, C.B.: Creating a web 2.0 government: Views and perspectives, pp. 648–651 (2010)
33. Turati, A., Cerizza, D., Celino, I., Valle, E.D.: Analyzing user actions within a web 2.0 portal to improve a collaborative filtering recommendation system. vol. 3, pp. 65–68 (2009)

34. Venkatesh, V., Davis, F.D.: A theoretical extension of the technology acceptance model: four longitudinal field studies. Management Science 46(2), 186–204 (2000)
35. Wan, L.: Application of web 2.0 technologies in e-learning context. In: Proc. of 2nd International Conference on Networking and Digital Society (ICNDS), vol. 1, pp. 437–440 (2010)
36. Wang, C.Y., Chou, S.C., Chang, H.C.: Emotion and motivation: Understanding user behavior of web 2.0 application, pp. 1341–1346 (2009)
37. Waqar, M., Khan, Z.: Web 2.0 content extraction (2010)
38. Wasco, M., Faraj, S.: Why should i share? Examining social capital and knowledge contribution in electronic networks of practice. MIS Quarterly 29(1), 35–57 (2005)
39. Wu, M.Y., Chou, H.P., Weng, Y.C., Huang, Y.H.: A study of web 2.0 website usage behavior using tam 2, pp. 1477–1482 (2008)

The Concept of Transformation of XML Documents into Quasi-Relational Model*

Robert Marcjan and Leszek Siwik

AGH University of Science and Technology, Kraków, Poland
{marcjan,siwik}@agh.edu.pl

Abstract. XML has evolved from a document markup language to a data model commonly and widely used for storing, exchanging and sharing hierarchically (semi)structured data. Simultaneously relational model is still a production standard for storing and querying data. Since there are on the market so many advanced and mature, relational– and sql–based tools, systems and solutions just loading xml documents into relational model (for instance as BLOBs) would be the easiest way for storing and then processing XML data.

Unfortunately, in such a way, hierarchical and (semi)structured data becomes flat and unstructured. That is why so many effort is made to provide efficient models and tools for storing and processing XML data and such solutions as XPath or XQuery have been proposed and become a standard in this field. It results however in developing heterogenous and not unified data storing and processing standards, models and tools. It is similar to well-known object-relational impedance mismatch.

Taking its power, scientific foundations, available tools and flexibility as well as „assumed and hidden knowledge"(everybody knows it) it would be great to process XML documents effectively with the use of SQL queries.

There are of course tools responsible for mapping between XML and relational model on the market. Using them however results often that either hierarchical data structure is lost and/or manual transformation (or at least transformation definition) is required.

In this paper the concept of transformation of XML data into quasi–relational model is proposed. The general assumption of this transformation is making it possible to process XML document with the use of SQL-like language without making manual transformation and with preserving hierarchical structure of XML data.

Keywords: XML databases, relational model, data transformation, SQL for XML.

1 Motivation

In [10,5] the comparison of relational and hierarchical data models is presented. According to this comparison the most important similarities and differences between these two models are as follows:

---

* The research reported in the paper was partially supported by grants No. 0008/R/ID1/2011/01 and No. DOBR-BIO4/060/13423/2013 from the Polish National Centre for Research and Development.

S. Kozielski et al. (Eds.): BDAS 2014, CCIS 424, pp. 569–580, 2014.

- „relational model is recommended to store flat, structured data since it uses flat tables to store data in column form". XML data model is „useful when it is required to preserve the document order or when the schema is flexible or unknown";
- „it is difficult to model semi-structured data using relational model" whereas XML data model „...provides excellent support for representing semi-structured data with variable or evolving schema";
- relational data model is „...not suitable for storing markup data beyond BLOB storage" whereas XML data model is „...excellent for storing markup data such as HTML, RTF and so on";
- relational data model „...supports nested data by using multiple tables and linking them with foreign keys, but the complexity of queries required for searching nested data stored in relational form increases when the depth of nesting increases or is unknown" whereas XML data model „...provides excellent support for expressing nested and hierarchical data";
- relational data model, in contrast to XML data model, doesn't not preserve the order of data;
- in relational data model input data is homogenous in contrast to XML data model;
- in relational data model result set is homogenous in contrast to XML data model.

The general conclusion coming from the above comparison is that XML data model is (much) more powerful, flexible and allows for storing much more classes of data structures. And even assuming that it is, as the matter of fact, slightly biassed and not precise comparison since its goal is to present advantages of XML documents—it is difficult to disagree with that in general.

One may ask so, why should we still respect relational model and, the more so, does it make sense to develop transformation tools allowing for transforming of „scalable and hierarchical" XML data into „flat and difficult in design" relational model?

Apart from classical and general features (and advantages) of relational model discussed for instance in [3,16] it can be answered as follows:

- flat data structures (sets) are ideologically closer to the organization of computer memory and in the consequence organization of (simple) data structures. Since data is processed in practice on linearly addressed memory, all data structures are „flat" by definition even if they are wrapped by classes, objects, collections etc. In the consequence, software taking heterogenous input data and producing heterogenous output data is working in fact on homogenous structures on the low level what is close to relational model;
- data modeled according to relational model is much more simple and easy to define, describe and understand. Each tuple is representing one single (real) entity in isolation from the rest of tuples and the rest of relations. One may focus so on processing and interpreting one single relation disregarding the rest of them. The rest of relations are included into interpretation with the use of appropriate operators and conditions only in required level. Only rarely it is required to operate on the complete relational model entirely;
- the schema understood here as the set of constraints and dependencies is much more simple to define in relational model since defined constraints can be considered— one by one—separately. Schemas used in XML documents define simultaneously

both the structure of the documents and constraints as well and regard the entire information always. In this model it is difficult to define constraints on relationships between elements as well;

- data in relational model can not be processed in isolation from its definition and constraints since they (names, types, dependencies) are the integral part of the model and are the basis of its functioning. XML document can be processed in the isolation from its schema. It is more difficult of course since constraints have to be defined and contained by the processing software—but it is possible;
- hierarchical data model imposes its interpretation since such data is processed top-down by definition;
- hierarchical data model may require redundancy whereas the relational model deals with that thanks to normalization. For instance, XML document with information about products belonging to at least one ore more groups of products can be defined in three possible ways, i.e.:
  - elements describing groups of products are the main parts containing particular products as subelements. In such an approach every single product has to be contained many times by every single group it belongs to;
  - elements describing products are the main parts and they contain the groups they belong to. In such a case every single group has to appear many times for every single product belonging to this group;
  - groups and products are processed as independent elements in the document but there are elements describing the relationship (membership) between products and groups. In such a case products and groups have to be identified uniquely and their identifiers have to be repeated in elements describing the membership—the model becomes so more relational than hierarchical[1].

Apart from the above at least equally important are some practical reasons such as available tools or „hidden and assumed knowledge" i.e. SQL is assumed as a fundamental and standard knowledge in IT. As one may see, there are some important reasons to consider storing XML data in relational models especially if it would be possible either use SQL directly or to provide SQL–like manipulation language to make it simple and easy to build queries and process XML data effectively. It's natural so, that there are on the market some tools allowing for mapping XML documents into relational or object-relational model [13]. Also particular vendors of RDBMS provide some mechanisms responsible for dealing with XML documents in relational model [14]. As the example „XML Query rewrite" technique in Oracle XML DB [6], pureXML in DB2 [9,2], XML capabilities in PostgreSQL [15], $ExtractValue()$ and UdateXML() functions in MySql[2] or implementation of XQuery and OPENXML function in MS SQL allowing for mapping particular nodes in XML document on defined relation [12,11] can be mentioned. There are some important limitations and shortcomings however, such as:

---

[1] It is important to note that in the first model it is impossible to define products not belonging to any group. It is possible only in the second or third approach or by providing artificial „not-in-the-group" group what is not intuitive and can be not in the accordance to the document's schema and its constraints.

[2] `http://dev.mysql.com/doc/refman/5.7/en/xml-functions.html`

- requirement for using XPath [4,1] to point out interesting nodes;
- building links by nodes identifiers can be done only between element and its direct parent. It is impossible to receive meta-data about any parental node what means for instance that one has to go through the whole document hierarchy to gain information about a customer and products it bought;
- accessing different nodes of the same document requires calling OPENXML function many times for the same document;
- every time one has to define the structure of output relation—otherwise OPENXML returns only nodes' names and identifiers;
- queries are difficult to define if document has not a fixed structure, i.e. if for instance the order of subelements is not fixed;
- manual user-driven definition of transformation as in [8] etc.

That is why research on more flexible and powerful mapping tools and approaches is highly required. The goal of this paper is to present interesting and promising idea of transformation of XML data into (quasi)-relational model.

2   XML-2-Relational Transformation Approach

The idea of transformation of XML documents into (quasi)–relational model proposed in this paper is presented in algorithmic (and simplified of course) way as Algorithm 1 and Algorithm 2 below. First the table is created for root element passed as first argument. Next, for each subelement specified as the second argument an empty row is created. Next each element is addressed appropriately and processed recursively in $elementTransformation()$ procedure where consecutive columns are added to the given row gradually. Processing is continued as long as all elements are transformed (simple textual values are terminals here).

Algorithm 1. transformationProcedure(RootPath root, ElementName e)

---
1 Create table T for $root$ element;
2 **foreach** $element\ with\ name\ e\ belonging\ to\ root$ **do**   /* e can be proceeded with special characters indicating if only direct or also indirect subelements in $root$ should be considered */
3 | create row r_i ;
4 | Set the address of e_i as the concatenation of $address(root).name(e)$;
5 | $elementTransformation(e_i, r_i)$;
6 **end**
7 Merge all r_i in table T;
                    /* Setting null as a default value for each tuple */

Finally all created rows are added to the table. If some columns have not been created during transformation process for particular row but they exists for different rows their values are set to null.

Algorithm 2. elementTransformation(element e, row r)

```
1  Add new column to row r ;
2  Set the name of added column equal to the address of e ;
3  Set the value of added column equal to the terminal (textual) value of e;
4  if e is a terminal textual element then
5  │   exit();
6  else
7  │   foreach subelement se_k of e do     /* including textual content and direct
   │   attributes */
8  │   │   switch typeOf(se_k) do
9  │   │   │   case terminal textual value
10 │   │   │   │   Set the address of se_k as the concatenation of address(e).#;
11 │   │   │   │   break;
12 │   │   │   end
13 │   │   │   case direct attribute
14 │   │   │   │   Set the address of se_k as the concatenation of address(e).#name(se_k);
15 │   │   │   │   break;
16 │   │   │   end
17 │   │   │   otherwise
18 │   │   │   │   Set the address of se_k as the concatenation of address(e).name(se_k);
19 │   │   │   end
20 │   │   endsw
21 │   │   if in r exists column with its name equal to address(se_k) then
22 │   │   │   continue
23 │   │   else
24 │   │   │   elementTransformation(se_k, r);
25 │   │   end
26 │   end
27 end
```

2.1 Sample Transformation

Since it's always easier to follow the algorithm on the basis of an example let's consider an XML document as presented in listing 1. It contains some pieces of information about particular persons described by some (more or less complex) attributes such as *name*, *age* but also *address* $=< street, state \dots >$, *hobbies* $=< hobby_1, \dots hobby_n >$ etc). So let's analyze what will be the result of running proposed transformation approach applied to sample document assuming that *transformationProcedure*() is called with *people* as root and *person* as interesting element(s).

First according to step 1 of *transformationProcedure*() the table for root element (*people*) is created. Next, in first iteration of loop in lines $2-6$ an empty row $r_1 = \emptyset$ is created. The address of first subelement is set to *person* and next *elementTransformation* procedure with *person* $< \cdots >^3$ and r_1 as arguments is called. According to lines $1-3$ of this procedure a new column is added to r_1. Its name is set to *person* and its value is set to *null* (since there is no anonymous textual content for this person). So now,

---

[3] *person* $< \dots >$ means that the whole element is passed as the argument.

$$r_1 = \{person = null\} \tag{1}$$

Since passed element is not a terminal textual value condition in line 4 is not fulfilled so processing is continued. In first iteration of loop from lines $7 - 26$ the attribute $id =' 34'$ is processed. So according to line 14 of *elementTransformation*() procedure its address is set to *person.#id* and next according to line 24 *elementTransformation*($id =' 34', r_1$) is called.

Again according to lines $1 - 3$ a new column is created and added to r_1 with its name set to *person.#id* and with value equal to 34. So, now:

$$r_1 = \{person = null, person.\#id = 34\} \tag{2}$$

Again, condition from line 4 is not fulfilled, so processing is continued. The first (and the only) subelement for element $id =' 34'$ is the terminal value 34. So according to line 10 its address is set to *person.#id.#* and *elementTransformation*($'34', r_1$) is called.

Listing 1. Sample XML document

```xml
<?xml version="1.0" encoding="utf-8"?>
<people>
    <person id="34">
        <firstname>John</firstname>
        <lastname>Foo</lastname>
        <age>32</age>
        <address>
            <city>NY</city>
            <street>PV</street>
            <house type="Apt">
                <bldg>34</bldg>
                <flat>12</flat>
            </house>
            <country></country>
        </address>
        <hobbies>
            <hobby>Dogs</hobby>
            <hobby>Cats</hobby>
            <hobby>Parrots</hobby>
        </hobbies>
    </person>
    <person id="35" missing="false" />
</people>
```

According to lines $1 - 3$ a new column is created and added to row r_1. Its name is set to *person.#id.#* and its value is set to 34. So now:

$$r_1 = \{person = null, person.\#id = 34, person.\#id.\# = 34\} \tag{3}$$

The condition from line 4 is fulfilled this time, so the further processing in this procedure call is finished. Since there are no further direct attributes for the first element the next subelement to be processed is $< firstname > John < /firstname >$.

According to line 18 of $elementTransformation(person < \cdots >, r_1)$ procedure the address of this element is set to $person.firstname$ and according to line 24 $elementTransformation(< firstname > John < /firstname >, r_1)$ is called.

In lines $1 - 3$ a new column with its name set to $person.firstname$ and with its value set to $John$ is created and added to row r_1. So now:

$$r_1 = \{person = null, person.\#id = 34, person.\#id.\# = 34, \\ person.firstname = John\} \tag{4}$$

Condition from line 4 is obviously not fulfilled. The first (and the only) subelement of element $< firstname > John < /firstname >$ is terminal textual value $John$. So in first (and the only this time) iteration in loop $7 - 26$, according to the line 10 the address of this subelement is set to $person.firstname.\#$ and then $elementTransformation(John, r_1)$ is called.

In lines $1 - 3$ the column $person.firstname.\#$ with value $John$ is created and added to r_1 so now:

$$r_1 = \{person = null, person.\#id = 34, person.\#id.\# = 34, \\ person.firstname = John, person.firstname.\# = John\} \tag{5}$$

In line 5 the procedure is recursively coming back to the next iteration. The process is continued until all subelements of $person_1$ are processed. Once finished, r_1 row contains the following columns (and their values):

$$r_1 = \{person = null, person.\#id = 34, person.\#id.\# = 34, \\ person.firstname = John, person.firstname.\# = John, \\ person.lastname = Foo, person.lastname.\# = Foo, \\ person.age = 32, person.age.\# = 32, \\ person.address = null, \\ person.address.city = NY, person.address.city.\# = NY, \\ person.address.street = PV, person.address.street.\# = PV, \\ person.address.house = null, person.address.house.\#type = Apt, \\ person.address.house.\#type.\# = Apt, person.address.house.bldg = 34, \\ person.address.house.bldg.\# = 34, person.address.house.flat = 12 \\ , person.address.house.flat.\# = 12, person.address.country = null, \\ person.hobbies = null, person.hobbies.hobby.\# = dogs\} \tag{6}$$

After finishing processing for $person_1$ in second iteration of the loop in lines $2 - 6$ of $transformationProcedure$ $person_2$ will be processed. As the result the row r_2 with the following columns (and their values) will be created:

$$r_2 = \{person = null, person.\#id = 35, person.\#id.\# = 35, \\ person.\#missing = false, person.\#missing.\# = false\} \tag{7}$$

Table 1. Final relation created on the basis of sample XML document for *people* as a root node and *person* as the element

Column name	Node type	Value for row r_1	Value for row r_2
person	Element	null	null
person.#id	Attribute	34	35
person.#id.#	Text	34	35
person.#missing	Attribute	null	false
person.#missing.#	Text	null	false
person.firstname	Element	John	null
person.firstname.#	Text	John	null
person.lastname	Element	Foo	null
person.lastname.#	Text	Foo	null
person.age	Element	32	null
person.age.#	Text	32	null
person.address	Element	null	null
person.address.city	Element	NY	null
person.address.city.#	Text	NY	null
person.address.street	Element	PV	null
person.address.street.#	Text	PV	null
person.address.house	Element	null	null
person.address.house.#type	Attribute	Apt	null
person.address.house.#type.#	Text	Apt	null
person.address.house.bldg	Element	34	null
person.address.house.bldg.#	Text	34	null
person.address.house.flat	Element	12	null
person.address.house.flat.#	Text	12	null
person.address.country	Element	null	null
person.hobbies	Element	null	null
person.hobbies.hobby	Element	Dogs	null
person.hobbies.hobby.#	Text	Dogs	null

Since there are no more subelements named *person* the procedure goes to line 7 i.e. both rows r_1 and r_2 are merged and added to table *people*. During this process the row r_1 will be completed with the columns

$$r_1 = r_1 \cup \{person.\#missing = null, person.\#missing.\# = null\} \qquad (8)$$

whereas row r_2 will be completed with all columns present in r_1 and not existing in r_2 and their values will be set to null.

The final product of proposed transformation procedure for sample XML document from listing 1 is presented in Tab. 1.

3   Quering XML Data Transformed into Quasi-Relational Model

The transformation described in previous sections is the basis for the construction of the relational query language called SQLxD [7].

SQLxD i.e. *SQL for XML Data*, is the query language for XML documents, with the syntax resembling indistinguishably the syntax known from ANSI–SQL.

There is also provided the engine responsible for transforming XML documents "as they come" on–the–fly according to the transformation model presented in section 2 on the basis of SQLxD queries and processing (querying) transformed hierarchical data as if it was a regular relational database.

What is important, it is performed in accordance to a general assumption that transforming the XML structure into the "flat" table model does not result in the loss of the XML document's "essence" i.e. the hierarchical relationship between the nodes. It is realized by providing functionality called the *natural join* which is explained in the second part of this section.

The detailed description of the SQlxD language is out of scope of the paper, so only a few examples, illustrating the idea and justifying the described method of transformation into quasi–relational model are presented in this section.

In listing 2 there is presented simple SQLxD query retrieving sample data regarding particular persons from XML document presented in listing 1.

On the basis of FROM clause processing engine is performing on-the-fly transformation presented in section 2 creating a virtual relational table containing information about particular persons. In fact on the basis of this FROM clause procedure *transformationProcedure()* presented in Algorithm 1 with *people* as a root node and *person* as interesting elements is performed. Assuming XML document from listing 1 virtual table exactly as presented in Tab. 1 is created. Next particular columns defined in SELECT clause are selected and data as presented in listing 2 is returned.

Listing 2. SQLxD simple example

```
SELECT
    person.#id,
    person.firstname,
    person.lastname,
    person.age,
    person.hobbies.hobby
FROM
    people.person AS person
```

Table 2. Results of applying simple SQLxD query

person.#id	person.firstname	person.lastname	person.age	person.hobbies.hobby
34	John	Foo	32	Dogs
35	null	null	null	null

It is important to note, that according to what has been previously written ("first and direct" principle) only one ("*dogs*") instead of three values of the column hobby for John Foo has been returned. All remaining hobbies ("*cats*", "*parrots*") are ignored (according to line 22 of *elementTransformation()* procedure). At this point, it could seem like a limitation, however it is eliminated by the natural join mechanism, which fits well into the assumptions of the relational transformation of the document.

In XML documents some important information is implicitly encoded "between the nodes" i.e. the information is not explicitly presented as attributes or textual values but is expressed by parent–child relationship between particular nodes.

On the contrary to other XML processing techniques, SQLxD, uses such information intensively by NATURAL JOIN clause. Although there is no strict relationship based on the primary/foreign keys, the natural join is performed on the basis of the hierarchy of the document, bringing the tree-structure back to the table-of-tables-of-objects structure.

The syntax of the natural join does not require any additional logical operators or precise data-to-column mapping. Generally speaking parent node addressed in NATURAL JOIN clause becomes the root in virtual subdocument containing this element and all its subelements only. Transformation procedure on such (restricted) subdocument is then performed producing table containing one single row for one single subelement and then typical NATURAL JOIN between rows of parent table and rows of child table is performed.

Table 3. Final relation created on the basis of sample XML document for *hobbies* as a root node and *hobby* as interesting the element

Column name	Node type	Value for row r_1	Value for row r_2	Value for row r_3
hobby	Element	Dogs	Cats	Parrots
hobby.#	Text	Dogs	Cats	Parrots

Listing 3. SQLxD natural join example

```
SELECT
    person.#id,
    person.firstname,
    person.lastname,
    hobby
FROM
    people.person AS person
NATURAL JOIN person.hobbies.hobby AS hobby
```

Table 4. Results of applying SQLxD query with natural join

person.#id	person.firstname	person.lastname	hobby
34	John	Foo	Dogs
34	John	Foo	Cats
34	John	Foo	Parrots

The sample query with NATURAL JOIN clause is presented in listing 3. Assuming XML document from listing 1, first on the basis of "people.person AS person" part of NATURAL JOIN clause transformation as in the previous example is performed. Next, according to the *person.hobbies.hobby AS hobby* part of NATURAL JOIN clause for each *person* element a virtual subdocument restricted to this element and all its subelements is created and then *transformationProcedure()* with hobbies as a root and hobby as interesting elements is called producing for John Foo a virtual table as presented in Tab. 3. Next a typical natural join operation is performed between JOHN FOO row(s) in parental relation and all rows in child table returning data as presented in listing 4.

The examples show just a draft of the SQLxD functionality however, it should be noted, that it provides almost everything that the programmer needs to work effectively with the XML documents, similarly as he would with conventional relational databases. All remaining language functionalities such as inner and outer joins, column expressions, filtering, sorting and aggregation etc come directly from the relational algebra and SQL. They simply operate on relational data produced by transformation model proposed in this paper performed on–the–fly on the basis of SQLxD statements.

4 Conclusions

There are of course on the market tools responsible for mapping between XML and relational model. When they are used however either hierarchical data structure is lost and/or manual transformation (or at least transformation definition) is required as for instance in explicit user-driven method proposed in [8].

In this paper the concept of quasi-relational transformation of XML data into relational model is proposed. The first impression taking a look into proposed transformation model is that it is a little bit strange producing sometimes redundant data and sometimes losing some information (for instance in relation from Tab. 1 produced for document from listing 1 there are some—seemingly—redundant records like *person.lastname = Foo* and *person.lastname.# = Foo*; and simultaneously only one hobby instead of three for person John Foo is there mapped.

This is only seeming shortcoming. As it is shown in section 3 it turns out that on the basis of such transformation model it is possible to propose SQL–like language (called SQLxD) and processing engine making it possible to:

- process effectively XML document with SQL-like queries,
- preserve hierarchical structure of XML data,
- transform data without any manual transformations, conversions, decompositions or even (fixed) transformation definition.

It can be said so for sure, that the main goals of research have been achieved and that proposed both: transformation quasi-relational model and SQLxD language are really promising and interesting achievements and they can effectively eliminate the impedance mismatch between XML and relational data.

References

1. The xpath 2.0 standard (2007), http://www.network-theory.co.uk/w3c/xpath/
2. Balmin, A., Özcan, F., Singh, A., Ting, E.: Grouping and optimization of xpath expressions in db2 purexml. In: Proceedings of the 2008 ACM SIGMOD International Conference on Management of Data, SIGMOD 2008, pp. 1065–1074. ACM, New York (2008), http://doi.acm.org/10.1145/1376616.1376722
3. Codd, E.F.: A relational model of data for large shared data banks. Commun. ACM 13(6), 377–387 (1970)
4. Hidders, J., Paredaens, J.: Xpath/xquery. In: Liu, L., Özsu, M.T. (eds.) Encyclopedia of Database Systems, pp. 3659–3665. Springer US (2009)
5. Kappel, G., Kapsammer, E., Retschitzegger, W.: Xml and relational database systems - a comparison of concepts. In: Proceedings of the 2001 International Conference on Internet Computing, pp. 199–205. CSREA Press, Las Vegas (2001)
6. Krishnaprasad, M., Liu, Z.H., Manikutty, A., Warner, J.W., Arora, V., Kotsovolos, S.: Query rewrite for xml in oracle xml db. In: Proceedings of the Thirtieth International Conference on Very Large Data Bases, VLDB 2004, vol. 30, pp. 1134–1145, VLDB Endowment (2004), http://dl.acm.org/citation.cfm?id=1316689.1316786
7. Marcjan, R., Wyrostek, J.: Processing xml documents on the basis of quasi-relational model and sqlxd language. Studia Informatica 32(2A), 111–120 (2011)
8. Mlynkova, I.: An xml-to-relational user-driven mapping strategy based on similarity and adaptivity. In: SYRCoDIS (2007)
9. Nicola, M., Kumar-Chatterjee, P.: DB2 pureXML Cookbook: Master the Power of the IBM Hybrid Data Server, 1st edn. IBM Press (2009)
10. Noh, S.Y., Gadia, S., Jang, H.: Comparisons of three data storage models in parametric temporal databases. Journal of Central South University 20(7), 1919–1927 (2013), http://dx.doi.org/10.1007/s11771-013-1691-8
11. Pal, S., Fussel, M., Dolobovsky, I.: Xml support in microsoft sql server 2005. TechNet Library (2005), http://technet.microsoft.com/en-us/library/ms345117(v=sql.90).aspx
12. Pal, S., Cseri, I., Seeliger, O., Rys, M., Schaller, G., Yu, W., Tomic, D., Baras, A., Berg, B., Churin, D., Kogan, E.: Xquery implementation in a relational database system. In: Proceedings of the 31st International Conference on Very Large Data Bases, VLDB 2005, pp. 1175–1186. VLDB Endowment (2005)
13. Runapongsa, K., Patel, J.M.: Storing and querying XML data in object-relational DBMSs. In: Chaudhri, A.B., Unland, R., Djeraba, C., Lindner, W. (eds.) EDBT 2002. LNCS, vol. 2490, pp. 266–285. Springer, Heidelberg (2002)
14. Rys, M., Chamberlin, D., Florescu, D.: Xml and relational database management systems: the inside story. In: Proceedings of the 2005 ACM SIGMOD International Conference on Management of Data, SIGMOD 2005, pp. 945–947 (2005), http://portal.acm.org/citation.cfm?id=1066157.1066298
15. Samokhvalov, N.: Xml support in postgresql. In: SYRCoDIS (2007)
16. Silberschatz, A., Korth, H.F., Sudarshan, S.: Database system concepts, 6th edn. McGraw-Hill, New York (2010), http://www.db-book.com/

Performance Issues in Data Extraction Methods of ETL Process for XML Format in Oracle 11g

Lukasz Wycislik

Silesian University of Technology, Institute of Informatics,
16 Akademicka St., 44-100 Gliwice, Poland
lukasz.wycislik@polsl.pl

Abstract. The article presents problems of data extraction and transmission as a parts of ETL process based on Oracle database example. Taking into attention more and more widespread cloud computing technologies it must be noticed that bandwidth of data links could be a serious limitation nowadays. The article describes technologies useful for data extraction in xml format from Oracle database. Further, the comparison of different methods taking into account both efficiency and volume of data to be transmitted, is presented. Finally several proposals on tuning these parts of ETL process are developed. The purpose of the article is to point the most efficient way of data extraction in XML format for Oracle database according to any existing nonfunctional requirements such us software portability, bandwidth limitation etc.

Keywords: Oracle, ETL, XML, EXI, efficiency, tuning.

1 Introduction

Information plays a key role in all areas of the world today. Almost everyone wants to know the weather forecast for the next few days or check fuel prices in one's city. Number of information stored and processed is growing and it doesn't seem that this trend would have changed. Today's information systems are rarely separated islands cut off the outside world. The data often are collected from multiple sources and only merged make usable logical unit. If data processing is performed only within a single datacenter, where internal data links are very fast, the process of data migration is typically a small fraction of the total processing time. However, taking into attention more and more widespread cloud computing technologies it must be noticed that bandwidth of data links could be a serious limitation. The reason for this is that not all data can be stored directly in the cloud. Some of them are forbidden by law, the other may be part of trade or corporate secrets that should not be stored or published outside. A broader discussion of the potential causes of legal limitations can be found in [3]. As we can see, after bringing the concept of cloud computing to life, the significance of different parts of ETL (Extract, Transform, Load) process has changed taking into consideration time-consumption. The whole basic concept of ETL process

S. Kozielski et al. (Eds.): BDAS 2014, CCIS 424, pp. 581–589, 2014.
© Springer International Publishing Switzerland 2014

is presented for example in [4,9]. These articles cover such aspects as data warehouse architectures, data mapping, ETL modeling, data quality issues including
consistency, validity, conformity, accuracy, integrity etc. Widely used for data
migration XML format, proved to have too much overhead on the transmitted
data volume. This resulted in situation where tuning of data extraction and migration is an integral part of the implementation of ETL solutions. This applies
in particular for systems such as real-time data warehousing [9] or continuous
monitoring systems [8]. In this article different technologies and strategies used
for data extraction in Oracle database are compared taking into account both
efficiency and volume of data to be transmitted. Information about Oracle tools
supporting other stages of the ETL process can be found in [11].

2   XML Technologies

XML [2] is data format developed as universal and human readable to exchange
information between systems. Its great advantage of being easy to understand
(considering both grammar and content) affected on one serious defect - a large
overhead on the size of transmitted/stored data. Over the years a number of
technologies and concepts to support the processing of XML documents were
developed. In particular, the Oracle company has introduced XML support of
database server in version 8i which was released in 1998. The support was limited
then only to XML document importing and exporting but not storing. The next
9i version, published in 2001, brought native support for XML. The new column
type - XMLType [5] was introduced that was designed to XML data storage.
Thanks to this type it is also possible to navigate the structure of the XML
document and read and write individual data. This version has also provided
SQL functions for creating and merging XML documents. Released in 2003,
version 10g brought a number of development tools such as hierarchical queries
to generate XML documents, support for XQuery language and added some new
functionality to XMLType. The latest version of Oracle Database 11g, released
in 2007, expands the functionality of binary XML storage. Also the new type of
index was implemented to support XMLType columns. There are several ways
to extract data in XML format from relational schema of Oracle database. The
trivial one that involves usage of select statements where XML document must
be formed 'manually'. If it is required to generate data from multiple tables
the all functionality may be encapsulated in PL/SQL [7] stored procedure or
(pipelined) function. The example is presented on Listing 1.1.

Listing 1.1. 'Manual' XML forming

```
FUNCTION GET_XML( pId Decimal )
RETURN TABLEOFVARCHAR pipelined AS
BEGIN
    pipe row ('<?xml_version="1.0"_encoding="UTF-8"?>');
    pipe row('<master>');
    for m in ( select attr1, attr2 from master m
```

```
                    where a.id = pId
) loop
   pipe row('<master_attr1="'||m.attr1||'"'||attr2="'||
      m.attr2||'">');
   for s in (select attr from slave s where m.id=s.id
          ) loop
              pipe row('<slave_attr="'||s.attr||'/>');
   end loop;
   pipe row('</master>');
   return;
END GET_XML;
```

The second method is based on XMLElement functionality. The trivial, analogical to Listing 1.1, example is presented on Listing 1.2.

Listing 1.2. 'XMLElement' based functionality

```
select xmlroot (
   xmlelement ( name "collection", xmlattributes
   ('http://example.com' as "xmlns"),
      xmlagg (
        xmlelement ( name "master", xmlattributes
            (m.attr1 as "attr1", m.attr2 as "attr2"),
          xmlagg( xmlelement (name "slave",
              xmlattributes(s.attr as "attr")))
        )
      )
   )
   ,
   version '1.0'
)
from master m, slave s where s.id=m.id group by m.attr;
```

The other methods are related to technologies derived from outside native database server software. These are for instance common java libraries that connect to database with standard JDBC driver or with object-relational mappers (e.g.Hibernate) for database connection and SAX (simple API for XML) for XML documents formatting. These processes can thus be realized inside database server (as that java virtual machine is embedded inside Oracle database), outside database server but on the same physical node or at separate client node. Considering process efficiency of formatting and transmitting XML documents as total time of document creation and transition it may be crucial to ensure that generated document is as small as possible. This could be achieved using several techniques which include shortening XML tags, general methods of lossless compression and also dedicated methods for XML compression such as EXI [1] (Efficient XML Interchange). While the first two are obvious the last one is not standard yet and exists in two versions - based on XML schema definition existence and the other - possible to use without the existence of a formal XML schema definition. Unfortunately, so far none of them are natively supported by the Oracle database, so deciding on the use of that technique the third-party software component is needed. The EXI format is so promising that it can not only ensure shorter XML documents but also speeds up the parsing process and reduces the amount of memory needed.

3   Survey

3.1   Research Environment and Methods of Measurement

The study was conducted on an environment virtualized by Oracle Virtual Box software and is shown on the Fig. 1. Both server and client CPU time and also the bandwidth of data link between them where arbitrary limited and controlled. Each virtual PC device (PC - server and PC - client) was 'pinned' to one core of Intel i5 processor and 'equipped' with 2GB RAM. The computing power of each device was controlled by 'execution cap' functionality of Virtual Box. Communication between PC server and PC client was implemented by 'internal network' functionality of Virtual Box. Unfortunately, it has no ability to control bandwidth so dedicated NetLimiter software was deployed both on server and client.

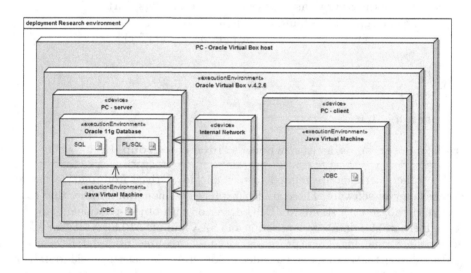

Fig. 1. The environment virtualized by Oracle Virtual Box software

The exact values of computing power and bandwidth have minor importance for considerations generality because production systems may have different parameters however some trends and conclusions may be regarded as general. Also the schema of relations in database, sample data and concrete XML format may affect individual scores. To measure the execution time of each operation the

ContiPerf library was applied. This is commonly used software for performance testing that allows for setting the repetition number of measured operation and collects execution time of each operation. It also calculates the mean time and median as well. As the differences in the measurement results for given operation were typically less than 5%, on figures with results only average values were presented.

3.2   Results

The first test case, shown on the Fig. 2, compares time needed to generate the XML document on the client side. It consists of finding data in database, sending content via the network and saving the resulting file at the client file system. In both situations the application creating XML document is running on client side but the first method uses only standard SQL and document formatting is done by SAX library. The second method, which appears to be almost four times faster, uses native XML support as presented on Listing 1.2. The second test case, presented on the Fig. 3, compares time needed to generate XML document on the server side only. The PL/SQL stored procedure proved to be the quickest way to format XML document. Somewhat surprising was the low efficiency of the method based on 'manual' formatting presented on Listing 1.1 - it appeared to be even slower then external application using SQL with native XML support. Compression is a common solution when one must operate with limited bandwidth. On the Fig. 4 different compression methods applied to the sample XML document are compared. As can be seen, the tag shortening method is much worse than the other. It is surprising that the EXI methods (which are developed especially for XML processing) proved to be less effective than ones based on zip - this is probably due to the specificity of this particular XML

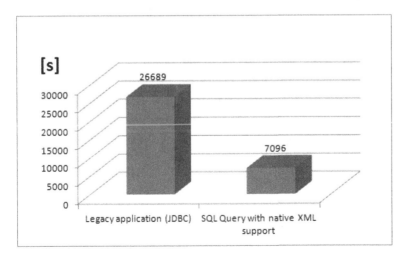

Fig. 2. Creation of XML document on a client side

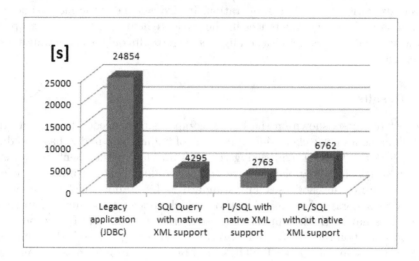

Fig. 3. Creation of XML document on a server side

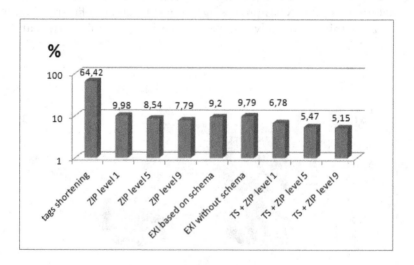

Fig. 4. Compression of XML document

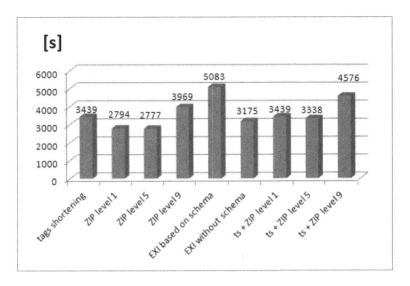

Fig. 5. Creation of compressed XML document on a client side

format as the results found in the literature [10] are generally slightly better for EXI. The best results are observed for compression mixed of tags shortening and zip but it's obviously the most time consuming. However the bandwidth is not the only limited resource. The second one is also processing time. Thus on the Fig. 5 the total time needed for both compression and transition via network is presented. What is the most surprising the EXI method based on XML schema definition appeared as the worst. The best ones are zip methods but neither the strongest nor the weakest of the methods proved to be the best. The best one is somewhere in the middle so the compression level should be tuned always when bandwidth or computational power changed. In addition, it is worth noting that using mixed methods is unjustified. Examples illustrating the efficiency of the generation and transmission of XML data process, depending on the bandwidth are shown on the Fig. 6. As can be seen, apparently not very interesting tags shortening method turns out to be better starting from bandwidth of 5 Mbit/s than an EXI based on schema that is claimed to be dedicated for XML processing. ZIP compression seems to be the undisputed winner but starting from bandwidth of 7 Mbit/s differences compared to EXI without schema are not significant.

4   Summary

This paper presents the aspects of XML documents formatting and transmitting that are the essential part of ETL process. Due to the popularization of the concept of cloud computing some data still need to be stored outside of data centers. Frequently data links to these centers have a limited bandwidth and

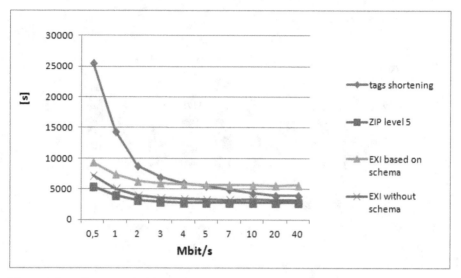

Fig. 6. The efficiency of the generation and transmission of XML data process, depending on the bandwidth

therefore an important part of the data integration process becomes tuning of that part of the ETL process. As shown in the examples, there is no single best way to configure the whole process. Each time, one should take into account specific conditions such as bandwidth or processing capacity. The experimental part of this article is limited to Oracle technology which is however presently the most common 'industrial' database [6]. When one is faced with the task of ETL tuning process this means that the whole database/data warehouse is already developed, deployed and in production phase and there is usually no economic reasons to change the technology. From this point of view the limitation only to Oracle technology is justified and practically useful. When using Oracle database the most efficient way to generate XML document is to use the native XML support inside PL/SQL procedure. The obvious price will be the lack of portability on different database platforms. Method of XML document generating presented on Listing 1.1 rather should not be used because is slower than method with native XML support and still does not give portability benefits. Taking into account data compression, it brings some disappointment that can be observed relatively low efficiency of method dedicated to XML processing - EXI. The zip format has proven to be more efficient and also provides greater flexibility in tuning process by possibility of setting different compression levels. It should be noted, however, that the EXI format makes it possible to parse XML documents without decompressing them, that does not allow universal zip format, so it may be considered when there is a need not only to transmit XML but to store it persistently as well. So far, Oracle database doesn't support EXI compression so that all tests were performed using external software component. It will be worth checking of performance of Oracle EXI implementation as soon as it integrate it into its database.

References

1. Efficient xml interchange (January 2014), http://www.w3.org/XML/EXI/
2. Extensible markup language (xml) (January 2014), http://www.w3.org/XML/
3. Balboni, P., Fontana, F.: Cloud computing: A guide to evaluate and negotiate cloud service agreements in the light of the actual european legal framework. ICT Law Review 1, 12–17 (2013)
4. El-Sappagh, S.H.A., Hendawi, A.M.A., Bastawissy, A.H.E.: A proposed model for data warehouse ETL processes. Journal of King Saud University - Computer and Information Sciences 23(2), 91–104 (2011), http://www.sciencedirect.com/science/article/pii/S131915781100019X
5. Geeta, A.: Oracle xml db: Choosing the best xml type storage option for your use case. Oracle Corporation (December 2009)
6. Graham, C., Correia, J., Coyle, D., Biscotti, F., Cheung, M., Contu, R., Dharmasthira, Y., Eid, T., Eschinger, C., Granetto, B., Swinehart, H.H., Mertz, S., Pang, C., Raina, A., Sommer, D., Sood, B., Wurster, M.D.L., Zhang, J.: Market share: All software markets (March 2012)
7. Harper, S.: Working with pl/sql, oracle Corporation (November 2007)
8. Jestratjew, A., Kwiecień, A.: Using cloud storage in production monitoring systems. In: Kwiecień, A., Gaj, P., Stera, P. (eds.) CN 2010. CCIS, vol. 79, pp. 226–235. Springer, Heidelberg (2010)
9. Kakish, K., Kraft, T.A.: Etl evolution for real-time data warehousing. In: Proceedings of the Conference on Information Systems Applied Research (2012) ISSN: 2167-1508
10. Snyder, S.L.: Efficient xml interchange compression and performance benefits: development, implementation and evaluation (March 2010)
11. Wiak, S., Drzymała, P., Welfle, H.: Using oracle tools to generate multidimensional model in warehouse. Przegld Elektrotechniczny (2012) ISSN: 0033-2097

Author Index